# Bioorganic Chemistry

Highlights and New Aspects

Edited by
Ulf Diederichsen, Thisbe K. Lindhorst,
Bernhard Westermann, and
Ludger A. Wessjohann

**WILEY-VCH**

Weinheim · New York · Chichester · Brisbane · Singapore · Toronto

Edited by:

Prof. Dr. Ulf Diederichsen
Institut für Organische Chemie
der TU München
Lichtenbergstraße 4
D-85747 Garching
Germany

Prof. Dr. Bernhard Westermann
Fachbereich 13
Organische Chemie
der Universität-GH Paderborn
Warburgerstraße 100
D-33098 Paderborn
Germany

Prof. Dr. Thisbe K. Lindhorst
Institut für Organische Chemie
der Universität Hamburg
Martin-Luther-King-Platz 6
D-20146 Hamburg
Germany

Prof. Dr. Ludger A. Wessjohann
Faculty of Chemistry
Vrije Universiteit Amsterdam
De Boelelaan 1083
NL-1081 HV Amsterdam
The Netherlands

Front cover: *Dendrobium nobile* Lindl., a southeast asian orchid rich in alkaloids such as dendrobine (back cover).

Library of Congress Card No. applied for

A catalogue record for this book is available from the British Library.

Die Deutsche Bibliothek – CIP-Einheitsaufnahme

Ein Titelsatz für diese Publikation ist bei
Der Deutschen Bibliothek erhältlich.

© WILEY-VCH Verlag GmbH, D-69469 Weinheim (Federal Republic of Germany), 1999

Printed on acid-free and chlorine-free paper.

Composition: Alden Bookset, Oxford, England
Printing: betz-druck, D-64291 Darmstadt
Bookbinding: W. Osswald, D-67433 Neustadt/Wstr.
Printed in the Federal Republic of Germany

# Preface

Think global – this tendency cannot only be observed in politics and the business market, but also more and more in science. The term bioorganic chemistry is certainly a good example. It represents the wide interface of organic chemistry, biochemistry, natural product chemistry, pharmaceutical and medicinal chemistry on the 'chemistry' side, and biotechnology, microbiology, molecular biology, pharmaceutical biology and medicine on the 'biology' side. Thus, bioorganic chemistry is a typical new branch of modern science, with scientists involved having crossed subject borders and developing a better understanding of complex interactions on a molecular level. These subjects often have more than the usual impact outside of the chemical community as applications, medical, social or even ethical issues are commonly involved. Bioorganic projects are typically initiated from phenomenons observed in nature. However, a deeper understanding of these observations often requires the development of a molecular tool which has to be provided by chemists. But also the opposite approach, the solution of chemical problems with 'biological' tools is part of bioorganic chemistry. In these collaborative interactions, scientists experience a new, completely different and stimulating view of a research topic. This way of open, modern scientific thinking results in broader scientific communication in suitable journals. It cannot be overlooked that traditional chemical top journals such as the *Journal of the American Chemical Society* or *Angewandte Chemie* publish more work which includes biochemistry, biotechnology, enzymology, and molecular biology, and typical biological journals like *Journal of Bacteriology* contain more chemistry than some years ago. Last but not least, prestigious new journals, such as *Chemistry and Biolology* or *Bioorganic Chemistry* were founded just for the purpose to publish science overlapping the traditional fields of chemistry and biology.

To stimulate discussions and establish a forum for young scientists representing these different aspects of bioorganic chemistry, the 'Clausthal meeting', in Clausthal-Zellerfeld, a small University town in the heart of the Harz mountains, was initiated. Organized at first by Werner Klaffke (formerly U. Hamburg, Germany, now U. Münster, Germany), Andreas Kirschning (TU Clausthal-Zellerfeld) and Jürgen Rohr (formerly U. Göttingen, Germany, now Charleston, SC, U.S.A.), then Stefan Schulz (formerly U. Hamburg, now TU Braunschweig, Germany) and Ludger Wessjohann (formerly U. München, now VU Amsterdam), this conference has been established as an intriguing meeting and an open-minded exchange of ideas. Young researchers (biologists, chemists, medical researchers, and pharmaceutical scientists) from academia and industry from Germany and various neighbouring European countries were invited to contribute to an open forum with as little constraints to free discussion as possible. In addition, contacts made in these 'three days of September' have led to a number of productive collaborations, some of which can be seen in this book. This meeting has been moved from its initial Clausthal-location to Paderborn, and will move on to Heidelberg. Also the team is constantly renewed, the baton passed to new assistant professors, some of them being editors of this book. The meeting has been generously sponsored, initially by the *Volkswagenstiftung*, and later exclusively by the chemical and pharmaceutical industry (Altus, AnalytiCon, Asta, BASF, Bayer, Boehringer Ingelheim/Thomae, Boehringer Mannheim, Christ, Degussa, Dragoco, DSM Research, Fonds der Chemischen Industrie, Hoechst, Hoffmann-LaRoche, E. Merck, Novartis, Schering, Spektrum-Verlag and Unilever). This broad support not only reveals the modern, open-minded thinking of the industry, moreover it enabled the organizers of the meeting to reimburse most traveling and logistic costs for active participants from the universities and state research institutions, an important factor for young scientists who often do not have any traveling grants available. A complementary European meeting (WEB) has been set up by L. Wessjohann, M. Kalesse (U. Hannover) and A. Young (U. Cambridge), establishing the same flair as its Clausthal-counterpart.

This book contains more than 50 articles, written by academic and industrial researchers from the Czech Republic, Germany, Great Britain, The Netherlands, Switzerland, and the U.S.A., most of whom contributed regularly to and enjoyed the free, enthusiastic spirit of the 'Clausthal meeting'. The topics chosen are from typical 'bioorganic' fields, covering Analytical Methods, Biochemistry, Biosynthesis, Biotransformation, Carbohydrates, Drug Research, Enzymes, Enzymatic Synthesis, Glycobiology, Immunology, Medicinal Chemistry and QSAR, Molecular Biology, Natural Products, Nucleic Acid Chemistry, Organic and Combinatorial Synthesis of Model Drugs, Peptide Chemistry, and Spectroscopic Methods. We believe that this book represents a good first retrospective and reflects some of the spirit of the bioorganic chemistry meeting of young scientists which started 1992 in Clausthal, Germany.

*Andreas Kirschning, Technical University Clausthal, Germany, and*
*Jürgen Rohr, Medical University of South Carolina, Charleston, SC, U.S.A.*

# List of Contributors

**Chris Abell**

University of Cambridge Department of Chemistry, Lensfield Road, Cambridge CB2 1EW, U.K.
ca26@cam.ac.uk

**Wolf-Rainer Abraham**

GBF – National Research Center for Biotechnology, Chemical Microbiology, Mascheroder Weg 1, D-38124 Braunschweig, Germany
wab@gbf.de

**Christoph Arenz**

Institut für Organische Chemie, Universität Karlsruhe, Richard-Willstätter-Allee 2, D-76128 Karlsruhe, Germany

**Hans-Georg Batz**

Roche Diagnostics GmbH, D-82372 Penzberg, Germany

**Andreas Bechthold**

Institut für Pharmazeutische Biologie, Universität Tübingen, Auf der Morgenstelle 8, D-72076 Tübingen, Germany
andreas.bechthold@uni-tuebingen.de

**Annette G. Beck-Sickinger**

Department of Pharmacy, Federal Institute of Technology, (ETH) Zürich, Winterthurerstr. 190, CH-8057 Zürich, Switzerland
beck-sickinger@pharma.ethz.ch

**Ludger Beerhues**

Institut für Pharmazeutische Biologie, Universität Bonn, Nußallee 6, D-53115 Bonn, Germany
beerhues@uni-bonn.de

**Claus Beninga**

Department of Pharmaceutical Sciences, Medical University of South Carolina, 171 Ashley Avenue, Charleston, SC 29425-2303, U.S.A.
Fax: ++49/843-9536615

**Uwe T. Bornscheuer**

Institute for Technical Biochemistry, University of Stuttgart, Allmandring 31, D-70569 Stuttgart, Germany
bornscheuer@po.uni-stuttgart.de

**Bernd Buchmann**

Institute of Medicinal Chemistry, Preclinical Drug Research, Schering AG, D-13342 Berlin, Germany

**Lars Burgdorf**

Department of Organic Chemistry, Swiss Federal Institute of Technology, ETH-Zentrum, Universitätstrasse 16, CH-8092 Zürich, Switzerland

**Jens Butenandt**

Department of Organic Chemistry, Swiss Federal Institute of Technology, ETH-Zentrum, Universitätstrasse 16, CH-8092 Zürich, Switzerland

**Chiara Cabrele**

Department of Pharmacy, Federal Institute of Technology, (ETH) Zürich, Winterthurerstr. 190, CH-8057 Zürich, Switzerland

**Thomas Carell**

Department of Organic Chemistry, Swiss Federal Institute of Technology, ETH-Zentrum, Universitätstrasse 16, CH-8092 Zürich, Switzerland
tcarell@org.chem.ethz.ch

**Ricardo Cortese**

Istituto di Ricerche di Biologia Moleculare, I-00040 Pomezia, Italy
cortese@irbm.it

**Rudolf Dernick**

Heinrich-Pette-Institute for Experimental Virology and Immunology, University of Hamburg (HPI), Martinistrasse 52, D-20251 Hamburg, Germany
dernick@t-online.de

**Marco-Aurelio Dessoy**

Institut für Organische Chemie, Ludwig-Maximilians-Universität München, Karlstraße 23, D-80333 München, Germany

**Ralf Dettmann**

Institut für Organische Chemie, Universität Köln, Greinstraße 4, D-50939 Köln, Germany

**Ulf Diederichsen**

Institut für Organische Chemie und Biochemie, Technische Universität München, Lichtenbergstr. 4, D-85747 Garching, Germany
diederichsen@ch.tum.de

**Nicole Diedrichs**

Universität-GH Paderborn, Chemie und Chemietechnik, D-33095 Paderborn, Germany
nd@chemie.uni-paderborn.de

**Michael Duszenko**

Institut für Physiologische Chemie, Universität Tübingen, Hoppe-Seyler-Straße 4, D-72076 Tübingen, Germany

**Hubert Dyker**

Bayer AG, Central Research, D-51368 Leverkusen, Germany
Hubert.Dyker.hd@bayer-ag.de

**Lothar Elling**

Institut für Enzymtechnologie, Heinrich-Heine-Universität Düsseldorf, Forschungszentrum Jülich, D-52426 Jülich, Germany
L.Elling@fz-juelich.de

**Magdalena Endová**

Institute of Organic Chemistry and Biochemistry, Academy of Sciences of the Czech Republic, Flemingovo nám. 2, 16610 Prague 6, Czech Republic

**Robert Epple**

Department of Organic Chemistry, Swiss Federal Institute of Technology, ETH-Zentrum, Universitätstrasse 16, CH-8092 Zürich, Switzerland

**Winfried Etzel**

Bayer AG Crop Protection Business Group, Chemical Research, Agricultural Centre Monheim, D-51368 Leverkusen, Germany
winfried.etzel.WE@bayer-ag.de

**Eduard R. Felder**

Novartis Pharma AG, Core Technology Area, Postfach, CH-4002 Basel, Switzerland
Fax: ++41/61-6968360

**Gianfranco Fragale**

Institut für Organische Chemie, Universität Basel, St. Johanns Ring 19, CH 4056 Basel, Switzerland

**Ina Gedrath**

Universität-GH Paderborn, Chemie und Chemietechnik, D-33095 Paderborn, Germany
ik@chemie.uni-paderborn.de

**Matthias Gehling**

Bayer AG, Life Science Center Natural Products, D-40789 Monheim, Germany
Matthias.Gehing.mg@bayer-ag.de

**Gerd Gemmecker**

Institut für Organische Chemie und Biochemie II, Technische Universität München, Lichtenbergstr. 4, D-85747 Garching, Germany
Gerd.Gemmecker@ch.tum.de

**Athanassios Giannis**

Institut für Organische Chemie, Universität Karlsruhe, Richard-Willstätter-Allee 2, D-76128 Karlsruhe, Germany
giannis@ochhades.chemie.uni-karlsruhe.de

**Steffen J. Glaser**

Institut für Organische Chemie und Biochemie, Technische Universität München, Lichtenbergstr. 4, D-85747 Garching, Germany
Glaser@ch.tum.de

**Susanne Grabley**
Hans-Knöll-Institut für Naturstoff-Forschung, Beutenbergstraße 11, D-07745 Jena, Germany
sgrabley@pmail.hki-jena.de

**Michael Grol**
Roche Diagnostics GmbH, D-82372 Penzberg, Germany

**Oliver Gutbrod**
Bayer AG, Central Research, D-51368 Leverkusen, Germany
Oliver.Gutbrod.og@bayer-ag.de

**Alfons Hädener**
Department of Pharmacy, University of Basel, Totengässlein 3, CH-4051 Basel, Switzerland
haedener@ubaclu.unibas.ch

**Sara Häuptli**
Institut für Organische Chemie, Universität Basel, St. Johanns Ring 19, CH 4056 Basel, Switzerland

**Thomas Henkel**
Institut für Organische Chemie, Universität Göttingen, Tammannstr. 2, D-37077 Göttingen, Germany

**Birgit Holz**
Procter & Gamble, European Services GmbH, Sulzbacher Str. 40, D-65823 Schwalbach, Germany

**Meike Holzenkämpfer**
Department of Pharmaceutical Sciences, Medical University of South Carolina,
171 Ashley Avenue, Charleston, SC 29425-2303, U.S.A.

**Henning Hopf**
Institut für Organische Chemie, Technische Universität Braunschweig, Hagenring 31,
D-38106 Braunschweig, Germany
H.Hopf@tu-bs.de

**Stefan Jaroch**
Schering AG, Institute of Medicinal Chemistry, Preclinical Drug Research, D-13342 Berlin, Germany

**Peter Jeschke**
Bayer AG, Crop Protection Business Group, Chemical Research & Animal Health Division,
Agricultural Centre Monheim, D-51368 Leverkusen, Germany
peter.jeschke.PJ@bayer-ag.de

**Stephan Jordan**
Bayer AG, Central Research, D-51368 Leverkusen, Germany
Stephan.Jordan.sj@bayer-ag.de

**Jochen Junker**

Institut für Organische Chemie, Johann-Wolfgang Goethe-Universität, Marie-Curie-Str. 11,
D-60439 Frankfurt, Germany
jj@org.chemie.uni-frankfurt.de

**Markus Kalesse**

Institut für Organische Chemie, Universität Hannover, Schneiderberg 1B, D-30167 Hannover,
Germany
kalesse@mbox.oci.uni-hannover.de

**Uli Kazmaier**

Organisch-Chemisches Institut, Universität Heidelberg, Im Neuenheimer Feld 270,
D-69120 Heidelberg, Germany
ck1@popix.urz.uni-heidelberg.de

**Eckhard Kirschning**

Heinrich-Pette-Institute for Experimental Virology and Immunology,
University of Hamburg (HPI), Martinistraße 52, D-20251 Hamburg, Germany
kirschni@uke.uni-hamburg.de

**Andreas Kirschning**

Institut für Organische Chemie, Technische Universität Clausthal, Leibnizstraße 6,
D-38678 Clausthal-Zellerfeld, Germany
andreas.kirschning@tu-clausthal.de

**Werner Klaffke**

Organisch-Chemisches Institut, Westfälischen Wilhelms-Universität, Corrensstraße 40,
D-48149 Münster, Germany
werner.klaffke@uni-muenster.de

**Matthias Köck**

Institut für Organische Chemie, Johann-Wolfgang Goethe-Universität, Marie-Curie-Str. 11,
D-60439 Frankfurt, Germany
km@org.chemie.uni-frankfurt.de

**Ulrich Koert**

Institut für Chemie, Humboldt-Universität zu Berlin, Hessische Straße 1-2, D-10115 Berlin,
Germany
koert@lyapunov.chemie.hu-berlin.de

**Šárka Králíková**

Institute of Organic Chemistry and Biochemistry, Academy of Sciences of the Czech Republic,
Flemingovo nám. 2, 16610 Prague 6, Czech Republic

**Ronald F.M. Lange**

DSM Research, PCM, P.O. Box 18, NL-6160 MD Geleen, The Netherlands
ronald.lange@dsm-group.com

**Michele Leuenberger**

Institut für Organische Chemie, Universität Basel, St. Johanns Ring 19, CH-4056 Basel,
Switzerland

**Radek Liboska**

Institute of Organic Chemistry and Biochemistry, Academy of Sciences of the Czech Republic, Flemingovo nám. 2, 16610 Prague 6, Czech Republic


**Folker Lieb**

Bayer AG, Central Research, D-51368 Leverkusen, Germany
Folker.Lieb.fl@bayer-ag.de


**Michael R. Linder**

Institut für Organische Chemie, Universität Stuttgart, Pfaffenwaldring 55, D-70569 Stuttgart, Germany
michael.linder@po.uni-stuttgart.de


**Thisbe K. Lindhorst**

Institut für Organische Chemie, Universität Hamburg, Martin-Luther-King-Platz 6, D-20146 Hamburg, Germany
tklind@chemie.uni-hamburg.de


**Torsten Linker**

Institut für Organische Chemie, Universität Stuttgart, Pfaffenwaldring 55, D-70569 Stuttgart, Germany
torsten.linker@po.uni-stuttgart.de


**Andreas L. Marzinzik**

Novartis Pharma AG, Core Technology Area, Postfach, CH-4002 Basel, Switzerland
Fax: ++41/61-6968360


**Michael Maurer**

Institut für Organische Chemie, Universität Stuttgart, Pfaffenwaldring 55, D-70569 Stuttgart, Germany
maurer@chemie.uni-wuerzburg.de


**E.W. Meijer**

Eindhoven University of Technology, P.O. Box 512, NL-5600 MB Eindhoven, The Netherlands
tgtobm@chem.tue.nl


**Rob Meloen**

Central Veterinary Institute, NL-8200 Lelystad, The Netherlands
r.h.meloen@id.dlo.nl


**Sabine Müller**

Institut für Chemie, Fachinstitut für Organische und Bioorganische Chemie, Humbodt-Universität zu Berlin, Hessische Str. 1-2, D-10115 Berlin, Germany
sabine.mueller@chemie.hu-berlin.de


**Carsten Oelkers**

Institut für Organische Chemie, Universität Göttingen, Tammannstraße 2, D-37077 Göttingen, Germany

**Thorsten Oost**

Institut für Organische Chemie, Universität Hannover, Schneiderberg 1B, D-30167 Hannover, Germany

**Miroslav Otmar**

Institute of Organic Chemistry and Biochemistry, Academy of Sciences of the Czech Republic, Flemingovo nám. 2, 16610 Prague 6, Czech Republic

**Stefan Peters**

Institut für Pharmazeutische Biologie, Universität Bonn, Nußallee 6, D-53115 Bonn, Germany

**Markus Pietzsch**

Institute of Biochemical Engineering, University of Stuttgart, Allmandring 31, D-70569 Stuttgart, Germany
pietzsch@ibvt.uni-stuttgart.de

**Joachim Podlech**

Institut für Organische Chemie, Universität Stuttgart, Pfaffenwaldring 55, D-70569 Stuttgart, Germany
joachim.podlech@po.uni-stuttgart.de

**Dominik Rejman**

Institute of Organic Chemistry and Biochemistry, Academy of Sciences of the Czech Republic, Flemingovo nám. 2, 16610 Prague 6, Czech Republic

**Monika Ries**

Institut für Organische Chemie, Technische Universität Clausthal, Leibnizstraße 6, D-38678 Clausthal-Zellerfeld, Germany

**Gilles Ritter**

Testex AG, Gotthardstrasse 61, CH-8027 Zürich, Switzerland
Fax: ++41/1-2025527

**Jürgen Rohr**

Department of Pharmaceutical Sciences, Medical University of South Carolina, 171 Ashley Avenue, Charleston, SC 29425-2303, U.S.A.
rohrj@musc.edu

**Ivan Rosenberg**

Institute of Organic Chemistry and Biochemistry, Academy of Sciences of the Czech Republic, Flemingovo nám. 2, 16610 Prague 6, Czech Republic
ivan@uochb.cas.cz

**Marco T. Rudolf**

Institut für Organische Chemie, Abt. Bioorganische Chemie, UFT, Universität Bremen, Leobener Str., D-28359 Bremen, Germany

**Patrick M. Schaeffer**

Research School of Chemistry, The Australian National University, Canberra ACT 0200, Australia
pms@rsc.anu.edu.au

**Werner Schmidt**

Institut für Pharmazeutische Biologie, Universität Bonn, Nußallee 6, D-53115 Bonn, Germany

**Andrew Schnaars**

Institut für Organische Chemie, Abt. Bioorganische Chemie, UFT Universität Bremen, Leobener Str., D-28359 Bremen, Germany
schnaars@chemie.uni-bremen.de

**Jens Schneider-Mergener**

Jerini BioTools GmbH, D-12489 Berlin, Germany
biotools@jerini.fta-berlin.de

**Andreas Schönberger**

Institut für Organische Chemie, Technische Universität Clausthal, Leibnizstraße 6, D-38678 Clausthal-Zellerfeld, Germany

**Andreas Schoop**

Bayer AG, Central Research, D-51368 Leverkusen, Germany
Fax: ++49/214-3056676

**Carsten Schultz**

Institut für Organische Chemie, Abt. Bioorganische Chemie, UFT, Universität Bremen, Leobener Str., D-28359 Bremen, Germany
schultz@chemie.uni-bremen.de

**Stefan Schulz**

Institut für Organische Chemie, Technische Universität Braunschweig, Hagenring 30, D-38106 Braunschweig, Germany
stefan.schulz@tu-bs.de

**Ulrich Schwarz-Linek**

Department of Biochemistry, University of Oxford, South Parks Road, Oxford OX1 3QU, U.K.
uli@bioch.ox.ac.uk

**Christoph Schwemler**

Bayer AG, Central Research, D-51368 Leverkusen, Germany
Christopf.Schwemler.cs@bayer-ag.de

**Anja Schwögler**

Department of Organic Chemistry, Swiss Federal Institute of Technology, ETH-Zentrum, Universitätstrasse 16, CH-8092 Zürich, Switzerland

**Christoph Seidel**

Roche Diagnostics GmbH, D-82372 Penzberg, Germany
christoph.seidel@roche.com

**Norbert Sewald**

Institut für Organische Chemie, Universität Leipzig, Talstraße 35, D-04103 Leipzig, Germany
sewald@organik.orgchem.uni-leipzig.de

**Hans-Christian Siebert**

Bijvoet Center, Department of Bioorganic Chemistry, P.O. Box 80075, NL-3508 TB Utrecht, The Netherlands
siebert@boc.chem.uu.nl

**Werner Skuballa**

Schering AG, Institute of Medicinal Chemistry, Preclinical Drug Research, D-13342 Berlin, Germany

**Richard Söll**

Department of Pharmacy, Federal Institute of Technology, ETH Zürich, Winterthurerstr. 190, CH-8057 Zürich, Switzerland

**Thomas Sommermann**

Institute of Organic Chemistry, University of Stuttgart, Pfaffenwaldring 55, D-70569 Stuttgart, Germany
Thomas.Sommermann@po.uni-stuttgart.de

**Bernd Sontag**

Institut für Organische Chemie, Ludwig-Maximilians-Universität München, Karlstraße 23, D-80333 München, Germany

**N. Patrick J. Stamford**

School of Chemical Sciences, University of East Anglia, Norwich NR4 7TJ, UK
n.stamford@uea.ac.uk

**Ralf Thiericke**

Hans-Knöll-Institut für Naturstoff-Forschung, Beutenbergstraße 11, D-07745 Jena, Germany

**Zdeněk Točík**

Institute of Organic Chemistry and Biochemistry, Academy of Sciences of the Czech Republic, Flemingovo nám. 2, 16610 Prague 6, Czech Republic

**Karl-Heinz van Pée**

Institut für Biochemie, Technische Universität Dresden, Mommsenstr. 4, D-01062 Dresden, Germany
Karl-Heinz.vanPee@chemie.tu-dresden.de

**Robert Velten**

Bayer AG, Central Research, D-51368 Leverkusen, Germany
Robert.Velten.rv@bayer-ag.de

**Marina Vogel**

Institut für Organische Chemie, Universität Leipzig, Talstrasse 35, D-04103 Leipzig, Germany
vogel@organik.orgchem.uni-leipzig.de

**Barillas Wagner**

Institut für Pharmazeutische Biologie, Universität Bonn, Nußallee 6, D-53115 Bonn, Germany

**Armin Walter**

Universität-GH Paderborn, Chemie und Chemietechnik, D-33095 Paderborn, Germany
aw@chemie.uni-paderborn.de

**Susanne Weber**

Institut für Organische Chemie, Universität Göttingen, Tammannstr. 2, D-37077 Göttingen, Germany

**Thomas Weimar**

Institut für Chemie, Medizinische Universität Lübeck, Ratzeburger Allee 160, D-23538 Lübeck, Germany
thio@chemie.uni-luebeck.de

**Elmar Weinhold**

Max-Planck-Institut für molekulare Physiologie, Abteilung Physikalische Biochemie, Rheinlanddamm 201, D-44139 Dortmund, Germany
elmar.weinhold@mpi-dortmund.mpg.de

**Ulrike Weißbach**

Institut für Organische Chemie, Universität Göttingen, Tammannstr. 2, D-37077 Göttingen, Germany

**Ludger A. Wessjohann**

Bio-organic Chemistry, FEW/OAC, Vrije Universiteit Amsterdam, De Boelelaan 1083, NL-1081 HV Amsterdam, The Netherlands
wessjohn@chem.vu.nl

**Bernhard Westermann**

Universität-GH Paderborn, Chemie und Chemietechnik, Organische Chemie, Warburger Str. 100, D-33095 Paderborn, Germany
bw@chemie.uni-paderborn.de

**Lucia Westrich**

Institut für Pharmazeutische Biologie, Universität Tübingen, Auf der Morgenstelle 8, D-72076 Tübingen, Germany

**Thomas Wirth**

Institut für Organische Chemie, Universität Basel, St. Johanns Ring 19, CH-4056 Basel, Switzerland
wirth@ubaclu.unibas.ch

**Sven-Eric Wohlert**

Institut für Organische Chemie, Universität Göttingen, Tammannstraße 2, D-37077 Göttingen, Germany

**Marion Zerlin**

Hans-Knöll-Institut für Naturstoff-Forschung, Beutenbergstraße 11, D-07745 Jena, Germany

**Thomas Ziegler**

Institut für Organische Chemie, Universität Köln, Greinstraße 4, D-50939 Köln, Germany
Fax.: ++49/221-4705057

# Contents

# 1 Natural Products and Drug Research

## 1.1 Structural diversity of surface lipids from spiders

*Stefan Schulz*

The surface lipids of insects are well characterized for a fair number of species [1–3], long chain hydrocarbons being the prominent compound class in most insects investigated. They usually occur as mixtures with varying chain length and often with methyl branches along the chain at different positions. The majority of them are saturated, but unsaturated hydrocarbons have also often been found. Despite their simple structure, the alkanes require a quite large energy input for their biosynthesis, two ATP for formation of the condensing malonyl-units, and two NADPH for removal of the oxygen of the $\beta$-ketocarbonyl intermediate. Oxidized derivatives of these hydrocarbons containing one or two additional functional groups are also common to many species, and might have been overlooked in other cases. Alcohols, aldehydes, esters, rarely epoxides, carboxylic acids, ketones, diols, oxoaldehydes, and oxoketones, as well as aliphatic ethers have all been identified in one or more species as components of the insect cuticle.

The function of the surface lipids are multitudinous. While designed primarily to protect against desiccation, they also function in chemical communication in many ways [1, 3, 4], and serve as a barrier against microorganisms or chemicals.

In contrast to insects, the lipids of spiders or other arachnids have not been received much attention [5]. While they could be expected to occur on the body of spiders, their presence on spider silk was a surprise. During our investigation of the sex pheromone of the European spider *Linyphia triangularis* (Linyphiidae) [6] we found considerable amounts of lipids on their webs. The main components of these silk lipids are unbranched alkanes with an odd number of carbons in the chain (termed odd-numbered compounds later on) and 2-methylalkanes such as 2-methylhexacosane with an even number of carbons in the chain (termed even-numbered compounds later on). They are accompanied by usual fatty acids such as palmitic acid and small amounts of other components such as fatty acid amides or wax-type esters. The amount of linear alkenes varies, being higher in juveniles than in adults. (Z)-9-Tricosene and the only unsaturated acetate present on the web, (Z)-11-octadecenyl acetate, are known pheromones of flies [1, 7]. It is tempting to speculate that these compounds are used to attract prey, but proof is lacking so far.

Besides these well-known compounds, considerable amounts of methyl-branched alkyl methyl ethers are present on the silk. In contrast to the compound classes mentioned before, these ethers (1-methoxyalkanes) are unique to spiders and have not been found elsewhere. The webs of ten different linyphiid species were investigated

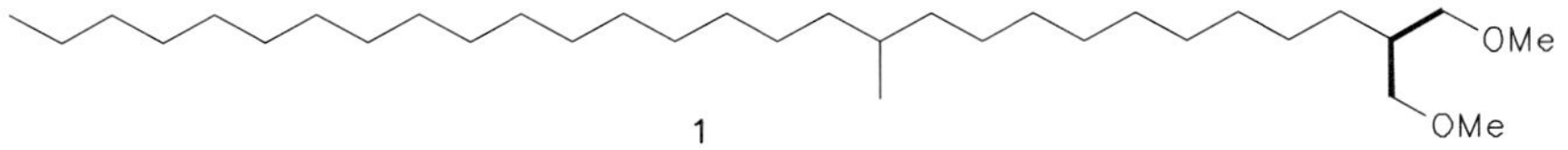

**Figure 1.1.1.** Bismethyl ether from *Labulla thoracica.*

for lipids. They contain between 5% and 40% ethers, while the majority of the lipids are hydrocarbons. The composition of these web hydrocarbons is quite similar within these species, showing pentacosane, heptacosane, 2-methylhexacosane, 2-methylocta-cosane, and 2-methyltriacontane as most abundant components. In contrast, the composition of the ethers varies widely. In *L. triangularis,* 1-methoxy-2,24-dimethyl-heptacosane is the major ether accompanied by other ethers which exhibit preferen-tially methyl branches at C-2, $\omega$-2 (even-numbered carbon chain), or $\omega$-3 (odd-numbered carbon chain) [8]. The related species *L. tenuipalpis* uses longer ethers like 1-methoxy-2,28-dimethylhentriacontane and 1-methoxy-2,30-dimethyltritriacontane, which is also the major ether of *Microlinyphia impigra.* In *Neriene emphana,* additional branchings in the middle of the chain occur, as in the major component 1-methoxy-2,14,24-trimethylheptacosane [9]. Another type of ether occurs on the web of the linyphiid *Labulla thoracica.* This species contains bismethyl ethers like 1-methoxy-2-(methoxymethyl)-12-methylnonacosane (**1**) in addition to the 1-methoxyalkanes of the type mentioned before. These bismethyl ethers can be regarded as carba-analogues of glycerol ethers.

Other spider families also contain these ethers. Thus, the lipids of webs from the golden orb weaver *Nephila clavipes* (Araneidae) comprise of more than 80% ethers. Besides lower amounts of 1-methoxy-2,($\omega$-3)-dimethylalkanes, ethers with up to four internal double bonds occur such as 1-methoxy-14,18,22,26-tetramethylhentria-contane [10]. As often found in arthropod hydrocarbons [1, 2], the methyl groups are separated by three methylene units. The webs of the black widow spider *Latro-dectus revivensis* (Theridiidae) also are covered with 1-methoxyalkanes and the usual hydrocarbons [11]. All the species containing the methoxyalkanes are members of the Araneoidea, but others like *Araneus diadematus* (Araneidae) lack them. A different type of lipid is used by the daddy longleg spider *Pholcus phalangoides* (Pholcidae) which belongs to a family not closely related to those mentioned so far. While the latter use complex mixtures of up to 150 compounds resembling chemical libraries, *P. phalangoides* web extracts contain up to 90% of one compound, the wax-type ester icosyl 2,4,6-trimethyltridecanoate. The remainder of the lipid layer consists of homologs, hydrocarbons, aldehydes, and methylketones [12]. Such highly branched wax esters are known especially from the uropygdial gland of waterfowl, mammals, or microorganisms [13].

The silk type on which the lipids are located is unknown. Nevertheless, silk from hunting spiders which produce dragline silk only, like *Cupiennius salei* (Ctenidae), do not contain lipids. Also dragline silk freshly drawn from *Nephila* sp., which is also dragline silk, shows only a very low abundance of lipids [14]. Therefore, the lipids must be associated with other silk types or are applied to the silk after the construction of the web.

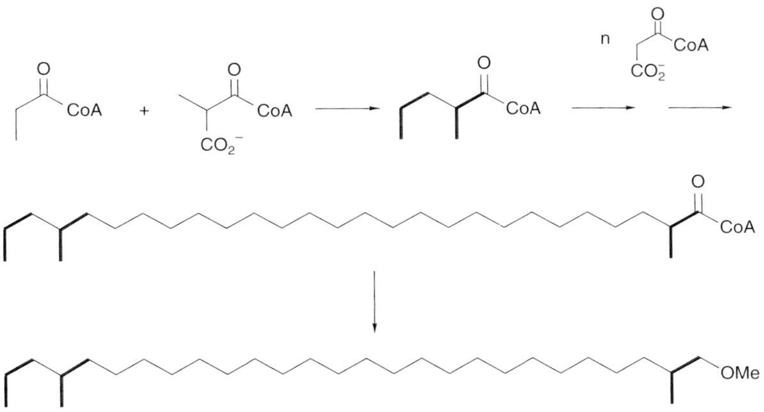

**Figure 1.1.2.** Suggested biosynthetic pathway to 1-methoxy-2,24-dimethylheptacosane (**2**). Propionate units are denoted in bold.

Interestingly, the branching pattern of the hydrocarbons and the ethers are different within one species. This points to paths of formation for hydrocarbons and ethers which diverge early. Hydrocarbons are normally formed by the elongation of fatty acids with malonyl-CoA to form long chain fatty acyl-CoA compounds. After decarboxylation, odd-numbered hydrocarbons are formed [1, 15]. Methyl branches are introduced by incorporation of methylmalonyl-CoA in the chain, giving rise to methyl branches at odd-numbered positions. Starting the biosynthesis of fatty acids with isobutyryl-CoA (derived from valine) instead of acetyl-CoA leads to even-numbered 2-methylalkanes, while 3-methylbutyryl-CoA (derived from leucine) yields odd-numbered 2-methylalkanes. The formation of 1-methoxy-2,24-dimethylheptacosane (**2**) is depicted in Figure 1.1.2. Because closely related even-numbered ($\omega$-2) and odd-numbered ($\omega$-3) branched ethers occur, a chain start with a amino acid-derived acyl-CoA intermediate seems unlikely. No common amino acid features an ($\omega$-3) branching. Instead, propionyl-CoA is elongated with methylmalonyl-CoA to form 2-methylpentanoyl-CoA. Further elongation with malonyl-CoA furnishes the long chain, on which in the final stage a methylmalonyl-CoA unit is attached. Reduction to the alcohol and methylation completes the ether. By use of acetyl-CoA as starter and elongation with methylmalonyl-CoA, 2-methylbutanoyl-CoA is formed, which finally will produce ($\omega$-2) even-numbered ethers. When acetyl-CoA is used as a stopper, the methyl group at C-2 is missing. The formation of the bismethyl ethers of *Labulla* might follow a similar role, using 3-hydroxypropionyl-CoA as final stopper. It has been shown that in insects 3-hydroxypropionate is an intermediate in the synthesis of acetyl-CoA from propionyl-CoA [1]. The characteristic starter units in this scheme might be sufficient to distinguish between the pathways leading to hydrocarbons and ethers. In no case could an unbranched ether be identified so far.

The silk lipids may have different functions. These include protection of the high-performance fibers against chemicals, particles or microorganisms, regulation of the water balance, which in some cases is essential for web function, and chemical communication (for a more detailed discussion see [16]).

Unusual 1,3-difunctionalized compounds have been found in webs of *N. clavipes* and *Tegenaria atrica* (Agelenidae). In both species 1,3-diols like 1,3-docosanediol occur, which are not known from other arthropods [10, 17], besides lower amounts of usual 1-alkanols. In webs of *T. atrica,* oxidized derivatives of such compounds could be identified, like vinylketones, 3-oxoalcohols and 3-oxoaldehydes. These compounds are known from insect defensive secretions, especially termites [18]. The soldiers brushing them onto the cuticle of their enemies. There they polymerize as a result of inherent chemical reactivity and thus lower the mobility of the enemy, or penetrate the cuticle. It is tempting to speculate that these compounds may enhance the 'catching' efficacy of the web.

The cuticular lipids of the spider itself resembles more the usual insect profiles. The ethers present on the silk can also be found on the cuticle of the linyphiids, but in markedly less abundance compared to the dominating hydrocarbons. The proportion of the ethers in *N. clavipes* falls from 90% of the whole lipids in the web to about 30% on the legs, to finally about 15% on the rest of the body. Obviously the ethers are mainly associated with the silk. Hydrocarbons and fatty acids have been identified in different *Tegenaria spp.* [19]. Novel *n*-propyl esters of long-chain methyl-branched fatty acids dominate the cuticular lipids of the social spider *Anelosimus eximus* (Theridiidae) [20]. Major components are propyl 6,20- and 4,20-dimethylhentriacontanoates. The major esters are odd-numbered, pointing to a starter unit of propionyl-CoA instead of acetyl-CoA in their biosynthesis.

The identified branched alkane derivatives can be characterized by a combination of GC and EI-GC-MS methods. They usually occur in complex mixtures with only a limited amount of extract available. This is due to the considerable effort necessary to obtain enough clean silk material for extraction. Therefore, isolation of pure compounds in amounts sufficient for NMR analysis is unrealistic. It is not always possible to distinguish whether a mass spectrum represents a mono-, di-, tri-, or tetramethyl-1-methoxyalkane or related compound, because methyl ethers easily lose methanol to form alkenes under EI-MS conditions [8]. Mass spectra of esters are dominated by fragments formed by the carboxyl group. The use of retention indices ($I$), which has been shown to be very helpful in the determination of branching positions and numbers of methyl groups along the chain in hydrocarbons [2], can be used for this purpose. Calculated $I$ values can be obtained ($I_c$) according to equation 1, by adding increments for a functional group ($FG$), and methyl groups ($Me_i$; $i$ indicating the position of the methyl group along the chain) to the base value of the respective unbranched hydrocarbon ($N$, *e.g.* 2700 for a $C_{27}$ carbon chain).

$$I_c = N + FG + \sum Me_i - S \qquad (1)$$

The values of $Me_i$ vary with the position of the methyl groups in the chain, and were obtained, as those for $FG$, by measuring $I$ of reference compounds (see Table 1.1.1 and [8]). Small deviations occur, especially when less than four methylene units separate adjacent methyl branches and with chain length. This is taken into account by a steric factor $S$, which increases with the number of such methyl groups. No general

**Table 1.1.1:** Experimental *I* increments of long-chain alkane derivatives, obtained on a BPX-5 stationary phase

| | R–CH₃ | R–CH₂–OMe | R–CO₂Me | R–CO₂Pr | R–CH₂–CN |
|---|---|---|---|---|---|
| *FG* | 0 | 232 | 348 | 476 | 453 |
| *(ω-1)-Me* | 63 | | 60 | 63 | 61 |
| *(ω-2)-Me* | 73 | 75 | 75 | 75 | 78 |
| *(ω-3)-Me* | 57 | 59 | 58 | | 40 |
| *2-Me* | 63 | 39 | 30 | | |
| *4-Me* | 52 | | 49 | 51 | 48 |
| *6-Me* | 38 | | 35 | 37 | 35 |
| *10-Me* | 34 | 31 | 32 | | |
| *12-Me* | 32 | 31 | 32 | | |
| *16-Me* | 32 | 31 | 27 | 28 | 29 |

rules for his value exist so far. Nevertheless, it is still possible to determine the number of branches in the chain in most cases.

Mass spectra of methyl branched 1-methoxyalkanes or esters do not allow determination of branching positions, particularly when present in complex mixtures often overlayed with alkanes, or when more than one methyl group is present in the chain. A way out of the dilemma is the transformation into derivatives, which allows easy determination of the branching positions. We generally use the microprocedures represented in Figure 1.1.3, which can be performed in high yields with low amounts of material [8, 21].

Key compounds are cyanides, which can be formed from iodides by tetraethylammonium cyanide. Their mass spectra clearly show the branches, because most of the ions above $m/z = 100$ still contain nitrogen, and hydrocarbon fragments are seen in low abundance only (see Figure 1.1.4). Thus fragments **A** and **B** are formed in higher abundance and allow location of methyl groups starting from C-5. Near the cyanide head methyl groups cannot easily be located, but this can be achieved by the distinct mass spectra of 2-, 3- and 4-methylalkyl methyl esters [22]. Methyl

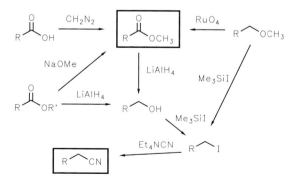

**Figure 1.1.3.** Microderivatizations used for location of methyl branches in long-chain mono functionalized compounds by EI-GC-MS.

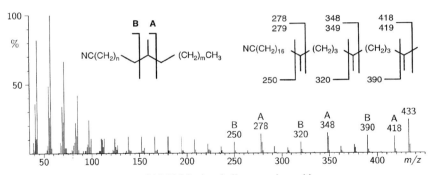

**Figure 1.1.4.** Mass spectrum of 17,21,25-trimethylhexacosyl cyanide.

ethers can be conveniently transformed into methyl esters by oxidation with $RuO_4$. Other transformations are also easily achieved. As an example, the propyl esters of *A. eximus* were first transesterified with 1% NaOMe in absolute methanol. The resulting methyl esters were then further transformed into alcohols with $LiAlH_4$, which in turn were transformed into iodides and finally into cyanides as discussed above. The mass spectra of alcohols or iodides did not allow determination of branching positions.

In conclusion, the lipids of spiders show a surprisingly broad spectrum of compound classes. They also contain unique compounds, the functions of which are not yet fully understood. Nevertheless, methods exists for their identification, even when only small amounts are available. By the synthesis of synthetic analogs, more information on the properties of these compounds will be obtained.

**Acknowledgements:** I like to thank all my coworkers for their enthusiastic input and especially the biologists which I had the pleasure to cooperate with: Yael Lubin, Harald Tichy, Sören Toft, Marie Trabalon, Gabriele Uhl, and Fritz Vollrath. I also like to thank the Deutsche Forschungsgemeinschaft for generous support of our work.

# References

[1]  R. Nelson, R. J. Blomquist, in *Waxes: Chemistry and Molecular Biology and Functions* (ed. R. J. Hamilton), The Oily Press, Dundee, **1995** pp. 1–90.

[2]  H. Lockey, *Comp. Biochem. Physiol.*, **1988**, *89B*, 595–645; G. J. Blomquist, D. R. Nelson, M. de Renobales, *Arch. Insect Biochem. Physiol.*, **1987**, *6*, 227–265; J. S. Buckner, in *Insect Lipids: Chemistry, Biochemistry and Biology* (eds. D. W. Stanley-Samuelson, D. R. Nelson), University of Nebraska Press, Lincoln, **1993**, pp. 227–270; D. R. Nelson, in *Insect Lipids: Chemistry, Biochemistry and Biology* (eds. D. W. Stanley-Samuelson, D. R. Nelson), University of Nebraska Press, Lincoln, **1993** pp. 271–315.

[3]  R. W. Howard, in *Insect Lipids: Chemistry, Biochemistry and Biology* (eds. D. W. Stanley-Samuelson, D. R. Nelson), University of Nebraska Press, Lincoln, **1993** pp. 179–226.

[4]  K. E. Espelie, E. A. Bernays, J. J. Brown, *Arch. Insect Biochem. Physiol.*, **1991**, *17*, 223–233.

[5]  E. C. Toolson, N. F. Hadley, *J. Comp. Physiol.*, **1979**, *129*, 319–325.

[6] S. Schulz, S. Toft, *Science*, **1993**, *260*, 1635–1637.

[7] J. L. Jackson, M. T. Arnold, G. J. Blomquist, *Insect Biochem.*, **1981**, *11*, 87; M. Schaner, R. J. Bartelt, J. L. Jackson, *J. Chem. Ecol.*, **1987**, *13*, 1777; K. Hedlund, R. J. Bartelt, M. Dicke, L. E. M. Vet, *J. Chem. Ecol.*, **1996**, *22*, 1835–1844.

[8] S. Schulz, S. Toft, *Tetrahedron*, **1993**, *49*, 6805–6820.

[9] S. Schulz, S. Toft, unpublished results.

[10] S. Schulz, unpublished results.

[11] S. Schulz, M. Papke, Y. Lubin, unpublished results.

[12] S. Schulz, G. Uhl, unpublished results.

[13] D. O'Hagan, *The Polyketide Metabolites*, Ellis Horwood, Chichester, **1991**.

[14] S. Schulz, F. Vollrath, unpublished results.

[15] P. von Wettstein-Knowles, in *Waxes: Chemistry, Molecular Biology and Functions* (ed. R. J. Hamilton), The Oily Press, Dundee, **1995** pp. 91–130.

[16] S. Schulz, *Angew. Chem., Int. Ed. Engl.*, **1997**, *36*, 314–326.

[17] O. Prouvost, M. Trabalon, M. Papke, S. Schulz, *Arch. Insect Biochem. Physiol.*, **1999**, *40*, 194–202.

[18] G. D. Prestwich, *Annu. Rev. Entomol.*, **1984**, *29*, 201–232; G. D. Prestwich, M. S. Collins, *J. Chem. Ecol.*, **1982**, *8*, 147–160.

[19] M. Trabalon, A. G. Bagnères, N. Hartmann, A. M. Vallet, *Insect Biochem. Molec. Biol.*, **1996**, *26*, 77–84; M. Trabalon, A. G. Bagnères, C. Roland, *J. Chem. Ecol.*, **1997**, *23*, 747–758.

[20] A.-G. Bagnères, M. Trabalon, G. J. Blomquist, S. Schulz, *Arch. Insect Biochem. Physiol.*, **1997**, *36*, 295–314.

[21] S. Schulz, *Chem. Commun.*, **1997**, 969–970.

[22] R. Ryhage, E. Stenhagen, *Arkiv för Kemi*, **1961**, *15*, 291–304.

## 1.2        Chemistry of marine pyrrole-imidazole alkaloids

*Thomas Lindel, Holger Hoffmann, and Matthias Hochgürtel*

The chemistry of life in the ocean is the fascinating subject of marine natural products research. Thirty years of worldwide investigations have led to the characterization of about 11 000 marine natural products [1, 2] and there is undoubtedly still an immense reservoir of unique, biologically active natural products to be explored, in particular from temperate zones. However, for some organisms such as Caribbean sponges, a quite detailed picture of their secondary metabolites now exists. Bioorganic marine chemistry has reached the point at which the identification of new metabolites has to continue and studies towards the chemical properties and biological potential of marine natural products have to be intensified.

Pyrrole-imidazole alkaloids are exclusive to marine sponges, the most primitive multi-cellular animals which account for about 40% of all characterized marine natural products. The cyclization and dimerization of the $C_{11}N_4$ skeleton of oroidin (**1**) [3] give rise to a variety of heterocyclic natural products with different geometrical orientations of functional groups and thereby structural diversity [4]. In total, five modes of cyclization (Figure 1.2.1) and five modes of dimerization of this key structural motif have been discovered in nature. Members of the branched family of oroidin alkaloids are found across sponge orders in Agelasida, Axinellida, and Halichondrida. The knowledge of their biological activities is still very incomplete, frequently because only preliminary studies were carried out in course of their isolation and structure determination. Considerable cytotoxicity has been observed for dibromophakellstatin (**2**, *e.g.* $ED_{50}$ 110 ng · ml$^{-1}$ against both the KM20L2 (colon) and the SK-MEL-5 (melanoma) cell lines) [5] and for agelastatin A (**6**, *e.g.* $IC_{50}$ 33 ng · ml$^{-1}$ against L1210, 75 ng · ml$^{-1}$ against KB cells) [6]. Recently, arthropod toxicity was discovered for agelastatin A (**6**).

Oroidin (**1**) itself is a major secondary metabolite (about 1.5% of the dry weight in Bahamian specimens of *Agelas clathrodes*) and was shown to deter the generalist predator reef fish *Thalassoma bifasciatum* from feeding on the sponge. The total soluble protein content in the same species was determined to about 20% [7]. The structural diversity of the cyclized and dimerized marine pyrrole-imidazole alkaloids in combination with the biosynthetic cost of their linear precursors leads to the hypothesis that this system of secondary metabolites serves biological functions. It is predicted that the exploration of their synthesis will lead to the availability of novel natural products in advance of their isolation from natural sources.

With respect to the discovery of new biologically active molecules, it is a promising approach to select naturally occurring, versatile key structures and to analyze their true potential. A parallel, multi-directed cyclization of a given pyrrole-imidazole core will lead to complex product mixtures. The computer program COCON ("Constitutions from CONnectivities") will facilitate and improve the constitutional analysis of the expected proton-poor reaction products [8].

Our research program includes the development of short synthetic routes to the basic, non-cyclized pyrrole-imidazole alkaloids, the analysis of their cyclization, and the investigation of their ecological activity.

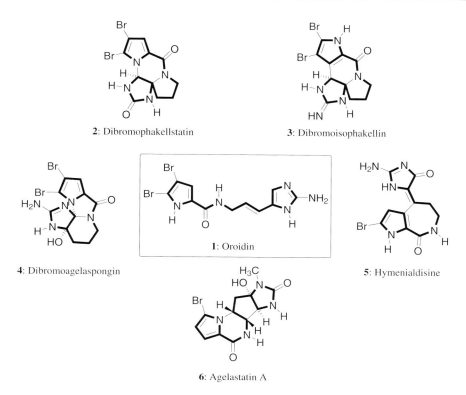

**Figure 1.2.1.** The pyrrole-imidazole substructure of the sponge metabolite oroidin (**1**) is the basis of a structurally versatile group of marine alkaloids. Five modes of cyclization and four modes of dimerization of the oroidin skeleton have so far been observed.

## 1.2.1 Synthesis of non-cyclized pyrrole-imidazole alkaloids

Syntheses designed to render gram quantities of any of the linear pyrrole-imidazole alkaloids must be regio- as well as stereoselective. We focused on the preparation of the metabolites keramadine (**9**) [9], dispacamide A (**10**) [10], and midpacamide (**14**) [11] which, unlike oroidin (**1**) itself [12], are not available in sufficient quantities by isolation (Figure 1.2.2). The linear pyrrole-imidazole alkaloids can be divided into two groups for which independent syntheses had to be developed. While oroidin (**1**), hymenidin (**7**), clathrodin (**8**), and keramadine (**9**) exhibit a 2-amino-5-vinylimidazole substructure, the dispacamides (**10**–**13**), the midpacamides (**14**–**15**), mauritamide A (**16**) [11c], and the tauroacidins (**17**–**18**) [13] are oxidized at C-4 of the imidazole ring.

Although the imidazol-4-ones are oxidation products of the corresponding 2-aminoimidazoles, their economic synthesis did not necessarily have to follow that pathway, but was developed starting from hydantoins. The 3-acylaminoaldehyde **19** is readily available in four steps from pyrrole [14] and served as the key synthetic

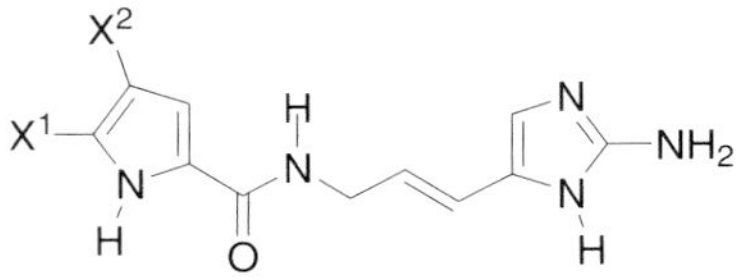

**1**: Oroidin ($X^1$=Br, $X^2$=Br)
**7**: Hymenidin ($X^1$=H, $X^2$=Br)
**8**: Clathrodin ($X^1$=H, $X^2$=H)

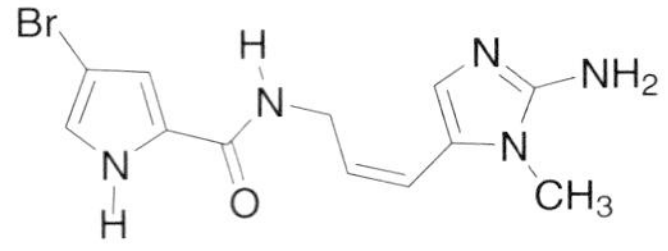

**9**: Keramadine

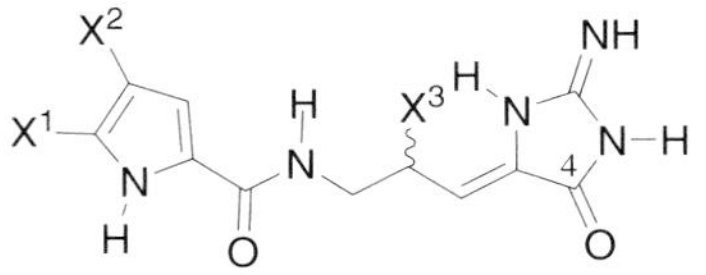

**10**: Dispacamide A ($X^1$=Br, $X^2$=Br, $X^3$=H)
**11**: Dispacamide B ($X^1$=H, $X^2$=Br, $X^3$=H)
**12**: Dispacamide C ($X^1$=Br, $X^2$=Br, $X^3$=OH)
**13**: Dispacamide D ($X^1$=H, $X^2$=Br, $X^3$=OH)

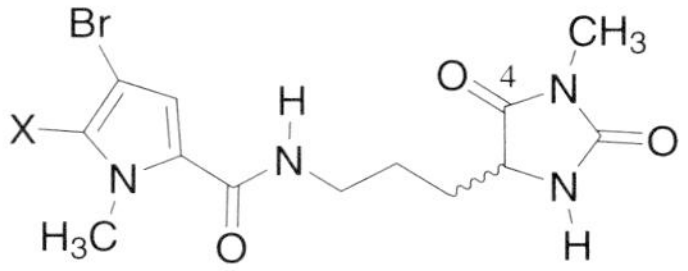

**14**: Midpacamide (X=Br)
**15**: Debromomidpacamide (X=H)

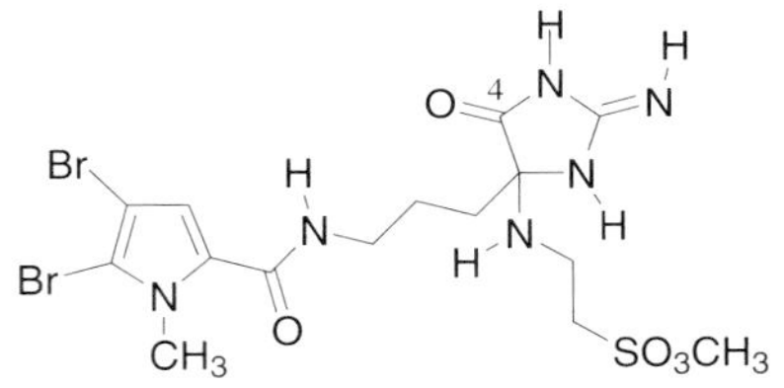

**16**: Mauritamide A

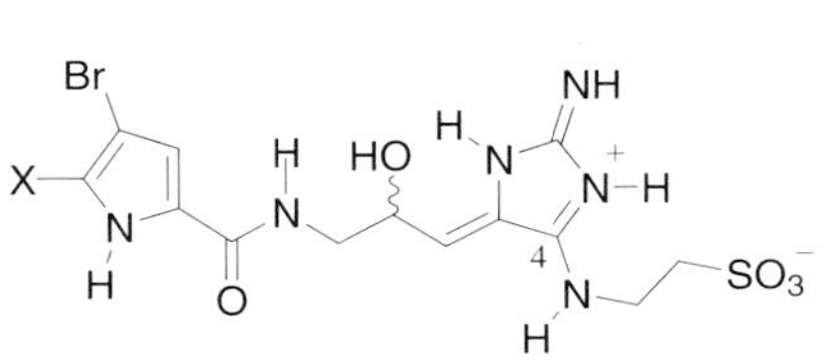

**17**: Tauroacidin A (X=Br)
**18**: Tauroacidin B (X=H)

**Figure 1.2.2.** Non-cyclized pyrrole-imidazole alkaloids isolated from marine sponges.

intermediate for the synthesis of both dispacamide A (**10**) and midpacamide (**14**) (Scheme 1.2.1).

Attempts to directly couple hydantoin or its 2-imino analogue glycocyamidine with the aldehyde **19** led to decomposition products [15]. When hydantoin was first converted to its 5-phosphonate **20** [16], a Horner-Wittig reaction conveniently gave access to 5-alkylidene hydantoin **21** in a yield of 83% ($E{:}Z \approx 2{:}3$). Midpacamide (**14**) was obtained in an overall yield of 40% after regioselective dimethylation of **21**, followed by chemoselective hydrogenation of the exocyclic double bond between C-10 and C-11 in the presence of 5% ruthenium on aluminum oxide as a catalyst. When 10% palladium on charcoal was used, undesired debromination occurred. Natural midpacamide (**14**) is probably racemic [11].

For the synthesis of the alkylidene glycocyamidine dispacamide A (**10**), 2-thiohydantoin (**22**) was chosen as a precursor, due to the lack of reactivity of glycocyamidine itself [17]. The nucleophilic reactivity of the 5-position of 2-thiohydantoin (**22**) is sufficient to achieve condensation with the aldehyde **19** in the presence of piperidine.

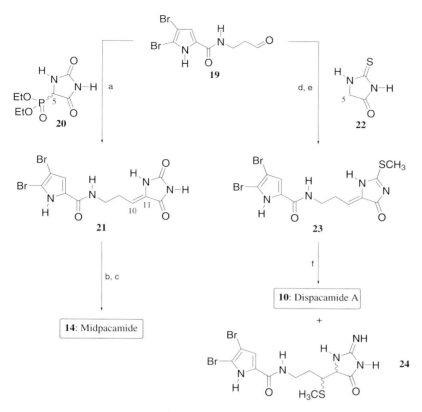

**Scheme 1.2.1.** Hydantoin pathways to dispacamide A (**10**) and midpacamide (**14**). (a) **20**, NaOEt, EtOH, r.t., 3 h, 83%, mixture of isomers ($E:Z \approx 2:3$); (b) MeI, $K_2CO_3$, DMF, r.t., 4 h, 58%, mixture of isomers ($E:Z \approx 1:3$); (c) $H_2$, 1 atm, 5% $Ru/Al_2O_3$, DMF, r.t., 3 d, 83%; (d) **22**, piperidine, $EtOH/H_2O$ (8:2), r.t., 4 h, 80%; (e) $CH_3I$, $K_2CO_3$, dry DMF, 0°C, 2 h, 75%; (f) $NH_3$ (sat.)/$NH_4Cl$ (2 equiv.), dry MeOH, sealed tube, 60°C, 7 h, 80% (**10**:**24** $\approx 60:40$).

Subsequent regioselective *S*-methylation of the product yielded the stereochemically pure mercaptoimidazolone **23**. On treatment of **23** with $NH_3/NH_4Cl$ in methanol under elevated pressure and temperature, a product mixture was obtained with dispacamide A (**10**) being the major constituent. Analysis by NMR spectroscopy and MS revealed that both diastereomers of the mercaptane **24** were formed as side products in substantial amounts (**10**:**24** $\approx 60:40$). Under the reaction conditions, methylmercaptane formed in the primary ammonolysis had to remain in solution and was available for a subsequent Michael addition to dispacamide A (**10**). Tedious separation through preparative reversed phase HPLC (RP-18, phosphate buffer pH 7.8/MeOH) allowed production of pure material of **10**.

The key to the chemoselective synthesis of dispacamide A (**10**) proved to be the electrophilic activation of C-14 through oxidation of the thioxo group (Scheme 1.2.2). We discovered that the use of *tert.*-butylhydroperoxide (TBHP) and aqueous

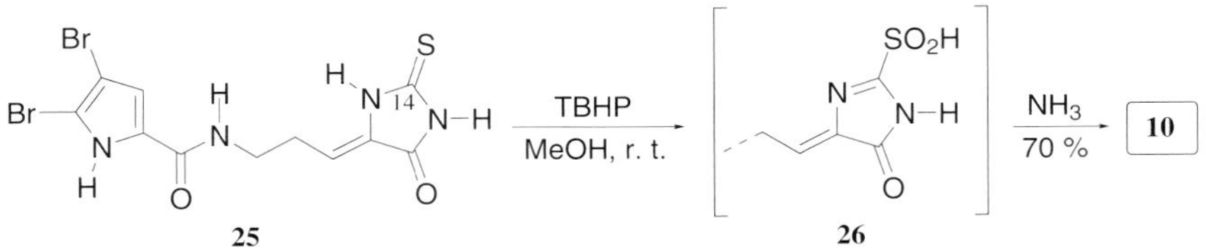

**Scheme 1.2.2.** One-pot transamination of alkylidene thiohydantoins to the corresponding glycocyamidines as the key step of the chemoselective synthesis of dispacamide A (**10**).

ammonia in methanol at room temperature gives efficient access to **10** without significant formation of side products. The aminoiminomethanesulfinic acid **26** can be proposed as probable reaction intermediate [18], because exactly two equivalents of TBHP are required to achieve complete conversion. The use of three equivalents of TBHP accelerates the reaction.

The convenient synthetic access to dispacamide A (**10**) could not be transferred to the synthesis of the oroidin alkaloids with a 2-amino-5-vinylimidazole unit, because both the alkylidene hydantoins and glycocyamidines resisted all attempts of selective reduction. Among the non-cyclized oroidin alkaloids, solely keramadine (**9**) possesses a Z-double bond in vinyl position of a trisubstituted imidazole ring. Our synthesis of keramadine (**9**) for the first time employs alkyne precursors to build up a trisubstituted (Z)-2-amino-5-vinylimidazole (Scheme 1.2.3) [19]. Keramadine (**9**) was isolated from

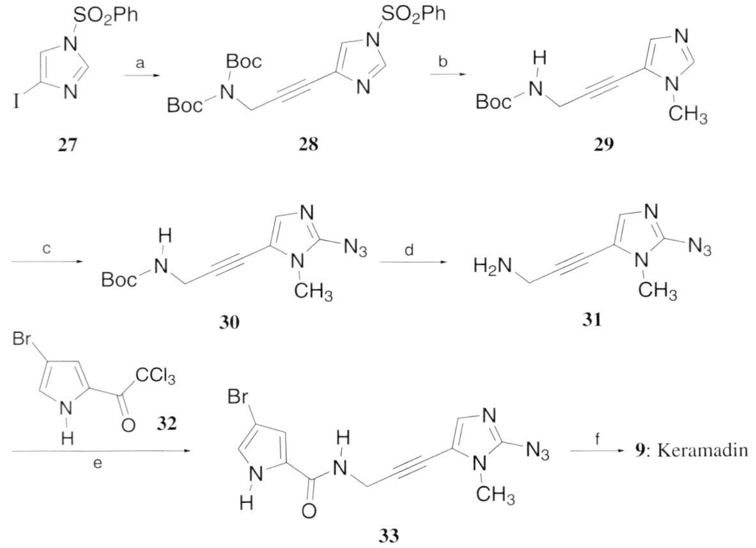

**Scheme 1.2.3.** Stereoselective alkyne pathway to keramadine (**9**). (a) bis-Boc-protected propargylic amine, $Pd(PPh_3)_2Cl_2$ (0.05 equiv.), CuI (0.1 equiv.), DIPA (3.0 equiv.), THF, r.t., 24 h, 90%; (b) $(CH_3)_3OBF_4$ (1.5 equiv.), $CH_2Cl_2$, r.t., 12 h, 80%; (c) n-BuLi (2.1 equiv.), THF, −75°C, TosN$_3$ (1.5 equiv.), 10 min, 60%; (d) TFA (40 equiv.), $CH_2Cl_2$, r.t., 24 h, quant.; (e) **32** (1.1 equiv.), DMF, r.t., 8 h, 60% from **30**; (f) $H_2$/Pd-Lindlar, THF/MeOH (5:1), r.t., 24 h, quant. conversion, 55% after chromatography.

*Agelas* sp. in low yields as an antagonist on serotonergic receptors of the rabbit aorta [9]. While *N*-unsubstituted analogs could undergo double bond isomerization through diazafulvene intermediates [20], the *N*-methylation of the natural product keramadine (**9**) seems to stabilize its configuration. Therefore, the methylation of its imidazole ring took place prior to the stereoselective generation of the vinyl double bond.

Pd-catalyzed coupling of 1-benzenesulfonyl-4-iodoimidazole (**27**) and fully *tert.*-butoxycarbonyl (Boc)-protected propargylic amine (**28**) was achieved in 90% yield in the presence of copper iodide (Sonogashira conditions [21]) providing regiochemically pure 4-alkynylimidazole **28** (Scheme 1.2.3). The benzenesulfonyl group serves the double purpose of both activating the imidazole ring for the carbon-carbon bond formation and protecting the reaction product against quaternization in the subsequent methylation. Treatment of **28** with trimethyloxonium tetrafluoroborate (Meerwein's salt) in dry dichloromethane, followed by methanolysis of the intermediate imidazolium salt led to the regiochemically pure 1-methyl-5-alkynylimidazole **29**. Simultanously, one of the two protecting groups was removed. Deprotonation of **29** with *n*-butyllithium and treatment with tosyl azide [22] gave the 2-azidoimidazole **30** in a yield of 60%. After quantitative removal of the carbamate (TFA), the skeleton of keramadine (**9**) was completed by treatment of **31** with the monobrominated pyrrolyltrichloromethyl ketone **32** [23]. In the final step, double hydrogenation of **33** (Lindlar catalyst) reduced the azide function to the amino group and the triple bond to the desired *Z*-double bond. It proved to be important to use a mixture of THF and methanol as solvent in order to avoid over-reduction. The ratio of isomers ($Z:E \approx 18:1$) could be determined by NMR spectroscopy only in [D$_4$]methanol, but not in [D$_6$]DMSO which was used as a solvent in course of the original structure elucidation of keramadine (**9**). The overall yield of our six-step synthesis is 14%. By keeping the double bond masked as a triple bond until the last step of the sequence, the risk of its isomerization was minimized.

## 1.2.2  Synthesis of the cyclized oroidin skeleton

The only synthesis of a marine pyrrole-imidazole alkaloid via biomimetic cyclization has been reported by Büchi et al. who obtained racemic dibromophakellin by treatment of dihydrooroidin with bromine in acetic acid [24]. Completed, presumably non-biomimetic syntheses were reported for hymenialdisine (**5**) and analogs [25]. Recently, Weinreb et al. reported the enantioselective synthesis of the skeleton of the agelastatins [26]. Overman et al. achieved the stereocontrolled synthesis of the tetracyclic core of the alkaloids palau'amine and styloguanidine [27].

Scheme 1.2.4 outlines our two independent approaches towards the ABCD ring system of the cytotoxic natural product dibromophakellstatin (**2**) of which originally only very small amounts were isolated from the sponge *Phakellia mauritiana* [5].

A convenient access to the chiral N,O-acetal **35** and to the *N*-vinylpyrrole **36** was discovered when the primary alcohol **34** was oxidized. **34** is directly accessible via condensation of the pyrrolyl trichloromethylketone **32** and *L*-prolinol. Swern

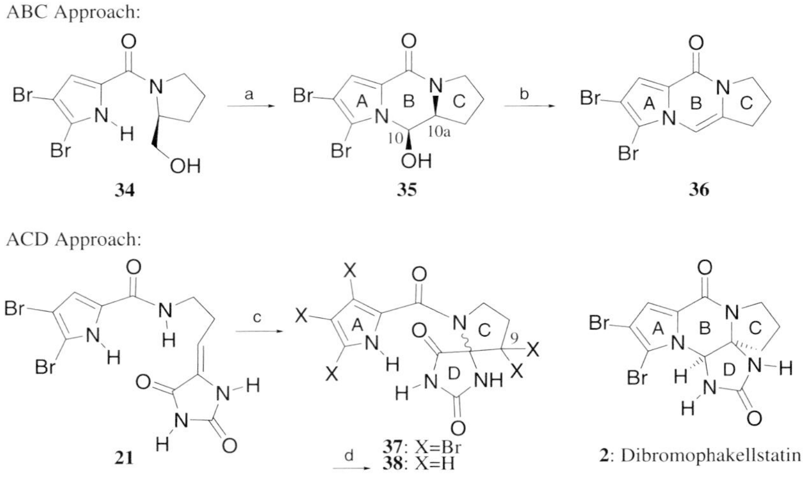

ABC Approach:

ACD Approach:

**Scheme 1.2.4.** Preparations of both the ABC and the ACD subunits of dibromophakellstatin (**2**). (a) DMSO, (COCl)$_2$, Et$_3$N, CH$_2$Cl$_2$, $-78°$C, 1 h, 55%; (b) TosCl, Et$_3$N, CH$_2$Cl$_2$, r.t., 48 h, 50%; (c) 1 eq. Br$_2$, HOAc, 100°C, 1 h, 62%; (d) H$_2$/Pd-C, NaOAc, MeOH, r.t., 15 h, 60%.

oxidation was immediately followed by intramolecular acetalization, leading to the clean formation of **35**. The $^1$H NMR spectrum showed the exclusive formation of one diastereomer with $^3J_{\text{10-H,10a-H}} \approx 2.6$ Hz being in agreement with a *cis*-arrangement of the two protons. The absolute configuration of **35** was confirmed via esterification with (*S*)- *resp.* (*R*)-2-methoxy-2-phenyl-2-(trifluoromethyl)acetic acid chloride and subsequent $^1$H NMR-chemical shift analysis of the resulting diastereomers (modified Mosher method [28]). Current studies aim at the enantioselective completion of the ABCD ring system of dibromophakellstatin (**2**), pursuing the pathways outlined in Scheme 1.2.4. The *N*-vinylpyrrole **36** was obtained via a tosylation/dehydration sequence. [2 + 3] Cycloaddition reactions of **36** are investigated as an alternative to the functionalization of the chiral N,O-acetal **35**.

In a biomimetic approach both the pyrrole and the amide nitrogen atoms would take part as nucleophiles. In the latter case, ring C should be formed first, because the initial formation of a strained 9-ring is unlikely. When the alkylidene hydantoin **21** was treated with excess of bromine in acetic acid at 80°C, the *spiro*-hydantoin **37** was formed with three bromine substituents at the pyrrole ring and two in geminal positions at C-9. A likely intermediate may result from the addition of one equivalent of bromine to the exocyclic double bond of **21**, followed by elimination of hydrogen bromide. The formation of **37**, which contains the ACD ring system of dibromophakellstatin (**2**), may have occurred via a bromonium ion formed from a vinylbromide. As soon as a tribenzylated alkylidene hydantoin is submitted to the same cyclization conditions, the analogous compound with a monobrominated ring C is formed. Hydrogenation in the presence of palladium on charcoal yielded the reduced compound **38** in which all five bromine substituents are removed.

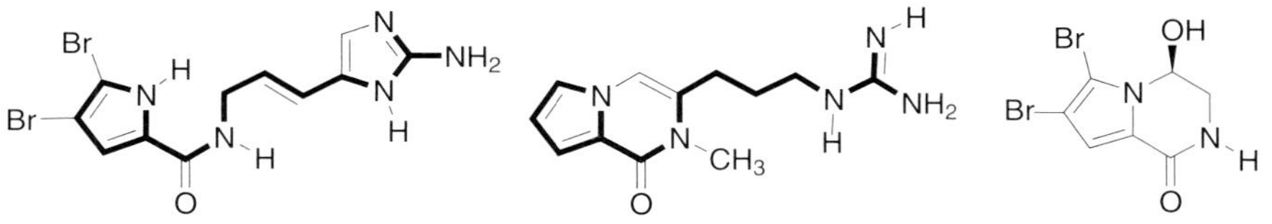

**1**: Oroidin            **39**: Peramine            **40**: Longamide A

**Figure 1.2.3.** The terrestrial pyrrole-guandinium alkaloid peramine (**39**) is an antifeedant against certain insects and resembles the structure of the marine fish feeding deterrent oroidin (**1**).

As a biological function of oroidin (**1**), we identified the fish feeding deterrence towards the reef fish *Thalassoma bifasciatum* [7]. Our synthetic program enables us to conduct a thorough study on structure-activity relationships in marine chemical ecology. Preliminary results indicate that the presence of the pyrrole ring is required and that the molecular size of oroidin (**1**) is beneficial for fish feeding deterrence. In aquarium assays, the alkylidene hydantoin **21** showed about the same activity as oroidin (**1**), while the free 3-aminopropylidene hydantoin part was not deterrent. The racemic form of the natural product longamide A (**40**) from *Agelas longissima* [29] was not fish feeding deterrent.

Interestingly, the structurally related pyrrole-guanidinium alkaloid peramine (**39**) from the terrestrial microfungus *Acremonium lolii* acts as a strong antifeedant against certain insects (Figure 1.2.3) [30]. In marine sponges of the genus *Agelas*, the incorporation of pyrrole and guanidinium substructures to induce feeding deterrence seems to be paralleled.

In conclusion, the pyrrole-imidazole alkaloids belong to the most prominent, exclusively marine secondary metabolites. If their structural versatility is mirrored by biological functions, their biogenetic relationship will define an economical biochemical system. Our newly developed, hydantoin and alkyne synthetic pathways to both groups of non-cyclized oroidin alkaloids form a basis for the structurally complex cyclized and dimerized family members. We could achieve the participation of both the amide and the pyrrole nitrogen atoms in intramolecular cyclizations and thereby solve two of the major problems on the way to dibromophakellstatin, but also to the agelastatins. Our synthetic program provides various derivatives of pyrrole-imidazole alkaloids in quantities sufficient to study structure-activity relationships in marine chemical ecology.

# References

[1] The scientific literature on marine natural products is compiled in the database MarinLit, developed by Murray H. G. Munro et al., see: http://www.marinlit.ac.nz.
[2] (a) For a complete documentation of marine natural products, see: D. J. Faulkner, *Nat. Prod. Rep.* **1998**, *15*, 113–158, and earlier reviews in this series; (b) A recent summary of the oroidin alkaloids is given in: G. W. Gribble in *Fortschr. Chem. Org. Naturst., Vol. 68* (Eds.: W. Herz, G. W. Kirby, R. E. Moore, W. Steglich, C. Tamm), Springer, Wien, **1996**, p. 137ff.

[3]  (a) S. Forenza, L. Minale, R. Riccio, E. Fattorusso, *J. Chem. Soc., Chem. Commun.* **1971**, 1129–1130; (b) E. E. Garcia, L. E. Benjamin, R. I. Fryer, *J. Chem. Soc., Chem. Commun.* **1973**, 78–79.

[4]  F. Balkenhohl, C. von dem Bussche-Hünnefeld, A. Lansky, C. Zechel, *Angew. Chem.* **1996**, *108*, 2437–2488; *Angew. Chem. Int. Ed.* **1996**, *108*, 2288–2337.

[5]  G. R. Pettit, J. McNulty, D. L. Herald, D. L. Doubek, J. C. Chapuis, J. M. Schmidt, L. P. Tackett, M. R. Boyd, *J. Nat. Prod.* **1997**, *60*, 180–183.

[6]  (a) M. D'Ambrosio, A. Guerriero, C. Debitus, O. Ribes, J. Pusset, S. Leroy, F. Pietra, *J. Chem. Soc., Chem. Commun.* **1993**, 1305–1306; (b) T. W. Hong, D. R. Jímenez, T. F. Molinski, *J. Nat. Prod.* **1998**, *61*, 158–161.

[7]  (a) B. Chanas, J. R. Pawlik, T. Lindel, W. Fenical, *J. Exp. Mar. Biol. Ecol.* **1996**, *208*, 185–196; (b) B. Chanas, J. R. Pawlik, *Oecologia* **1996**, *107*, 225–231.

[8]  (a) T. Lindel, J. Junker, M. Köck, *J. Mol. Model.* **1997**, *3*, 364–368; (b) T. Lindel, J. Junker, M. Köck, *Eur. J. Org. Chem.* **1999**, 573–577; (c) M. Köck, J. Junker, W. Maier, M. Will, T. Lindel, *Eur. J. Org. Chem.* **1999**, 579–586.

[9]  H. Nakamura, Y. Ohizumi, J. Kobayashi, Y. Hirata, *Tetrahedron Lett.* **1984**, *25*, 2475–2478.

[10] F. Cafieri, E. Fattorusso, A. Mangoni, O. Taglialatela-Scafati, *Tetrahedron Lett.* **1996**, *37*, 3587–3590.

[11] (a) L. Chevolot, S. Padua, B. N. Ravi, P. C. Blyth, P. Scheuer, *Heterocycles* **1977**, *7*, 891–894; (b) R. Fathi-Afshar, T. M. Allen, *Can. J. Chem.* **1988**, *66*, 45–50; (c) C. Jiménez, P. Crews, *Tetrahedron Lett.* **1994**, *35*, 1375–1378.

[12] Oroidin has also been synthesized: (a) G. de Nanteuil, A. Ahond, C. Poupat, O. Thoison, P. Potier, *Bull. Soc. Chim. Fr.* **1986**, 813–816; (b) T. L. Little, S. E. Webber, *J. Org. Chem.* **1994**, *59*, 7299–7305; (c) A. Olofson, K. Yakushijin, D. A. Horne, *J. Org. Chem.* **1998**, *63*, 1248–1253; (d) T. Lindel, M. Hochgürtel, unpublished results.

[13] J. Kobayashi, K. Inaba, M. Tsuda, *Tetrahedron* **1997**, *53*, 16679–16682.

[14] (a) T. Lindel, H. Hoffmann, *Liebigs Ann./Recueil* **1997**, 1525–1528; (b) Y. Xu, K. Yakushijin, D. A. Horne, *J. Org. Chem.* **1997**, *62*, 456–464.

[15] E. Ware, *Chem. Rev.* 1950, *46*, 403–470 and ref. cited therein. Aryl aldehydes react with glycocyamidines in boiling piperidine, see: Guella, G.; Mancini, I.; Zibrowius, H.; Pietra, F. *Helv. Chim. Acta* **1988**, *71*, 773–782.

[16] N. A. Meanwell, H. R. Roth, E. C. R. Smith, D. L. Wedding, J. J. K. Wright, *J. Org. Chem.* **1991**, *56*, 6897–6904.

[17] T. Lindel, H. Hoffmann, *Tetrahedron Lett.* **1997**, *38*, 8935–8938.

[18] Aminoiminomethanesulfinic and sulfonic acids react with amines to give the corresponding guanidines: C. A. Maryanoff, R. C. Stanzione, J. N. Plampin, J. E. Mills, *J. Org. Chem.* **1986**, *51*, 1882–1884.

[19] (a) T. Lindel, M. Hochgürtel, *Tetrahedron Lett.* **1998**, *39*, 2541–2544. The only known, recent synthesis of keramadine (**9**) employs a Wittig-Schweizer reaction to generate the (Z)-double bond; (b) S. Daninos-Zeghal, A. Al Mourabit, A. Ahond, C. Poupat, P. Potier, *Tetrahedron* **1997**, *53*, 7605–7614.

[20] M. Braun, G. Büchi, *J. Am. Chem. Soc.* **1976**, *98*, 3049–3050.

[21] K. Sonogashira, Y. Tohda, N. Hagihara, *Tetrahedron Lett.* **1975**, *16*, 4467–4470.

[22] N. S. Narasimham, R. Ammanamanchi, *Tetrahedron Lett.* **1983**, *23*, 4733–4734.

[23] D. M. Bailey, R. E. Johnson, *J. Med. Chem.* **1973**, *16*, 1300–1302.

[24] L. H. Foley, G. Büchi, *J. Am. Chem. Soc.* **1982**, *104*, 1776–1777.

[25] H. Annoura, T. Tatsuoka, *Tetrahedron Lett.* **1995**, *36*, 413–416. See also ref [14b]..

[26] G. T. Anderson, C. E. Chase, Y. Koh, D. Stien, S. M. Weinreb, *J. Org. Chem.* **1998**, *63*, 7594–7595.

[27] L. E. Overman, B. N. Rogers, J. E. Tellew, W. C. Trenkle, *J. Am. Chem. Soc.* **1997**, *119*, 7159–7160.

[28] Unpublished results: (a) J. A. Dale, D. L. Dull, H. S. Mosher, *J. Org. Chem.* **1969**, *34*, 2543–2549; (b) I. Ohtani, T. Kusumi, Y. Kashman, H. Kakisawa, *J. Am. Chem. Soc.* **1991**, *113*, 4092–4096.
[29] F. Cafieri, E. Fattorusso, A. Mangoni, O. Taglialatela-Scafati, *Tetrahedron Lett.* **1995**, *36*, 7893–7896.
[30] D. D. Rowan, M. B. Hunt, D. L. Gaynor, *J. Chem. Soc., Chem. Commun.* **1986**, 935–936.

## 1.3    Natural products and their role in pesticide discovery

*Robert Velten, Hubert Dyker, Matthias Gehling, Oliver Gutbrod,*
*Folker Lieb, and Andreas Schoop*

The discovery of new lead structures is one of the most crucial steps in life science research. Even today, when technologies like combinatorial chemistry and high-throughput screening have accelerated the screening process, the identification of innovative lead structures remains a difficult task.

In general, natural compounds have a broad chemical and structural diversity which is an excellent complement to synthetic libraries. Additionally, they have often provided insight into potential drug target mechanisms in the past. To become a suitable lead structure, a compound must possess an interesting biological activity or a novel pharmacophoric pattern. Since its native biological properties are usually not sufficient for commercial product development, it is essential to identify the pharmacophoric or toxophoric regions for further optimization steps. This procedure is time-consuming and may include both total synthesis and microderivatization strategies and support from structure-based design studies. In the following, we will present some examples of the different approaches.

### 1.3.1    Cripowellin−microderivatization

In our screening program for new bioactive substances, extracts of the Amaryllidaceae plant *Crinum powellii* showed strong insecticidal activity. In a bioassay-guided isolation, we identified the two novel insecticidal alkaloids named cripowellin A (**1**) and B (**3**) [1].

R = H, **1**
R = Ac, **2**

**3**

Their structures were elucidated by positive ion FAB-HRMS analysis and interpretation of two-dimensional NMR spectra, including data from COSY, HMQC and HMBC experiments. In order to confirm the structures proposed by spectroscopic methods, as well as to assign the absolute configuration, X-ray analysis of cripowellin A diacetate (**2**) was carried out. The absolute stereochemistry shown in Figure 1.3.1 can be proposed based on the assumption that the carbohydrate moiety is biologically derived from $\beta$-D-glucose.

Upon treatment of cripowellin B (**3**) with hydrochloric acid, carboxylic acid **4** and corresponding ester **5** could be isolated as a result of amide bond cleavage. Interestingly, the glycosidic bond was found to be stable under these acidic conditions.

Aglycone monoacetate **7** was prepared in three steps from cripowellin B (**3**). Selective iodination of the primary C6′-hydroxy group and acylation of the remaining secondary C14-alcohol provided iodide **6**. Fragmentation of **6** according to Vasella [2] afforded aglycone monoacetate **7** and aldehyde **8**. Aglycone monoacetate **7** is as active as cripowellin A (**1**) against insects.

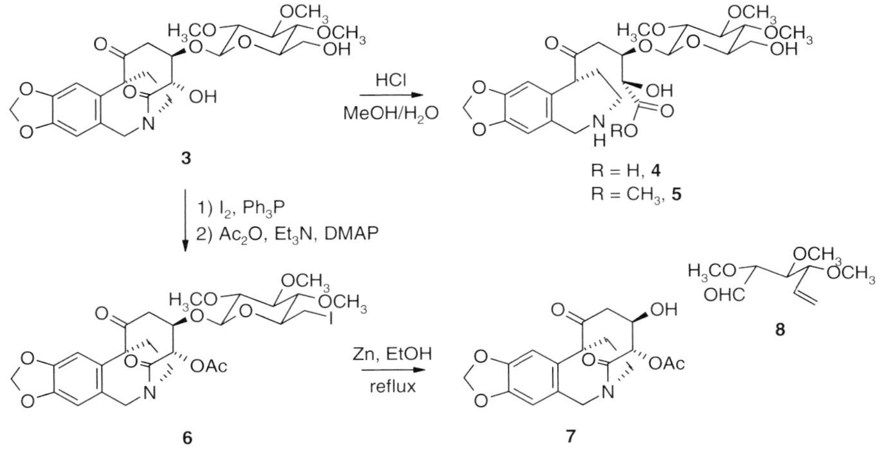

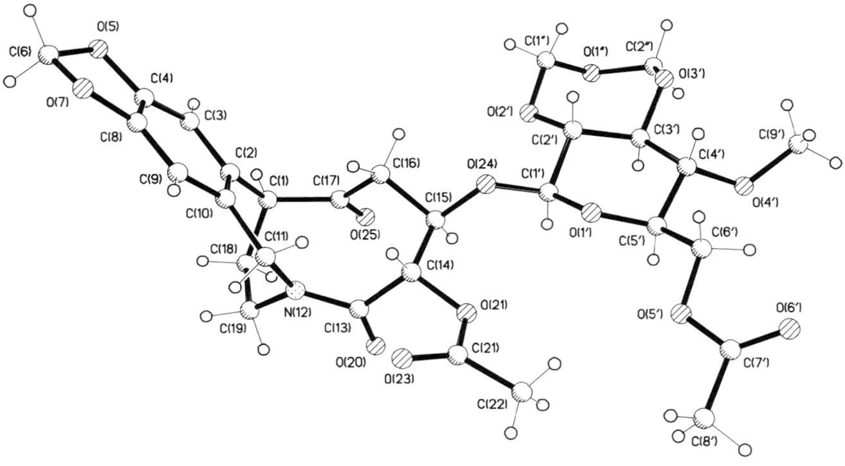

**Figure 1.3.1.** Perspective view of cripowellin A diacetate (**2**).

## 1.3.2    Rocaglamide – synthesis towards new analogs

In 1982, the natural product rocaglamide (**9**) was isolated from *Aglaia elliptifolia* [3] and found to exhibit antileukemic activity. About ten years later, this compound class was rediscovered in an insecticidal screening and indicated a potent activity comparable to that of the natural product azadirachtin [4–6]. Since then, various groups have isolated rocaglamide derivatives, including analogs of the pyrimidinone type **10** [7] from various *Aglaia species* [8–10] with strong insecticidal activity ($LC_{50} = 0.9$ ppm against the larvae of the test organism *Spodoptera littoralis* for **9**).

9                    10

Though different syntheses towards the natural product rocaglamide (**9**) have already been reported, we have been interested in a facile access to the cyclopenta-benzofuran core structure for further structure–activity relationship studies. The synthetic approach described here is based on a strategy to construct the tricyclic core structure in a [3 + 2] addition from an olefin and a 1,3-dicarbonyl compound.

Starting from bromoketone **11a** [11], the five-membered carbocycle was prepared by an intramolecular Wittig reaction utilizing allylidenenetriphenylphosphorane **12** [12]. Treatment of the enolether **13a** with HCl followed by reduction with NaBH$_4$ yielded alcohol **14a** as a single diastereoisomer. This olefin **14a** was then subjected to an oxidative dihydrofuran synthesis under conditions reported in the literature [13, 14]. In the presence of cerium ammonium nitrate (CAN) and 1,3-cyclohexanedione, compound **15a** was formed. In spite of rather moderate reaction yields of 35% the compound could be isolated as a single isomer.

When the CAN-oxidation was applied to the olefin **14b** in the presence of dimedone, the derivative **15c** was obtained in 44% yield. This compound shown in Figure 1.3.2 was easily crystallized to give an X-ray structure to establish the relative configuration. It confirmed that the cyclohexanedione radical adds to the less-hindered face of the double bond and cyclization takes place to give the *cis* annelated ring system with the two aryl substituents in the desired *cis* orientation.

Compound **15c** as well as the derivatives **15a** and **15b** did not show any insecticidal activity. Since 1-*O*-acetyl rocaglamide [8] is less potent than rocaglamide (**9**) itself, it seemed reasonable that the configuration at C-1 accounts for the loss of activity.

Oxidation of compound **15a** by iodine in methanol allowed the formation of the methyl ether and the dehydrogenation in one step to give the aromatized derivative

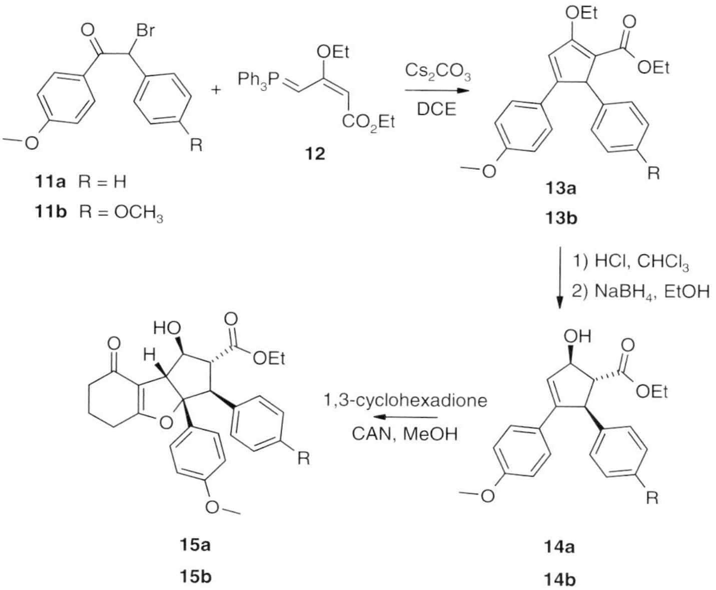

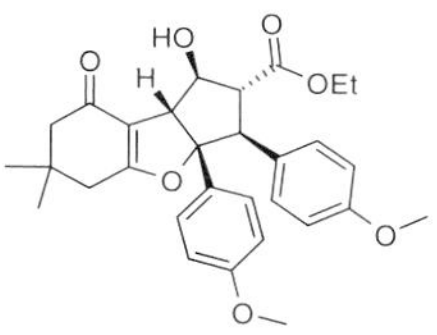

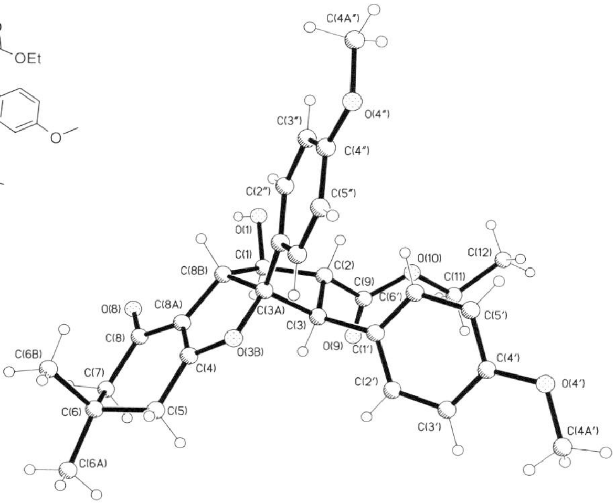

Figure 1.3.2. Perspective view of rocaglamide analog **15c**.

**16**. The configuration at C-1 was then inverted in an oxidation reduction sequence. Oxidation was most successful using Dess-Martin periodinane to give the desired keto ester **17**. Finally reduction of **17** was accomplished by borohydride reaction. The best conditions were found using tetrabutylammonium borohydride in dichloromethane, yielding the desired diastereoisomer **18** with a selectivity of 85:15.

Though compound **18** represents the complete carbon skeleton of rocaglamide (**9**) as well as most of the functional groups with correct configurations, **18** proved to be inactive as well as its intermediates **16** and **17**. It is important to note that **18** differs from **9** not only in the ester moiety and a missing methoxy substituent, but also in the missing C-8b hydroxy group. Analogs of rocaglamide (**9**) with different substitution pattern in the aryl moiety or the amide side chain are active. Therefore, one can assume that the activity of rocaglamide (**9**) depends on the C-8b hydroxy group which is omitted in our synthetic approach.

## 1.3.3   Pyrenocine – total synthesis of analogs

Phytotoxines have the potential to serve as structural leads for new herbicides. Pyrenocine A (**19**) and B (**20**) are produced by the phytopathogenic fungus *Pyrenochaeta terrestris*, which causes the onion pink root disease [15, 16]. Tal reported that **19** and **20** are also lethal to the leaves of *Helianthus* sp. [17].

The synthesis of the pyrenocines was carried out by Ichihara [18]. Condensation of malonyl chloride and acetylacetone provided the pyrone as starting material for the synthesis of **19** and **20**. It seemed convenient to synthesize 3-aryl-4-hydroxy-pyrone analogs **24** to enhance the herbicide activity. Therefore aryl-chlorocarbonyl-ketenes

**22** were used as building blocks. Nakanishi and Butler described a proper synthetic route of phenyl-chlorocarbonyl-ketene **22a** via phenyl-malonic acid **21a** [19]. As a result of this work a large number of different substituted aryl-chlorocarbonyl-ketenes are now accessible (*e.g.* **22b**). The reaction of **22b** with different ketones **23** provides the 3-aryl-4-hydroxy-pyrones **24** in very good yields, thus allowing broad structural variations of the pyrenocines. Structure–activity relationship studies revealed pyrones of the type **24a** as the most active compounds. The pyrones **24a,b** exhibit herbicidal activity against various weed species in post- and pre-emergence treatment. The mode of action is the inhibition of acetyl-CoA carboxylase, an important target of grass herbicides [**24a**: $pI_{50} = 5, 7$ (maize)].

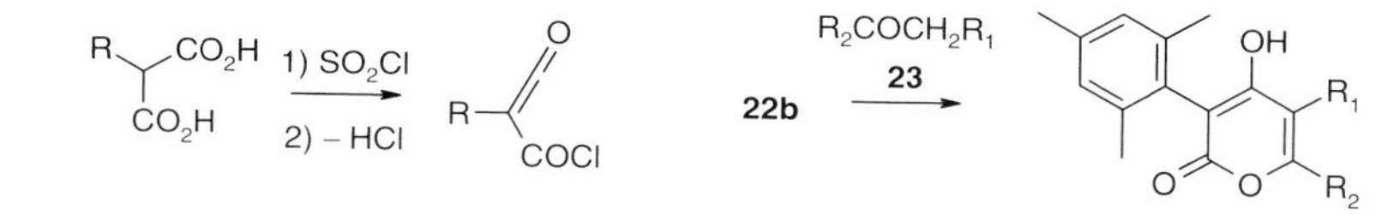

R=Ph, **21a**

R=2,4,6-Me$_3$C$_6$H$_2$, **21b**

**22a**

**22b**

R$_1$ = Me, R$_2$ = 4-F-C$_6$H$_4$, **24a**

R$_1$ = H, R$_2$ = Me, **24b**

The scope and limitation of the reaction of aryl-chlorocarbonyl-ketenes **22** was studied. The reaction of alkyl cylopentylketone **26** with **22b** gave the pyrone **25**, whereas the reaction of **22b** with the silylenolether **27** produced the 6-cyclopentyl-pyrone **28**.

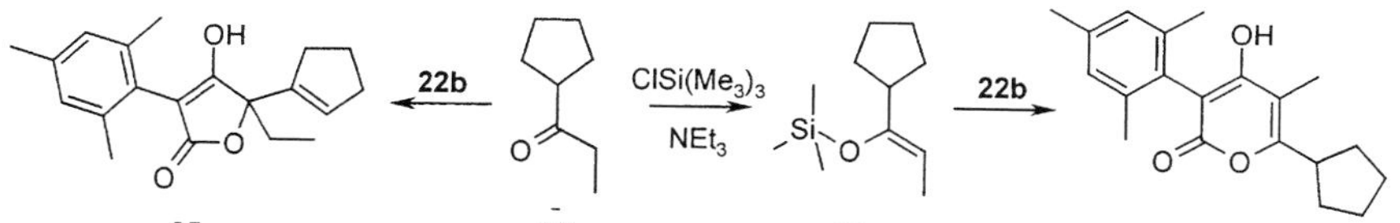

**25**     **26**     **27**     **28**

2-Methyl-cyclopentanone **29** reacts with chlorocarbonyl-ketene **22b** to give the expected pyrone **30** and additionally spiro-tetronic acid **31**. The formation of the products **25** and **31** may be explained by a carbocationic-rearrangement. Homologs and the acyclic ketones react in an analogous way to these two examples. Mechanistic studies were not undertaken due to a complete lack of herbicidal activity.

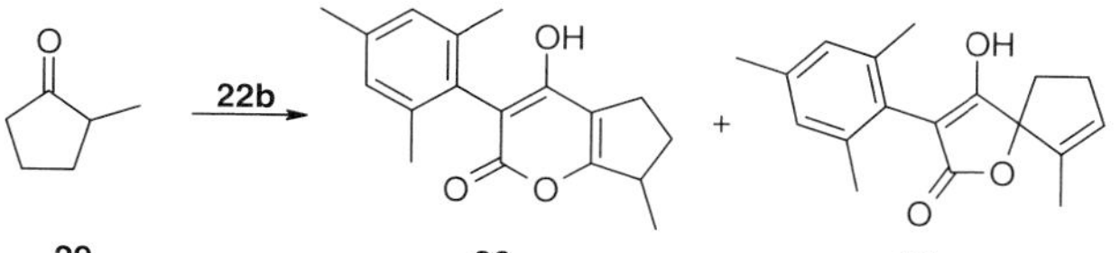

**29**     **30**     **31**

## 1.3.4   PF 1022A – synthesis of an aza analog of the cyclodepsipeptide

The cyclo-octadepsipeptide PF 1022A (**32**) was first isolated from *Mycelia sterilia PF1022* found in the microflora of the plant *Camellia japonica* [20]. It represents a promising new class of antiparasitics with broad-spectrum anthelmintic activity against a number of economically relevant intestinal nematodes in livestock and domestic animals, and it is unique both structurally and in its mode of action. To date, five total syntheses of this cyclodepsipeptide have been reported in the literature [21].

In order to investigate the structure–activity relationship of PF 1022A (**32**) we devised a synthesis of aza analogs of this cyclodepsipeptide in which the $\alpha$-carbons of the leucin residues are substituted by nitrogen. The primary goal in developing a synthesis of these analogs was to elucidate the functional role of the backbone and the importance of the chiral centers by this method of spatial screening. Although the alteration of the backbone by replacement of $\alpha$-carbons by nitrogen is a common manipulation in peptide chemistry (aza peptides) [22], the concept of aza analogs of depsipeptides to our knowledge has not been established in the literature.

We describe here the synthesis of the first aza analog of PF 1022A (**33**) in which two of the four leucin $\alpha$-carbons are substituted by nitrogen. The synthetic strategy is a convergent one in which both parts of the molecule, fragment A, bearing the natural part of the molecule, and fragment B, which contains the bis aza portion, are pre-assembled separately, then combined and finally cyclized. Throughout the synthesis, BOP-Cl was chosen as an efficient coupling reagent to form the *N*-methyl amide linkages, including the final cyclization.

Fragment A (**34**) was made by dimerization of the two didepsipeptide units MeLeu-PhLac and MeLeu-Lac as described in the literature [21]. The didepsipeptides were constructed from benzyl 3-phenyl-D-lactate, benzyl D-lactate, and Boc-*N*-methyl-D-leucine using Boc/Bn as protecting group. The appropriately protected aza tetradepsipeptide fragment B (**35**) was constructed by coupling the two *N*-Boc and *O*-Bn aza didepsipeptides, Me-aza-Leu-PhLac and Me-aza-Leu-Lac. These two subunits were synthesized by joining Boc protected *N*-methyl-*N'*-isobutyl hydrazine with benzyl 3-phenyl-D-lactate and benzyl D-lactate, respectively *via* phosgene.

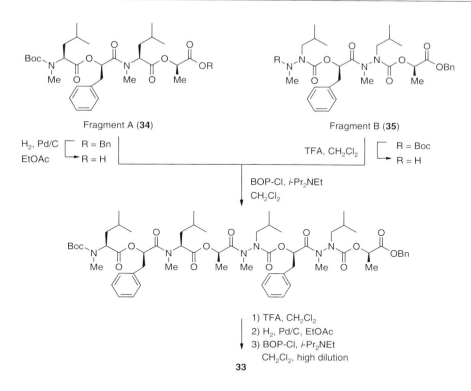

Fragment A (**34**)                Fragment B (**35**)

H₂, Pd/C  ⌐ R = Bn                      ⌐ R = Boc
EtOAc     └→ R = H          TFA, CH₂Cl₂  └→ R = H

BOP-Cl, *i*-Pr₂NEt
CH₂Cl₂

1) TFA, CH₂Cl₂
2) H₂, Pd/C, EtOAc
3) BOP-Cl, *i*-Pr₂NEt
   CH₂Cl₂, high dilution
**33**

Condensation of the two fragments **34** and **35** after protecting group manipulation furnished the linear precursor of **33**. Following the deprotection step, macrocyclization with BOP-Cl as the coupling reagent under high dilution conditions afforded the bis aza PF 1022 (**33**) in excellent yield. The X-ray crystal structure analysis of **33** shown in Figure 1.3.3 revealed that the introduction of nitrogen in the backbone of

**Figure 1.3.3.** Comparison of the X-ray structure of PF 1022A (**32**, open bonds, C in position 10 and 16) and bis aza PF 1022 (**33**, solid bonds, N in position 10 and 16). BOP-Cl = bis(2-oxo-3-oxazolidinyl)phosphonic chloride, Boc = *tert*-butoxycarbonyl.

PF 1022A (**32**) results in almost complete conservation of the 3D structure [23] of the natural product with only minor deviations at the new nitrogen positions.

However, when tested *in vivo* in sheep infected with the two nematodes *Haemonchus contortus* and *Trichostrongylus colubriformis*, a significant loss in biological activity of **33** was observed as compared to the natural product.

In summary, we have established a convenient synthesis for aza analogs of the natural cyclodepsipeptide PF 1022A (**32**) and of aza depsipeptides in general. The bis aza PF1022 (**33**) is the first example of this interesting class of compounds. By subtly changing the synthetic route and by choosing differently substituted hydrazines, a wide range of analogs can be generated in a facile manner.

**Acknowledgements:** We are grateful for contributions and additional support from our colleagues in the Central Research, the Agrochemical, and the Animal Health Divisions.

# References

[1]  R. Velten, C. Erdelen, M. Gehling, A. Göhrt, D. Gondol, J. Lenz, O. Lockhoff, U. Wachendorff, D. Wendisch, *Tetrahedron Lett.* 1998, 39, 1737–1740.

[2]  B. Bernet, A. Vasella, *Helv. Chim. Acta* **1984**, *67*, 1328–1347.

[3]  M. L. King, C.-C. Chiang, H.-C. Ling, E. Fujita, M. Ochiai, A. T. McPhail, *J. Chem. Soc., Chem. Commun.*, **1982**, 1150–1151.

[4]  C. Satasook, M. B. Isman, P. Wiriyachitra, *Pestic. Sci.* **1992**, *36*, 53–58.

[5]  J. Janprasert, C. Satasook, P. Sukumalanand, D. E. Champagne, M. B. Isman, P. Wiriyachitra, G. H. N. Towers, *Phytochemistry*, **1993**, *32*, 67–69.

[6]  F. Ishibashi, C. Satasook, M. B. Isman, G. H. N. Towers, *Phytochemistry*, **1993**, *32*, 307–310.

[7]  U. Kokpol, B. Venaskulchai, J. Simpson, R. T. Weavers, *J. Chem. Soc., Chem. Commun.*, **1994**, 773–774.

[8]  B. W. Nugroho, R. A. Edrada, B. Güssregen, V. Wray, L. Witte, P. Proksch, *Phytochemistry*, **1997**, *44*, 1455–1461.

[9]  B. W. Nugroho, B. Güssregen, V. Wray, L. Witte, G. Bringmann, P. Proksch, *Phytochemistry*, **1997**, *45*, 1579–1585.

[10]  B. Güssregen, M. Fuhr, B. W. Nugroho, V. Wray, L. Witte, P. Proksch, Z. Naturforsch. **1997**, *52c*, 339–344.

[11]  I. Lantos, P. E. Bender, K. A. Razgaitis, B. M. Sutton, M. J. DiMartino, D. E. Griswold, D. T. Wald, *J. Med. Chem.* **1984**, *27*, 72–75.

[12]  M. Hatanaka, Y. Himeda, R. Imashiro, Y. Tanaka, I. Ueda, *J. Org. Chem.* **1994**, *59*, 111–119.

[13]  E. Baciocchi, R. Ruzziconi, *J. Org. Chem.* **1991**, *56*, 4772–4778.

[14]  V. Nair, J. Mathew, *J. Chem. Soc. Perkin Trans* 1, **1995**, 187–188.

[15]  H. Sato, K. Konoma, and S. Sakamura, *Agric. Biol. Chem.* **1979**, *43*, 2409–2411.

[16]  H. Sato, K. Konoma, S. Sakamura, A. Furusaki, T. Matsumoto, and T. Matsuzaki, *Agric. Biol. Chem.* **1981**, *45*, 795–797.

[17]  B. Tal, and D. J. Robeson, *Z. Naturforsch.* **1986**, *41c*, 1032–1036.

[18]  A. Ichihara, K. Murakami, and S. Sakamura, *Tetrahedron* **1987**, *43*, 5245–5250.

[19]  S. Nakanishi, and K. Butler, *Organic Preparations and Procedures Int.* **1975**, *7*, 155–158.

[20]  T. Sasaki, M. Takagi, T. Yaguchi, S. Miyadoh, T. Okada, M. Koyama, *J. Antibiot.* **1992**, *45*, 692–697.

[21] (a) J. Scherkenbeck, A. Plant, A. Harder, N. Mencke, *Tetrahedron* **1995**, *51*, 8459–8470; (b) F. E. Dutton, S. J. Nelson, *J. Antibiot.* **1994**, *47*, 1322–1327; (c) M. Ohyama, K. Iinuma, A. Isogai, A. Suzuki, *Biosci. Biotech. Biochem.* **1994**, *58*, 1193–1194; (d) M. Kobayashi, T. Nanba, T. Toyama, A. Saito, *Annu. Rep. Sankyo Res. Lab.* **1994**, *46*, 67–75; (e) *Jap. Patent* JP 05–229997-A, CA **1994**, *120*, 135–149.

[22] J. Gante, *Synthesis* **1989**, 405–413.

[23] Y. Kodama, Y. Takeuchi, A. Suzuki, *Sci. Reports of Meiji Seika Kaisha* **1992**, *31*, 1–8.

## 1.4      Common principles in marginolactone (macrolactone) biosynthesis

*Marion Zerlin and Ralf Thiericke*

In the course of our chemical screening approach [1,2] we discovered new poly-hydroxylated macrolactones, named oasomycins, as well as the already known structurally related metabolite desertomycin (**1**) in the culture broth of *Streptoverticillium baldacci* subsp. *netropse* (FH-S 1625) [4, 5]. Most of these compounds bear a 42-membered macrolactone ring as a typical element. However, oasomycin C (**6**) and its aglycon oasomycin D (**7**) are the first representatives of 44-membered macrolactones. The oasomycins and desertomycins vary in the side chain located at C-41 or C-43, respectively, as well as in the presence of an α-linked D-mannose moiety attached to 22-OH (Scheme 1.4.1).

In the large group of macrocyclic lactones biosynthetic studies have already been highlighted in several classes, *e.g.* for macrolides [6], or for polyene antibiotics [7]. However, only little is known about polyhydroxylated macrolactones of the non-polyene type. Because of the structural characteristics, the oasomycins and desertomycins embody useful model substances to study the biosynthesis of polyhydroxylated macrolactones in detail. In the course of our studies with the oasomycin/desertomycin family we discovered unexpected analogies and common principles in the formation of a number of structurally related macrocyclic lactones [5, 8]. Due to the obvious biosynthetic relationships we introduce the term 'marginolactones' [8] for this class of macrolactones originating from *Actinomyces* species. The term **marginolactone** arises from the common structural element of these metabolites, the **macrolactone** ring of more than 31 carbon atoms. In addition, all members of this group usually bear a side chain in the α-position to the lactone carbonyl with a terminal amino or guanidino functionality as the second typical characteristic. On the basis of biosynthetic studies the origin of the side chains is derived from the amino acid **argin**ine or the related **orn**ithine.

This article is intended to summarize the information available on the biosynthesis and biosynthetic relationships of marginolactones.

### 1.4.1      Biosynthetic building blocks

Our working hypothesis for the biosynthesis of the carbon skeleton of the oasomycins envisioned a similar polyketide-type pathway as in the case of monazomycin [9], guanidylfungin A [10], and azalomycin $F_{4a}$ [11], resulting from the already reported feeding experiments with acetate as $C_2$- and propionate as $C_3$-building blocks. Feeding of sodium [1-$^{13}$C]- and [1,2-$^{13}$C$_2$]acetate, and [1-$^{13}$C]propionate as typical polyketide precursors resulted in the expected labeling pattern in oasomycin B (**5**) (Scheme 1.4.2) [8]. The propionate feeding experiment clearly showed that the methyl branches of the macrocycle derive from propionate (methylmalonyl-CoA). No incorporation of acetate and propionate was observed in the α-D-mannose (C1′–C6′) and in the γ-lactone moiety (C43–C46) of oasomycin B (**5**).

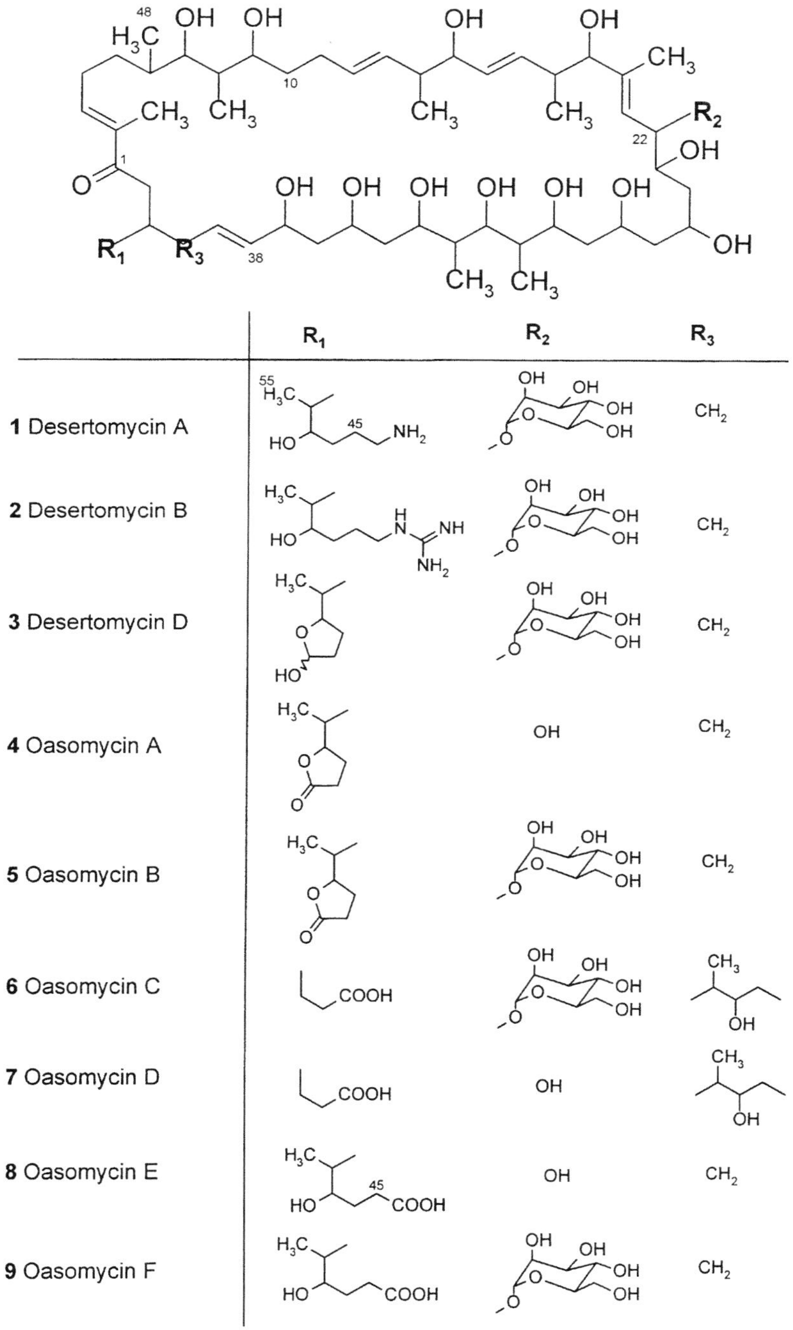

|   | R₁ | R₂ | R₃ |
|---|-----|-----|-----|
| **1** Desertomycin A | | | CH₂ |
| **2** Desertomycin B | | | CH₂ |
| **3** Desertomycin D | | | CH₂ |
| **4** Oasomycin A | | OH | CH₂ |
| **5** Oasomycin B | | | CH₂ |
| **6** Oasomycin C | | | |
| **7** Oasomycin D | | OH | |
| **8** Oasomycin E | | OH | CH₂ |
| **9** Oasomycin F | | | CH₂ |

**Scheme 1.4.1.**  Secondary metabolites of the oasomycin/desertomycin family.

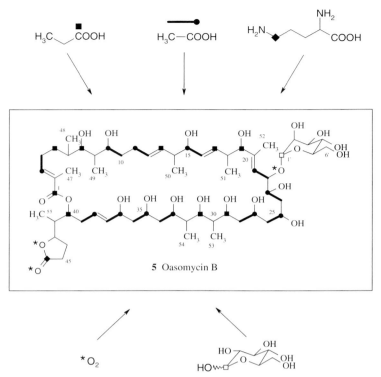

**Scheme 1.4.2.** Positional and bond-labeling patterns of oasomycin B (5) biosynthesized from [$^{13}$C]- and [$^{18}$O]-labeled precursors.

The origin of the mannose unit in oasomycin B (**5**) was studied by feeding D,L-[1-$^{13}$C]glucose which, as expected, resulted in a high specific enrichment at C1′ and therefore proved the direct incorporation of glucose. However, significant label transfer into the methyl group (C2) of each acetate polyketide building block was observable pointing to an efficient glucose degradation process *via* the Embden–Meyerhof–Parnas pathway and oxidative decarboxylation of pyruvate leading to acetate.

The previous results revealed the four carbon atoms of the γ-lactone of oasomycin B (**5**) to be the polyketide starter. Considering biosynthetic relationships of the oasomycins and desertomycins deduced from fermentation time curves and pH-static fermentations [5] (see below) the polyketide starter was sought in the amino acid pool. On the basis of several incorporation studies with [$^{13}$C]-, and [$^{15}$N]-labeled precursors [8], ornithine and the related arginine have been shown to be incorporated into the polyketide starter units for oasomycin/desertomycin biosynthesis (Scheme 1.4.2). In particular, a feeding experiment with D,L-[5-$^{13}$C]ornithine resulted in both, a significant yield (60 mg/l) of the main fermentation product of this experiment, desertomycin D (**3**), and in high $^{13}$C-signal enhancements of the hemiacetal C46 in both stereoisomers. Thus, an activation of the amino acid ornithine towards a CoA-activated C$_4$-starter unit involves a deamination of the α-amino group and a decarboxylation step.

Because oxygen atoms in the polyketide biosynthesis frequently originate from their $C_2$- or $C_3$-building blocks, the hydroxy groups at C-9, C-15, C23, C-25, C27, C33, C35, and C37 in oasomycin B (**5**) should derive from the carboxy group of acetate. However, a feeding experiment with [1-$^{13}$C, $^{18}O_2$]acetate gave significant carbon label incorporation, but the oxygen label was not detectable. This was assumed to be correlated to a high metabolic turnover of acetate at the time of the oasomycin production (logarithmic growing phase). As our main interest focused on the origin of the glycosidic oxygen at C22, which is typically not located at a C1 position of an acetate or propionate building block, a fermentation in a closed vessel system under an atmosphere of 20% [$^{18}O_2$] and 80% $N_2$ was performed. Analysis of the obtained oasomycin B sample was done by both, FAB mass spectrometry, and $^{13}$C NMR spectroscopy by observation of [$^{18}$O]-isotope induced shifts of $^{13}$C signals. As expected, C22 and C1' showed label incorporation into the glycosidic oxygen atom. In addition, $^{18}$O label could be observed in the $\gamma$-lactone ring at both, the carbonyl, and the ring oxygen atom (Scheme 1.4.2).

Together, the results of the feeding experiments have provided a detailed picture of the biosynthetic origin of oasomycin B (**5**). The polyketide biosynthesis is initiated by ornithine or the related arginine, whose carbon skeleton is located in the $\gamma$-lactone side chain (C43 to C46). Therefore, an activation of these amino acids involving oxidative deamination with an introduction of an oxygen atom from molecular $O_2$ is presumed which explains $^{18}O_2$ label incorporation. After an oxidative decarboxylation the activated CoA-building block initiates polyketide biosynthesis using methylmalonyl-CoA located in the methyl group C55 and C41/C42 for the first elongation step. In summary, twelve $C_2$ (acetate) and nine $C_3$ (propionate units are used to form the macrolactone moiety of oasomycin B (**5**). The characteristic oxygenation pattern and the existence of four double bonds illuminate the enzymatic processes at the level of the oasomycin polyketide synthase, obviously of a PKS I-type.

The formed 42-membered polyketide intermediate is attacked by an oxygenase at position 22, introducing a hydroxy group from molecular oxygen. To date, it is not possible to distinguish between the two plausible pathways: a direct introduction of OH groups catalyzed by hydroxylases, or the introduction *via* a monooxygenase with an epoxide intermediate. The latter requires a double bond for attack and should be followed by stereoselective oxirane ring opening to yield the *trans*-diol. A following biosynthetic step involves a glycosylation reaction using $\alpha$-D-mannose as a building block. Remarkably, the OH group at C22 is favored for the mannosyltransferase, indicating that this position seemed to be spatially exposed because of folding of the 42-membered macrolactone ring. Analogous observations were made in the case of chemical derivatization reactions in the oasomycin family [12].

## 1.4.2 Biosynthetic relationships

A detailed analysis of the fermentation time-course of the producing organism *Streptoverticillium baldacci* subsp. *netropse* (FH-S 1625), pH-static fermentation studies and *in vitro* conversions of the purified metabolites resulted in a complete picture of the biosynthetic relationships of the marginolactones of the oasomycin/ desertomycin family [5]. From a typical fermentation time-course the fermentation process can be divided into three parts. During the logarithmic growing phase the mannosylated products desertomycin A (**1**), the first product detected in the bio-synthetic sequence, and oasomycin B (**5**) were biosynthesized, whereas parallel to the beginning of the stationary phase a demannosylation led to the aglycon oasomycin A (**4**) as the typical main product in part II. In the later fermentation stage the appearance of both, oasomycin D (**7**), and E (**8**) correlated with an increase of the pH-value in the culture broth pointing to alkaline-catalyzed pH-dependent conversions (part III).

Considering the typical pH time-course in correlation with the metabolite pattern during cultivation, pH-static fermentations were performed. At a constant pH of 5.0, desertomycin A (**1**) was the only metabolite detectable during the whole fermentation procedure. A fermentation at pH 8.0 resulted in a nearly analogous metabolite sequence as a non-influenced fermentation starting with desertomycin A (**1**), followed by the glycosylated oasomycins B (**5**), C (**6**), and F (**9**), to the corresponding aglyca oasomycin A (**4**), D (**7**), and E (**8**). As a consequence of the fermentation studies, the production of the different oasomycins seems mainly to be dependent on the actual pH-value of the fermentation.

In order to analyze appropriate instabilities and interconversion reactions, the various purified oasomycins were stirred in mixtures of water/methanol (1:1) at pH-values of 1–12 at 30°C for 24 h. For example, oasomycin A (**4**) was found to be stable under acidic conditions, while at a pH-value of 8 ring-opening of the $\gamma$-lactone moiety was observed yielding oasomycin E (**8**). Stronger alkaline conditions favor this reaction. Additionally, at pH >8.0 oasomycin A (**4**) is also converted into oasomycin D (**7**), which exhibits the already opened $\gamma$-lactone as well as the ring-enlarged 44-membered macrolactone. Oasomycin D (**7**) itself appeared to be stable under slightly acidic conditions (down to pH 3.0). At pH <3.0, ring contraction to the 42-membered macrolactone as well as $\gamma$-lactone formation in the side chain is favored, giving rise to oasomycin A (**4**) at pH 1.0 in nearly quantitative yield. Under alkaline conditions (pH 8–9) both compounds, oasomycin D (**7**) and E (**8**), resulted in an equilibrium (ca. 2:3) with traces of oasomycin A (**4**). In parallel studies, the corresponding glycosylated metabolites showed analogous interconversion reactions. The results with the aglyca of the oasomycins as well as with the mannosylated metabolites are summarized in Scheme 1.4.3 [5].

In combination, these results led to the conclusion that, under acidic fermentation conditions, oasomycins A (**4**) and B (**5**), which bear a $\gamma$-lactone side chain and a 42-membered macrolactone ring, are the thermodynamically favored compounds. At pH-values of 6.4–7.0 in the early fermentation stage the main metabolite oasomycin B (**5**) is formed *via* the biosynthetic intermediate oasomycin F (**9**). Obviously, the formation of the $\gamma$-lactone in the side chain is not necessarily catalyzed by an enzyme. It is remarkable that oasomycin B undergoes a $\gamma$-lactone opening reaction leading to oasomycin F (**9**) under alkaline fermentation conditions. In addition, under alkaline conditions oasomycin E (**8**) and the corresponding mannosyl-derivative oasomycin F (**9**) undergo ring extension reactions to yield the 44-membered macrolactones oasomycin D (**7**) and C (**6**), respectively. Thus, in between the formation of the

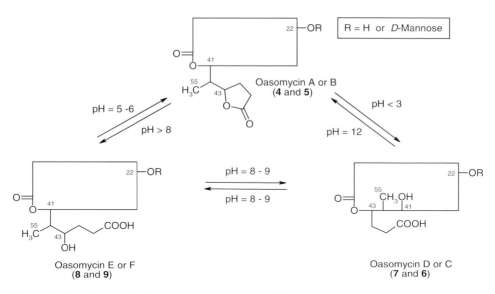

**Scheme 1.4.3.** Acid- and alkaline-catalyzed conversions of the oasomycins.

various oasomycins starting from oasomycin F (**9**), only the demannosylation reaction is enzymatically catalyzed, while $\gamma$-lactone formation and opening reactions as well as the macrolactone enlargement and contraction processes seem to be non-enzymatic.

The results from the analysis of the fermentation time-course, pH-static cultivations and *in vitro* conversions of the different oasomycins illuminate the biosynthetic relationships of the oasomycin/desertomycin family, depicted in Scheme 1.4.4 [5].

## 1.4.3   Common principles

The results described from the biosynthesis of the marginolactones of the oasomycin/desertomycin family can be transferred to a number of structurally related secondary macrolactones (selected examples are shown in Scheme 1.4.5) [8]. As a consequence, the following common principles can be given for marginolactones:

(a) All marginolactones derive from the polyketide pathway with ornithine or arginine as polyketide starters.
(b) The starter amino or guanidino group can be methylated.
(c) The CoA-activated polyketide starters are usually elongated by acetate or propionate building blocks.
(d) The length of the polyketide chains and the sequence of the $C_2$ or $C_3$ units are individually determined by the different polyketide synthases of the particular producing organism.
(e) In general, the linkage to the macrolactone ring is formed by the hydroxy group of the first propionate unit, which is preserved during the polyketide biosynthesis.

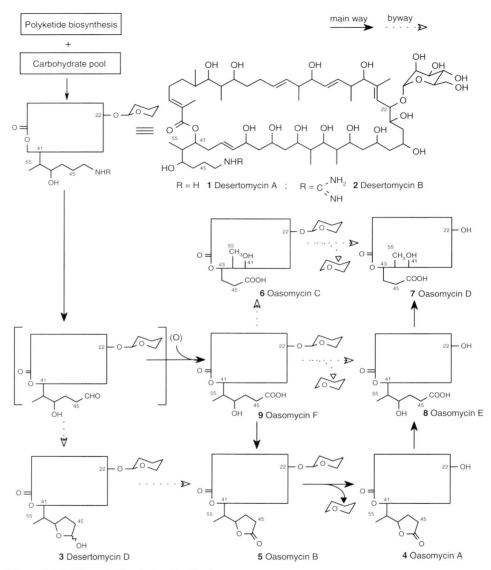

**Scheme 1.4.4.** Biosynthetic relationships in the oasomycin/desertomycin family.

(f) Each metabolite of the marginolactone group bears one hydroxy group (in the middle region between $A_7$ and $A_{10}$), which obviously is introduced from molecular oxygen.

(g) In the aglycona of some marginolactones the oxygen atom deriving from molecular $O_2$ is found as a hydroxy group located at the 6-membered hemiketal ring.

(h) Glycosylated marginolactones bear this particular oxygen atom as the glycosidic O-atom of their mannose or arabinose residues.

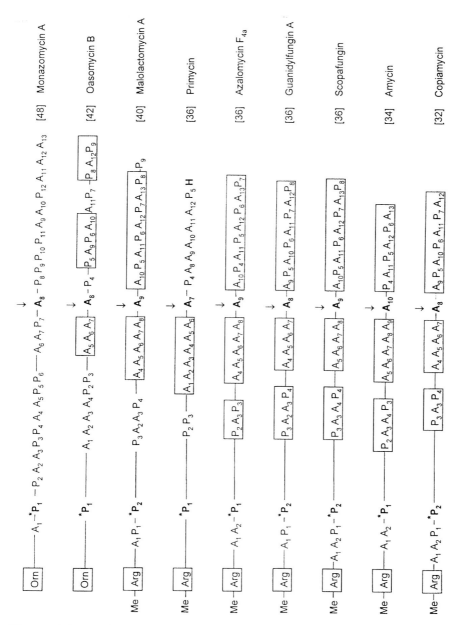

[Me = methyl group, A = acetate, p = propionate, Arg = arginine, Orn = ornithine, H = $C_6$-building block, *P = linkage position to form the macrolactone ring, ↓ O-atom from molecular $O_2$, [ ] = size of the macrolactone ring.

**Scheme 1.4.5.** Illustration of the common principles in the biosynthesis of marginolactones (selected examples).

Besides the obviously evolutionarily related polyketide synthases of different *Actinomycetes* strains, further post-polyketide modifications such as the oxygenation and glycosylation reactions reflect analogous processes for the functionalization of polyhydroxylated macrolactones of the marginolactone group. Thus, the marginolactones can be subdivided as a separate class of macrocyclic lactones, grouped by their close structural and biosynthetic relationships.

### 1.4.4   Manipulation of the metabolite pattern

The knowledge of the late biosynthesis in the oasomycin/desertomycin family has been used successfully to manipulate the metabolite pattern of *Streptoverticillium baldacci* subsp. *netropse* (FH-S 1625), directing the fermentation process to produce desired metabolites [5]. Variations of the fermentation time led either to glycosylated compounds such as oasomycin B (**5**), or to the corresponding aglycons such as oasomycin A (**4**). On the other hand, the enzymatic deamination step can be completely blocked by pH-static fermentations at pH 5.0 resulting in desertomycin A (**1**). The carboxylic acid derivatives of the oasomycins can be obtained *via* pH-static fermentations at pH 8.0. However, a convenient chemical derivatization starting from oasomycin A (**4**) or B (**5**), respectively, seems to be the more elegant route.

Furthermore, manipulation of the secondary metabolite pattern was achieved by feeding amino acids into cultures of the producing organism [13]. Feeding of 10 g/l of L-tryptophan initiated the biosynthesis of metabolites of the streptovaricin-complex {*e.g.* streptovaricin C (**10**) and G (**11**)} [14] as well as of tryptophan-related metabolites such as tryptophan-dehydrobutyrin-diketopiperazine (**12**, Scheme 1.4.6) [15].

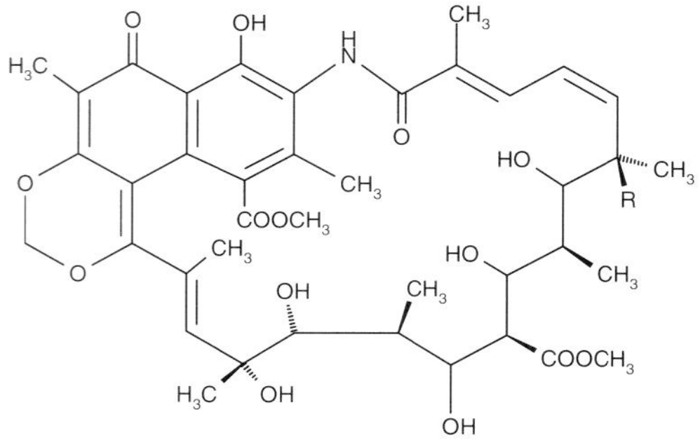

**10** Streptovaricin C   R = H

**11** Streptovaricin G   R = OH

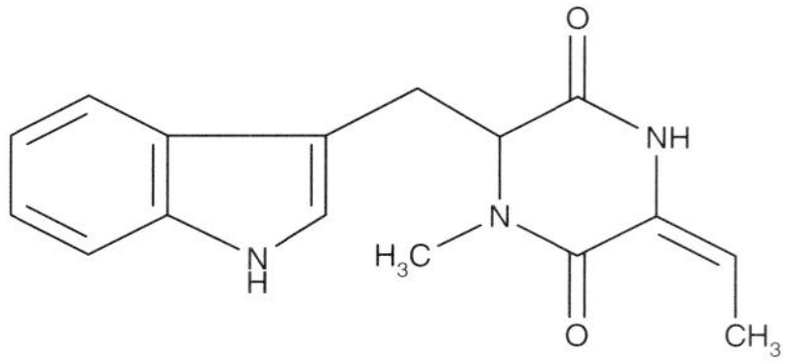

**12** Tryptophan-dehydrobutyrin-

diketopiperazin

**Scheme 1.4.6.** Streptovaricin C (**10**) and G (**11**), and tryptophan-dehydrobutyrin-diketopiperazine (**12**) isolated after manipulation of the metabolite pattern of *Streptoverticillium baldacci* subsp. *netropse* (FH-S 1625).

# References

[1]  S. Grabley, R. Thiericke, A. Zeeck, *Drug Discovery from Nature*, Springer Verlag, Heidelberg, **1999**.

[2]  S. Grabley, R. Thiericke in *New Aspects in Bioorganic Chemistry*, (U. Diederichsen, T. K. Lindhorst, L. Wessjohann, B. Westermann eds.), Wiley-VCH, Weinheim, **1999**, 409.

[3]  A. Bax, A. Aszalos, Z. Dinya, K. Sudo, *J. Am. Chem. Soc.* **1986**, *108*, 8056.

[4]  S. Grabley, G. Kretzschmar, M. Mayer, S. Philipps, R. Thiericke, J. Wink, A. Zeeck, *Liebigs Ann. Chem.* **1993**, 573.

[5]  M. Mayer, R. Thiericke, *J. Chem. Soc. Perkin Trans. 1* **1993**, 2525.

[6]  D. O'Hagan, *Nat. Prod. Rep.* **1992**, 9, 447.

[7]  R. B. Herbert, *The Biosynthesis of Secondary Metabolites*, Chapman and Hall, London, **1989**.

[8]  M. Zerlin, R. Thiericke, *J. Org. Chem.* **1994**, 59, 6986.

[9]  H. Nakayama, K. Furihata, H. Seto, N. Otake, *Tetrahedron Lett.* **1981**, *22*, 5217.

[10]  K. Takesako, T. Beppu, *J. Antibiot.* **1984**, *37*, 1170.

[11]  S. Iwasaki, K. Sasaki, M. Namikoshi, S. Okuda, *Heterocycles* **1982**, *17*, 331.

[12]  G. Kretzschmar, M. Krause, L. Radics, *Tetrahedron* **1997**, *53*, 971.

[13]  M. Zerlin, R. Thiericke, P. Henne, A. Zeeck, *Nat. Prod. Lett.* **1997**, *10*, 217.

[14]  K. L. Rinehart, Jr., L. S. Shield, *Prog. Chem. Org. Nat. Prod.* **1976**, *33*, 231.

[15]  K. Kakinuma, K. L. Rinehart, Jr., *J. Antibiot.* **1974**, *27*, 733.

## 1.5    Synthesis and biological activity of leukotriene derivatives

*Stefan Jaroch, Bernd Buchmann, and Werner Skuballa*

Leukotriene $B_4$ (LTB$_4$) is a biologically active eicosatetraenoic acid derivative formed by oxidation of arachidonic acid (1) via the 5-lipoxygenase pathway (Scheme 1.5.1) [1]. LTB$_4$ plays an important role during inflammatory processes.

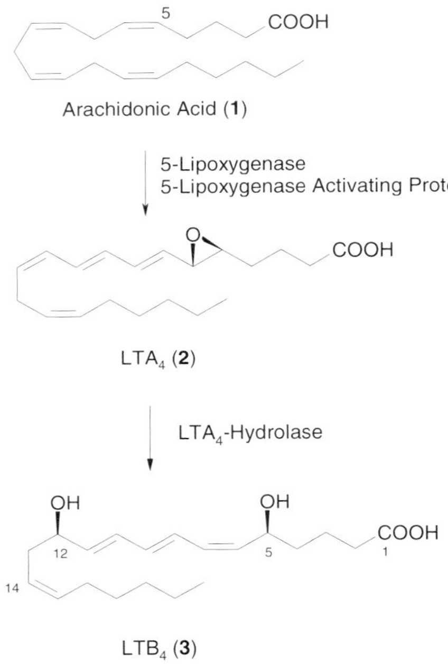

**Scheme 1.5.1.** LTB$_4$ biosynthesis.

Leukocytes produce LTB$_4$ (**3**) to fight invading pathogens in a process known as phagocytosis. LTB$_4$ is a mediator of inflammation in that it recruits (chemotaxis) and activates neutrophils which respond by releasing lysosomal enzymes and superoxide radicals. Though beneficial under certain conditions elevated levels of LTB$_4$ seem to play a pathophysiological role in chronic inflammatory diseases such as psoriasis, colitis ulcerosa, Crohn's disease, and rheumatoid arthritis [1]. A remedy against these ailments could rely on blocking LTB$_4$'s detrimental action in the inflamed tissue [2]. There are two options to achieve this goal. The first approach is directed towards the inhibition of the enzymes 5-lipoxygenase activating protein (FLAP) [2], 5-lipoxygenase [2, 3], or LTA$_4$-hydrolase [4] which are involved in LTB$_4$ biosynthesis. Alternatively, since binding of LTB$_4$ to a G-protein-coupled 7TM receptor [5] on the surface of neutrophils triggers its biological effects, blocking this receptor by antagonists should shut down LTB$_4$-mediated events. We decided to follow this second approach and set out to find an LTB$_4$-R antagonist considering agonistic LTB$_4$ as a lead structure (Scheme 1.5.2).

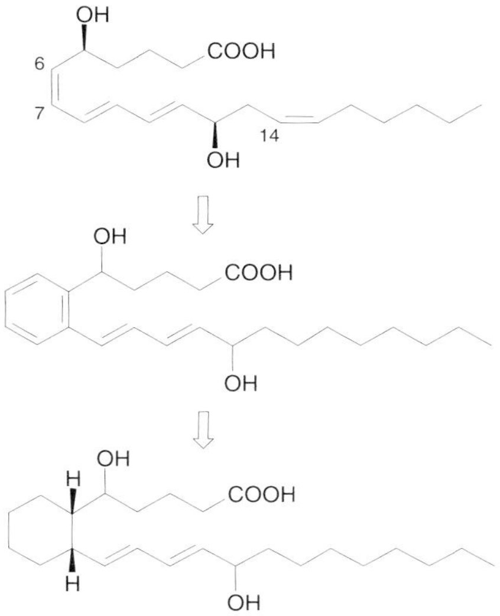

LTB$_4$ (**3**)
(agonist)

BLTB$_3$ (**4**)
(agonist/antagonist)

*cis*-CLTB$_2$ (**5**)
(antagonist)

**Scheme 1.5.2.** Lead and design.

Our design aimed to incorporate the labile *cis*-6,7-double bond into a ring system leading to benzoleukotriene B$_3$ (BLTB$_3$, **4**) [6] and cyclohexanoleukotriene B$_2$ (CLTB$_2$, **5**). In the latter case, the substituents at the cyclohexane core were arranged in a *cis*-fashion to reflect the geometry of the LTB$_4$ 6,7-double bond. While BLTB$_3$ displayed a mixed agonistic/antagonistic profile of action [7], cyclohexanoleukotriene B$_2$ (CLTB$_2$, **5**) acted as a pure LTB$_4$-R antagonist.

The synthesis of CLTB$_2$(**5**) started with *meso*-cyclohexenedicarboxylic acid (**6**), which was diesterified and subjected to an enantioselective PLE saponification to yield monoester (**7**) (Scheme 1.5.3) [8]. DIBAH reduction of (**7**), protecting group transformations, and Swern oxidation gave aldehyde (**10**) [9]. Horner–Wadsworth–Emmons homologization and octylmagnesiumbromide addition accomplished the construction of the $\omega$-chain. The resulting epimeric alcohols (**12**) and (**13**) were separated by chromatography and were processed separately.

The appendage of the $\omega$-chain started after acetylation of the secondary hydroxyl group and cleavage of the silyl ether in (**12**) (Scheme 1.5.4). Oxidation to aldehyde (**14**) set the stage for a Grignard addition providing a mixture of alcohols (**15**) and (**16**) which were readily separated by chromatography. Epimer (**15**) was transformed to *cis*-(6$S$,7$S$)-CLTB$_2$ (**18**) by acetylation, silylether cleavage, stepwise oxidation of the primary alcohol to the acid, and deacetylation [10].

The binding affinitiy [11] of the stereoisomeric CLTB$_2$s to the LTB$_4$-R relies to a great extent on the configuration at C-12. While both (*12R*)-compounds (**18**) and

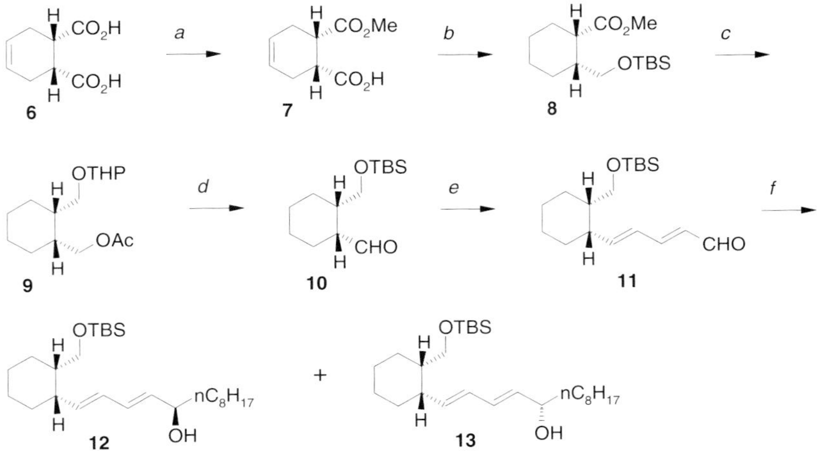

**Scheme 1.5.3.** Construction of the $\omega$-chain: (a) MeOH, $H_2SO_4$; PLE (98% ee); (b) $SOCl_2$; $NaBH_4$, MeOH; TBSCl, imidazole, DMF; $H_2$, Pd-C; (c) DIBAH, PhMe; DHP, *p*TsOH; TBAF, THF; $Ac_2O$, Py; (d) HOAc, THF-$H_2O$; TBSCl, imidazole, DMF; DIBAH, PhMe; $(COCl)_2$, DMSO, $Et_3N$, $CH_2Cl_2$; (e) $(EtO)_2P(O)CH_2CH=CHCO_2Et$, DBU, LiCl, MeCN; DIBAH, PhMe; $MnO_2$, PhMe; (f) $nC_8H_{17}MgBr$, THF.

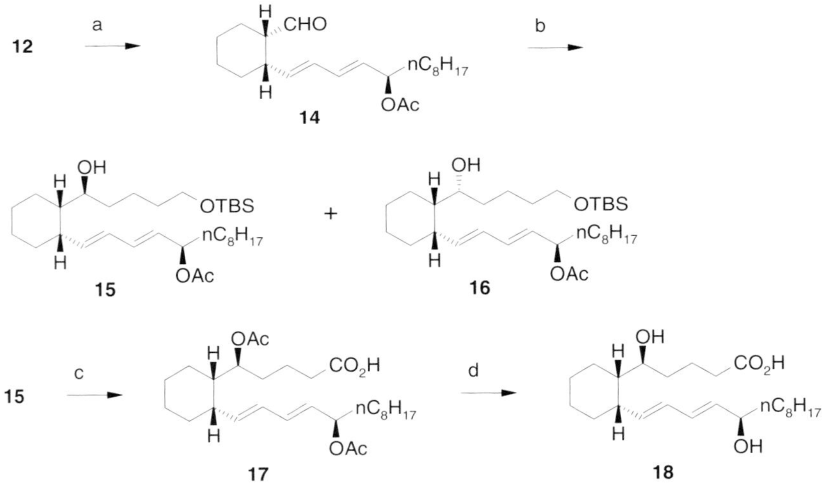

**Scheme 1.5.4.** Construction of the $\omega$-chain and completion of the synthesis: (a) $Ac_2O$, Py; TBAF, THF; $CrO_3Py_2$, $CH_2Cl_2$; (b) $ClMg(CH_2)_4OTBS$, THF; (c) $Ac_2O$, Py; TBAF, THF; $CrO_3Py_2$, $CH_2Cl_2$; $CrO_3$, $H_2SO_4$; (d) LiOH, THF-MeOH.

**(19)** bind equally well despite different C-5-stereochemistry, both (*12S*)-epimers display a strongly reduced affinity. Hence, we conclude that the hydroxyl group at C-12 plays a more dominant role in receptor recognition than the one at C-5.

Cyclohexanoleukotriene $B_2$ (**18**) represented a new template for the design and synthesis of high-affinity LTB$_4$-R antagonists.

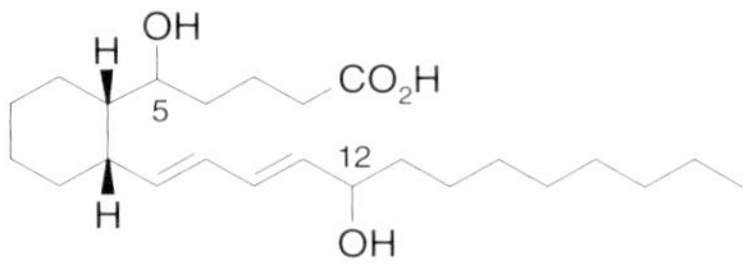

| 18 | (5S, 12R), | CF = 85 |
| 19 | (5R, 12R), | CF = 79 |
| 20 | (5R, 12S), | CF = 272 |
| 21 | (5S, 12S), | CF = 244 |

**Scheme 1.5.5.** Configuration and binding affinities [11].

**Acknowledgements:** We thank Dr. Claudia Giesen, Dr. Roland Ekerdt and Dr. Wolfgang Fröhlich (Experimental Dermatology) for the pharmacological testing of the compounds and subsequent discussions. Dr. Günter Michl's (Department Strukturanalyse) support in elaborating the chemical structure of various stereoisomers is gratefully acknowledged.

# References

[1]  B. Samuelsson, *Angew. Chem.* **1982**, *94*, 881–962; *ibid.* **1983**, *95*, 854–864; J. Rokach (Ed.), *Leukotrienes and Lipoxygenases*, Elsevier, Amsterdam, **1989**; M. J. Müller, *Pharm. Unserer Zeit* **1995**, *24*, 264–272.

[2]  C. D. W. Brooks, J. B. Summers, *J. Med. Chem.* **1996**, *39*, 2829–2654.

[3]  J. H. Musser, A. F. Kreft, *J. Med. Chem.* **1992**, *35*, 2501–2524.

[4]  M.-Q. Zhang, *Current Med. Chem.* **1997**, *4*, 67–78.

[5]  S. W. Djuric, D. J. Fretland, T. D. Penning, *Drugs of the Future* **1992**, *17*, 819–830; T. Yokomizo, T. Izumi, K. Chang, Y. Takuwa, T. Shimizu, *Nature* **1997**, *387*, 620–624.

[6]  A related approach was pursued by Searle scientists: S. W. Djuric, R. A. Haack, S. S. Yu, *J. Chem. Soc. Perkin Trans. 1* **1989**, 2133–2134. New Synthesis: F. Babudri, V. Fiandanese, G. Marchese, A. Punzi, *Synlett* **1995**, 817–818.

[7]  R. Ekerdt, B. Buchmann, W. Fröhlich, C. Giesen, J. Heindl, W. Skuballa in *Advances in Prostaglandin, Thromboxane, and Leukotriene Research, Vol. 21*, (Ed.: B. Samuelsson), Raven Press, New York, **1990**, S. 565–568.

[8]  H.-J. Gais, K. L. Lukas, W. A. Ball, S. Braun, H. J. Lindner, *Liebigs Ann.* **1986**, 687–716.

[9]  The enantiomer of **10** is readily available by DIBAH reduction of **8** and subsequent Swern oxidation. These transformations offer a convenient entry into the class of the *cis*-(6R,7R)-CLTB$_2$ stereoisomers which, however, proved to be less active than the *cis*-(6S,7S)-CLTB$_2$ derivates.

[10] The stereochemistry at C-5 was determined by cyclizing CLTB$_2$ (H$_2$SO$_4$ in THF) and $^1$H NMR (NOE) experiments with the resulting tetrahydrofurans:

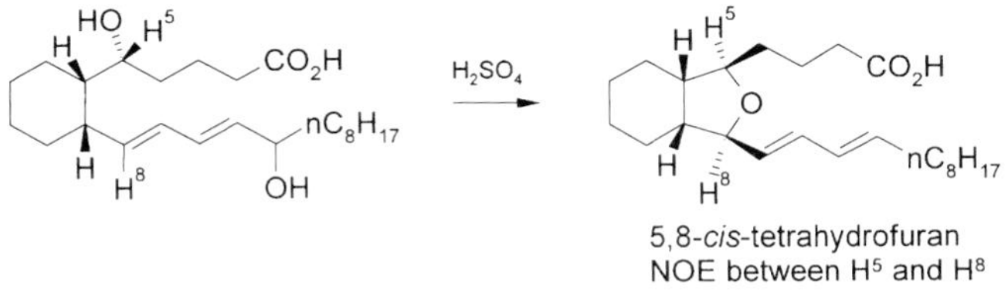

5,8-*cis*-tetrahydrofuran
NOE between H$^5$ and H$^8$

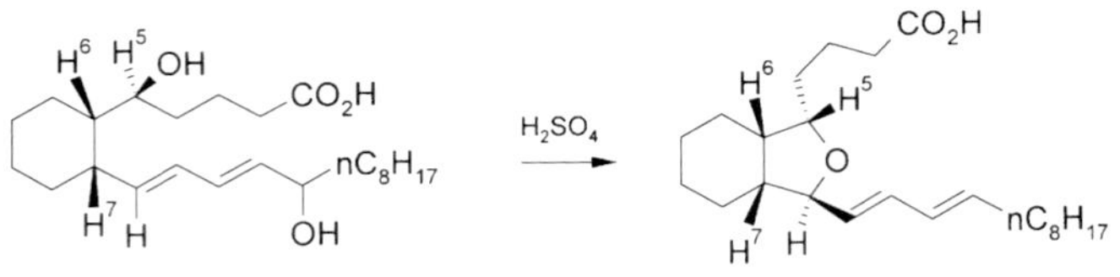

5,8-*trans*-tetrahydrofuran
NOE between H[5], H[6] and H[7]

The stereochemistry at C-12 was proved by asymmetric syntheses of chiral intermediates according to *Corey* and *Depezay* (E. J. Corey, H. Niwa, J. Knolle, *J. Am. Chem. Soc.* **1978**, *100*, 1942–1943; Y. Le Merrer, C. Gravier-Pelletier, D. Micas-Languin, F. Mestre, A. Dureault, J.-C. Depezay, *J. Org. Chem.* **1989**, *54*, 2409–2416):

[11] The competition factor CF quantifies how well or poorly the test compounds bind to the $LTB_4$-R compared to $LTB_4$. It is defined as $IC_{50}$ (test compound)/$IC_{50}$ ($LTB_4$) which means the lower CF the better the binding. $IC_{50}$s were determined with cell membrane fragments of human polymorphonuclear neutrophilic granulocytes (PMN).

## 1.6  Diazoketones – versatile starting materials for the diastereoselective synthesis of β-lactams

*Joachim Podlech and Michael R. Linder*

A dramatic increase of bacterial resistances against antibiotics has been observed over the past 20 years [1]. Several attempts have been made to confront this problem. Among other strategies, the development of new, resistance-stable antibiotics is one promising method. β-Lactam antibiotics are still a dominant substance class in this area; in particular, thienamycin [2] and the recently discovered trinems [3] are such resistance-stable substrates (Figure 1.6.1). They differ from most other β-lactam antibiotics (*e.g.* penicillins and cephalosporins) by their being *trans*-3,4-substituted β-lactam ring and having hydroxyalkyl substituent at position C-3. We are interested in the development of new routes to β-lactams.

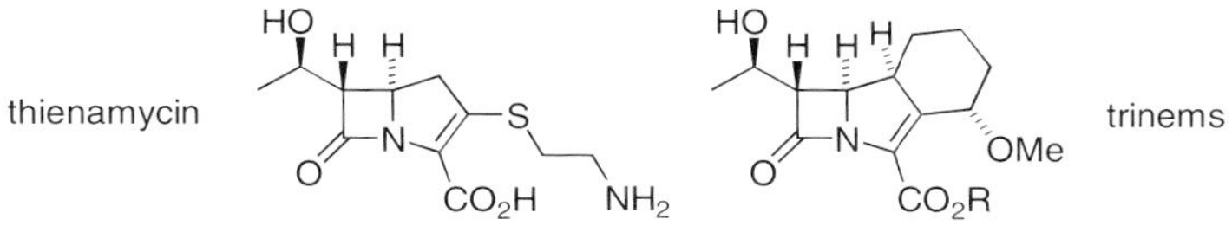

**Figure 1.6.1.**

### 1.6.1  Preparation of β-lactams

The Arndt-Eistert homologation of suitably protected amino acids was predominantly used for the preparation of β-amino acid derivatives (Scheme 1.6.1) [4].

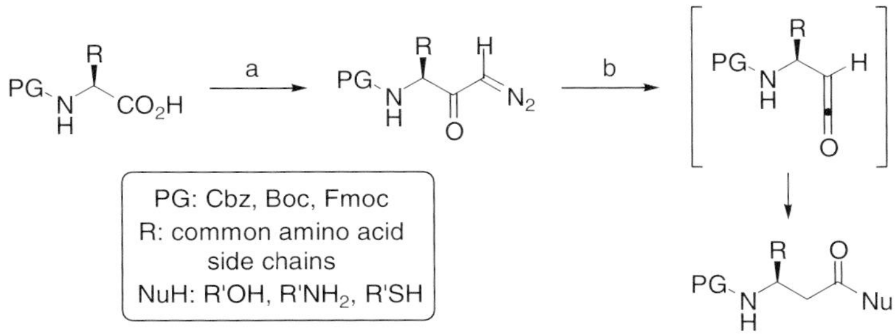

**Scheme 1.6.1.** Homologation of amino acids. (a) 1. NEt$_3$, THF; 2. ClCO$_2$Et, 3. CH$_2$N$_2$; (b) 3 Eq. NEt$_3$, NuH, 0.11 eq. silver benzoate, THF (PG = protective group, Cbz = PhCH$_2$OCO, Boc = *t*BuOCO, Fmoc = 9-Fluorenylmethoxycarbonyl).

The herein generated ketenes are also suitable intermediates in the reaction with imines to form β-lactams [5] in analogy to a protocol developed by Staudinger at the beginning of the century [6]. The diazoketones used as starting materials in this

**Table 1.6.1.** Decomposition of amino acid-derived diazoketones in the presence of imines leading to $\beta$-lactams (see Scheme 1.6.2)

| Entry | PG | R | Amino acid | Yield | d.r. |
|---|---|---|---|---|---|
| 1 | Boc | Me | Ala | 70% | 71:29 |
| 2 | Cbz | $^i$Bu | Leu | 89% | 70:30 |
| 3 | Cbz | $^i$Pr | Val | 89% | 82:18 |
| 4 | Cbz | *sec*Bu | Ile | 90% | 83:17 |
| 5 | Cbz | *tert*Bu | Tle | 88% | 93:7 |
| 6 | Cbz | Bn | Phe | 58% | 59:41 |
| 7 | Boc | CbzNH–(CH$_2$)$_3$– | Orn(Cbz) | 73% | 70:30 |
| 8 | Boc | CbzNH–(CH$_2$)$_4$– | Lys(Cbz) | 71% | 65:35 |
| 9 | Boc | BnOCH$_2$– | Ser(Bn) | 63% | 70:30 |

reaction sequence (Scheme 1.6.2) can be prepared without racemization from amino acids [4]. The subsequent Wolff rearrangement cannot – as usually (Scheme 1.6.1) – be performed in the presence of a silver catalyst, since this catalyst is obviously complexed by the imine and is rendered ineffective [7]. Therefore, we tried photochemical reaction conditions for the rearrangement and obtained exclusively two of the four possible diastereoisomers, in which the hydrogens attached to positions C-3 and C-4 are *trans*-arranged. The diastereoselectivity in this reaction is dominantly influenced by the steric bulk of the side chain introduced with the amino acid. They range from 2:1 (starting with alanine, R = CH$_3$) to 13:1 (*tert*-leucine, R = $t$Bu; Scheme 1.6.2, Table 1.6.1) [5]. Suitably protected trifunctional amino acids, *e.g.* lysine, aspartic acid or serine derivatives can also be used.

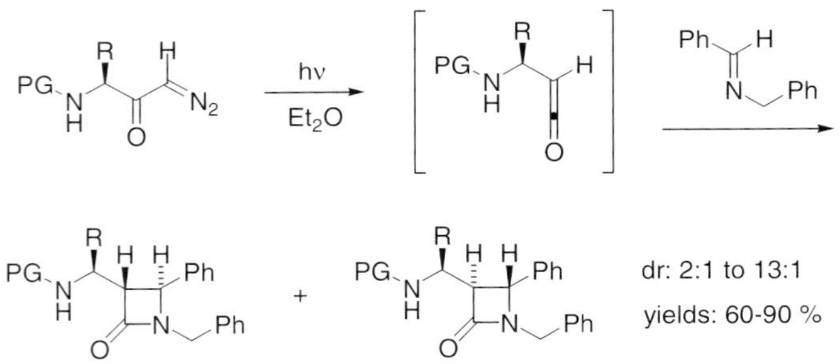

**Scheme 1.6.2.** Synthesis of $\beta$-lactams starting from diazoketones (See Table 1.6.1).

To improve the product ratio in this reaction, we utilized chiral imines derived from phenethylamines [5, 8], providing the possibility for match-mismatch interaction [9, 10]. The use of (*S*)-*N*-phenethyl benzaldimine led to an increased selectivity (rising from 2.5:1 to 7:1, R = Me, PG = Boc; Table 1.6.2). Consequently, the selectivity dropped almost to 1:1, when the analogous (*R*)-configurated imine was used [5, 8].

Both the *N*-benzyl- and the *N*-phenethyl-substituent bear a great disadvantage: These protecting groups at the lactam-nitrogen introduced with the imine are difficult

**Table 1.6.2.** Decomposition of diazoketones in the presence of chiral imines

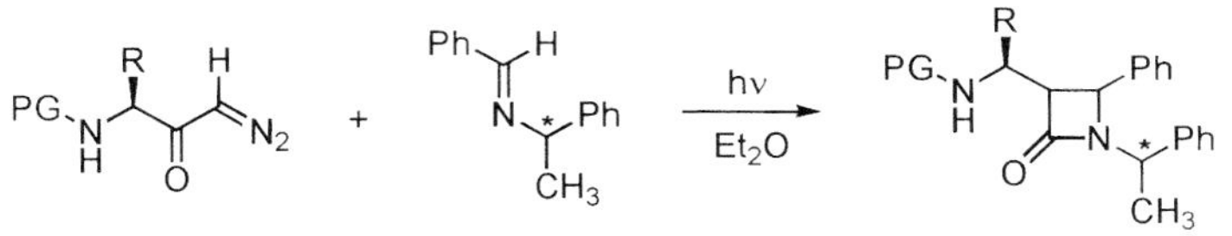

| Entry | PG | R | Amino acid | Configuration of imine | Yield | d.r. |
|---|---|---|---|---|---|---|
| 1 | Boc | Me | Ala | (R) | 72% | 55:45 |
| 2 | Boc | Me | Ala | (S) | 88% | 88:12 |
| 3 | Cbz | secBu | Ile | (R) | 72% | 73:27 |
| 4 | Cbz | secBu | Ile | (S) | 56% | 86:14 |

to remove [6], which is unfavorable with respect to further reactions. Therefore we checked, whether similar imines with other *N*-substituents can be used. It transpired that virtually any carbon substituent can be introduced when suitable imines are used. The synthesis of a *p*-methoxybenzyl (PMB)-substituted β-lactam is depicted in Scheme 1.6.3; the PMB-group can easily be cleaved using oxidative conditions [5, 8].

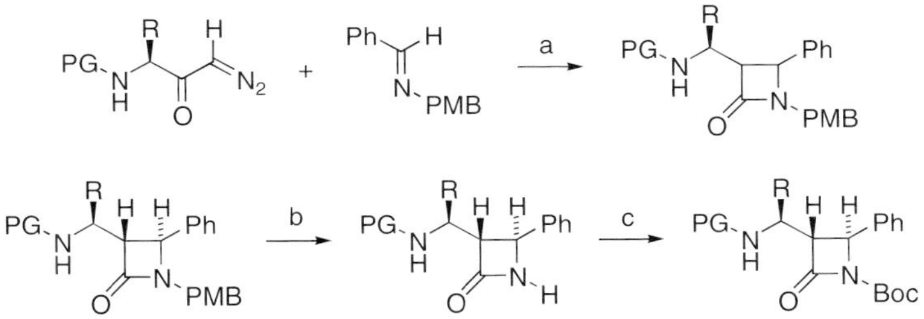

**Scheme 1.6.3.** PMB-substituted imines and cleavage of the PMB group. (a) 1. hν, Et₂O, 2. Separation of isomers; (b) K₂S₂O₈, buffer; (c) Boc₂O, NEt₃, DMAP.

The hitherto described β-lactams bear a phenyl group at position C-4. Again, this is unsuitable, since a replacement or modification of the phenyl ring is difficult. Surprisingly, all attempts using imines which are not derived from benzaldehyde failed. Neither aliphatic aldimines nor imines or iminoesters prepared from acrolein, glyoxalates or formates led to the formation of β-lactams. Consequently, we made a virtue from necessity and used imines prepared from substituted benzaldehydes in our reaction. Both, electron-rich (*e.g.* 4-methoxy-substituted) and electron-deficient (*e.g.* 4-nitro-substituted) benzaldimines, and especially heteroaromatic (2-thienyl or 2-furyl) carbaldimines can be used successfully (Scheme 1.6.4).

We were especially interested in compounds with electron-rich aromatic substituents, since these can be easily degradated oxidatively to carboxylic acids with ruthenium oxide (prepared *in situ* from RuCl₃/H₅IO₆) [11]. This works very well with

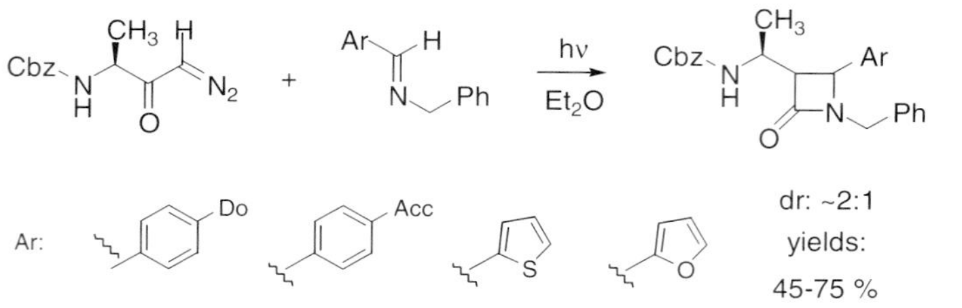

CH₃ H

Cbz–N ... N₂  +  Ar–CH=N–Ph  →(hv, Et₂O)→  Cbz–N ... Ar, N–Ph lactam

**Scheme 1.6.4.** Synthesis of C-4-aryl- and heteroaryl-substituted β-lactams.

dr: ~2:1
yields:
45–75 %

Ar: (p-substituted aryl with Do; p-substituted aryl with Acc; thienyl; furyl)

furyl-substituted β-lactams: The corresponding carboxylic acid is formed within 10 minutes in 80% yield (Scheme 1.6.5).

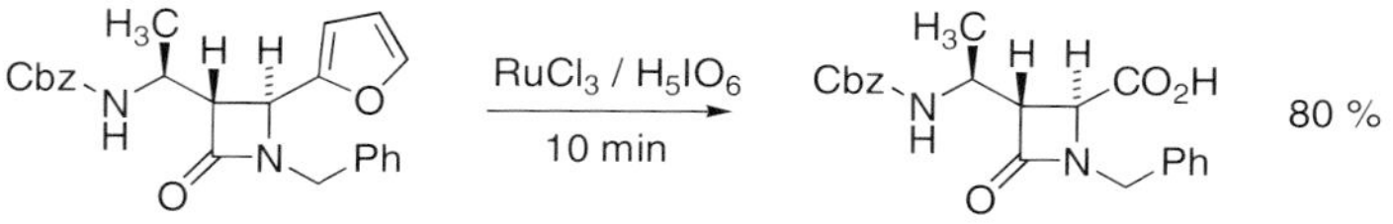

Cbz-furyl-β-lactam  →(RuCl₃ / H₅IO₆, 10 min)→  Cbz-CO₂H-β-lactam   80 %

**Scheme 1.6.5.** Degradation of a furyl-substituent to the corresponding carboxylic acid.

The configuration of several β-lactams could be unequivocally assigned by X-ray crystallographic analysis. Comparision of the NMR spectra elucidated the relative configuration of all prepared β-lactams. The supposed easy modification of the carboxylic acid functionality together with the variability in the choice of the substituent at the β-lactam nitrogen allows the required flexibility with respect to further transformations. This should enable ring closure reactions leading to substrates which are analogs of thienamycin (*trans*-substitution and an aminoalkyl- instead of a hydroxyalkyl-substituent, *vide supra*) and eventually exhibit similar biological activities.

## 1.6.2  Mechanistic Considerations

Since the *trans*-selectivity observed in our reaction is quite unusual, it appears necessary to examine the mechanism of this reaction. One might argue that a change of the reaction conditions (from thermal to photochemical conditions) would lead to a change in the substitution pattern (from *cis* to *trans*) similar to ring closures of butadienes [12]. Semi-empirical calculations [13] showed that in fact this reaction is not a pericyclic reaction, but is rather an intramolecular attack of an enolate to an imine. Nevertheless, the photochemical reaction conditions might be responsible for the observed selectivity: The reaction mechanism for a thermal Staudinger reaction (without irradiation) leading to *cis*-substituted β-lactams is depicted in Scheme 1.6.6 (top) [6]. With irradiation an isomerization either of the imine (**A → B**) or of the acyliminiumion (**C → D**) would lead consequently to *trans*-substituted β-lactams. This photochemical-induced isomerization (**A → B**) is well known for imines [14]. Further investigations to explain the unusual selectivity are ongoing in our laboratories.

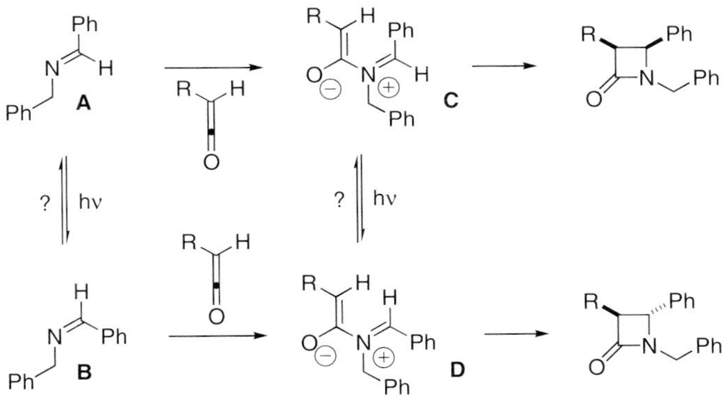

**Scheme 1.6.6.** Mechanistic considerations: Possible explanations for the *trans*-selectivity.

**Acknowledgements:** The help of all members of the Institut für Organische Chemie, especially of Prof. Dr. V. Jäger and Prof. Dr. Dr. h. c. F. Effenberger is gratefully acknowledged. We also thank Mrs. S. Henkel and Dr. W. Frey for the X-ray crystallographic analyses. This work was further supported by the Fonds der Chemischen Industrie, the Deutsche Forschungsgemeinschaft, the BASF AG (donation of phenethylamines), and the Degussa AG (amino acids).

# References

[1] *The World Health Report 1996*, World Health Organization, Genf, **1996**.

[2] U. Gräfe, *Biochemie der Antibiotika*, Spektrum Akademischer Verlag, Heidelberg, **1992**.

[3] C. Ghiron, T. Rossi, R. J. Thomas, *Tetrahedron Lett.* **1997**, *38*, 3569–3572.

[4] J. Podlech, D. Seebach, *Liebigs Ann. Chem.* **1995**, 1217–1228; J. L. Matthews, C. Braun, C. Guibourdenche, M. Overhand, D. Seebach in *Enantioselective Synthesis of β-Amino Acids* (Ed.: E. Juaristi), Wiley-VCH, New York, **1997**, pp. 105–126 and cited literature therein.

[5] J. Podlech, *Synlett* **1996**, 582–584; J. Podlech, M. R. Linder, *J. Org. Chem.* **1997**, *62*, 5873–5883.

[6] *The Organic Chemistry of β-Lactams* (Ed.: G. I. Georg), VCH, New York, **1993**.

[7] E. Hoyer, V. V. Skopenko, *Russ. J. Inorg. Chem.* **1966**, *11*, 436–439.

[8] J. Podlech, S. Steurer, *Synthesis* **1999**, 650–654.

[9] See for example: B. Krämer, T. Franz, S. Picasso, P. Pruschek, V. Jäger, *Synlett* **1997**, 295–297.

[10] S. Masamune, W. Choy, J. S. Petersen, L. R. Sita, *Angew. Chem.* **1985**, *97*, 1–31; *Angew. Chem. Int. Ed. Engl.* **1985**, *24*, 1–30.

[11] P. H. J. Carlsen, T. Katsuki, V. S. Martin, K. B. Sharpless, *J. Org. Chem.* **1981**, *46*, 3936–3938; see as well: V. Jäger, H. Grund, V. Buß, W. Schwab, I. Müller, R. Schohe, R. Franz, R. Ehrler, *Bull. Soc. Chim. Belg.* **1983**, *92*, 1039–1054.

[12] R. B. Woodward, R. Hoffmann, *The Conservation of Orbital Symmetry*, Verlag Chemie, Weinheim, **1970**.

[13] Unpublished results; see also: F. P. Cossio, J. M. Ugalde, X. Lopez, B. Lecea, C. Palomo, *J. Am. Chem. Soc.* **1993**, *115*, 995–1004.

[14] G. Wettermark in *The Chemistry of the Carbon-Nitrogen Double Bond* (Ed.: S. Patai), Interscience, London, **1970**, pp. 565–596.

# 1.7     Selenium compounds in chemical and biochemical oxidation reactions

*Gianfranco Fragale, Sara Häuptli, Michele Leuenberger, and Thomas Wirth*

Stereoselective synthesis has gained much attraction during the last decades: The insight to biochemical and medical processes on a molecular level has gone hand in hand with developments in the synthesis of complex target molecules by methods of synthetic organic chemistry.

Very efficient and advanced methods have been established for some synthetic transformations to highly enriched stereoisomers. But only a few useful methods are known for the stereoselective functionalization ofunactivated or only weakly activated C–H bonds or C=C bonds. In our research projects we are investigating and developing stoichiometric as well as catalytic reactions with such compounds leading to products with new stereogenic centers.

Stereoselective functionalizations of unactivated double bonds has been performed mainly with chiral alkenes. Chiral electrophiles have been used rarely in these reactions. It can be shown, that addition reactions of chiral electrophiles to alkenes can be performed with good stereoselectivities. C=C bonds can be attacked by chiral selenium electrophiles. The three-membered seleniranium ions of type **1** formed by this reaction can then undergo a subsequent ring-opening reaction with a nucleophile. This reaction usually leads to addition products **2** by an *anti* addition of electrophile and nucleophile (Scheme 1.7.1).

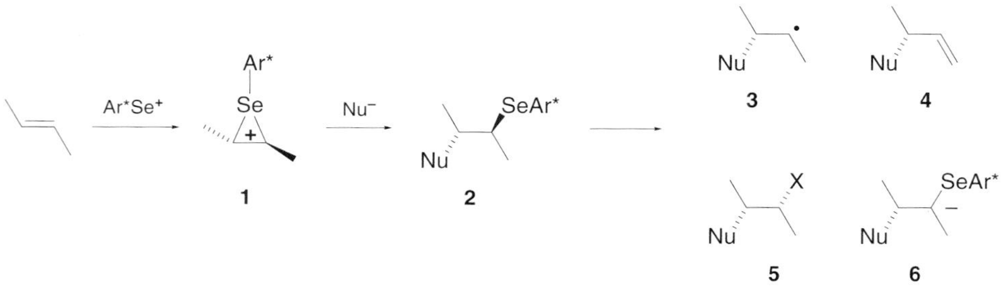

**Scheme 1.7.1** Reaction of double bonds with chiral selenium electrophiles and subsequent reactions.

The compounds **2** obtained are versatile building blocks for various subsequent reactions and have been employed in natural product syntheses. Furthermore, we are investigating the use of small organoselenium compounds as mimetics for the enzyme glutathione peroxidase. The selenium – carbon bond can be cleaved homolytically which opens the broad field of radical chemistry (**3**). Oxidation to the selenoxide and subsequent $\beta$-elimination introduces again a double bond (**4**), which is then functionalized in the allylic position. Further oxidation to the selenone generates an excellent leaving group, which can be substituted by a second nucleophile X (**5**). Deprotonation in $\alpha$-position can be the entry to the chemistry of carbanions (**6**).

We designed chiral diselenides, which were prepared by a short synthesis from read-ily available starting materials. After transformation into the electrophilic selenium species **7** by reaction with bromine and then with silver triflate, we observed a good transfer of chirality in addition reactions to various alkenes. An interaction between an electron lone pair of the hydroxy group in the chiral side chain with the $\sigma^*$-orbital of the selenium is a prerequisite for a reaction with high stereoselectivity. The addition products **9** can be obtained in diastereomeric ratios up to 95:5 [1].

We have investigated the mechanism of the methoxyselenenylation in detail. We have proved the assumption, that the formation of the intermediate diastereomeric selenir-anium ions **8a** and **8b** is reversible. Furthermore, we have shown experimentally as well as by calculations, that these two diastereomeric seleniranium ions resulting from a *Re*- and from a *Si*-attack of the chiral selenium electrophile to the alkene are different in energy. Each of them has been synthesized independently and their differing reac-tions with methanol is further proof for the face selectivity observed in these addition reactions (Scheme 1.7.2) [2]. To increase the energetic difference between **8a** and **8b** we substituted the second *ortho*-position of the selenium with a methoxy-group. With the diselenide **10** it is now possible to obtain addition products of type **9** with diastereo-selectivities up to 98% [3].

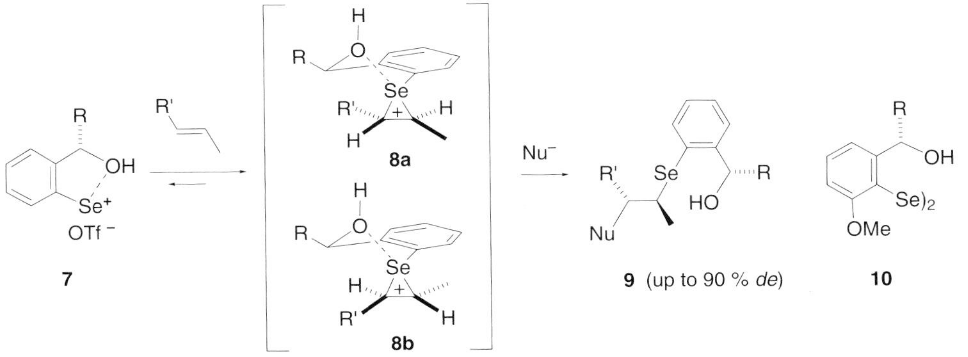

**Scheme 1.7.2.**   Mechanistic course of the stereoselective methoxyselenenylation reaction.

The seleniranium-adducts can be trapped with external nucleophiles as shown above, but intramolecular selenocyclizations can also be performed. This is demonstrated by the synthesis of lactone **11a** and tetrahydrofuran derivative **11b** with asymmetric tetrasubstituted carbon atoms [4]. We used the intramolecular aminoselenenylation for the stereoselective synthesis of tetrahydroisoquinoline alkaloids like salsolidin **12** as shown in Scheme 1.7.3 [5].

Nucleophiles bearing other functionalities can also be used in the addition reactions. Even alcohols with double or triple bonds can be added. The selenium functionality in these products can then be used for a subsequent radical cyclization. Addition pro-ducts **13** are obtained in about 55% yield with diastereomeric ratios of 15:1. After radical cyclization, oxidative cleavage of the double bond and deprotection cyclization

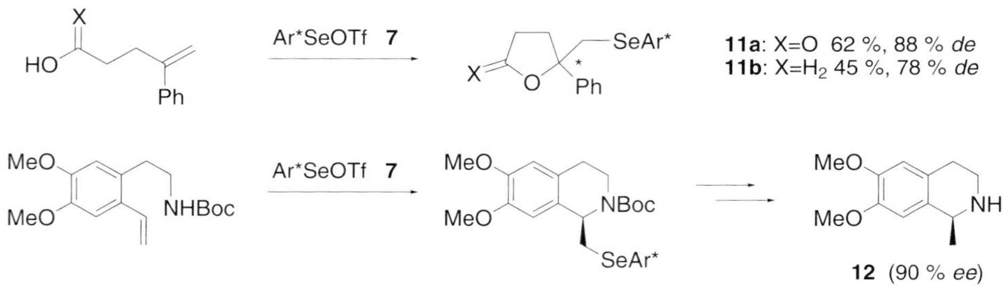

**Scheme 1.7.3.** Stereoselective selenocyclizations.

to the hemiacetal occured spontaneously and a short total synthesis of the furofuran lignan (+)-samin **14a**, was accomplished [6]. Grignard addition to **14b** lead to the first synthesis of (+)-membrin **15** (Scheme 1.7.4) [7]. Samin **14a** is a component of the sesame oil [8] and a known precursor for the synthesis of a variety of furofuran lignans. Membrin **15** was isolated in 1993 from the grains of *Rollina membranacea*, a plant growing in the area of Antioquia in Columbia [9].

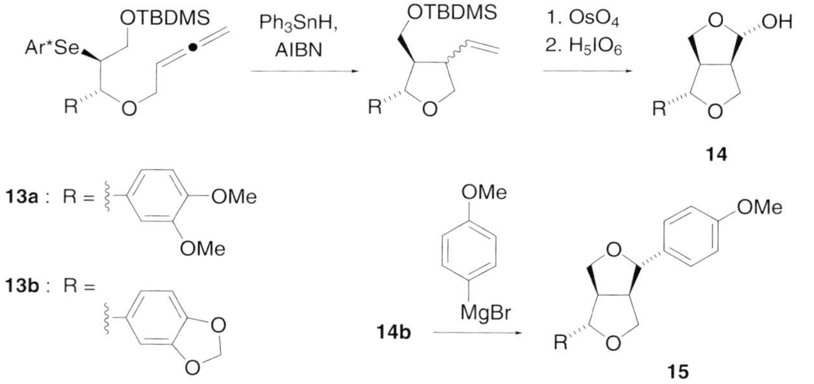

**Scheme 1.7.4.** Stereoselective synthesis of the furofuran lignans samin and membrin.

Stereoselective selenenylation reactions can be followed by a $\beta$-elimination to regenerate a double bond which is functionalized in the allylic position. A one-pot procedure should be possible if selenenylation and elimination are perfomed without isolation of the initially formed selenenylated product. Furthermore, if the liberated selenium compound can be used again for a selenenylation reaction, a process of addition and subsequent elimination with only catalytic amounts of chiral diselenide can be achieved [10]. Recently we could show, that chiral selenium compounds can indeed be used for an asymmetric version of this reaction. With 10 mol% of the diselenide **16** the chiral ether **17** can be obtained in 51% yield and in up to 75% *ee*, as shown in Scheme 1.7.5 [11].

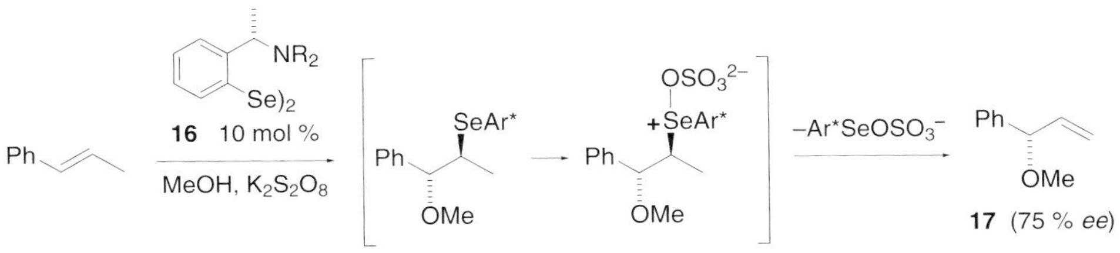

**Scheme 1.7.5.** Catalytic stereoselective selenomethoxylation.

The enzyme glutathione peroxidase, an important enzyme in thiol metabolism, posseses a selenium moiety in the active site. After this discovery, there has been a growing interest in the biochemistry of selenium. In cooperation with OXIS International, France, we have investigated the biological activity of some diselenides as glutathione peroxidase mimics. Glutathione peroxidases (GPx) are antioxidant selenoenzymes which protect various organisms from oxidative stresses by catalyzing the reduction of hydroperoxides at the expense of glutathione [12]. Based on the observation, that the selenium in glutathione peroxidase is stabilized by nitrogen from other amino acid residues, various organoselenium compounds containing other heteroatoms have been synthesized to mimic the active site of GPx. The antioxidant properties (GPx activities) and the pro-oxidant properties (glutathione oxidase [GOx] activities) of diselenides **18** and **19** were investigated (Scheme 1.7.6).

**18** R = H, Me, CF$_3$, CH(OH)Et       **19**

**Scheme 1.7.6.** Possible glutathione peroxidase mimics.

The GPx activities were measured with the protocol of Paglia *et al.* using the glutathione reductase coupled assay with either hydrogen peroxide or *tert*-butyl hydroperoxide as oxidant [13]. The diselenides were tested at 20 μM selenium equivalents. The GOx activities were derived from the kinetics of NADPH oxidation without peroxides in air-saturated buffer with catalase. In the presence of thiol, oxygen is reduced by the diselenides in a one-electron transfer process leading to the production of reactive oxygen species. The GOx activities are therefore an estimation of the cytotoxicity of these molecules.

As shown in Table 1.7.1, we found that diselenides of type **18** have reasonably high GPx values with relatively low GOx values [14]. Diselenide **19** with less possibility for a strong interaction between the oxygen and the selenium has a lower GPx value. The compounds of type **18** are therefore interesting compounds to design more efficient glutathione peroxidase mimetics.

**Table 1.7.1.** GPx and GOx activities of diselenides **18** and **19**

| Diselenide | GPx activities (20 µM Se-equivalents) [nmol of NADPH · min⁻¹] 100 µM H₂O₂ | GPx activities (20 µM Se-equivalents) [nmol of NADPH · min⁻¹] 200 µM *t*-BuOOH | GOx activities (20 µM Se-equivalents) [nmol of NADPH · min⁻¹] |
|---|---|---|---|
| **18a** (R = H) | 26.6 | 13.9 | 1.1 |
| **18b** (R = Me) | 20.2 | 13.0 | 0.3 |
| **18c** (R = CF₃) | 8.7 | 4.7 | 0[a] |
| **18d** (R = CH(OH)Et) | 18.8 | 9.5 | 0.3 |
| **19** | 14.5 | 7.8 | 0.2 |

(*a*) not detectable.

**Acknowledgements:** This work was supported by the Schweizer Nationalfonds and by the Treubel Fonds. We thank J.-C. Yadan and I. Erdelmeier, Oxis International S.A. (Bonneuil-sur-Marne, France), for the measurement of the GPx and GOx activities and for many fruitful and stimulating discussions and Prof. B. Giese for continuous support.

# References

[1]  Reviews: T. Wirth, *Liebigs Ann./Recueil* **1997**, 2189–2196. T. Wirth, *Tetrahedron* **1999**, *55*, 1–28. *Organoselenium Chemistry*; Ed.: T. Wirth, *Top. Curr. Chem.* **2000**, *Vol. 208*.

[2]  T. Wirth, G. Fragale, M. Spichty, *J. Am. Chem. Soc.* **1998**, *120*, 3376–3381.

[3]  G. Fragale, T. Wirth, *Chem. Commun.* **1998**, 1867–1868.

[4]  G. Fragale, T. Wirth, *Eur. J. Org. Chem.* **1998**, 1361–1369.

[5]  T. Wirth, G. Fragale, *Synthesis* **1998**, 162–166.

[6]  T. Wirth, K. J. Kulicke, G. Fragale, *J. Org. Chem.* **1996**, *61*, 2686–2689.

[7]  T. Wirth, *Liebigs Ann./Recueil* **1997**, 1155–1158.

[8]  W. Adriani, *Z. Unters. Lebensm.* **1928**, *56*, 187–194. P. Budowski, *J. Am. Oil Chem. Soc.* **1964**, *41*, 280–285.

[9]  J. Saez, S. Saphaz, L. Villaescusa, R. Hocquemiller, A. Cavé, D. Cortes, *J. Nat. Prod.* **1993**, *56*, 351–356.

[10]  K. Fujita, M. Iwaoka, S. Tomoda, *Chem. Lett.* **1994**, 923–926. R. Kaur, H. B. Singh, R. P. Patel, *J. Chem. Soc., Dalton Trans.* **1996**, 2719–2726. S. Fukuzawa, K. Takahashi, H. Kato, H. Yamazaki, *J. Org. Chem.* **1997**, *62*, 7711–7716. K. Fujita, *Rev. Heteroatom Chem.* **1997**, *16*, 101–117.

[11]  T. Wirth, S. Häuptli, M. Leuenberger, *Tetrahedron: Asymmetry* **1998**, *9*, 547–550.

[12]  Reviews: T. C. Stadtman, *J. Biol. Chem.* **1991**, *266*, 16257–16260. F. Ursini, in *Oxidative Processes and Antioxidants*; Ed.: R. Paoletti, Raven Press: New York, 1994; pp. 25–31.

[13]  D. E. Paglia, W. N. Valentine, *J. Lab. Clin. Med.* **1967**, *70*, 158–169.

[14]  T. Wirth, *Molecules* **1998**, *3*, 164–166.

# 1.8 Synthetic routes to oligo-tetrahydrofurans and oligo-pyrrolidines

*Ulrich Koert*

Biological membranes are impermeable to ions like $K^+$, $Na^+$, $Ca^{2+}$, and $Cl^-$. A difference in ion concentration of the external cell enviroment compared to the internal cytoplasmic compartment is generated and maintained by membrane located ion pumps [1]. Gated ion channels open and close resulting in a change in specific ion concentration, in turn leading to a change in the transmembrane potential [2]. The control of the transmembrane potential by ion channels is a key function, and molecular science aims to explain such functions on the molecular level by understanding the structures of the active species involved in the process. The goal is to understand and control the active conformation of the molecule in order to regulate its function.

Bioorganic Synthesis permits the design, preparation, and testing of molecules at the interface of molecular form (conformation) and molecular function (*e.g.* ion channel activity) [3].

As part of a project directed towards the synthesis of artificial ion channels we choose oligo-tetrahydrofurans (oligo-THFs) of type **1** [4], oligo-THF-amino acids of type **2** [5], and oligo-pyrrolidines of type **3** [6] as target structures (Figure 1.8.1). While the polyether structures are designed for cation transport, the polyamines should be suitable for the binding and transport of anions.

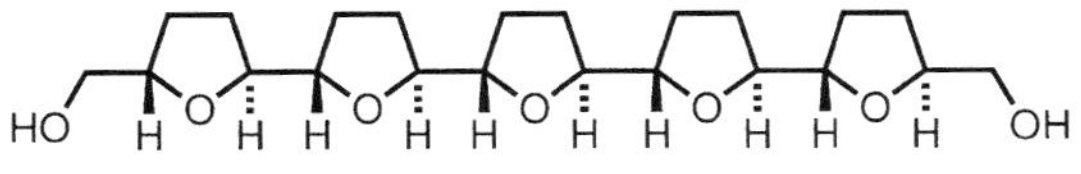

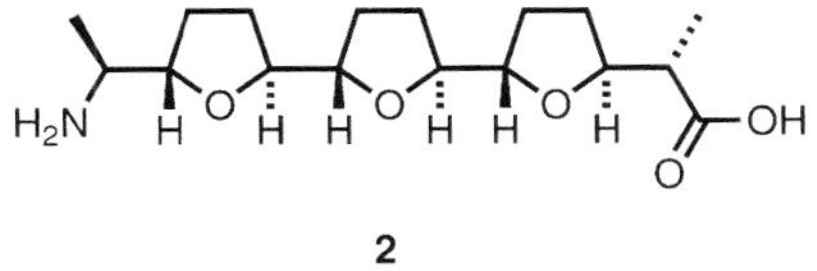

**Figure 1.8.1.** Synthetic targets: oligo-THF **1**, oligo-THF amino acid **2**, and oligo pyrrolidine **3**.

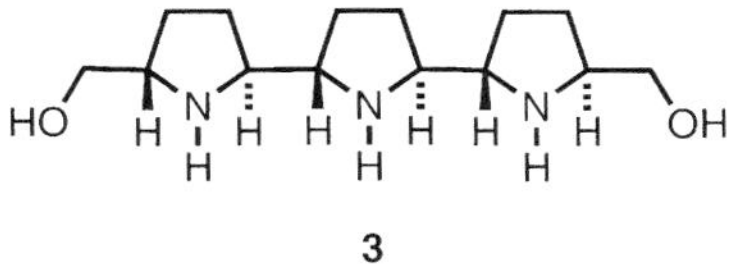

Two different approaches are available for the synthesis of the 2,5-linked oligo-THF framework with the *all-trans-anti* configuration: First, the acid-catalyzed, intramolecular opening of an oligo-epoxide (**4** → **5**). This route – sometimes called epoxide cascade – is related to the biosynthesis of several polyethers [7]. Second, a multiple five-membered-ring-selective Williamson reaction (**6** → **7**) [8, 9].

The introduction of the stereocenters necessary for the multiple Williamson reaction is possible by use of the Sharpless dihydroxylation. Scheme 1.8.2 summarizes the combination of the Sharpless dihydroxylation and the multiple Williamson reaction for the synthesis of a THF-Tetramer [8].

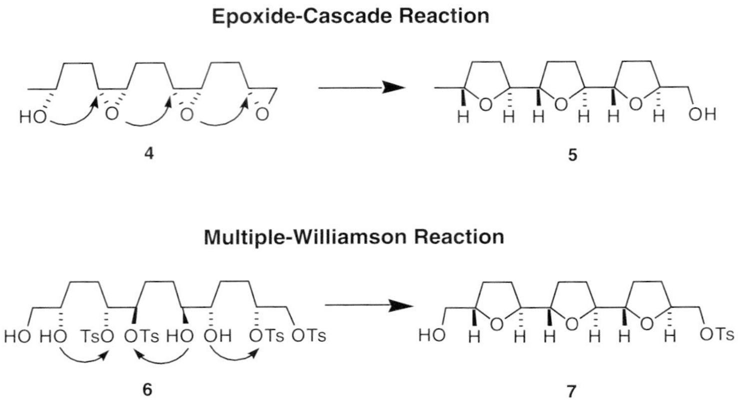

**Epoxide-Cascade Reaction**

**4** → **5**

**Multiple-Williamson Reaction**

**6** → **7**

**Scheme 1.8.1.** Epoxide cascade and multiple Williamson reaction, two different routes to *all-trans anti* oligo-THFs.

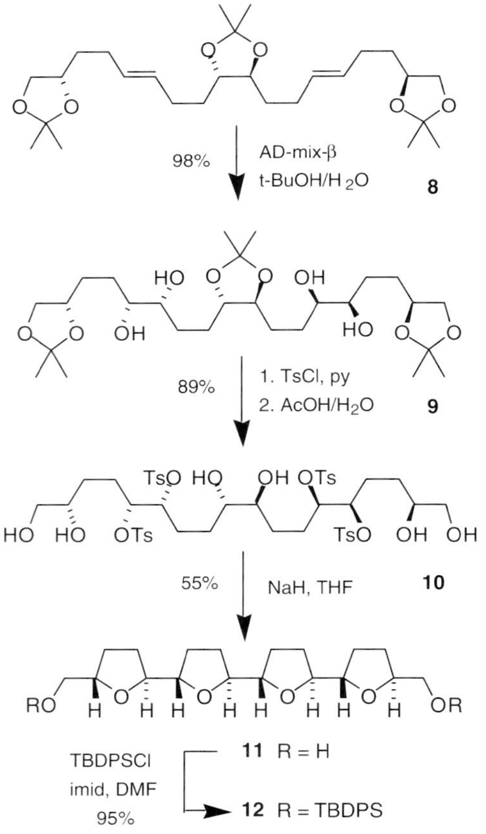

**Scheme 1.8.2.** Synthesis of the THF-tetramer **12** using a multiple Williamson reaction as a key step.

Dihydroxylation of the enantiomerically pure diene **8** gave the tetraol **9** (9:1 diastereoselectivity per double bond). A quadruple tosylation and a subsequent triple acetonide cleavage resulted in the hexahydroxytetratosylate **10**. All four THF rings were formed in a single step by the following multiple Williamson reaction. The resulting water-soluble diol **11** was converted to the di-TBDPS-ether, which is soluble in organic solvents.

Oligo-THFs such as **1** or **11** are to short to span a membrane. Thus, the synthesis of molecules containing 30 or more THF-rings required an additional structural extension: the use of amide linkages. This type of linkage required THF-amino acids such as **2**.

The starting point for the synthesis of the Ter-THF amino acid **2** was *N*-Tosyl-L-alanine **13** (Scheme 1.8.3) [5].

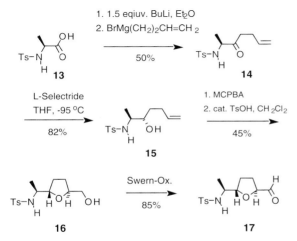

**Scheme 1.8.3.** Synthesis of the ter-THF amino acid; part I: construction of the THF-aldehyde **17**.

Reaction of the lithium salt of **13** with 2-butenylmagnesium bromide afforded the ketone **14**. L-Selectride reduction gave way to the alcohol **15**, the latter being epoxidized with MCPBA to yield a 1:1 mixture of both diastereomers. An acid-catalyzed intramolecular opening of the epoxide function by the hydroxyl group gave (after chromatography) 45% of the *trans*-THF-alcohol **16** and 45% of the corresponding *cis*-THF-alcohol. A Swern-oxidation converted alcohol **16** into the mono-THF aldehyde **17**.

The synthesis of the sulfone **26** started with the aldehyde **18**, which was allowed to react with allyltrimethylsilane in a chelate-controlled addition reaction (Scheme 1.8.4). The resulting alcohol **19** was transformed via the diol **20** into the acetonide **21**.

Ozonolysis of the double bond in **21** followed by $NaBH_4$-reduction gave the alcohol **22**, which was converted into the bromide **23**. Alkylation of **23** yielded the alcohol **24**. After an *E*-selective reduction of alkyne **24** to afford alkene **25** and conversion of the

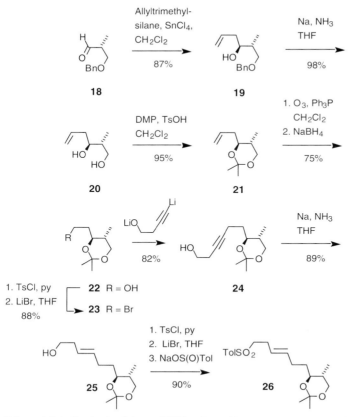

**Scheme 1.8.4.** Synthesis of the ter-THF amino acid; part I: construction of the sulfone **26**.

primary hydroxyl group into an aryl sulfone, the desired building block **26** was obtained.

The coupling of the aldehyde **17** with the sulfone **26** was addressed next (Scheme 1.8.5). The lithiated sulfone was added smoothly to the aldehyde, resulting in a secondary alcohol which was oxidized directly to the ketosulfone **27**, following the Dess-Martin protocol. Reductive desulfonation gave the ketone **28** which was transformed into the olefin **29** by stereoselective ketone reduction and subsequent TMS-protection.

The two THF-units still missing were installed simultaneously by a Sharpless dihydroxylation (**29** → **30**) followed by a multiple Williamson reaction (**31** → **32**). A change in the N-protective group and a two-step oxidation of the primary alcohol to the corresponding carboxylic acid provided the Boc-protected Ter-THF amino acid **33**, which was transformed into oligo-THF-peptides of type **34**. Compound **34** was incorporated into planar lipid bilayers and the change in conductance by transport of $K^+$-ions was analyzed [5]. A rapid change in conductance on the order of milliseconds was observed, which could not be resolved as a series of discrete single

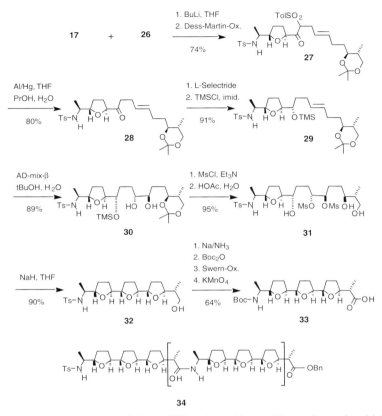

**Scheme 1.8.5.** Synthesis of the ter-THF amino acid; part III: coupling and multiple Williamson reaction.

events. The observed current peaks of varying intensity (up to 100 pA) are indicative of a membrane-spanning, porelike structure rather than a carrier mechanism.

The terpyrrolidine skeleton of **3** was assembled from two molecules of aldehyde **35** and an acetylene linker (Scheme 1.8.6) [6].

Addition of the cer-acetylide to **35** gave the alcohol **36**. After TMS-protection the terminal acetylene function was deprotonated and allowed to react with a second molecule of the aldehyde **35**. The resulting product **37** was TMS-deprotected and a subsequent hydrogenation of the triple bond gave the saturated 1,4-diol **38**.

Finally, the formation of center-pyrrolidine ring was addressed (Scheme 1.8.7).

Therefore, the diol **38** was converted into the cyclic sulfate **39**. Reaction of **39** with lithium azide followed by hydrolysis of the sulfate group gave the azido alcohol **40**. Mesylation of the hydroxyl group and a subsequent reduction of the azide to the amine afforded the desired terpyrrolidine derivative **41** [6].

The synthetic routes to oligo-THFs, oligo-THF amino acids and peptides as well as oligo-pyrrolidines presented here lay ground for further studies of their conformational properties and their membrane-modifying activities.

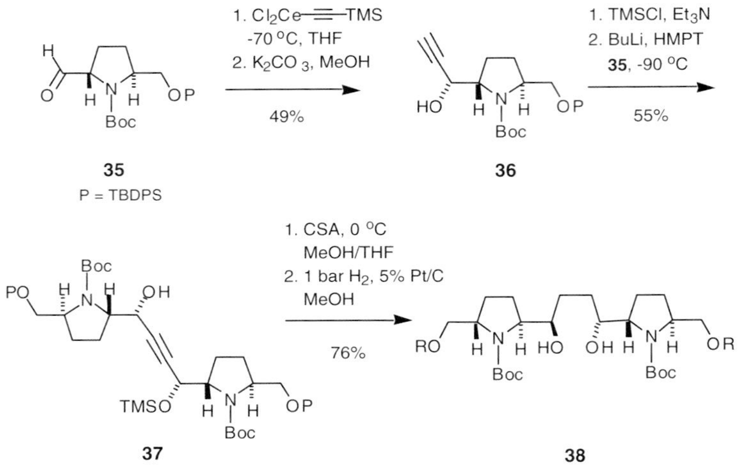

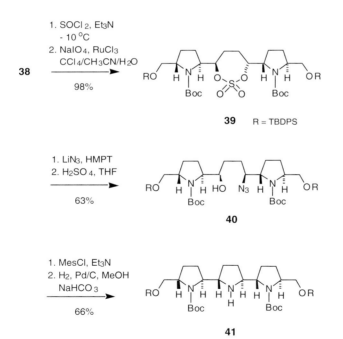

**Scheme 1.8.6.** Synthesis of the terpyrrolidine; part I.

**Scheme 1.8.7.** Synthesis of the terpyrrolidine; part II.

**Acknowledgments:** We thank the Deutschen Forschungsgemeinschaft, the Fonds der Chemischen Industrie, the Pinguin Foundation and the Schering AG for financial support.

# References

[1]  R. B. Gennis *Biomembranes, Molecular Structure and Function,* Springer, New York 1989.

[2]  B. Hille *Ionic Channels of Excitable Membranes* Sinauer, Sunderland MA, 2. ed. 1992.

[3]  N. Voyer, *Top. Curr. Chem.* 1996, *184,* 1; G. W. Gokel, O. Murillo, *Acc. Chem. Res.* 1996, *29,* 425; Y. Kobuke in *Advances in Supramolecular Chemistry* (ed. G. W. Gokel), Vol. 4, Greenwich, 1997, p. 163; G. G. Cross, T. M. Fyles, T. D. James, M. Zojaji, *Synlett* 1993, 449.

[4]  U. Koert, M. Stein, K. Harms, *Angew. Chem.* 1994, 106, 1238; *Angew. Chem.. Int Ed. Engl.* 1994, *33,* 1180. Related work in the oligo-THP-series: X. Wang, S. D. Erickson, T. Iimori, W. C. Still, *J. Am. Chem. Soc.* 1992, *114,* 4128. For a poly-THFs see: W. J. Schultz, M. C. Etter, A. V. Pocius, S. Smith, *J. Am. Chem. Soc.* 1980, *102,* 7984.

[5]  H. Wagner, K. Harms, U. Koert, S. Meder, G. Boheim, *Angew. Chem.* 1996, *108,* 2836; *Angew. Chem. Int Ed. Engl.* 1996, *35,* 2643.

[6]  H.-D. Arndt, K. Polborn, U. Koert, *Tetrahedron Lett.* 1997, *38,* 3879.

[7]  D. Cane, W. D. Celmer, J. W. Westley, *J. Am. Chem. Soc.* 1983, *105,* 3594.

[8]  H. Wagner, U. Koert, *Angew. Chem.* 1994, *106* , 1939; *Angew. Chem. Int Ed. Engl.* 1994, *33,* 1873.

[9]  U. Koert, H. Wagner, M. Stein, *Chem. Eur. J.* 1997, *3,* 1170.

## 1.9     Glycosphingolipids of myelin: Potential target antigens of demyelinating antibody activity in multiple sclerosis

*Eckhard Kirschning and Rudolf Dernick*

One of the most severe diseases affecting the human central nervous system (CNS) is multiple sclerosis (MS). The hallmark features of this disease are myelin depletion and the loss of the myelin-forming cells, called oligodendrocytes (OLs), at multiple sites of the CNS. Although both the etiology and the pathogenesis remains to be established, experimental data indicate that autoimmune mechanisms might be involved in myelin destruction [1]. In this context certain glycosphingolipids (GSL) being expressed in myelin and on the OL surface were suggested to serve as target structures of demyelinating antibody activity [2]. The most abundant GSL occurring in myelin are galactosylceramide (GalC) and galactosyl (3-O-sulfate) ceramide (sulfatide), which in human CNS myelin constitute 24% and 7% of the lipids, respectively [3]. The structure of sulfatide is shown in Figure 1.9.1. The fatty acids found in amide linkage to the sphingosine range from C-16 to C-28. Some unsaturation may occur, particularly in the longer-chain fatty acids. In the case of GalC and sulfatide, 2-D hydroxy fatty acids also occur. Until now, demyelinating anti-GalC or anti-sulfatide antibodies have been obtained exclusively from presensitized animals, but not from humans. In order to evaluate the human anti-glycolipid immune repertoire at the clonal level we established monoclonal antibody (mAb)-secreting B-cell lines from patients with MS and control persons [4]. In a liposome agglutination assay, four mAbs showed a positive reaction with GalC and/or sulfatide. These autoantibodies were exclusively obtained from MS patients. Autoantibodies are generally divided into two subgroups: (i) pathological auto-antibodies (PAA); and (ii) natural autoantibodies (NAA). The former occur exclusively in patients with autoimmune diseases, while the latter are also present in healthy individuals [5]. Both PAA and NAA can be directed against a wide range of self antigens. PAA generally are monospecific and exhibit a high affinity towards the antigen, while NAA are polyreactive and of low affinity. In order to find out whether these mAbs resemble PAA or NAA with regard to their binding properties, we determined their reaction with antigens of neuronal and non-neuronal origin.

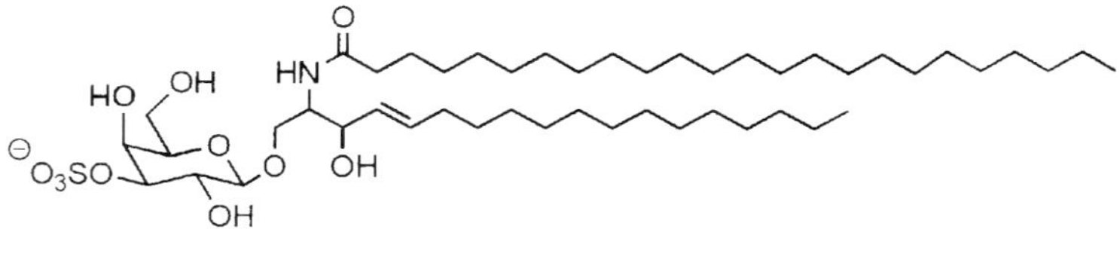

**Figure 1.9.1.** Structure of sulfatide.

## 1.9.1 Binding of human mAbs to surface antigens of cultured brain cells

Using immunogold labeling and indirect immunofluorescence methods, we determined the binding pattern of our myelin glycolipid-reactive mAbs with cultures of cells grown from dissociated brain tissue of newborn rats. Cultured brain cells provide a suitable model for this type of study, since (a) cells from the OL lineage divide and differentiate on a schedule very similar to that observed in vivo, and (b) OL differentiation is accompanied by the specific expression of GalC and sulfatide on their surfaces [6]. At a later stage, these cells express the myelin basic protein (MBP), which is another specific marker for Ols. Living cultures were stained with human mAb and anti-human immunoglobulin (Ig)-specific gold-conjugates. For light microscopy, the specific gold label was visualized by silver enhancement. One of these mAbs, the sulfatide-reactive mAb DS1F8, bound to cells with typical OL morphology, i.e. with round cell bodies and multiple arborized processes and large-sized, myelin-like elaborated membrane sheets (Figure 1.9.2). The three other mAbs introduced in this study did not exhibit any immunoreactivity. The OL specificity of human mAb DS1F8 was revealed by immunofluorescence double-labeling of aldehyde-fixed cultures with human mAb DS1F8 (visualized with anti-human Ig Texas Red-conjugates), and a panel of glial cell-specific marker mAbs (made visible with anti-mouse Ig Fluorescein-conjugates) [7]. DS1F8 selectively bound to a subpopulation of GalC + Ols. This OL-specificity of mAb DS1F8 was confirmed by double-labeling experiments carried out with mAb anti-MBP. As shown in Figure 1.9.3, the DS1F8-label was exclusively confined to MBP + Ols.

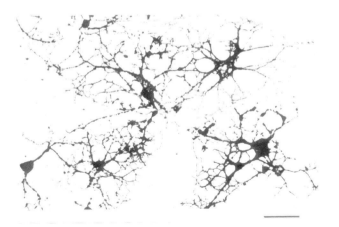

**Figure 1.9.2.** Human autoantibody DS1F8 binds to surface structures of cells with typical oligodendrocyte (OL) morphology. Scale bar = 25 μm.

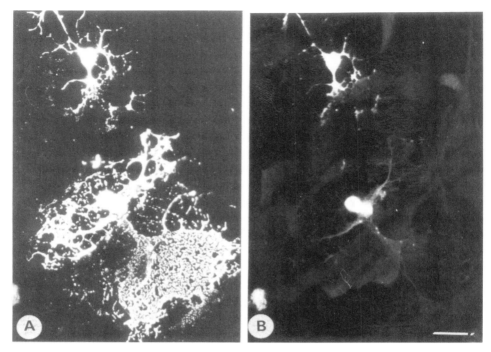

**Figure 1.9.3.** Human autoantibody DS1F8 (A) binds specifically to MBP + Ols (B) in mixed brain cell cultures. Scale bar = 25 μm.

## 1.9.2 Reaction pattern of autoantibody DS1F8 with intracellular antigens

Immunolabeling of opened brain cell cultures revealed some additional reaction of mAb DS1F8 with fibrous antigens in cells with epithelioid morphology [8]. Double-labeling studies carried out with mAb DS1F8 and the murine GalC-reactive mAb O1 in opened HeLa cells (human cervix carcinoma origin) revealed a very similar reaction of both mAbs with microtubuli-like structures (Figure 1.9.4). This cross-reaction with internal antigens of fibrous structure seems to be a common feature of GalC- and sulfatide-reactive antibodies of different species including mouse and rabbit [9]. The fact that HeLa cells were shown to express GalC [10], together with our observation that any intracellular staining could be abolished by pretreatment of cells with organic solvents, argues in favor of a lipidic nature of these internal antigens [8].

## 1.9.3 Reaction pattern of mAb DS1F8 with purified lipids in ELISA

Because of the low sensitivity of a a liposome agglutination assay, we tested the specificity of mAb DS1F8 with lipid antigens in an enzyme-linked immunosorbent

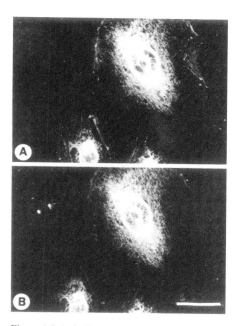

**Figure 1.9.4.** Sulfatide-reactive mAb DS1F8 (A) and GalC-reactive mAb O1 (B) exhibit a similar staining pattern of microtubuli-like structures in opened HeLa cells. Scale bar = 25 μm.

assay (ELISA) [11]. Commercially available purified lipid standards were coated to the wells of an ELISA plate. Wells were incubated with mAb DS1F8 followed by anti-human Ig alkaline phosphatase-conjugates. The enzyme reaction was performed using *p*-nitrophenylphosphate as chromogenic substrate. Enzyme activity was recorded at 414 nm and 540 nm wavelength in an automatic reader.

As shown in Figure 1.9.5, the observed reaction of DS1F8 with sulfatides in a liposome agglutination assay could be confirmed in an anti-lipid ELISA. Besides a pronounced reaction with a glycosphingolipid mixture, a significant albeit minor reaction with cholesterol was observed, while GalC, lecithine and some other lipids introduced in this assay did not exhibit any immunoreactivity with DS1F8. This binding of DS1F8 to cholesterol is difficult to interpret, but might be due to an intrinsic property of sulfatide-reactive antibodies, since cholesterol reactivity was also described for a murine sulfatide-reactive mAb [12].

Lecithin (Lec), GalC, sulfatides (Sulf), cholesterol (Chol), sphingomyelin (SM), monogalactosyldiglyceride (MG) and a mixture of gangliosides (G-Mix) and sphingo-lipids (S-Mix).

## 1.9.4 Conclusions

In this study we for the first time describe the existence of a B-cell occuring in the human autoimmune repertoire which has specificity for surface structures of

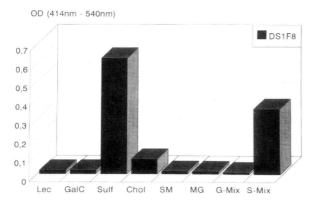

**Figure 1.9.5.** Pronounced reaction of human mAb DS1F8 with sulfatides and a sulfatides containing glyco-sphingolipid-mixture in an ELISA with purified lipids.

oligodendrocytes. With regard to its phenotypic properties, this sulfatide-reactive mAb DS1F8 closely resembles PAA. From these data we conclude that autoantibodies with a similar specificity as DS1F8 might react with oligodendrocytes and/or myelin in vivo and thus might be involved in demyelinating processes in the CNS. A recently started molecular biology study aimed at identifying the genes encoding for the antigen-binding domains of DS1F8 will provide additional information as to whether mAb DS1F8, which was isolated from a patient with MS, is due to a non-specific site effect or is a type of disease-specific part of the autoimmune network leading to MS [13].

**Acknowledgements:** Part of this study was supported by grants from the Gemeinnützige Hertie-Stiftung (GHS 132/86, GHS 316/94), Frankfurt/Main. The preparation of this manuscript was made possible by generous financial support by the Verein zur Förderung von Ursachen auf dem Gebiet der Multiplen Sklerose und anderer chronischer Krankheiten (MUCK e.V.), Hamburg. We thank Dr. A. Kirschning for providing one of the figures.

# References

[1]  C. Ewing, C. C. Bernard, *Immunol. Cell. Biol.*, **1998**, 76, 47–54.
[2]  R. C. Sergott, M. J. Brown, D. Silberberg, R. P. Lisak, *J. Neurol. Sci.*, **1984**, 64, 297–303; C. Ffrench-Constant, *Lancet*, **1994**, 343, 271–275.
[3]  G. Nonaka, Y. Kishimoto, *Biochim. Biophys. Acta*, **1979**, 572, 432–441.
[4]  H. Uhlig, R. Dernick, *Autoimmunity*, **1989**, 5, 87–99.
[5]  S. Avrameas, *Immunol. Today*, **1995**, 12, 154–159.
[6]  M. C. Raff, *Science*, **1989**, 243, 1450–1455.
[7]  E. Kirschning, G. Rutter, H. Uhlig, R. Dernick, *J. Neuroimmmunol.*, **1995**, 56, 191–200; E. Kirschning, *Dissertation*, Universität Hamburg, **1995**.
[8]  E. Kirschning, G. Rutter, R. Dernick, unpublished results.

[9]  A. E. Warrington, S. E. Pfeiffer, *J. Neurosci. Res.*, **1992**, 33, 338–353; R. Reynolds, R. Hardy, *J. Neurosci. Res.*, **1997**, 47, 455–470.

[10]  K. Sakakibara, T. Momoi, T. Uchida and Y. Nagai, *Nature*, **1981**, 293, 76–79.

[11]  E. Kirschning, G. Rutter, H. Hohenberg, *J. Neurosci. Res.*, **1998**, 53, 465–474.

[12]  R. Bansal, A. E. Warrington, A. L. Gard, B. Ranscht, S. E. Pfeiffer, *J. Neurosci. Res.*, **1989**, 24, 548–557.

[13]  E. Kirschning, K. Jensen, H. Will, unpublished results.

## 1.10    Combinatorial chemistry for the identification of novel bioactive compounds

*Andreas L. Marzinzik and Eduard R. Felder*

### 1.10.1    Combinatorial chemistry in drug discovery

Although combinatorial chemistry has been used in material science [1], catalysis [2] and polymer research [3], it is mostly disseminated in agrochemical and pharmaceutical research [4]. The methods of lead finding in the research-oriented pharmaceutical industry are continuously reviewed in order to evaluate their innovative potential for identifying compounds with unique mechanisms of action or outstanding activity profiles. If a new technology becomes relevant for a more efficient identification of drug candidates, the implementation of this technology would rearrange resources. The influence of functional genomics or molecular modeling on recent research exemplifies the effects of methodologies on drug development. Currently, the multidisciplinary field of combinatorial chemistry is also developing rapidly. Investment in combinatorial technologies is being driven by the need dramatically to increase the evaluation efficiency of new chemical entities. The validation of high-throughput assays with minimal compound consumption in connection with combinatorial methods is expected to enhance the possibilities for the identification or optimization of novel lead structures. In its simplest expression, combinatorial chemistry may be reduced to the utilization of a set of validated reactions and compatible building blocks for generating an array of discrete compounds in a single experimental run that covers all combinations of desired reactants.

Combinatorial chemistry or multiparallel synthesis have also been applied to traditional organic chemistry simply to increase the throughput of traditional processes: solid-phase catalysts [5], scavengers on polymer [6] or assistance of fluorous-phase [7] enhance the efficiency of solution-phase chemistry without adaptation of the reaction conditions to another format. Nevertheless, compared with solution-phase chemistry, solid-phase chemistry has several considerable advantages. Complex work-up or purification procedures are unnecessary because of the immobilization of products or intermediates. A substantial excess of reagents accelerates the reaction and can drive the reaction to completion. In addition, the one-bead-one-compound method allows for the preparation of millions of physically separated individual compounds simultaneously on microcarriers, *e.g.* beads [8]. Another substantial benefit of solid-phase chemistry is offered by the more extensive automation of a multi-step synthesis. In addition to generic diversity systems, combinatorial chemists work with focused, thematic libraries, where the synthetic pathway has to be customized to an already existing lead structure, originally identified from the compound archive. Realizing the time investment needed to adapt the solution synthesis to a solid-phase protocol, industry also engages in major efforts to increase speed and automation of solution-phase chemistry. This is the naturally more straightforward way to produce analogs of leads originally synthesized in solution.

At present, the access to partial or full automation in solution- and solid-phase synthesis is a key factor in preventing mistakes by handling a large number of samples and enhancing reproducibility rather than speeding up the whole process. The impact of combinatorial technologies will be greatly enhanced by synergies with ongoing parallel developments in solid-phase chemistry, informatics and computing (compound logistics), high-throughput screening, and automation.

## 1.10.2    Molecular diversity on rigid scaffolds

Recently, scientists have focused on the design of synthetic schemes which allow for the greatest possible diversity in the synthesis of small-molecule libraries. Low-molecular weight compounds of a certain physical property profile (*e.g.* 'rule of five') [9] are attractive lead structures for medicinal chemistry programs in terms of their likelihood of sufficient absorption or permeation. In the discovery setting, the 'rule of five' predicts that poor absorption or permeation is more likely when there are more than five H-bond donors, 10 H-bond acceptors, the molecular weight is >500, and the calculated log p is >5.

Various assessment factors have to be considered before a reaction sequence should be transferred to the solid-phase. First of all, the modular scaffold generated in a low number of reaction steps should display pharmacophoric moieties in a conformationally constrained way. Preferred are reactions utilizing readily accessible or commercially available building blocks, and reactions which are suitable for automation. Especially useful are reaction schemes in which an intermediate is utilized for various pathways [10]. Prerequisite for all solid-phase syntheses is the compatibility of cleavable linker moieties that anchor the reaction products to the solid-support during the template assembly in a reversible manner. Ideally, linkers are cleaved without leaving a trace of reagents in the test samples. For lead finding, compatibility with the split-and-mix [8] synthesis is desired, in order exponentially to raise the number of synthesized compounds compared to an array synthesis. As an example of a low-molecular weight library preparation we were able to validate the synthesis of pyrazoles and isoxazoles on solid-phase [11]. A scope and limitation study was performed utilizing reagents which represent the spectrum of reactivity in order to grasp the system's breadth of applicability.

To obtain **1**, a carboxylic acid acetyl component bearing the $R^1$ residue was immobilized by the acid labile Rink linker [12] under standard conditions. All samples were analyzed and identified by liquid chromatography/mass spectrometry (LC/MS) of the cleaved product. Optimization of formation of the Claisen condensation product **2** revealed that heating for 1 h at 90°C was suitable to condense deactivated benzoates as well as various heterocyclic esters, without an appreciable amount of side products. On the other hand, carboxylic esters with $\alpha$-hydrogens and weakly acidic heteroaromatics are not suited. With regard to the $\alpha$-alkylation, the best results to give **3** were obtained in the presence of tetrabutylammonium fluoride (TBAF). The hydrogen bond acceptor TBAF increases the nucleophilicity and inhibits *O*-alkylation by shielding the oxygen atom of the enol intermediate. The reaction provides reasonable yields for a variety of alkylating agents, but is restricted to substrates **2** without acidic or basic functionalities in $R^1$ and $R^2$. Numerous monosubstituted hydrazines and hydroxylamine afforded regioisomeric pyrazoles **4** and isoxazoles **5**, respectively, from highly functionalized **3** (Scheme 1.10.1). However, there are also drawbacks in solid-phase chemistry in addition to the limited possibilities of monitoring a reaction on solid-phase. If using a vast excess of reagent, the absence of highly reactive impurities has to be ensured. It transpired that some aromatic hydrazines contain hydrazine itself as an impurity, which has a much higher tendency to cyclize than aromatic hydrazines, leading to *N*-unsubstituted pyrazoles almost exclusively. We used acetylacetone as a scavenger to eliminate those impurities. In selecting the hydrazines, we had to consider that they would decompose if kept at higher temperatures.

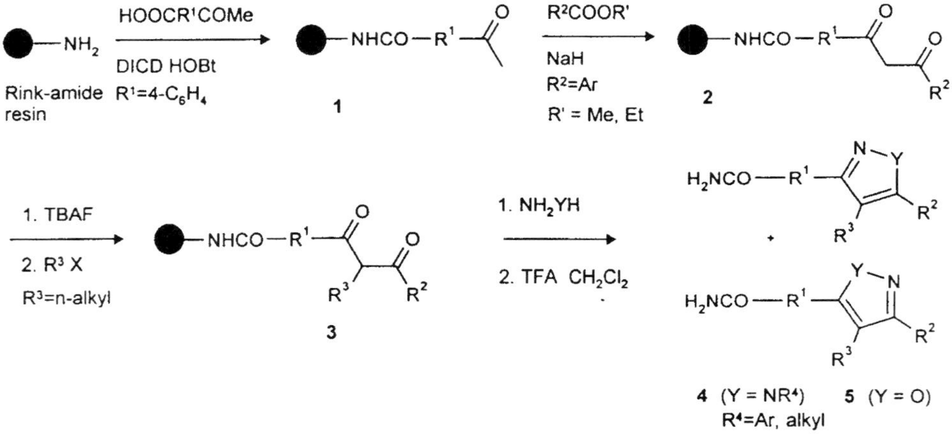

HOOCR¹COMe — DICD HOBt — Rink-amide resin — $R^1$=4-$C_6H_4$ — NHCO—R¹ **1** — $R^2$COOR' NaH $R^2$=Ar R' = Me, Et — NHCO—R¹ $R^2$ **2**

1. TBAF  2. R³ X  R³=n-alkyl — NHCO—R¹ $R^3$ $R^2$ **3** — 1. NH₂YH  2. TFA CH₂Cl₂ — $H_2NCO$—R¹ (N—Y, $R^2$, $R^3$) **4** (Y = NR⁴) + $H_2NCO$—R¹ (Y—N, $R^2$, $R^3$) **5** (Y = O)  R⁴=Ar, alkyl

**Scheme 1.10.1.** Solid-phase synthesis of pyrroles and isoxazoles.

The collective data on the conversion rates of a variety of reactants within our proposed synthetic scheme provide an information basis sufficient for the planning of combinatorial libraries by split-and-mix [8], sort-and-combine [13] or by multiparallel synthesis [14]. After a decision was made as to which kind of building blocks were to be included, a library of more than 10 000 compounds in sublibraries of 140 regioisomers was synthesized [15]. The achieved progress in miniaturization of biological in vitro testing permits an economical handling of samples from combinatorial chemistry, which enables varied usage in several projects for many years.

## 1.10.3    Solid-phase synthesis toward multiple core structures

Without precise knowledge of the structural and physico-chemical parameters involved in a particular ligand–target interaction, efforts should be spent on planning an efficient process of exhaustive exploration of space availability at the target site, based on experimental approaches. The de novo ligand design has revealed insufficient predictive capability so far. Intellectual efforts in combinatorial chemistry deal with the selection and development of modular scaffolds for high-throughput synthesis. Besides the novelty of the core structure and patentabilty of the final compounds, valuable criteria may be the similarity to known biologically active compound classes, or more pragmatic factors such as ease of synthesis and number of diversity points. For lead finding, the production of highly diverse molecules demands a large variety of commercially available or proprietary building blocks. Computational studies assessing diversity measures are used not only for selecting building blocks, but also for predicting the complementarity of the scaffold to existing libraries. Taking into consideration the medicinal chemistry aspects, scaffold selection may also comprise an estimation of pharmacokinetic aspects such as absorption and metabolism, or pharmacodynamic factors such as toxicity.

Bearing these criteria in mind, the application of combinatorial chemistry within modern lead finding processes not only demands the combination of numerous building blocks on a particular template, but also the ability to provide a wide variety of such scaffolds. It is evident that the core structure of a compound class contributes to the pharmacological profile. In addition to its intrinsic effects, it mediates activity by directing the spacial arrangement of pharmacophoric substituents.

Besides the examples of solid-phase synthesis of a particular template for lead optimization, a few studies developed pathways for the synthesis of several scaffolds from a particular intermediate, *e.g.* quinones, phenols, naphthylfurans, pyrones, and indoles from squaric acid [16] or dioxopiperazines and dioxomorpholines from α-bromo-substituted dipeptides [17].

We were able to validate α,β-unsaturated ketones as intermediates for the generation of molecular diversity on small heterocycles, as shown in Scheme 1.10.2. Our investigation concentrated, in a first stage, on the solid-support synthesis of pyrimidines **8** which are the core structure of various pharmaceutically relevant compounds.

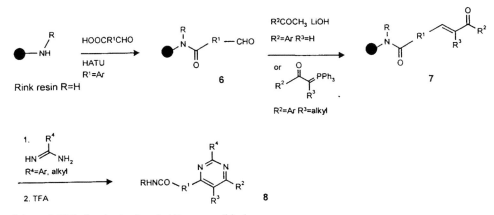

**Scheme 1.10.2.** Synthesis of pyrimidines on solid-phase.

The Claisen–Schmidt reaction met the requirements for a transfer to a combinatorial solid-phase synthesis. Various acetophenones indicated a satisfactory conversion to **7** in the presence of LiOH in dimethoxyethane for the immobilized carboxyaldehydes **6** on Rink resin. On the other hand, the simultaneous introduction of an $R^3$ residue (*e.g.* $R^3 = CH_3$) was not possible by the usage of propiophenone instead of acetophenone ($R^2 = Ph$, $R^3 = H$) under the same or similar reaction conditions. However, those substitution patterns were easily achieved by the broadly applicable Wittig reaction. In turn, pyrimidines exhibiting various substituent patterns were obtained upon condensation of intermediates **7** with various amidines at 100°C in presence of air. Likewise, we could achieve uniform reactions independently from steric and electronic properties of the amidines by using excess of reagent.

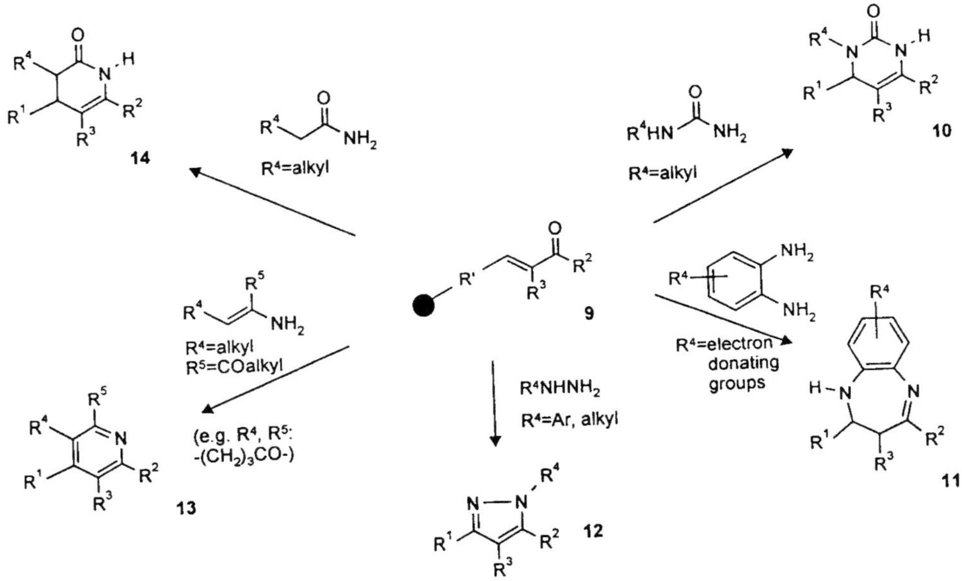

**Scheme 1.10.3.** Multiple core structures from $\alpha,\beta$-unsaturated ketones.

Subsequently, the synthesis of heterocycles other than pyrimidines from **9** was studied (Scheme 1.10.3). As in solution phase chemistry, $\alpha,\beta$-unsaturated ketones can act as three-carbon components and a primary enamine stabilized with an electron-withdrawing group can be utilized to complement the system for ring closure. For instance, the reaction of 3-amino-2-cyclohexen-1-one with **9** in the presence of potassium *tert*-butoxide afforded a pyridine **13** with an annelated ring ($R^4$, $R^5$: $-(CH_2)_3CO-$). Primary enamines are easily obtained, *e.g.* by the Thorpe–Ziegler reaction.

The aim of preparing compound arrays in multiparallel synthesis encourages researchers to synthesize the known pharmacophore dihydropyrimidine in a combinatorial fashion. One approach has been reported by the Biginelli multicomponent reaction [18]. Our stepwise reaction scheme allows to apply the split-and-mix methodology and, therefore, to generate diversity with high efficiency. Ring closure to form a heterocyclic scaffold has been tested successfully with carboxamides (providing dihydropyridone **14**), with monosubstituted hydrazines (to pyrazoles **12**) and also with 1,2-phenylenediamine to generate 2,3-dihydro-1,5-benzodiazepines **11**, of which the secondary amine and the azomethine moiety could be the starting point for further diversity generation.

The collected data on the conversion rates of a variety of reactants within our proposed synthetic scheme provide an information basis sufficient for the planning and designing of combinatorial libraries with regard to the selection of pharmacophores for diversity generation. The development of intermediates towards multiple core structures can play a pivotal role in the synthesis of non-thematic libraries for the identification of novel lead structures. The combinatorial synthesis of new molecular entities, meeting the criteria of a drug-like template, will pave the way for more efficient and comprehensive follow-up activities in the lead optimization phase.

# References

[1] X.-D. Xiang, X. Sun, G. Briceno, Y. Lou, K.-A. Wang, H. Chang, W.G Wallace-Freedman, S.-W. Chen, P. G. Schultz, *Science* **1995**, *268*, 1738–1740.

[2] M. S. Sigman, E. N. Jacobsen, *J. Am. Chem. Soc.* **1998**, *120*, 4901–4902.

[3] D. J. Gravert, K. D. Janda, *Tetrahedron Lett.* **1998**, *39*, 1513–1516.

[4] Reviews: (a) L. A. Thompson, J. A. Ellman, *Chem. Rev.* **1996**, *96*, 555–600; (b) F. Balkenhohl, C. von dem Bussche-Hünnefeld, A. Lansky, C. Zechel, *Angew. Chem. Int. Ed. Eng.* **1996**, *108*, 2288–2337; *Angew. Chem.* **1996**, *108*, 2437–2488.

[5] S.J. Shuttleworth, S.M. Allin, P.K. Sharma, *Synthesis* **1997**, 1217–1239.

[6] R.J. Booth, J.C. Hodges, *J. Am. Chem. Soc.* **1997**, 119, 4882–4886.

[7] A. Studer, P. Jeger, P. Wipf, D.P. Curran, *J. Org. Chem.* **1997**, *62*, 2917–2924.

[8] (a) K. S. Lam, S. E. Salmon, E. M. Hersh, V. J. Hruby, W. M. Kazmierski, R. J. Knapp, *Nature* **1991**, *354*, 82–84; (b) A. Furka, F. Sebestyen, M. Asgedom, G. Dibo, *Abstr. 14th Int. Congr. Biochem. Prague* **1988**, *5 (Abstr FR:013)*, 47.

[9] C. A. Lipinski, F. Lombardo, B. W. Dominy, P, J. Feeney *Adv. Drug Delivery Rev.* **1997**, *23*, 3–25.

[10] A. L. Marzinzik, E. R. Felder, *J. Org. Chem.* **1998**, *63*, 723–727.

[11] A. L. Marzinzik, E. R. Felder, *Tetrahedron Lett.* **1996**, *37*, 1003–1006.

[12] H. Rink, *Tetrahedron Lett.* **1987**, *28*, 3787–3790.

[13] K. C. Nicolaou, X. Y. Xiao, Z. Parandoosh, A. Senyei, M. Nova, *Angew. Chem. Int. Ed. Eng.* **1995**, *34*, 2289–2291; *Angew. Chem.* **1995**, *34*, 2476–2479.

[14] B. A. Bunin, J. A. Ellman, *J. Am. Chem. Soc.* **1992**, *114*, 10997–10998.

[15] A. L. Marzinzik, E. R. Felder, *Molecules* **1997**, *2*, 17–30.

[16] P. A. Tempest, R. W. Armstrong, *J. Am. Chem. Soc.* **1997**, *119*, 7607–7608.

[17] B. O. Scott, A. C. Siegmund, C. K. Marlowe, C. K.; Pei, Y.; K. L. Spear, *Mol. Diversity* **1996**, *1*, 125–134.

[18] P. Wipf, A. Cunningham, *Tetrahedron Lett.* **1995**, *36*, 7819–7822.

## 1.11       Supramolecular polymer chemistry based on multiple hydrogen bonding

*Ronald F.M. Lange and E.W. Meijer*

Inspired by the wealth of biomacromolecules, that have often well-defined architectures through multiple hydrogen bonding interactions, polymer chemists are pursuing the translation of the principles of supramolecular chemistry into novel synthetic polymeric materials. With the making and breaking of covalent bonds, chemists have achieved overwhelming results, as is demonstrated by the synthesis of, *e.g.* vitamin $B_{12}$ and Taxol [1, 2], and by the control over polymer tacticity in the polymerization of, *e.g.* isotactic polypropylene [3]. However, the limits have been reached, and the synthesis of even larger and more complicated molecules is likely to be impossible.

In nature, large and complicated systems are formed by the defined organization of macromolecules using secondary interactions, like ion–ion, dipole–dipole (including hydrogen bonding) or van der Waals forces [4]. This feature was the inspiration for the development of an area within the field of organic chemistry, *i.e.* supramolecular chemistry, which is chemistry aiming to control the secondary interactions between molecules. The subtleties which determine the balance between enthalpic and entropic factors make the complexes of relative small molecules often unstable. The co-operativity is restricted due to the limited size of the molecules, which is circumvented in nature by using molecules possessing a higher molecular weight as, *e.g.* polypeptides or polynucleotides.

Although the importance of co-operative secondary interactions and chirality is recognized, *e.g.* in nylons and polypropylene respectively, the use of pre-designed secondary interactions is generally neglected by the majority of polymer chemists. This work focuses on the existing gap within the triangle based on: (i) organic chemistry, which uses well-defined molecules possessing a low molecular weight, and from which supramolecular chemistry has developed; (ii) polymer chemistry, which uses high molecular weight compounds but generally neglects the importance of secondary interactions and chirality; and (iii) natural materials, which uses high molecular weight, chiral compounds (Figure 1.11.1).

In this paper our results will be summarized regarding the use of multiple hydrogen bonding interactions in order to arrive at well-defined, supramolecular polymer architectures, which are all based on commercially available compounds.

Complexation of melamine via triple hydrogen bonding to various imide-containing polymers resulted in well-defined 3:1 imide–melamine complexes, as was determined by extrapolation of powder diffraction X-ray data in combination with differential scanning calorimetry (DSC) [5]. The 3:1 imide-melamine complexation obtained, as depicted in Figure 1.11.2, is based on supramolecular interactions since the melamine could be quantitatively recovered by addition of hydrogen bond-breaking agents. Structural evidence of the 3:1 imide-melamine complexation was obtained by $^{13}$C MAS TRAPDOR NMR experiments, and the 3:1 complexation was supported by model studies. The model studies also showed the possibility of influencing the

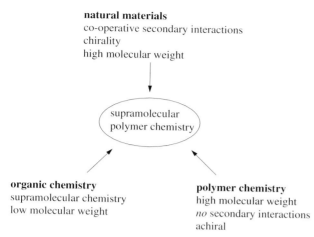

**Figure 1.11.1.** Supramolecular polymer chemistry bridging the gap between organic chemistry, polymer chemistry, and natural materials.

complexation ratio. Single crystal X-ray analysis of model compounds revealed that a 1:1, 2:1 or 3:1 imide-melamine complex ratio was obtained, in which the availability of the imide carbonyl acceptor sites forms a tool to tune the supramolecular crystal structure with melamine [6].

The introduction of imide and 2,4-diaminotriazine structures into various polymeric materials resulted in homogeneous, molecularly miscible polymer blends. The miscibility is attributed to the formation of specific, unidirectional imide-2,4-diaminotriazine triple hydrogen bonds. The prominent role of these specific triple hydrogen bonds was demonstrated by the use of *N*-methyl-substituted polyimides which prevent the triple hydrogen bond formation. This resulted in inhomogeneous, phase-separated polymer blends.

The imide-2,4-diaminotriazine triple hydrogen bond was also introduced into commercially available polymers. Due to the specific triple hydrogen bond formation,

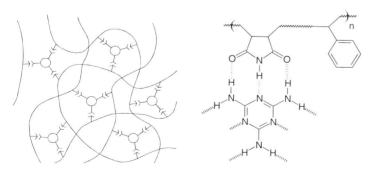

**Figure 1.11.2.** The formation of a well-defined polymeric supramolecular 3:1 imide–melamine complex.

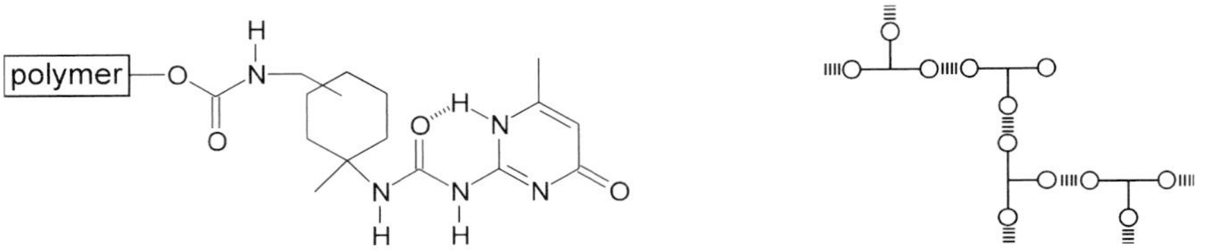

**Figure 1.11.3.** Hydrogen-bonded polymer network using the strong dimerizing ureido-pyrimidone unit (polymer is a block co-polymer of ethylene oxide and propylene oxide possessing three hydroxyl endgroups and a molecular weight of about $6 \times 10^3$ g·mol$^{-1}$).

an otherwise immiscible ABS-SMI blend (acrylonitrile-butadiene-styrene and styrene-maleimide, respectively) became miscible. The latter demonstrates the general applicability of the complementary couple used.

Unlike most of the biomaterials, no higher order of the supramolecular polymeric complexes was obtained using this triple hydrogen bonded imide – 2,4-diaminotriazine interaction. Attempts to synthesize stereoregular polymers to increase the overall imide – 2,4-diaminotriazine interaction strength due to co-operativity, were less successful. Therefore, the strong dimerizing, hydrogen-bonded ureido-pyrimidone unit, which has a dimerization constant exceeding $10^6$ l·mol$^{-1}$ in CHCl$_3$ [7], is used to obtain reversible polymer networks (Figure 1.11.3).

A new synthetic route from commercially available starting materials that has been developed is described [8]. The quadruple hydrogen-bonding ureido-pyrimidone unit is prepared using selectively 3(4)-isocyanatomethyl-1-methylcyclohexyl-isocyanate (IMCI) in the coupling reaction of the multi-hydroxy functionalized polymers with isocytosines. $^1$H- and $^{13}$C-NMR, IR, MS and ES-MS analysis, performed on a model reaction using butanol, demonstrated the formation of the hydrogen bonding ureido-pyrimidone unit in a yield of more than 95%.

In order to determine the properties of the undissolved, pure hydrogen-bonded polymer network in bulk, dynamic mechanical measurements at various temperatures were performed. To our surprise, it proved possible to employ the principle of time-temperature superposition. In Figure 1.11.4, master curves of the storage and loss moduli, as well as the viscosity and phase angle are plotted against the frequency at a reference temperature of 30°C. At 30°C, the hydrogen-bonded polymer network shows typical viscoelastic behavior: liquid-like behavior at low frequencies and solid-like behavior at high frequencies. Furthermore, a clear rubber-to-glass transition is visible as a maximum in the phase angle around $10^8$–$10^9$ rad·s$^{-1}$. The plateau moduli, determined at the minimum of the phase angle $\delta$, is $5 \times 10^5$ Pa, which corresponds to a molecular weight between crosslinks (M$_c$) of $5.1 \times 10^3$ g·mol$^{-1}$ [9]. Noticeable is the defined viscoelastic transition. This very sharp viscoelastic transition can be described to a good approximation with a simple relaxation time $\tau$. For a single Maxwell element $\omega\tau = 1$ at 45° (or G$''$ is a maximum) [10], so that a relaxation time is obtained of around $\tau = 0.1$ s. Interestingly, this three-dimensional, hydrogen-bonded network behaves as a viscoelastic material. This can be explained by the cleavage and reassembling of the hydrogen bonds. It has been shown for similar systems that when the characteristic breakage time of the dissociating molecules is short compared to the

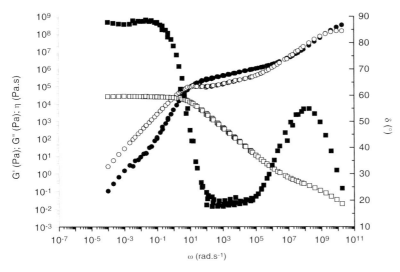

**Figure 1.11.4.** Master curves of the storage (●) and loss modulus (○), as well as viscosity (□) and phase angle (■) versus the frequency for the hydrogen-bonded polymer network.

relaxation time of the associated molecules, a single relaxation time is to be expected [11]. The capability of the hydrogen-bonding units to reassemble was shown for this hydrogen-bonded network by means of a steady-state measurement, in which the viscosity at lower frequency is unaffected compared to the values obtained by the dynamic measurements. This proves the existence of hydrogen-bonding at these lower frequencies.

The behavior of this hydrogen-bonded polymer network, possessing the well-defined, strong dimerizing hydrogen-bonding ureido-pyrimidone unit is unique. The quadruple hydrogen-bonded ureido-pyrimidone unit is able to construct a strong polymer network consisting of low-molecular weight compounds without the need of additional stabilization such as phase separation. In that way, it resembles a covalently crosslinked polymer network. To the best of our knowledge, no other reversible interacting unit displays this unique behavior. Although sugar-based natural materials are also able to form multiple hydrogen-bonding interactions resulting in high viscosities, *e.g.* a concentrated syrup solution, these systems do not show a viscoelastic behavior. The rheological behavior of this hydrogen-bonded polymer network, which show a well-defined viscoelastic transition, differs from the rheological behavior of the reversible, phase-separated polymer networks, which show a broadening of the transitions. The rheological behavior of this hydrogen-bonded polymer network resembles a linear, high-molecular weight polymer possessing a very small molecular weight distribution. However, since we deal here with a polymer network, the relaxation time is not caused by chain reptation, as is the case for high-molecular linear polymers, but is caused by disruption of the hydrogen-bonded ureido-pyrimidone units. The abrupt transition from an elastic hydrogen-bonded polymer network to an almost perfect Newtonian liquid is unique, and should offer many exciting possibilities and applications.

In conclusion, it can be said that the above-described results demonstrate that exciting possibilities exist to develop new polymer architectures, based on commercially available building blocks.

**Acknowledgements:**   The authors acknowledge the fruitful discussions with many colleagues at DSM Research and Eindhoven University of Technology. The Bayer Company is acknowledged for providing the 3(4)-isocyanatomethyl-1-methylcyclohexyl-isocyanate (IMCI).

# References

[1]  (a) R. B. Woodward, *Pure Appl. Chem.*, **1968**, *17*, 519; (b) A. Eschenmoser, *Quart. Rev.*, **1966**, *24*, 366; (c) A. Eschenmoser, *Chem. Soc. Rev.*,**1976**, *5*, 377; (d) A. Eschenmoser, *Nova Acta Leopoldina*, **1982**, *55*, 5.

[2]  (a) R. A. Holton, C. Somoza, H. B. Kim, F. Liang, R. J. Biediger, P. D. Boatman, M. Shindo, C. C. Smith, S. Kim, H. Nadizadeh, Y. Suzuki, C. Tao, P. Vu, S. Tang, P. Zhang, K. K. Murthi, L. N. Gentile, J. H. Liu, *J. Am. Chem. Soc.*, **1994**, *116*, 1597; (b) R. A. Holton, H. B. Kim, C. Somoza, F. Liang, R. J. Biediger, P. D. Boatman, M. Shindo, C. C. Smith, S. Kim, H. Nadizadeh, Y. Suzuki, C. Tao, P. Vu, S. Tang, P. Zhang, K. K. Murthi, L. N. Gentile, J. H. Liu, *J. Am. Chem. Soc.*,**1994**, *116*, 1599; (c) K. C. Nicolaou, Z. Yang, J. J. Liu, H. Ueno, P. G. Nantermet, R. K. Guy, C. F. Claiborne, J. Renaud, E. A. Couladouros, K. Paulvannan, E. Sorensen, *Nature*, **1994**, *367*, 630; (d) K. C. Nicolaou, P. G. Nantermet, H. Ueno, R. K. Guy, E. A. Couladouros, E. Sorensen, *J. Am. Chem. Soc.*,**1995**, *117*, 624; (e) K. C. Nicolaou, J. J. Liu, Z. Yang, H. Ueno, E. Sorensen, C. F. Claiborne, R. K. Guy, C. K. Hwang, M. Nakada, P. G. Nantermet, *J. Am. Chem. Soc.*, **1995**, *117*, 634; (f) K. C. Nicolaou, Z. Yang, J. J. Liu, P. G. Nantermet, C. F. Claiborne, J. Renaud, R. K. Guy, K. Shibayama, *J. Am. Chem. Soc.*, **1995**, *117*, 645; (g) K. C. Nicolaou, H. Ueno, J. J. Liu, P. G. Nantermet, Z. Yang, J. Renaud, K. Paulvannan, R. Chadha, *J. Am. Chem. Soc.*, **1995**, *117*, 653; (h) J. J. Masters, J. T. Link, L. B. Schneider, W. B. Young, S. J. Danishefsky, *Angew. Chem. Int. Ed. Engl.*, **1995**, *34*, 1723; S. J. Danishefsky, J. J. Masters, W. B. Young, J. T. Link, L. B. Schnyder, T. V. Magee, D. K. Jung, R. C. A.Isaacs, W. G. Bornmann, C. A. Alaimo, C. A. Coburn, M. J. Di Grandi, *J. Am. Chem. Soc.*, **1995**, *118*, 2843.

[3]  (a) G. W. Coates, R. M. Waymouth, *Science*, **1995**, *267*, 217; (b) H. H. Brintzinger, D. Fischer, R. Mülhaupt, B. Rieger, R. M. Waymouth, *Angew. Chem.*, **1995**, *107*, 1255; *Angew. Chem. Int. Ed. Engl.*, **1995**, *34*, 1143.

[4]  L. Stryer, *Biochemistry*, 4th Ed., W. H. Freeman and Co., New York, **1995**.

[5]  (a) R. F. M. Lange, E. W. Meijer, *Macromolecules*, **1995**, *28*, 782; (b) R. F. M. Lange, E. W.Meijer, *Macromol. Symp.*, **1996**, *102*, 301.

[6]  R. F. M. Lange, F. H. Beijer, R. P. Sijbesma, R. W. W. Hooft, H. Kooijman, A. L. Spek, J. Kroon, E. W. Meijer, *Angew. Chem.*, **1997**, *109*, 1006; *Angew. Chem. Int. Ed. Engl.*, **1997**, *36*, 696.

[7]  F. H. Beijer, H. Kooijman, A. L. Spek, R. P. Sijbesma, E. W. Meijer, *J. Am. Chem. Soc.*, submitted.

[8]  R. F. M. Lange, *Multiple Hydrogen Bonding in Reversible Polymer Networks*, Thesis, **1997**; R. P. Sijbesma, F. H. Beijer, L. Brunsveld, B. J. B. Folmer, K.Hirschberg, R. F. M. Lange, J. Lowe, E. W. Meijer, *Science*, **1997**, *278*, 1601.

[9] The entanglement molecular weight (Me) is determined using the formula $Me = \rho RT/Gd$, where $\rho$ is the density, R the gasconstant and T the temperature. The density of the linear poly(ethylene glycol)-*b*-poly(propylene glycol)-*b*-poly(ethylene glycol), which is $1020 \, \text{kg} \cdot \text{m}^{-3}$, is used as an estimate for the density of the hydrogen-bonded polymer network.

[10] J. D. Ferry, *Viscoelastic Properties of Polymers*, J. Wiley & Sons New York, 3rd Ed., **1980**.

[11] M. E. Cates, *Macromolecules*, **1987**, *20*, 2289.

# 2  Enzymatic Synthesis and Biotransformation

## 2.1  Enzymatic C–C coupling: The development of aromatic prenylation for organic synthesis

*Ludger Wessjohann, Bernd Sontag, and Marco-Aurelio Dessoy*

Enzymes offer advantages over synthetic catalysts because they commonly provide extraordinary chemo-, regio- and stereoselectivity, high catalytic activity, mild reaction conditions, use in water, and reproducibility on a microscale basis. This is especially important for reactions which are not easily achieved by classical methods [1]. The basic requirements for the successful introduction of an enzymatic method into a synthetic organic laboratory are:

1. Accessibility and/or stability of the biocatalyst.
2. Lack of cofactor problems.
3a. Knowledge of the substrate specificity [2], which should be as broad as possible if pure substrates can be applied, or alterable, *e.g.* by directed mutagenesis. (This requirement can be fulfilled by a single enzyme or, alternatively, by a group of available enzymes catalyzing the same transformation.)
3b. For industrial applications, aiming at a single product, the latter requirement may be substituted by process stability, high turnover, and cost effectiveness.

On one hand, hydrolases (lipases, proteases, esterases, amidases, etc.) fulfil all of the above requirements and, consequently, have become part of the standard repertoire of organic synthetic methods [1]. On the other hand, oxido-reductases belong to the most promising future biocatalysts, despite the fact that enzymatic oxidation reactions often require cofactor regeneration or whole-cell systems and demand many development cycles [3]. Both these 'big' groups of currently applied or developed biocatalysts are limited to the alteration or introduction of carbon-hetero bonds, whereas C–C bond formations have not yet found widespread application [4–9]. The most successful applications of enzymatic C–C couplings to date have been achieved with cyanohydrin reactions catalyzed by oxynitrilases and similar lyases [4, 7]. However, the cyanide addition can be regarded as special, as it does not share the major drawback of most other C–C-forming reactions of being energetically and at the same time kinetically less favorable (due to the requirement of an intermediate with a charged carbon and the lack of free electron pairs at the uncharged reaction centers). Quite to the contrary, enzyme catalysis in the cyanohydrin reaction has to compete with the general acid/base-catalyzed process.

More complex C–C-bond formations can be achieved with aldolases and similar enzymes catalyzing aldol reactions [1, 6, 8]. If acetyl- and malonyl-S-CoA-reactions are included, aldol reactions are certainly the most important C—C coupling reactions, crucial *e.g.* for sugar, lipid, and polyketide synthesis. For dihydroxyacetone phosphate and aldehydes as substrates, a set of enzymatic reactions allowing the selective synthesis of each of the four possible diastereomers has been developed, including elaborate schemes to generate the phosphate substrate [9, 1]. However, enzymatic aldol reactions in organic syntheses have fierce competition from classic chemical procedures.

Polyprenyl compounds (and therefrom steroids) and terpenoids comprise a group which also requires extensive C—C coupling. After synthesis of the $C_5$-isoprenoid units, C—C coupling is no longer based on aldol-type reactions, but on intra- or intermolecular prenyl transfer (= dimethylallyl-transfer) to C-nucleophiles [11]. The transfer of (poly)prenyl units happens exclusively from their diphosphates (pyrophosphates, **2**) and might be the second most important natural C—C-coupling reaction. Other C—C-coupling reactions, *e.g.* methylations, are less commonly found in nature.

## 2.1.1   Prenyl transferases

Prenyl transferases have gained little recognition in chemistry after Feodor Lynen's fundamental work on the mevalonate route was completed. Only recently, interest in isoprenoid chemistry has received an immense boost, almost simultaneously from various areas: The discovery of a new terpenoid pathway by Rohmer, Arigoni and Sahm; [*cf.* 11–13] the possibilities of molecular biology and X-ray crystallography to study enzymes like squalene synthase in detail; [14] and the importance of prenylation for the regulation and targeting of bioactive compounds in the cell, *e.g.* the farnesylation of proteins in signal transduction cascades important in carcinogenesis [15, 16]. Prenylated aromatic compounds constitute a special group (Figure 2.1.1), playing a crucial role in redox processes, *e.g.* the ubiquinones (coenzymes Q) as part of the respiratory chain [17a]. Other prenylated aromatics are precursors of vitamin E (tocopherols), of the Japanese drug and dye shikonin [17b], of anti-HIV compounds, of alkaloids and of many fungal metabolites. Prenylation often leads to activation or amplification of the biological effects of the aromatic moiety [18].

Much chemical work in prenylation has been directed towards the monosubstrate terpene cyclases, which are usually highly specific [19]. Bisubstrate enzymes have received less attention, although they are more interesting for synthetic purposes. Unfortunately, but in agreement with their often crucial role, many transferases appear to be highly specific, especially those of higher organisms. In addition, the majority of the phenol-prenyltransferases seem to be membrane-dependent, which may be an evolutionary consequence of the lipophilicity of the products formed. These properties make prenyltransferases less accessible, less stable and more difficult to characterize. Nevertheless, we tried to prove that suitable candidates for biocatalysis can be found [18, 20].

In concentrating on the bisubstrate C—C-coupling prenyl transferases, we excluded prenyl synthases [11], because the various oligoprenols are available either from nature

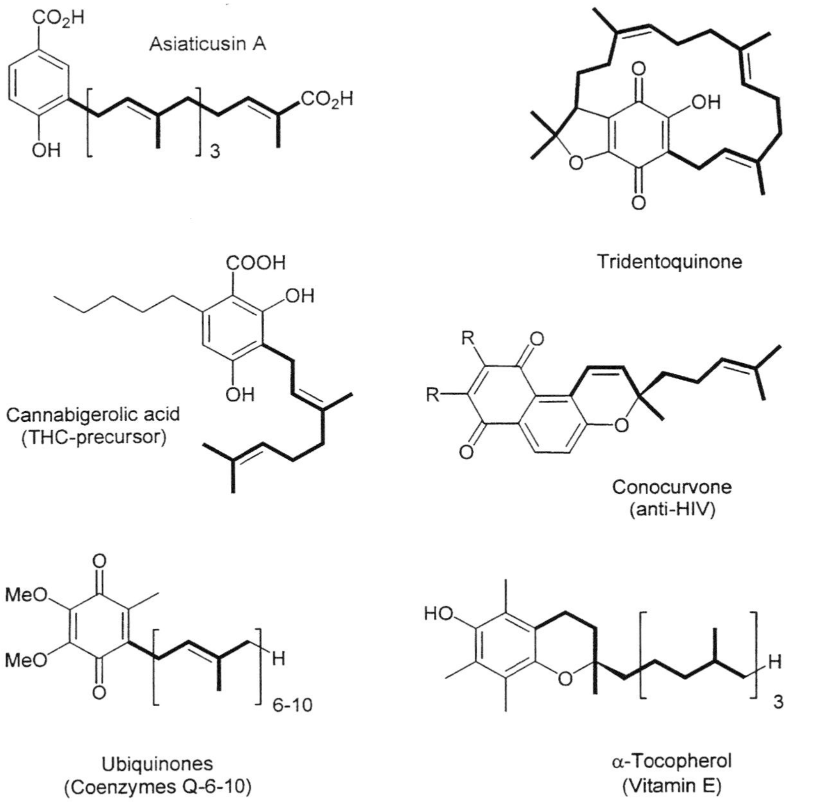

**Figure 2.1.1.** Some natural products derived from the prenylation of hydroxybenzoic acids.

or through classic synthesis. We focused on phenol-oligoprenyldiphosphate-prenyl-transferases as potential biocatalysts (Scheme 2.1.1) [20–27]. They allow the prenylation of aromatic compounds in a Friedel–Crafts-like reaction, most notably of hydroxybenzoic acids and hydroxyphenylketones. Both classes of prenylated products include important metabolites, but they are difficult to obtain chemically, especially in the natural *all-trans* configuration. While simple phenols and hydroquinones react readily in Lewis acid (or base) -catalyzed chemical prenylations, the acylated or carboxylated phenols usually do not give the desired regioselectively core-prenylated products at all. Depending on the procedure and substrate used, *O*-alkylation, over-alkylation (as the prenylated aromatics are better nucleophiles), isomerizations and addition reactions of double-bonds, and poor regioselectivity are observed. The most common chemical routes require a protective group strategy to allow regio-selective *ortho*-metallation, followed by prenylation and deprotection. Alternatively, palladium-based methods can be applied if suitably substituted precursors can be prepared. Both routes are cumbersome, and the prenylation is not always reliable [18, 28].

## 2.1.2 Production, stability and substrate specificity of 'ubiA-prenyltransferase'

Sources of phenol-prenyltransferases range from mammals (rat liver), plants and yeast to bacteria [25–27]. Several have been cloned in recent years [21, 29]. 4-Hydroxybenzoic acid (PHB, **1**) is the most common natural aromatic substrate. As already stated, most transferases – as far as has been tested – appear to be very substrate-specific. However, in the 1970s a prenyltransferase from *E. coli* required for the ubiquinone biosynthesis had been shown to possess activity for a variety of prenyl chain lengths [24]. This is in accordance with the variable side chain lengths of ubiquinones found in the bacterium. However, from these early works, the existence of isoenzymes could not be excluded.

Recently, Heide et al. cloned a gene (*ubiA*) encoding for such a transferase and generated overproducing *E. coli* strains (ca. 600 times overproduction). [21, 25, 30] The enzyme could be enriched ('isolated') in the membrane fraction, though solubilization has been unsuccessful so far. A purified membrane fraction from the best strain, K12-pALMU3, was used to establish a substrate model [18]. From this point, this membrane fraction will be called 'ubiA-transferase'. Apart from the two substrates [PHB (**1**) and oligoprenyldiphosphate (**2**)] only magnesium ions are required. The native reaction (probably with $n = 8$) is shown in Scheme 2.1.1; standard assays were done with geranyldiphosphate (GPP, **2**, $n = 2$) to yield GBA (**3**, $n = 2$). Recently enzyme production could be increased considerably over previously [18] reported values [LB-medium, growth to optical density $>7$, specific activity $>20\,\mathrm{mU/mg}$ protein from cell suspension, $>50\,\mathrm{mU/mg}$ protein ($>90$ rel.%) in the membrane fraction]. For the assay conditions reported previously [18], these preparations have to be diluted 1/10 or 1/30 [31].

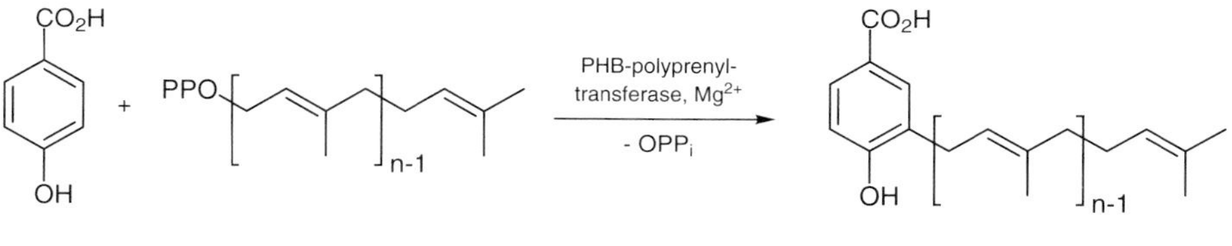

**Scheme 2.1.1.** Prenylation of 4-hydroxybenzoate (PHB, **1**; OPP = diphosphate; $n = 1$–12).

Despite the fact of being a membrane-bound enzyme, ubiA-tranferase can be stored for years without notable loss of activity at $-25^\circ$C. At $4^\circ$C it slowly loses activity and should not be kept at that temperature for more than two weeks. At room and reaction temperatures (optimum $37^\circ$C) it is stable for several hours [32]. However, under reaction conditions activity deteriorates within a few hours. The decrease is neither caused by mechanical agitation, buffer, magnesium ions or by oxygen, nor has severe product inhibition been found [18, 20, 31].

From the bacterial enzyme(s) we expected and found broader specificity than from plant enzymes. This was based on the hypothesis that a somewhat less exact ubiA-enzyme in the facultatively anaerobic *E. coli* would produce less selection pressure than a similar enzyme in mammals or plants which are highly dependent on the exact production of a specific coenzyme Q. Also, the availability of bacterial enzymes is better and production easier.

Knowing that we have a single enzyme, we could show that the broad acceptance of various prenyl diphosphates with different chain length is inherent in the ubiA-transferase and not the result of a set of isoenzymes. Interestingly, the best results and fastest reaction is obtained with geranyldiphosphate, whereas with longer chains (farnesyl, geranylgeranyl, solanosyl) activity decreases rapidly. This contrasts the finding of predominantly longer chain length products in the wild-type bacterium, *e.g.* Q-8 (octaprenyl ubiquinone). Reasons for this discrepancy are likely a result of the increased lipophilicity of the prenyl substrates and products. Among other effects, solubility problems of the prenyl substrate and retention of the products in the membrane are likely causes, both being more relevant factors at the unnatural high concentrations used in synthetic batches. As a consequence, under current assay conditions the real enzymatic activity is not measured (including $K_M$-data, etc.), but a reflection of secondary effects is obtained. For synthetic purposes this indicates that enzyme activity will be higher than expected and has not yet been fully exploited. Solutions of the problem are described below. As expected, prenylmonophosphates and cinnamyldiphosphate were not substrates.

The most important factor for a 'useful' phenol-prenyltranferase is a broad substrate acceptance in the aromatic substrate paired with high regioselectivity. UbiA-transferase exclusively prenylates the *meta*-position of the least substituted half of benzoic acids. Double prenylation of the second *meta*-position was never observed. A large set of possible substrates has been tested. Some are shown in Figure 2.1.2, and a substrate model could be deduced (Figure 2.1.2) which so far has proved to be valid for all compounds tested [18, 20].

The presence of a carboxylate group in the aromatic substrate is crucial for the ubiA-transferase. In *para*-hydroxybenzoate (PHB) the COOH-group cannot be omitted, nor can it be substituted by $CONH_2$, CHO, $CH_2OH$, $B(OH)_2$ or $SO_3H_2$. Also, a nitro-group ($NO_2$) was not accepted despite the fact that it is isolobal, isoelectronic and isoster to a carboxylate $COO^-$. Nitro-substituents in general do not give any (or extremely poor) activity, probably by lowering the electron density of the $\pi$-sextet too much. In the 4-position of benzoic acid an activating substituent with a hydrogen atom (OH, NH) has to be present. One rim of the PHB-derivative, *i.e.* the 5- and 6-position (Figure 2.1.2) can be substituted with a large variety of functional groups. Probably it is not possible to have both *ortho*-positions substituted at the same time, although such a case has been checked for one example only. The relative activity of PHBs with other substituents than nitro-groups, however, does not seem to be related to the 'textbook'-effects of aromatic substitution. Thus 3-acetyl-PHB (**10**) or 3-chloro-PHB (**8**) give better conversion than the electron rich 2,4-dihydroxybenzoic acid (**5**, ca. 20% rel. activity), 3-methoxy-PHB is even worse. On the other hand, the bulky, lipophilic 3-cyclohexyl-PHB (**9**) is prenylated with similar activity as the native substrate, but the less bulky 3-geranyl-PHB (GBA, **3**, n = 2) is not prenylated a second time, although lipophilicity and electronic influence on the aromatic ring

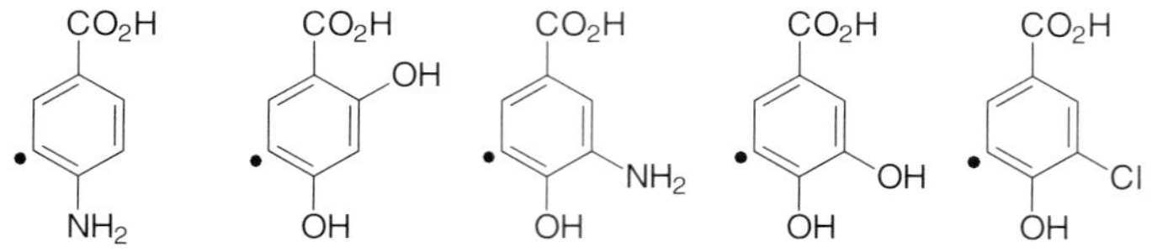

**4**  **5**  **6**  **7**  **8**

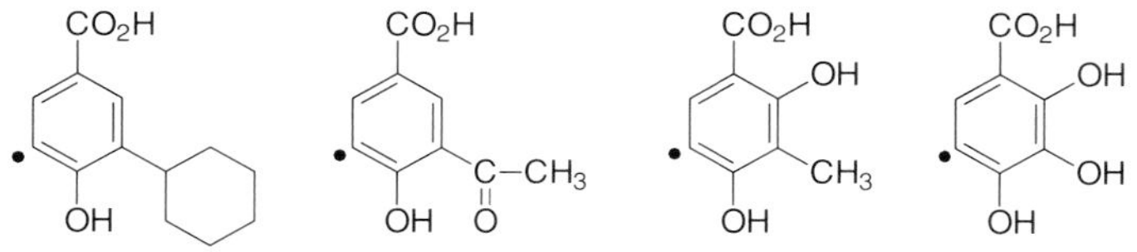

**9**  **10**  **11**  **12**

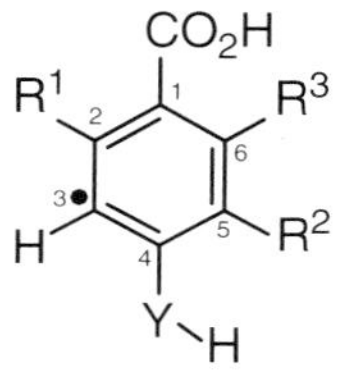

$R^1$ = H  [other substituents are probably not allowed]

$R^2$ = H, OH, NH$_2$, OCH$_3$, Cl, COCH$_3$, CH$_3$, cyclohexyl, etc.; not polyprenyl or very strongly electron withdrawing substituents like NO$_2$.

$R^3$ = H, OH, etc.

Y = O, NH  [other substituents may be possible]

**Figure 2.1.2.** A selection of benzoic acid derivatives acting as non-natural substrates of ubiA-transferase (with geranyldiphosphate, the dot ● marks the geranylated position), and a preliminary model for the aromatic substrate of ubiA-transferase.

should be similar to that of a cyclohexyl group. However, it must be clarified how double prenylation is excluded if simple steric exclusion is not the selection mechanism.

## 2.1.3   Parametrization and scale-up

The optimum temperature for reactions with ubiA-transferase is 37°C, and optimum pH is ~7.5. High sodium concentrations inhibit catalysis, magnesium ions are obligatory; they can be substituted by Mn$^{2+}$ ions (<15% rel. activity), but not by other divalent metal ions.

Most organic solvents destroy the activity of the membrane-bound ubiA-transferase irreversibly, as do most common detergents. Exceptions are DMF in very small amounts and especially DMSO, which can enhance activity (Figure 2.1.3). DMSO most likely causes an enhancement of the solubility of the amphiphilic geranyl-diphosphate and of the product, and thus increases availability and reaction rates. Increasing DMSO concentrations at the same time will give increased membrane destruction and protein denaturation. The two contrary effects will result in the

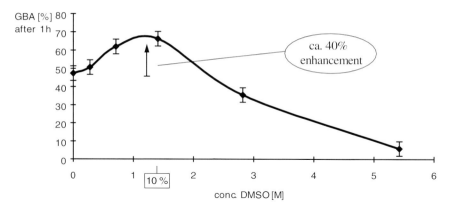

**Figure 2.1.3.** The influence of DMSO concentration on the enzyme-catalyzed reaction of PHB with geranyl diphosphate (GBA = 3-geranyl-4-hydroxybenzoic acid).

correlation observed, with a peak at around 10% DMSO addition and a ca. 40% activity increase for standard assay concentrations. The optimum DMSO concentration also depends on the concentration of the membrane fraction in the assay.

This fact prompted us to test cyclodextrins, especially 2,6-di-*O*-methyl-*β*-cyclodextrin, as modifiers. These host molecules have a lipophilic interior to accomodate the lipophilic end of the substrate and a hydrophilic exterior so as not to cause problems in the membrane. Indeed, increasing amounts of 2,6-di-*O*-methyl-*β*-cyclodextrin give faster conversion without any noteworthy negative effect at higher concentration, but unfortunately the maximum effect is limited by the low water solubility of this modifier. Nonetheless we could observe almost a doubling of the yield in a modified assay (reaction time reduced to 30 min, dilution of enzyme vs. standard assay 1:19, >90% increase in yield), which could be boosted further with DMSO (ca. 115% increase).

UbiA-transferase is not easily inhibited under the reaction conditions. No noteworthy effect was observed with inorganic ammonium hydrogencarbonate (used as buffer) or fluoride (up to 10 mM); inorganic diphosphate has only a small inhibitory effect up to ca. 5 mM. Fluoride and diphosphate thus can be used to block substrate-deteriorating phosphatases that potentially are present in technical enzyme preparations. Competitive inhibition was tested for some non-substrate aromatic compounds similar to PHB without much success. Iodoacetamide, an inhibitor acting predominantly at thiol positions, is only active after a certain incubation period, thus even electrophiles are not excluded as substrates if reacted quickly.

Of the various fractions in the purification process only whole cells did not show any noteworthy activity. Ruptured cells (ultrasound), supernatant (5 000 *g*) and membrane fraction (140 000 *g* pellet) all showed similar total activity (Figure 2.1.4) [20]. The storability and reproducibility of these preparations likely will be lower than that of the washed membrane fraction. The crude preparations will contain considerable amounts of proteases, as well as phosphatases that may act on the prenyldiphosphate. For direct synthetic application and production, however, the possible omission of all centrifugation steps is a great advantage and far outweighs the disadvantages.

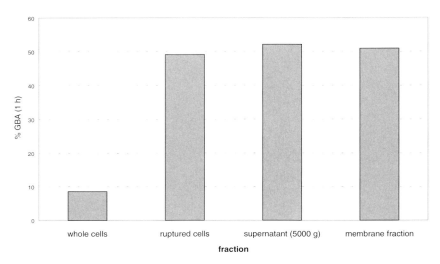

**Figure 2.1.4.** Influence of the enzyme fraction (equivalent proportions of the total amount isolated) of K12-pALMU3 on the yield of GBA in a standard assay [18].

Isolation of the products is equally easy as with the purified membrane fraction: Simple extraction with ethyl acetate/formic acid gives the product together with excess aromatic substrate almost exclusively. Separation of the much more lipophilic prenylated product from the educt is extremely simple, *e.g.* by chromatography. Several milligrams of compound, *e.g.* for NMR spectroscopy, are easily prepared from a batch of less than 100 ml of K12-pALMU3-culture.

## 2.1.4   Conclusion and Outlook

The first example of a membrane-bound enzyme suitable for organic C–C coupling was presented. UbiA-transferase is easily accessible in any amount, storable and does not require cofactors. The strain used is not problematic with regard to legislative requirements. The enzyme preparation can also be used without isolation of the membrane fraction. It allows the regioselective *meta*-prenylation of a large variety of benzoic acids with oligoprenyldiphosphates of all chain lengths, *i.e.* Friedel–Crafts-type allylations were performed regio- and chemoselectively in water. Up to 5-fold substituted prenylated aromatics have been synthesized, with a wide variety of substituents and functional groups. Products and unconverted starting material can be isolated by simple extraction with ethyl acetate. Yields are easily doubled by optimizing the conditions, and batches were scaled-up 1000-fold without problems. The chemical synthesis of the substrates will likely be the limiting factor for a further scale-up to produce multigram or kilogram quantities of prenylated phenols.

Prenyltransferases with other regioselectivities and without the limitation of requiring a 1-carboxylate group in the phenol are known [25–27]. It is likely that these will become available for biocatalysis in the future, providing a set of enzymes allowing access to almost any substitution pattern desired.

**Acknowledgements:** We thank Prof. Dr. Lutz Heide and his coworkers (U. Tübingen), especially Dr. K. Severin and Dr. M. Melzer, for providing us with the bacterial strains and numerous helpful discussions, Dr. Markus Pietzsch and P. Hörsch (U. Stuttgart) for optimization of enzyme production, and Prof. Dr. W. Steglich and Prof. Dr. T. Kutchan (LMU München) for the use of their facilities and instruments. This work was supported by the Fonds der Chemischen Industrie, BASF AG and Bayer AG. M.A.D. thanks the German Academic Exchange Service (DAAD) for a grant.

# References

[1]  K. Drauz, H. Waldmann (Eds.), *Enzyme Catalysis in Organic Synthesis – A Comprehensive Handbook*, VCH, Weinheim, **1995**; C.-H. Wong, G. M. Whitesides, *Enzymes in Synthetic Organic Chemistry*, Pergamon (Elsevier Science Ltd.), Oxford, **1994**; N. J. Turner, *Nat. Prod. Rep.* **1994**, *11*, 1–15; B. Berger, A. De Raadt, H. Griengl, W. Hayden, P. Hechtberger, N. Klempier, K. Faber, *Pure Appl. Chem.* **1992**, *64*, 1085; K. Faber, S. Riva, *Synthesis* **1992**, 895–910; W. Boland, C. Froessl, M. Lorenz, *Synthesis* **1991**, 1049.

[2]  J. B. Jones, M. Gold, J. K. Hogan, V. Martichonok, R. Sackowicz, T. Lee, *Acta Chem. Scand.* **1996**, *50*, 697–706.

[3]  H. L. Holland, *Organic Synthesis with Oxidative Enzymes*, VCH, New York, **1991**; R. Csuk, B. I. Glaenzer, *Chem. Rev.* **1991**, *91*, 49; T. Hudlicky, R. Fan, H. Luna, H. Olivo, J. Price, *Pure Appl. Chem.* **1992**, *64*, 1109.

[4]  F. Effenberger, *Angew. Chem.* **1994**, *106*, 1609–1619; *Angew. Chem. Int. Ed. Engl.* **1994**, *33*, 1555–1565.

[5]  H. Waldmann, *Nachr. Chem. Tech. Lab.* **1991**, *39*, 1408–1414.

[6]  M. D. Bednarski in *Comprehensive Organic Synthesis, Vol. 2* (Eds.: B. M. Trost, I. Fleming), Pergamon Press, Oxford, **1991**, p. 455.

[7]  F. Effenberger, A. Schwammle, *Biocatal. & Biotransf.* **1997**, *14*, 167–179.

[8]  C. F. I. Barbas, R. A. Lerner, *Acta Chem. Scand.* **1996**, *50*, 672–678.

[9]  W.-D. Fessner, O. Eyrisch, *Angew. Chem.* **1992**, *104*, 76–78; *Angew. Chem. Int. Ed. Engl.* **1992**, *31*, 56–58.

[10]  W.-D. Fessner, G. Sinerius, *Angew. Chem.* **1994**, *106*, 217–220; *Angew. Chem. Int. Ed. Engl.* **1994**, *33*, 209–212.

[11]  K. Ogura, T. Koyama, *Chem. Rev.* **1998**, *98*, 1263–1276.

[12]  S. Takahashi, T. Kuzuyama, H. Watanabe, H. Seto, *Proc. Natl. Acad. Sci. USA* **1998**, *95*, 9879–9884.

[13]  T. J. Bach, *Lipids* **1995**, *30*, 191–202.

[14]  B. A. Kellogg, C. D. Poulter, *Current Opinion in Chemical Biology* **1997**, *1*, 570–578.

[15]  F. L. Zhang, P. J. Casey, *Ann. Rev. Biochem.* **1996**, *65*, 241–269.

[16]  F. Tamanoi, I. Sattler, in *Molecular Biology Intelligence Unit: Regulation of the RAS Signaling Network*, (Eds.: H. Maruta, A. W. Burgess), R.G. Landes Company, Springer-Verlag GmbH & Co. KG (Intern. Distributor), Heidelberg, **1996**, pp. 95–137.

[17]  (a) L. Heide, M. Melzer, M. Siebert, A. Bechthold, J. Schröder, K. Severin, *J. Bacteriol.* **1993**, *175*, 5728–5729; R. E. Olson, H. Rudney, *Vitam. Horm. (N.Y.)* **1983**, *40*, 1–43; and Refs. [21, 25, 30].
(b) V. P. Papageorgiou, A. N. Assimopoulou, E. A. Couladouros, D. Hepworth, K. C. Nicolaou, *Angew. Chem.* **1999**, *111*, 280–311.

[18] L. A. Wessjohann, B. Sontag, *Angew. Chem.* **1996**, *108*, 1821–1823; *Angew. Chem. Int. Ed. Engl.* **1996**, *35*, 1697–1699.

[19] R. Croteau, *Chem. Rev.* **1987**, *87*, 929–954.

[20] L. Wessjohann, B. Sontag, *GIT Fachz. Lab.* **1998**, *42*, 229–230.

[21] M. Melzer, M. Siebert, A. Bechthold, L. Heide, *Planta Medica* **1992**, *58*, A596.

[22] M. Siebert, A. Bechthold, M. Melzer, U. May, U. Berger, G. Schröder, J. Schröder, K. Severin, L. Heide, *FEBS Lett.* **1992**, *307*, 347–350.

[23] K. Momose, H. Rudney, *J. Biol. Chem.* **1972**, *247*, 3930–3940.

[24] Z. El Hachimi, O. Samuel, R. Azerad, *Biochimie* **1974**, *56*, 1239–1247; *Chem. Abstr.* **1975**, *82*, 121392f.

[25] M. Melzer, *Untersuchungen zu 4-Hydroxybenzoat Polyprenyltransferasen der Ubichinonbiosynthese aus Escherichia coli und der Shikoninbiosynthese aus Lithospermum erythrorhizon SIEB. et ZUCC.*, Dissertation, Eberhard-Karls-Universität Tübingen, Tübingen, **1995**.

[26] L. Heide, M. Melzer, S. M. Li, R. Boehm, *Phytochemistry* **1997**, *44*, 419–424.

[27] R. Ibrahim, P. Gulick, H. Khouri, P. Laflamme, *Phytochemistry* **1993**, *34*, 147–151; C. L. Scharld, C. D. Poulter, J. C. Gebler, H. Wang, H.-F. Tsai, *Biochem. Biophys. Res. Commun.* **1995**, *216*, 119–125; K. Inoue, M. Senda, J. Kimata, H. Yamamoto, *Phytochemistry* **1997**, *44*, 23–28; M. Fellermeier, M. Zenk, *FEBS Lett.* **1998**, *427*, 283–285.

[28] A. Mühlbauer, *Untersuchungen zur Biosynthese des Pilzfarbstoffes Tridentochinon und verwandter Meroterpenoide*, Dissertation, Ludwig-Maximilians-Universität München, München, **1998**; S. Lang-Fugmann, *Untersuchungen zur Synthese und Biosynthese prenylierter Chinone in Höheren Pilzen*, Dissertation, Rheinische Friederich-Wilhelms-Universität Bonn, Bonn, **1987**.

[29] G. Wu, H. D. Williams, F. Gibson, R. K. Poole, *J. Gen. Microbiol.* **1993**, *139*, 1795–1805.

[30] L. Heide, M. Melzer, *Biochim. Biophys. Acta* **1994**, *1212*, 93–102.

[31] P. Hörsch, M. Pietzsch, M. A. Dessoy, L. A. Wessjohann, unpublished results, **1999**; cf. also: Hörsch, P., *diploma thesis*, Universität Stuttgart **1998**.

[32] The inherent stability of the enzyme above 0°C might be much higher than indicated, because loss of activity might be caused by residual proteases. Other preparations or protease inhibitors may provide increased stability also in unfrozen preparations.

## 2.2 Are there enzyme-catalyzed Diels–Alder reactions? An investigation into the polyketide-synthase system required for the biosynthesis of cytochalasans

*Alfons Hädener, Gilles Ritter, Patrick M. Schaeffer, and N. Patrick J. Stamford*

Enzyme-catalyzed reactions of a Diels–Alder type have been proposed to occur during the biosynthesis of various secondary metabolites including the cytochalasans [1], the brevianamides [2], nargenicin $A_1$ [3], mevinolin [4], the solanapyrones [5], the artonines C and D [6], chalcomoracin, and kuwanon J [7]. However, it has not been possible to identify and purify the enzymes involved in most of these cases [8]. The clearest evidence for an enzyme-catalyzed Diels–Alder reaction so far is established in the case of the biosynthesis of solanapyrone A: Oikawa et al. have been able to show that a cell-free extract of *Alternaria solani* can catalyze a cyclization of the Diels–Alder type [9].

In a related area of research, Schultz and coworkers initially succeeded in isolating an antibody against a bicyclic compound which supposedly resembles the transition state of a bimolecular Diels–Alder reaction. Remarkably, it was found that this antibody can catalyze the corresponding Diels–Alder reaction [10]. Several additional examples show that this approach can be broadly applied [11].

Our work on the present topic is intended to answer the title question for the biosynthesis of the antibiotic cytochalasin D (**1**) in *Zygosporium masonii*. The cytochalasans are a class of ca. 60 microbial secondary metabolites, most of which exhibit cytostatic activity [1]. Some cytochalasans also inhibit proteinase from HIV [12,13]. The carbon skeleton of these antibiotics not only includes a polyketide-derived moiety but also a substructure derived from an amino acid and a number of *S*-adenosyl methionine-derived methyl groups. The key step in the biosynthesis of the carbon skeleton of cytochalasans was repeatedly postulated to be an enzyme-catalyzed intramolecular Diels–Alder reaction in which a polyketide-derived compound **2** is converted into a tricyclic system **3** (R represents the side chain of an amino acid) [14,15] (Scheme 2.2.1).

### 2.2.1 Screening of blocked mutants of *Zygosporium masonii*

First, we intended to identify the putative Diels–Alder substrate **2** by analyzing compounds accumulated by ca. 150 blocked mutants of *Z. masonii* (*i.e.* mutants that are no longer capable of producing **1** but are instead expected to accumulate a compound of type **2**) [16]. These mutants had initially been identified in a screening over ca. 10 000 UV-generated mutants. Although compounds of type **2** were not identified among the products accumulated by the mutants, we did find a metabolite of type **3**, zygosporin J (**4**, Scheme 2.2.1), a new cytochalasan that is likely to be identical with the product of the Diels–Alder reaction [17]. A number of other new cytochalasans, the zygosporins H, K, L, and M, as well as some metabolites that are presumably not related to the

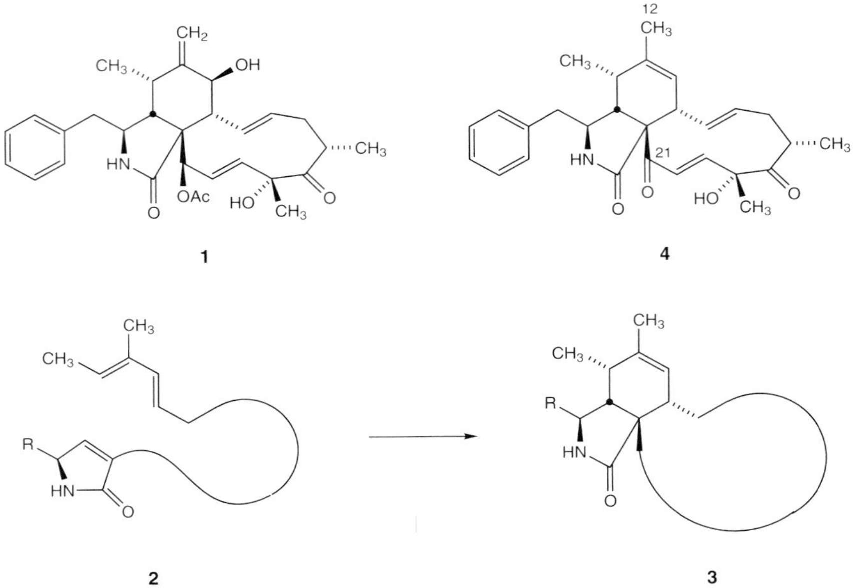

**Scheme 2.2.1.** Structures of cytochalysin D (**1**), zygosporin J (**4**), and an outline of the putative Diels–Alder reaction during the biosynthesis of **1** and **4**.

biosynthesis of cytochalasin D, were also identified in the course of this study [18,19].

The structure of **4** gave us some hints towards a more detailed depiction of **2**. For instance, the presence of both the C(12) methyl group and a carbonyl group at C(21) supports the idea that **2** is properly furnished for a smooth Diels–Alder reaction in exhibiting both an electron-rich diene and an electron-deficient dienophile moiety. Whereas the C(12) methyl group or a derivatized form thereof is present in all known cytochalasans, the carbonyl function at positions corresponding to C(21) is a structural feature that was not seen before in metabolites from *Z. masonii*, although it is common within the majority of other cytochalasans.

The project was continued, as detailed below, by following two independent approaches that complement each other. Both approaches are based on the fact that the Diels–Alder substrate **2**, if it exists, must be biosynthesized *via* polyketide synthesis. An enzyme that would catalyze the Diels–Alder cyclization of a compound of type **2** is therefore expected to be closely related to the polyketide synthase (PKS) multienzyme complex responsible for the assembly of an octaketide moiety.

## 2.2.2   Synthesis of derivatives of putative early polyketide intermediates and incorporation experiments

Recent insights into the mechanism of the biosynthesis of complex polyketides support the idea that acetate/malonate building blocks are assembled in a processive manner such that most of the functional groups seen in the final metabolite are individually established after each elongation step [20–22]. Since **2** is polyketide-derived and the putative Diels–Alder reaction can only occur after completion of polyketide synthesis, it is very likely that the functionalities exhibited by **2** are in close agreement with those found in the macrocycle of **4**. A possible version of **2** could be the depicted structure **5** (Scheme 2.2.2).

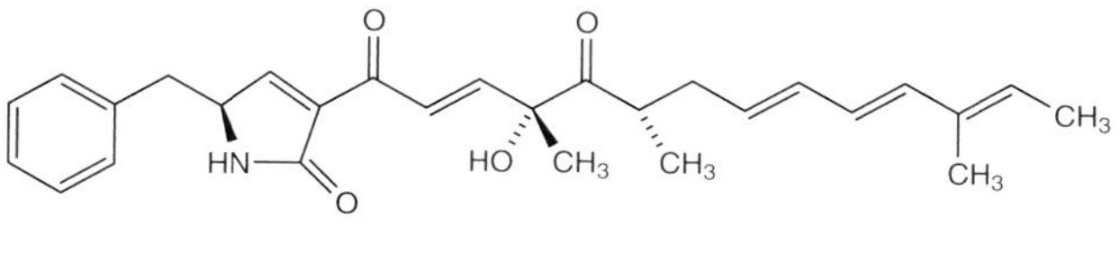

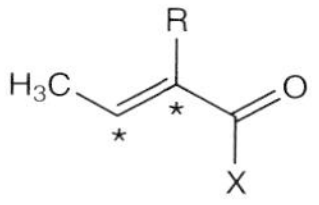

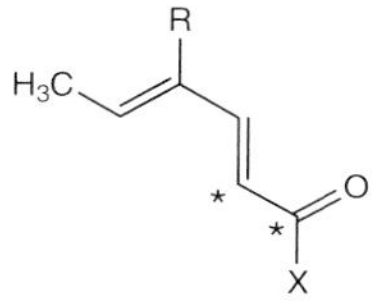

**Scheme 2.2.2.** Putative substrate **5** for the Diels–Alder reaction and postulated early polyketide intermediates **6–9** during the biosynthesis of **5**. The symbols * denote $^{13}$C-labeling positions.

The diene substructure shown in compound **2** would be expected to be present in early polyketide intermediates such as **6** or **7** (X = coenzyme A). However, the C(12) methyl group of **1** is not part of the polyketide chain but is derived from *S*-adenosyl methionine [23]. From a mechanistic point of view, the Diels–Alder reaction should be facilitated by the presence of this methyl group, but the question as to whether the methyl group is introduced early or late in the biosynthesis also needed to be clarified.

By synthesizing doubly $^{13}$C-labeled putative polyketide intermediates of both the versions with (**6, 7**) and without (**8, 9**) the methyl group and administering these in suitably derivatized form (X = SCH$_2$CH$_2$NHAc) to cultures of *Z. masonii*, we hoped not only to find support for the proposal that **1** is biosynthesized *via* an intramolecular Diels–Alder reaction, but also to decide whether the methyl group is introduced before the third, between the third and the fourth, or after the fourth polyketide chain-elongation step.

To prepare the 2,3-$^{13}$C$_2$-labeled *N*-acetyl cysteaminethioesters **6** and **8** (X = SCH$_2$CH$_2$NHAc) from commercially available [$^{13}$C$_2$]acetylene, a synthetic route was envisaged in which the key conversion is the stereoselective methylation of prop-inoic acid with (CH$_3$)$_2$CuLi [24]. The intermediates (*E*)-2-butenoic acid (crotonic acid, **8**, X = OH) and (*E*)-2-methyl-2-butenoic acid (tiglic acid, **6**, X = OH) were obtained by quenching the reaction with either H$_3$O$^+$ or CH$_3$I. For the synthesis of the 1,2-$^{13}$C$_2$-labeled *N*-acetyl cysteaminethioesters **7** and **9** (X = SCH$_2$CH$_2$NHAc), the intermediates (*E,E*)-2,4-hexadienoic acid (sorbic acid, **9**, X = OH) and (*E,E*)-4-methyl-2,4-hexadienoic acid (**7**, X = OH) were first prepared in Reformatsky reactions from ethyl bromo-[1,2-$^{13}C_2$]acetate and (*E*)-2-methyl-2-butenal or (*E*)-2-butenal, respectively. Following an established protocol [25], the doubly $^{13}C$-labeled carboxylic acids **6–9** (X = OH) were subsequently converted to the corresponding *N*-acetyl cysteaminethioesters (X = SCH$_2$CH$_2$NHAc) with overall yields between 46% and 68% with respect to the labeled starting materials [26].

To optimize the conditions for the administration of the labeled *N*-acetyl cysteaminethioesters **6–9** (X = SCH$_2$CH$_2$NHAc) to growing cultures of *Z. masonii*, the fungus was first grown in minimal media in the presence of 2,6-*di-O*-methyl-β-cyclodextrin (an activator of polyketide synthases), any one of the unlabeled *N*-acetyl cysteaminethioesters, and tetradecyl thiopropanoic acid (TDTP, an inhibitor of β-oxidation processes [27]). The time-courses of pH, of the amount of crude extract and mycelial mass, and of the production of cytochalasin D were registered and the optimal regimen for pulse feeding of the labeled precursors determined. It was found that **9** (both in the presence and absence of TDTP) and **7** (only in the presence of TDTP) inhibited the growth of the fungus to such an extent that **1** could not be isolated.

The labeled *N*-acetyl cysteaminethioesters were then administered independently under the conditions that had been found to be optimal. These incorporation experiments were carried out by using both continuously growing cultures and replacement cultures, in conjunction with a pulse-feeding regime, both in the presence and absence of TDTP. In all cases, no signs of intact incorporation of the ketide moieties of the labeled precursors into cytochalasin D were detected by $^{13}C$- and $^{1}H$-NMR spectroscopy. However, a detailed analysis of the $^{13}C$-NMR spectra of **1** revealed conclusively that $^{13}C$ labels from the precursors **6** and **8** had been incorporated specifically *via* acetate after breakdown of the precursors by β-oxidation processes [26].

Based on these findings, conclusions as to whether methylation occurs before the third, between the third and the fourth, or after the fourth chain-elongation step in polyketide biosynthesis, or whether or not an enzyme-catalyzed Diels–Alder reaction could occur in the course of cytochalasin D biosynthesis, are not possible at present.

### 2.2.3    Towards the characterization of the PKS gene cluster associated with cytochalasin D biosynthesis in *Z. masonii*

Among the known PKS systems, the PKS from *Z. masonii* will most probably have to be classified as type I modular system similar to the PKS responsible for the biosynthesis of 6-deoxyerythronolide B [22]. The method employed to probe *Z. masonii* genomic DNA for genetic elements of the PKS responsible for the biosynthesis of cytochalasin D was to construct degenerate DNA primers for use in a PCR strategy. With recent publication of the DNA gene clusters encoding the erythromycin [20,22], rapamycin [28], 6-methyl salicylic acid [29] and sterigmatocystin [30, 31] PKS, enough data have accrued for comparative DNA and protein alignments.

Protein and then DNA sequences of PKS abstracted from these literature sources were examined, and in each case it was noted that certain fragments of the β-ketosynthase were best conserved among the various PKS domains. Degenerate oligonucleotides were designed and synthesized for PCR against the two most similar $\beta$-ketosynthase sequences. These were located towards the 5′-phosphate of the $\beta$-ketosynthase gene alignments which correspond to the N-terminus of their respective protein products.

Following a standard procedure, genomic DNA was then prepared from cells of *Z. masonii* and PCR was performed on aliquots of the genomic DNA using *Thermus aquaticus* (Taq) DNA polymerase and the primers mentioned above. As a result of gene amplification, linear double-stranded DNA fragments of between about 220 to 260 nucleotides in length were expected. The visualization of the products of the PCR on an agarose gel containing ethidium bromide indicated that: (i) all available primer had been utilized in the extension reaction; (ii) essentially all of the available primer was unutilized in a negative control (using genomic DNA of *Escherichia coli* XL1-Blue MRF′ [32]); (iii) two distinct DNA fragments of nearly the expected length (ca. 240 bp and ca. 300 bp) had been amplified from *Z. masonii*; (iv) no amplification from *E. coli* XL1-Blue MRF′ had occurred; and v) a fragment of the expected length (ca. 250 bp) had been amplified in a positive control. For the latter control, genomic DNA from *Saccharopolyspora erythraea* (ATTC 11635) was used due to the fact that the nucleotide sequence coding for the PKS that is responsible for the biosynthesis of deoxyerythronolide B in this organism is known [22].

We were able to demonstrate that our PCR strategy was mainly successful by identifying a fragment of a PKS in *S. erythraea*. It is most notable, however, that this fragment does not form part of the known PKS of *S. erythraea* mentioned above, but represents a novel PKS, of putative type I and of hitherto unknown function, in this organism.

On the other hand, the 240-bp and 300-bp fragments from *Z. masonii* mentioned above were identified as being not part of a PKS. One of them was instead recognized as being part (241 bp) of a gene coding for a putative protein kinase. Inspection of the sequence complementary to the 241-bp fragment shows that this fragment had been amplified as a result of the utilization of one of the two primers only. This primer was thus used on the coding strand but, most probably due to its high G- and C-content, the complementary strand also allowed hybridization for reverse amplification.

The future course of the current project will be focused on the establishment of a modified strategy that would allow us to identify PKS gene fragments in *Z. masonii* more specifically. In addition, efforts to further characterize the novel PKS from *S. erythraea*, involving the sequencing of regions of DNA adjacent to the newly identified region, are currently underway in our laboratory.

**Acknowledgements:** The authors are grateful to K. Kirschner, Biozentrum of the University of Basel, for helpful discussions and support, and to the Swiss National Science Foundation for financial support.

# References

[1] S. W. Tanenbaum (Ed.) *Cytochalasins, Biochemical and Cell Biological Aspects*, North-Holland Publishing Company, Amsterdam, **1978**.

[2]   R. M. Williams, E. Kwast, H. Coffman, T. Glinka, *J. Am. Chem. Soc.* **1989**, *111*, 3064–3065.

[3]   D. E. Cane, W. Tan, W. R. Ott, *J. Am. Chem. Soc.* **1993**, *115*, 527–535.

[4]   R. N. Moore, G. Bigam, J. K. Chan, A. M. Hogg, T. T. Nakashima, J. C. Vederas, *J. Am. Chem. Soc.* **1985**, *107*, 3694.

[5]   H. Oikawa, Y. Suzuki, A. Naya, K. Katayama, A. Ichihara, *J. Am. Chem. Soc.* **1994**, *116*, 3605–3606.

[6]   Y. Hano, M. Aida, T. Nomura, *J. Natural Prod.* **1990**, *53*, 391.

[7]   Y. Hano, T. Nomura, S. Ueda, *J. Chem. Soc.; Chem. Comm.* **1990**, 610–613.

[8]   S. Laschat, *Angew. Chem. Int. Ed. Engl.* **1996**, *35*, 289–291.

[9]   H. Oikawa, K. Katayama, Y. Suzuki, A. Ichihara, *J. Chem. Soc.; Chem. Comm.* **1995**, 1321–1322.

[10]  A. C. Braisted, P. G. Schultz, *J. Am. Chem. Soc.* **1990**, *112*, 7430–7431.

[11]  D. Hilvert, *Acc. Chem. Res.* **1993**, *26*, 552–558.

[12]  R. B. Lingham, A. Hsu, K. C. Silverman, G. F. Bills, A. W. Dombrowski, M. E. Goldman, P. L. Darke, L. Huang, G. Koch, J. G. Ondeyka, M. Goetz, *J. Antibiotics* **1992**, *45*, 686–691.

[13]  J. L. Rösel, M. Poncioni, G. Ritter, A. Hädener, *J. Antibiotics* **1999**, manuscript in preparation.

[14]  C. Tamm, in *The Biosynthesis of Mycotoxins. A Study in Secondary Metabolism* (Ed.: P. S. Steyn), Academic Press, New York, **1980**, p. 269.

[15]  H. Oikawa, Y. Murakami, A. Ichihara, *J. Chem. Soc.; Perkin Trans. I* **1992**, 2949–2959.

[16]  G. Ritter, *Zur Frage der Existenz enzymkatalysierter Diels–Alder-Reaktionen. Untersuchungen zur Biosynthese von Cytochalasin D in Zygosporium masonii*, Dissertation, Universität Basel, **1995**.

[17]  A. Hädener, M. Poncioni, J. Schneider, G. Ritter, *Helv. Chim. Acta* **1999**, manuscript in preparation.

[18]  A. Hädener, M. Poncioni, G. Ritter, M. Neuburger, M. Zehnder, *J. Chem. Soc.; Chem. Comm.* **1999**, manuscript in preparation.

[19]  G. Ritter, A. Hädener, *Helv. Chim. Acta* **1999**, manuscript in preparation.

[20]  J. Cortes, S. F. Haydock, G. A. Roberts, D. J. Bevitt, P. F. Leadlay, *Nature* **1990**, *348*, 176–178.

[21]  D. A. Hopwood, D. H. Sherman, *Annu. Rev. Genet.* **1990**, *24*, 37–66.

[22]  S. Donadio, M. J. Staver, J. B. McAlpine, S. J. Swanson, L. Katz, *Science* **1991**, *252*, 675–679.

[23]  C.-R. Lebet, C. Tamm, *Helv. Chim. Acta* **1974**, *57*, 1785–1801.

[24]  E. J. Corey, J. A. Katzenellenbogen, *J. Am. Chem. Soc.* **1969**, *91*, 1851–1852.

[25]  D. E. Cane, R. H. Lambalot, P. C. Prabhakaran, W. R. Ott, *J. Am. Chem. Soc.* **1993**, *115*, 522–526.

[26]  P. M. Schaeffer, *Zur Polyketidbiosynthese in Zygosporium masonii und Saccharopolyspora erythraea*, Dissertation, Universität Basel, **1998**.

[27]  Ø. Spydevold, J. Bremer, *Biochim. Biophys. Acta* **1989**, *1003*, 72–79.

[28]  T. Schwecke, J. F. Aparicio, I. Molnar, A. Koenig, L. E. Khaw, S. F. Haydock, M. Oliynyk, P. Caffrey, J. Cortes, J. B. Lester, G. A. Boehm, J. Staunton, P. F. Leadlay, *Proc. Nat. Acad. Sci. USA* **1995**, *92*, 7839–7843.

[29]  J. Beck, S. Ripka, A. Siegner, E. Schiltz, E. Schweizer, *Eur. J. Biochem.* **1990**, *192*, 487–498.

[30]  D. W. Brown, J.-H. Yu, H. S. Kelkar, M. Fernandes, T. C. Nesbitt, N. P. Keller, T. H. Adams, T. J. Leonhard, *Proc. Nat. Acad. Sci. USA* **1996**, *93*, 1418–1422.

[31]  N. P. Keller, S. Segner, D. Bhatnagar, T. H. Adams, *Appl. Environ. Microbiol.* **1995**, *61*, 3628–3632.

[32]  W. O. Bullock, J. M. Fernandez, J. M. Short, *Bio-Techniques* **1987**, *5*, 376.

## 2.3 The truth about enzymatic halogenation in bacteria

*Karl-Heinz van Pée*

In nature, the formation of halometabolites is not unusual. To date, more than 3000 different naturally produced halogenated metabolites, with very diverse structures, have been isolated from bacteria, fungi, plants, marine invertebrates, and mammals [1].

Investigations on the biosynthesis of halogen-containing, bioactive metabolites from bacteria, such as 7-chlorotetracycline, chloramphenicol, or fluoroacetate have shown that halogenation occurs at defined biosynthetic steps and that the halogenating enzymes must have substrate specificity [2]. Feeding experiments have shown that only the feeding of metabolites before a certain biosynthetic step allowed the formation of halometabolites. Feeding of intermediates lying behind a certain step led only to the formation of non-halogenated analogs. In addition, comparison of different halometabolites suggests that halogenases must also be regiospecific. From the halogenation pattern of the halogenated indole derivatives shown in Figure 2.3.1, it may be concluded that the halogenating enzymes involved in the biosynthesis of these metabolites must be selective for a certain position in the indole ring system.

Surprisingly, this is not consistent with what has been known about halogenating enzymes for the last 35 years. What is the reason for this discrepancy between halogenation *in vivo* and *in vitro*?

### 2.3.1 Haloperoxidases

At the end of the 1950s, Hager's group detected the first halogenating enzyme during their work on the biosynthesis of the antibiotic caldariomycin produced by the fungus *Caldariomyces fumago* (Figure 2.3.2) [6]. This enzyme requires hydrogen peroxide, chloride, bromide or iodide, and an organic substrate susceptible to electrophilic attack for the formation of carbon–halogen bonds, and was thus named chloroperoxidase. Hager's group developed a spectrophotometric assay using monochlorodimedone as the organic substrate, inspired by the structural similarity of monochlorodimedone with 2-chloro-1,3-cyclopentanedione, an intermediate in caldariomycin biosynthesis, (Figure 2.3.2) [7]. This monochlorodimedone assay was subsequently used by all workers seeking for halogenating enzymes, regardless of the structure of the halogenated metabolite produced. During the following 20 years a number of iodo-, bromo- and chloroperoxidases were isolated which, like chlorperoxidase from *C. fumago*, contained heme as the prosthetic group. In 1984, a vanadium-dependent bromoperoxidase was isolated by Vilter from a marine alga [8]. Similar bromo- and chloroperoxidases were later isolated from other marine algae, lichen, and fungi [9]. In 1986, we detected a non-heme chloroperoxidase in the soil bacterium *Pseudomonas pyrrocinia*. This non-heme chloroperoxidase required neither metal ions nor any other cofactor for halogenating activity [2], and was only active in acetate or propionate buffer. During the following years cofactor- and metal-free haloperoxidases were also isolated from other bacteria [2].

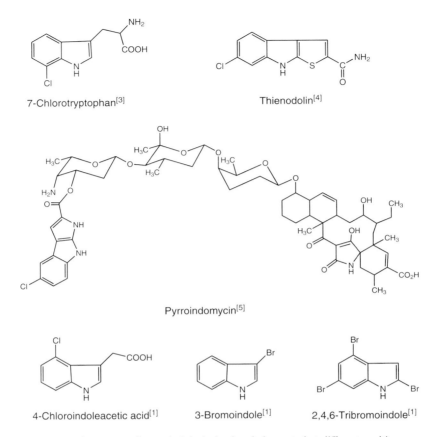

7-Chlorotryptophan[3]

Thienodolin[4]

Pyrroindomycin[5]

4-Chloroindoleacetic acid[1]    3-Bromoindole[1]    2,4,6-Tribromoindole[1]

**Figure 2.3.1.** Structures of some indole derivatives halogenated at different positions.

The mechanism by which these bacterial non-heme haloperoxidases catalyze the halogenation of organic substrates was difficult to understand. Halogenation using hydrogen peroxide was supposed to be a redox reaction, and these kinds of reactions require either metal ions or other cofactors, or both together. However, bacterial

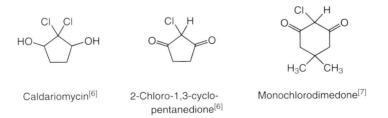

Caldariomycin[6]    2-Chloro-1,3-cyclo-    Monochlorodimedone[7]
                    pentanedione[6]

**Figure 2.3.2.** Structures of caldariomycin, 2-chloro-1,3-cyclopentanedione, a substrate for the halogenating enzyme involved in caldariomycin biosynthesis, and monochlorodimedone, the substrate used for the search for haloperoxidases.

non-heme haloperoxidases contained none of these! Instead, elucidation of the three-dimensional structure of chloroperoxidase from the 7-chlorotetracycline producer *Streptomyces aureofaciens* revealed the existence of a catalytical triad, as it is known from hydrolases [2]. This catalytical triad is conserved in all known bacterial non-heme haloperoxidases. Initially, the connection between the catalytical triade of hydrolases and halogenating activity could not be explained. However, when during the elucidation of the three-dimensional structure an electron density at the serine residue was detected that suggested the presence of an acetate molecule, the pieces of the puzzle fell into place. Obviously, acetate was necessary for halogenating activity, because acetate reacts with the serine residue to form an ester. In the presence of hydrogen peroxide this ester was perhydrolyzed to yield peracetic acid. This could be demonstrated by incubation of the enzyme with hydrogen peroxide in acetate buffer and subsequent removal of the enzyme by ultrafiltration. When monochlorodimedone and sodium bromide were added to the enzyme-free solution, bromination of monochlorodimedone could be observed [10]. This led to a hypothesis for the reaction mechanism of bacterial non-heme haloperoxidases involving the catalytical triade and acetate (Figure 2.3.3). The first step in the catalytic cycle is the formation of an ester between the active site serine and acetate, the ester being hydrolyzed by water or hydrogen peroxide. Hydrolysis with the better nucleophil hydrogen peroxide leads to the formation of peracetic acid which, as a strong oxidant, can oxidize halide

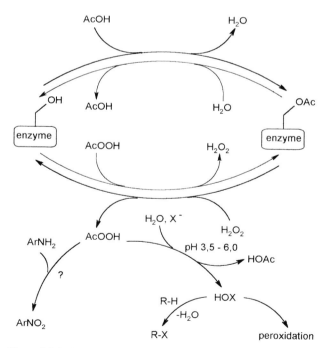

**Figure 2.3.3.** Suggested mechanism for the array of reactions catalyzed by bacterial non-heme haloperoxidases.

ions leading to the formation of hypohalite which acts as the actual halogenating agent. The formation of peracetic acid explains how bacterial haloperoxidases can catalyze the oxidation of thioethers to sulfoxides and the oxidation of aromatic amino to nitro groups [10], and why these enzymes show peroxidase activity only in the presence of bromide (Figure 2.3.3). According to this reaction mechanism, the term haloperoxidase is not correct for these enzymes and they should be called hydrolases or perhydrolases.

## 2.3.2   Haloperoxidases and halometabolite biosynthesis

Since haloperoxidases had been isolated from organisms producing halogenated compounds, they were believed to be involved in the biosynthesis of these halometabolites [2]. However, the fact that haloperoxidases in general did lack substrate specificity and regioselectivity, was not consistent with the specific role in halometabolite biosynthesis appointed to them. In 1995 it became clear that, at least in 7-chlorotetracycline biosynthesis, the halogenating enzyme was not a haloperoxidase. Dairi et al. [11] had cloned the biosynthetic genes for 7-chlorotetracycline biosynthesis and they detected the gene that coded for the enzyme which catalyzed the chlorination step in 7-chlorotetracycline biosynthesis. Sequencing of the gene and comparison of the deduced amino acid sequence with the sequences of haloperoxidases clearly showed that this enzyme was not a haloperoxidase. Final proof that haloperoxidases are not involved in halometabolite biosynthesis was obtained by gene replacement experiments. A bromoperoxidase–catalase gene that had been isolated from the chloramphenicol producer *Streptomyces venezuelae* [2] and a chloroperoxidase gene isolated from a pyrrolnitrin-producing *Pseudomonas fluorescens* strain [12] were inactivated by gene disruption. Exchange of the chromosomal gene in the wilde-type strain against the disrupted one, did not have any effect on chloramphenicol or pyrrolnitrin biosynthesis, respectively. After these results had been obtained, it was beyond any doubt that the haloperoxidase concept had come to an end, and it was necessary to develop a new concept for the detection and isolation of the specific and selective halogenating enzymes present in living organisms.

## 2.3.3   The search for new, specific halogenases

Until then, only monochlorodimedone which was not a natural compound had been used for the detection of halogenating enzymes. It was very unlikely that this compound could be a substrate for a specific halogenase. Therefore, we decided to use the natural substrates for the detection of specific halogenases. Only very few biosynthetic pathways for halometabolites had been elucidated to reveal the structure of the natural substrate for the halogenating enzyme. Often, even if the substrate is known, it is either unstable or difficult to obtain. Luckily, just at that time the genes for pyrrolnitrin biosynthesis were detected and cloned by Hammer et al. [13]. Pyrrolnitrin biosynthesis starts from tryptophan and only four genes (*prnA–prnC*) are

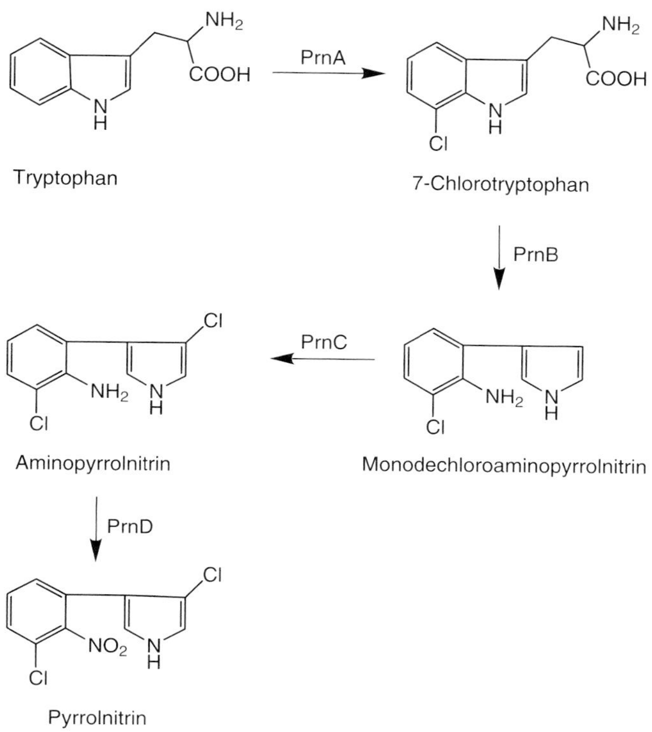

**Figure 2.3.4.** Pathway of pyrrolnitrin biosynthesis according to Kirner et al. (1998) [14].

required for the formation of pyrrolnitrin (Figure 2.3.4). Using a mutant with a deletion in the gene coding for the enzyme that catalyzes the second step, 7-chloro-L-tryptophan could be isolated, suggesting that the first step is the specific chlorination of L-tryptophan to 7-chloro-L-tryptophan. This could be verified by the isolation of 7-chlorotryptophan from a mutant from which all four genes for pyrrolnitrin biosynthesis had been deleted, but which contained the first gene (*prnA*) on a plasmid [3, 14].

Thus, for the first time a natural substrate for a specific halogenating enzyme was known that is stable and easily available. The second halogenation step in pyrrolnitrin biosynthesis is catalyzed by the enzyme PrnC, the substrate for which, monodechloro-aminopyrrolnitrin, can be obtained by directed biosynthesis [2]. Using the two natural substrates, tryptophan and monodechloroaminopyrrolnitrin, we tried to detect the halogenating activity of PrnA and PrnC. Since we did not know the cofactor requirements of the enzymes, a number of different cofactors and metal ions were tested, from which only NADH was found to be absolutely necessary for activity. When the amino acid sequences were screened for nucleotide binding sites, they were found near the N-termini; however, they are different for PrnA and PrnC, whose sequences do not show any homology to each other. On the other hand, PrnC is homologous to the halogenase involved in 7-chlorotetracycline biosynthesis, including the nucleotide binding site which can be found after correction of the sequencing error present in the sequence published by Dairi et al. [11]. Recent cloning of the genes for chloroeremomycin bio-synthesis and sequencing of the cluster have shown that the halogenase responsible for the incorporation of chloride into chloroeremomycin is homologous to PrnC and is probably also a NADH-dependent halogenase [15]. A gene for a non-heme haloper-oxidase was also detected in this cluster. Since bacterial non-heme haloperoxidases are actually hydrolases, we believe that this non-heme haloperoxidase is not involved in halogenation. However, the detection of a non-heme haloperoxidase gene in a bio-synthetic gene cluster provides a good possibility to elucidate the natural function of these enzymes. In other bacteria producing chlorinated or brominated pyrrole deriva-tives such as pyoluteorin [1], pentachloro- [16] or pentabromopseudilin [1], genes with a high homology to *prnC* were detected. Other hybridization experiments using *prnA* as the probe have shown that a number of bacteria that produce halometabolites with a halogenated moiety derived from tryptophan such as rebeccamycin [1], thienodolin [4], or pyrroloindomycin [5] contain genes highly homologous to *prnA*. These results suggest that NADH-dependent halogenases are involved in the biosynthesis of many halometabolites, if not all, and that they are widely spread in nature. They might even be involved in the formation of halometabolites in mammals, for example the iodide-containing thyroid hormone, thyroxine.

Only very little is known about the mechanism by which NADH-dependent halo-genases catalyze the incorporation of halide ions into organic molecules. However, incorporation definitely does not proceed *via* the oxidation of the halide ions, but rather must proceed *via* a change in the organic substrate, thus making it susceptible to nucleophilic attack by the halide ion. Figure 2.3.5 shows a hypothetical reaction mechanism for NADH-dependent halogenases involving the requirement of NADH, oxygen, and FAD. Further investigations of these novel halogenases will have to be done, to show whether this hypothetical mechanism is correct. The detection of specific, NADH-dependent halogenation is an important step towards both, the understanding of biological halogenation and the possibility of using specific enzymatic halogenation steps in the production of halogenated organic compounds.

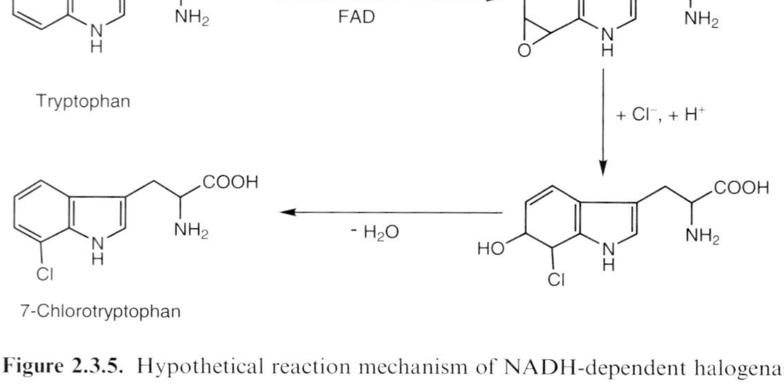

**Figure 2.3.5.** Hypothetical reaction mechanism of NADH-dependent halogenases.

# References

[1]  G. W. Gribble, *Prog. Chem. Org. Nat. Prod.*, **1996**, *68*.

[2]  K.-H. van Pée, *Annu. Rev. Microbiol.*, **1996**, *50*, 375–399.

[3]  K. Hohaus, A. Altmann, W. Burd, I. Fischer, P. E. Hammer, D. S. Hill, J. M. Ligon, K.-H. van Pée, *Angew. Chem. Int Ed. Engl.*, **1997**, *36*, 2012–2013.

[4]  K. Kanabe, H. Naganawa, K. T. Nakamura, Y. Okami, T. Takeuchi, *Biosci. Biotech. Biochem.*, **1993**, *4*, 636–637.

[5]  W. Ding, D. R. Williams, P. Northcote, M. M. Siegel, R. Tsao, J. Ashcroft, G. O. Morton, M. Alluri, D. Abbanat, W. M. Maiese, G. A. J. Ellestad, *J. Antibiotics*, **1994**, *47*, 1250–1257.

[6]  P. D. Shaw, J. R. Beckwith, L. P. Hager, *J. Biol. Chem.*, **1959**, *234*, 2560–2564.

[7]  L. P. Hager, D. R. Morris, F. S. Brown, H. Eberwein, *J. Biol. Chem.*, **1966**, *241*, 1769–1777.

[8]  H. Vilter, *Phytochemistry*, **1984**, *23*, 1387–1390.

[9]  M. C. R. Franssen, *Biocatalysis*, **1994**, *10*, 87–111.

[10]  M. Picard, J. Gross, E. Lübbert, S. Tölzer, S. Krauss, K.-H. van Pée, A. Berkessel, *Angew. Chem. Int. Ed. Engl.*, **1997**, *36*, 1196–1199.

[11]  T. Dairi, T. Nakano, K. Aisaka, R. Katsumata, M. Hasegawa, *Biosci. Biotech. Biochem.*, **1995**, *59*, 1099–1106.

[12]  S. Kirner, S. Krauss, G. Sury, S. T. Lam, J. M. Ligon, K.-H. van Pée, *Microbiology*, **1996**, *142*, 2129–2135.

[13]  P. E. Hammer, D. S. Hill, S. T. Lam, K.-H. van Pée, J. M. Ligon, *Appl.Environ. Microbiol.*, **1997**, *63*, 2147–2154.

[14]  S. Kirner, P. E. Hammer, D. S. Hill, A. Altmann, I. Fischer, L. J. Weislo, M. Lanahan, K.-H. van Pée, J. M. Ligon, *J. Bacteriol.*, **1998**, *180*, 1939–1943.

[15]  A. M. A. van Wageningen, P. N. Kirkpatrick, D. H. Williams, B. R. Harris, J. K. Kershaw, N. L. Lennard, M. Jones, S. J. M. Jones, P. J. Solenberg, *Chem. Biol.*, **1998**, *5*, 155–162.

[16]  B. Cavalleri, G. Volpe, G. Tuan, M. Berti, F. Parenti, *Curr. Biol.*, **1978**, *1*, 319–324.

## 2.4 Enzymatic Baeyer–Villiger oxidation with cyclohexanone monooxygenase: A new system of cofactor regeneration

*Marina Vogel and Ulrich Schwarz-Linek*

The enantioselective Baeyer–Villiger oxidation is an important reaction for the synthesis of lactones as precursors of natural products such as pheromones and antibiotics. Enantioselective chemical Baeyer–Villiger oxidations using chiral metal catalysts have been known only for a short time [1]. However, they are inferior to microbial and enzymatic reactions regarding the variety of substrates and enantio-selectivity, respectively.

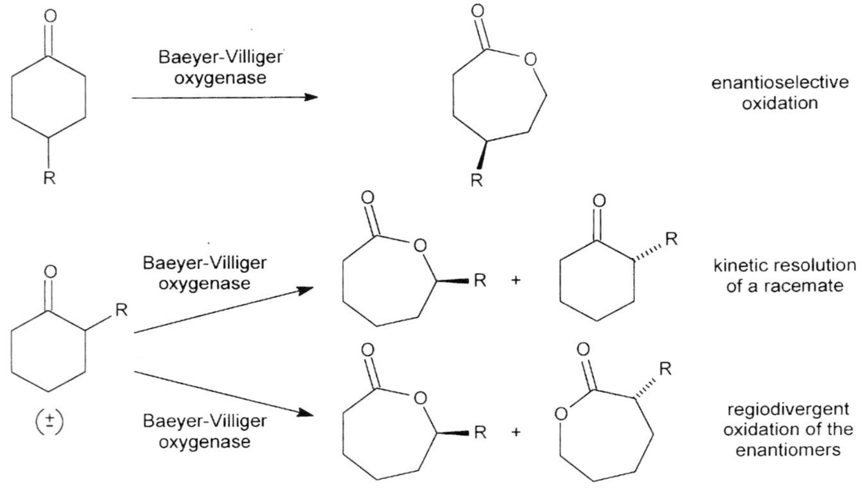

**Scheme 2.4.1.** General reactions of Baeyer–Villiger oxidases.

Microbial and enzymatic Baeyer–Villiger oxidations have been investigated inten-sively for the past 15 years [2, 3]. Some interesting applications are, *e.g.* the syntheses of precursors of azadirachtin [4], sarkomycin [5], (+)-lipoic acid [6], and (*R*)-(−)-baclofen [7]. There are two possibilities to carry out enzymatic Baeyer–Villiger oxidations on principle: with whole cells, or with isolated and enriched enzymes, respectively.

Working with whole cells the problem of cofactor regeneration does not exist. Unfortunately, competing enzymes and overmetabolism of the desired product have a negative influence on the reaction. This is also true if a recombinant strain like *e.g.* a cyclohexanone monooxygenase expressing baker's yeast is used [8]. Employment of isolated enzymes has some advantages. Generally, the reaction process can be executed in a simpler manner using higher and exactly defined concentrations of enzyme and substrate. Additionally, reaction conditions can be optimized by means of kinetic measurements. As a result of the conversion, purer products can be isolated easier and in higher yields. In some cases it is possible that the same enzyme can be used several times. However, application of isolated Baeyer–Villiger enzymes has a considerable disadvantage. All of these enzymes are cofactor-dependent, and most of them require NADPH [9]. Cofactor regeneration for NADPH-dependent monooxygenases has mostly been realized using glucose-6-phosphate/glucose-6-phosphate dehydrogenase [10]. Yet, this method is not applicable for syntheses of chiral lactones on a large scale because of the requirement of equimolar amounts of the expensive substrate glucose-6-phosphate. In the following, Baeyer–Villiger reactions of cyclohexanones with cyclohexanone monooxygenase from *Acinetobacter* NCIMB 9871 using a new cofactor regeneration system will be presented [11].

## 2.4.1 Cyclohexanone monooxygenase

*Acinetobacter calcoaceticus* NCIMB 9871 and the cyclohexanone monooxygenase (CHMO) [EC 1.14.13.22] [12] isolated from this strain received most attention and have been successfully employed for the Baeyer-Villiger oxidation, especially of substituted cyclohexanones [13]. Cyclohexanone monooxygenase is a yellow monomer flavoprotein which has a molecular weight of about 60 kDa [12]. The pH optimum is 9.0. It requires molecular oxygen and is absolutely specific regarding the prosthetic group (FAD) and the cofactor NADPH [12]. In contrast, the monooxygenase accepts as substrates, besides cyclohexanones, a wide range of cyclic four- to eight-membered ketones as well as bi- and tricyclic ketones as substrates [2]. By employing known biochemical methods like desintegration of the cells, fractionated ammonium sulfate precipitation, ion exchange chromatography, and a dye affinity chromatography the enzyme can easily be isolated and enriched, respectively [12, 14]. The step of chromatography is especially expensive and for that reason the known procedures have been reduced to the first three steps in a manner that enzyme fractions which can be isolated do not contain $\varepsilon$-caprolactone hydrolase and cyclohexanol dehydrogenase [15], both of which strongly impair the Baeyer–Villiger oxidation. The stability of the monooxygenase is noteworthy. Walsh et al. performed an oxidation successfully with a cell paste which had been stored at $-20\,°C$ for 9 months [16]. Comparable results could be found with enzyme samples which had been stored at $-70\,°C$ for one year [15]. (However, kinetic parameters have not yet been determined until now.) Under biotransformation conditions ($30\,°C$), the enzyme could be used three times (Figure 2.4.2). A small decrease of conversion was only obtained in the third batch. All these properties predestine the enzyme to be used as biocatalyst on a preparative scale.

## 2.4.2 A new method of NADPH regeneration

Like most of the known Baeyer–Villiger enzymes cyclohexanone monooxygenase from *Acinetobacter* NCIMB 9871 is dependent on NADPH [9]. Hence, when employing the enzyme on a larger scale, an effective cofactor regeneration is necessary. The cofactor $NADP^+$ can be reduced chemically [17], electrochemically [18], and photo-chemically [19], respectively. However, these reductions proceed under conditions which are not compatible with those of enzymatic reactions. For that reason, several enzymatic methods for the regeneration of reduced nicotinamide cofactors have been developed alternatively (Scheme 2.4.2) [20]. The system formate/formate dehydrogenase (EC 1.2.1.2) is one of the most effective cofactor regeneration systems [20, 21], though unfortunately, all known isoenzymes of this dehydrogenase are specific to $NAD^+$. This system was used successfully in Baeyer–Villiger oxidations with the NADH-dependent 2,5-diketocamphane monooxygenase from *Pseudomonas putida* [22]. Furthermore, alcohol dehydrogenases catalyze the oxidation of alcohols to ketones through which NADPH can be regenerated [20d]. Willetts et al. published an example of a Baeyer–Villiger reaction using cyclohexanone monooxygenase from *Acinetobacter* NCIMB 9871 and alcohol dehydrogenase from *Thermoanaerobium brockii* starting from norborneol [22b, 23]. The advantage of this reaction is that a substrate-coupled cofactor regeneration occurs. However, this reaction can only be used in special cases because of the necessity to find enzymes which possess approximately the same substrate specificities.

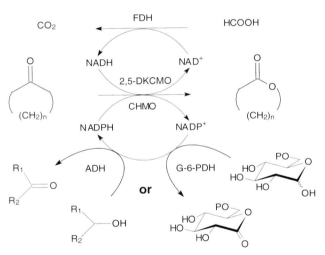

**Scheme 2.4.2.** Different methods of enzymatic cofactor regeneration employed in Baeyer–Villiger oxidations. (CHMO = cyclohexanone monooxygenase, 2,5-DKCMO = 2,5-diketocamphane monooxygenase, FDH = formate dehydrogenase, ADH = alcohol dehydrogenase, G-6-PDH = glucose-6-phosphate dehydrogenase).

Until recently, the oxidation of glucose-6-phosphate to 6-phosphogluconolactone with glucose-6-phosphate dehydrogenase from *Leuconostoc mesenteroides* [24] has been the most commonly used method to regenerate NADPH in Baeyer–Villiger reactions [2, 6]. However, it suffers from the application of stoichiometric amounts of the expensive substrate glucose-6-phosphate, which, therefore, has to be generated from glucose, with hexokinase including ATP regeneration for larger-scale conversions

[24]. In addition, phosphates act as acidic catalysts in the decomposition of NADPH. Moreover, a continuous correction of the pH is necessary because of the spontaneous formation of 6-phosphogluconic acid by hydrolysis of the 6-phosphogluconolactone [24]. Therefore, this regeneration system is not applicable for syntheses of chiral lactones on a large scale.

Tishkov et al. recently described a new $NADP^+$-dependent formate dehydrogenase (FDH) which was obtained by multipoint site-directed mutagenesis of the relevant gene from *Pseudomonas* sp. 101 [25]. The recombinant enzyme possesses all advantageous properties of the $NAD^+$-specific enzyme, and requires only formate, which is one of the cheapest hydrogen sources. The equilibrium of the oxidation of formate is strongly shifted towards $CO_2$ and NADPH formation [26]. $CO_2$ normally does not inhibit other oxidoreductases and can easily be removed from the reaction medium. The formate dehydrogenase possesses a broad pH optimum of activity and, additionally, a remarkable high stability. A first example of NADPH regeneration with the formate dehydrogenase was recently described [27]. Acetophenone was successfully reduced with alcohol dehydrogenase from *Lactobacillus* sp. using a continuously working enzyme-membrane reactor in the course of which a total turnover number (ttn) of about 85 could be achieved [27]. (The total turnover number, ttn, is defined as mol product per mol cofactor consumed.)

## 2.4.3 Baeyer–Villiger oxidation with the system cyclohexanone monooxygenase/formate dehydrogenase

### 2.4.3.1 Optimization of the reaction conditions

The coupling of cyclohexanone monooxygenase from *Acinetobacter* with the $NADP^+$-dependent formate dehydrogenase was actually the first application of this recombinant enzyme together with a monooxygenase [11]. Prochiral 4-methylcyclohexanone (**1**) was chosen as a model substrate which was oxidized by cyclohexanone monooxygenase, yielding the 5-methyl-oxepane-2-one (**2**) (Scheme 2.4.3).

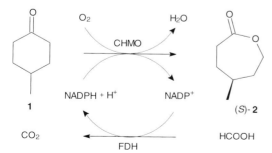

**Scheme 2.4.3.** Conversion of ketone **1** with the coupled enzyme system. (CHMO = cyclohexanone monooxygenase, FDH = formate dehydrogenase).

**Table 2.4.1.** Optimization of the reaction parameters.

|  | FDH | CHMO | Two-enzyme system |
|---|---|---|---|
| **Temperature** | optimum >40°C | optimum 30°C | 30°C |
| **pH-value** | optimum 6.0 | optimum 9.0 | 8.0 |
| **Formate** | $K_m = 40$ mM; optimal >200 mM | partial loss of activity >200 mM | 200 mM |
| **NADP$^+$** | $K_m = 0.25$ mM; optimal >0.5 mM | partial loss of activity >0.25 mM | 0.25 mM |
| **4-Methyl-cyclohexanone** | no influence at 40 mM | partial loss of activity >2 mM | 40 mM |
| **Activity under reaction conditions** | about 30% of optimum activity | about 20% of optimum activity | |

Reaction parameters of two-enzyme systems of course turn out to be a compromise between the optimal conditions of both enzymes. Temperature and pH value bear on enzymatic activity as well as on the concentration of substrates and cofactors. For that reason, the activities of both enzymes have been determined by kinetic measurements, varying the relevant reaction parameters [11, 15]. As a result, conversions were carried out under conditions which guaranteed the maximum activity of both cyclohexanone monooxygenase and formate dehydrogenase (Table 2.4.1).

The reaction temperature of 30°C is a compromise between stability and enzyme activity of cyclohexanone monooxygenase. At a pH of 8.0, the oxygenase has still a good activity while the activity of formate dehydrogenase remains acceptable. A two-fold excess of formate dehydrogenase was used in order to ensure a sufficiently high NADPH concentration, because the activity of the monooxygenase is drastically enhanced by the presence of NADPH. However, at a concentration of 40 mM of ketone **1,** no irreversible inhibition could be detected and the substrate had been oxidized completely. Of course, optimal concentrations have to be determined for any other ketone. At a NADP$^+$ concentration of 0.25 mM a maximum total turnover number (ttn) of 160 could be obtained in the oxidation of **1.** Further decrease of cofactor concentration below the $K_m$ value of formate dehydrogenase at 0.4 mM is not reasonable. Oxygen delivery turned out to be the limiting factor of the reaction. Standard methods of oxygen feeding such as bubbling of air or intensive stirring cannot be applied due to the deactivation of the enzymes caused by gas–liquid interfaces. This problem could be solved by continuous pumping of the reaction mixture through a thin-walled silicon tube for aeration. This technique of bubble-free aeration, which has been successfully used, *e.g.* in cell culture technology [28], was integrated in a simple reactor [11] which is shown in Figure 2.4.1.

### 2.4.3.2 Application of the new coupled enzyme system

Due to the sufficiently high stability of both cyclohexanone monooxygenase and formate dehydrogenase, the Baeyer–Villiger oxidation of ketone **1** (Scheme 2.4.3)

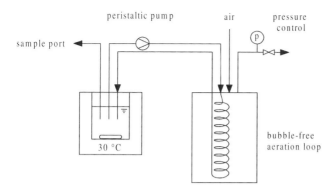

**Figure 2.4.1.** Reactor set-up with integrated bubble-free aeration.

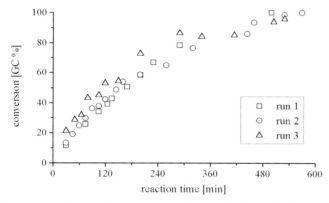

**Figure 2.4.2.** Conversion–time curves of the synthesis of **2** in three consecutive batches.

could be carried out in a repetitive batch mode [11]. After complete conversion of **1**, the enzymes were separated by ultrafiltration and reused after adding fresh reaction solution additionally twice, without a decrease in activity (Figure 2.4.2).

Lactone **2** could be isolated in good yield both with high purity (>99%, GC) and with high enantiomeric excess (>99%, GC).

Furthermore, ketone **3** could be oxidized in two consecutive batches successfully (Scheme 2.4.4) [15]. Kinetic resolution of the racemate **3** proceeded nearly completely yielding lactone (*R*)-**4** with high enantiomeric purity (>99%, GC, E = 217). A third batch was not successful, probably due to the irreversible inhibition of the mono-oxygenase by the lactone.

4-Phenylcyclohexanone (**5**) was also accepted by the cyclohexanone monooxygenase (Scheme 2.4.5) [15] which is in contrast to published results [13c]. However, the trans-formations which proceeded very slowly were only possible by the use of an organic solvent because of the insolubility of ketone **5** in the reaction medium. In order to solve this problem, addition of ethanol, ethylene glycol, and diethylene glycol mono-methylether was investigated in concentrations up to 10%. For example, using ethanol

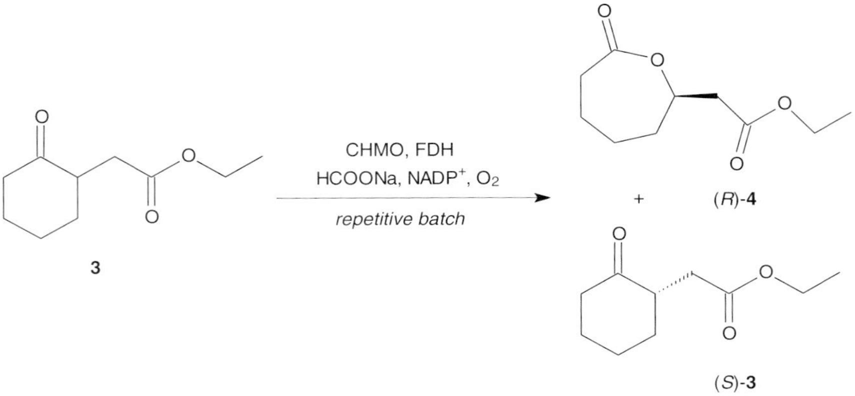

**Scheme 2.4.4.** Kinetic resolution of ketone **3** in repetitive batch.

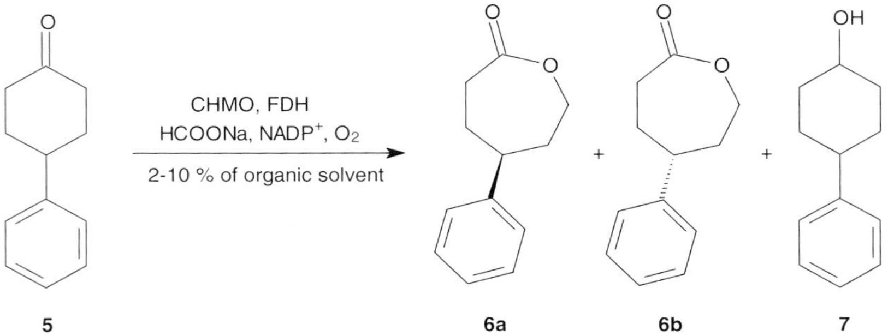

**Scheme 2.4.5.** Baeyer–Villiger oxidation of 4-phenylcyclohexanone (**5**).

as solvent the conversion stopped at 14%, but an enantiomeric excess of one of the enantiomers **6a** or **6b** of about 95% could be achieved. The conversion increased to 80% by using 5% ethylene glycol, whereas the enantiomeric excess decreased to 60%. These results show the strong influence of solvents on both the activity and probably the selectivity of the monooxygenase. The formation of the side product **7** was probably caused by traces of cyclohexanol dehydrogenase still present in the monooxygenase solution.

In conclusion, a new efficient method for cofactor regeneration could be developed combining a NADPH-dependent monooxygenase and a $NADP^+$-dependent formate dehydrogenase. This permits the application of the enzymatic Baeyer–Villiger oxidation of cyclohexanones on preparative scale using the repetitive batch technique. Furthermore, it could be shown that also ketones of low solubility can be successfully converted to the lactones by addition of organic solvents.

**Acknowledgements:** The authors wish to thank Dr. U. Kragl and S. Rissom (Research Center Jülich) for the fruitful cooperation in the development of the cofactor regeneration method, and Prof. Dr. V. I. Tishkov (Lomonossov State University, Moscov) for the generous gift of the formate dehydrogenase. We thank the Fonds der Chemischen Industrie and the Deutsche Forschungsgemeinschaft for financial support.

## References

[1] (a) Review: G. Strukul *Angew. Chem.* **1998**, *110*, 1256–1267; (b) C. Bolm, G. Schlingloff, K. Weickhardt *ibid.* **1994**, *106*, 1944–1946; (c) A. Gusso, C. Baccin, F. Pinna, G. Strukul *Organometallics* **1994**, *13*, 3442–3451.

[2] Review: S. M. Roberts, P. W. H. Wan *J. Mol. Catal. B: Enzymatic* **1998**, *4*, 111–136.

[3] Review: V. Alphand, R. Furstoss in: *Enzyme Catalysis in Organic Synthesis*, K. Drauz, H. Waldmann (Eds.) VCH, Weinheim, **1995**, vol. 2, 745–773.

[4] R. Gagnon, G. Grogan, S. M. Roberts, R. Villa, A. J. Willetts *J. Chem. Soc. Perkin Trans. 1* **1995**, 1505–1511.

[5] K. Königsberger, H. Griengl *Bioorg. Med. Chem.* **1994**, *2*, 595–604.

[6] (a) B. M. Adger, R. McCague, S. M. Roberts, *PCT Int. Appl.* WO 9638,437, 05.12.1996; *C.A.* **1997**, *126*, 89208q; (b) M. T. Bes, R. Villa, S. M. Roberts, P. W. H. Wan; A. J. Willetts *J. Mol. Catal. B: Enzymatic* **1996**, *1*, 127–134; (c) B. M. Adger, M. T. Bes, G. Grogan, R. McCague, S. Pedragosa-Moreau, S. M. Roberts, R. Villa, P. W. H. Wan, A. J. Willetts *Bioorg. Med. Chem.* **1997**, *5*, 253–261.

[7] C. Mazzani, J. Lebreton, V. Alphand, R. Furstoss *Tetrahedron Lett.* **1997**, *38*, 1195–1196.

[8] J. D. Stewart, K. W. Reed, C. A. Martinez, J. Zhu, G. Chen, M. M. Kayser *J. Am. Chem. Soc.* **1998**, *120*, 3541–3548.

[9] A. J. Willetts *Trends Biotechnol.* **1997**, *15*, 55–62.

[10] K. Faber *Biotransformations in Organic Chemistry*; Springer, Berlin, 2nd edn., **1995**.

[11] S. Rissom, U. Schwarz-Linek, M. Vogel, V. I. Tishkov, U. Kragl *Tetrahedron: Asymmetry* **1997**, *8*, 2523–2526.

[12] N. A. Donoghue, D. B. Norris, P. W. Trudgill *Eur. J. Biochem.* **1976**, *63*, 175–192.

[13] (a) M. J. Schwab, W. Li, L. P. Thomas *J. Am. Chem. Soc.* **1983**, *105*, 4800–4808; (b) M. J. Taschner, D. J. Black *ibid.* **1988**, *110*, 6892–6893; (c) O. Abril, C. C. Ryerson, C. Walsh, G. M. Whitesides *Bioorg. Chem.* **1989**, *17*, 41–52; (d) V. Alphand, R. Furstoss *Tetrahedron: Asymmetry* **1992**, *3*, 379–382; (e) V. Alphand, R. Furstoss, S. Pedragosa-Moreau, S. M. Roberts, A. J. Willetts *J. Chem. Soc. Perkin Trans. 1* **1996**, 1867–1872.

[14] (a) J. A. Latham, C. Walsh *J. Am. Chem. Soc.* **1987**, *109*, 3421–3427; (b) P. W. Trudgill *Methods Enzymol.* **1990**, *188*, 70–77.

[15] U. Schwarz-Linek, *PhD thesis*, University of Leipzig, **1998**; unpublished.

[16] C. C. Ryerson, D. P. Ballou, C. Walsh *Biochemistry* **1982**, *21*, 2644–2655.

[17] R. Ruppert, S. Herrmann, E. Steckhan *J. Chem. Soc., Chem. Comm.* **1988**, 1150–1151.

[18] R. Wienkamp, E. Steckhan *Angew. Chem.* **1982**, *94*, 786.

[19] R. Wienkamp, E. Steckhan *Angew. Chem.* **1983**, *95*, 508–509.

[20] (a) H. K. Chenault, G. M. Whitesides *Appl. Microbiol. Biotech.* **1987**, *14*, 147–197; (b) H. K. Chenault, E. S. Simon, G. M. Whitesides *Biotechnol. Genet. Eng. Rev.* **1988**, *6*, 221–270; (c) C.-H. Wong, G. M. Whitesides *Enzymes in Synthetic Organic Chemistry* Pergamon Press Elsevier, Oxford, **1994**, p. 131–136; (d) W. Hummel, M.-R. Kula *Eur. J. Biochem.* **1989**, *184*, 1–13.

[21] Z. Shaked, G. M. Whitesides *J. Am. Chem. Soc.* **1980**, *102*, 7104–7105.

[22] (a) G. Grogan, S. M. Roberts, A. J. Willetts *Biotechnol. Lett.* **1992**, *14*, 1125–1130; (b) R. Gagnon, G. Grogan, M. S. Levitt, S. M. Roberts, P. W. H. Wan, A. J. Willetts *J. Chem. Soc. Perkin Trans. 1* **1994**, 2537–2543; (c) G. Gagnon, S. M. Roberts, A. J. Willetts *J. Chem. Soc., Chem. Comm.* **1993**, 699–701; (d) S. M. Roberts, A. J. Willetts *Chirality* **1993**, *5*, 334–337.

[23] A. J. Willetts, C. J. Knowles, M. S. Levitt, S. M. Roberts, H. Sandey, N. F. Shipston *J. Chem. Soc. Perkin Trans. 1* **1991**, 1608–1610.

[24] C.-H. Wong, G. M. Whitesides *J. Am. Chem. Soc.* **1981**, *103*, 4890–4899.

[25] V. I. Tishkov, A. G. Galkin, G. N. Marchenko, Y. D. Tsygankov, H. M. Egorov *Biotechnol. Appl. Biochem.* **1993**, *18*, 201–207.

[26] H. Schütte, J. Flossdorf, H. Sahm, M.-R. Kula *Eur. J. Biochem.* **1976**, *62*, 151–160.

[27] K. Seelbach, B. Riebel, W. Hummel, M.-R. Kula, V. I. Tishkov, A. M. Egorov, C. Wandrey, U. Kragl *Tetrahedron Lett.* **1996**, *37*, 1377–1380.

[28] (a) B. Hambach, M. Biselli, P. W. Runstadler, C. Wandrey, in *Animal Cell Technology, Developments, Processes and Products* R. E. Spier, J. B. Griffiths, C. MacDonald (Eds.), Butterworth-Heinemann Ltd., Oxford, **1992**, 381–385; (b) M. Schneider, F. Reymond, I. W. Marison, U. von Stockar *Enzyme Microb. Technol.* **1995**, *17*, 839–847.

# 2.5    The synthesis of optically active [2.2]paracyclophanes by biotransformations

*Markus Pietzsch and Henning Hopf*

Interest in the synthesis of optically pure planar chiral ligands for transition metal-catalyzed asymmetric synthesis is rapidly growing since they are inter alia used in chiral catalysts for the preparation of enantiomerically pure compounds on an industrial scale. Recently, the multi-ton synthesis of the herbicide (*S*)-metolachlor by asymmetric hydrogenation using a new class of planar chiral iridium ferrocenyldiphosphine complexes has been reported [1]. Since the selectivity of this reaction [80% enantiomeric excess (*ee*)] is relatively low, the enormous activity and productivity of the catalyst seems to be more important than the optical purity of the product, in this particular case. The synthesis of the disubstituted planar chiral catalyst was achieved by diastereoselective *ortho*-lithiation of optically pure *N*,*N*-(1-ferrocenylethyl)dimethylamine and subsequent reaction with a substituted phoshine as an electrophile. The synthesis of optically active planar chiral ferrocenes has been reviewed recently by Togni [2].

Like ferrocene derivatives, planar chiral paracyclophanes are of growing interest because of their chemical stability against oxidation and racemization under acidic or basic conditions [3]. Hence considerable effort has been undertaken recently to show, that paracyclophanes are useful ligands in asymmetric synthesis, as well. For example, it has been shown by Vögtle et al. that dendrimers bearing terminal planar chiral [2.2]paracyclophanes can be employed for complexation of various metal cations and therefore have been foreseen to serve as asymmetric homogeneous catalysts [4]. Disubstituted enantiomerically pure [2.2]paracyclophanes have been used as chiral auxiliaries in the transition metal-catalyzed asymmetric synthesis of $\beta$-hydroxy-$\alpha$-amino acids [5] and in asymmetric hydrogenation reactions [6].

## 2.5.1    Preparation of optically active planar chiral paracyclophanes

Former attempts to prepare optically active mono- and disubstituted [2.2]paracyclophanes included resolution via diastereomeric derivatives [5–8] and chiral HPLC [9]. However, the methods published so far are lacking in generality and efficiency: while optical purities range from 85 to 99% *ee*, the yields obtained were generally too low to be useful in preparative terms.

## 2.5.2    Biocatalysis in the enantioselective preparation of cyclophanes

During the late 1980s the enormous potential for employing enzymes to resolve non-natural organic compounds was realized [10]. As a result, enzymes have captured an important place in organic synthesis [11, 12].

In contrast to central and axial chirality, planar chirality has not been found in nature so far. Nevertheless, some isolated enzymes as well as micro-organisms can act on planar chiral substrates and convert them not only regioselectively but also enantioselectively. For substituted ferrocenes and other metalloorganic planar chiral substrates, bioreductions of formyl-substituents, catalyzed by alcoholdehydrogenases [13], hydrolysis of esters and esterification of alcohols using lipases and esterases [14, 15] have been reported.

When we started our investigations, the use of chiral cyclophanes for enantioselective synthesis was limited due to the limited access to optically active material, and little was known about the stereoselective synthesis of cyclophanes using biocatalysis. Two attempts to synthesize optically pure [2.2]meta- and [2.2]paracyclophanes by microbial reduction and transesterification reactions resulted only in low enantiomeric excess [16, 17].

The aim of the present contribution is to review the efforts made in the synthesis of optically active mono- and disubstituted [2.2]paracyclophanes (Figure 2.5.1) using biotransformations, and to recount some general problems in applying biocatalysts to organic syntheses.

Above all, a few considerations on the terminology used in enzymatic kinetic resolution reactions are given. As commonly known, measurement of the *ee* of the product or substrate is not sufficient to describe the enantioselectivity of a kinetic resolution reaction and to compare biocatalysts properly, since the *ee* changes during the reaction course and is a function of the conversion. To compare biocatalysts, a mathematical model was developed by Chen et al., which defines an '*E*-value' of the enzyme under defined reaction conditions [18]. The '*E*-value' may be regarded as an enantioselectivity parameter, and there is some confusion in the literature about the correct terminology. The original work of Chen et al. introduced the '*E*-value' as an enantiomeric ratio. This is however rejected by some authors, since there is not a ratio of enantiomers calculated but the relation of $k_{cat}/K_M$ of the two enantiomers. Other expressions as 'enantioselectivity', '*E*-value', 'enantioselectivity-value', 'resolution selectivity', or 'resolution factor' have been recommended as a substitute. Although the present contribution refers to the original work of Chen et al [18]., the expression '*E*-value' is used instead of 'enantiomeric ratio', since this is in accordance with the terminology used in synthetic organic chemistry. However, it should be kept in mind that the limits of the mathematical model of Chen et al., *i.e.* irreversible reaction at the same active site and no substrate or product inhibition, are still valid.

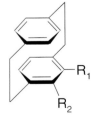

| Substrate | R$_1$ | R$_2$ | Reaction type |
|---|---|---|---|
| 1 | H | CHO | Bioreduction (substrate) |
| 2 | H | CH$_2$OH | Bioreduction (product) |
| 3 | H | OH | Esterification |
| 4 | H | OAc | Hydrolysis |
| 5 | CHO | OAc | Hydrolysis |
| 6 | CHO | OH | Bioreduction |

**Figure 2.5.1.** Biotransformations in the synthesis of optically pure mono- and disubstitued planar chiral [2.2]paracyclophanes.

## 2.5.3     Biotransformation of monosubstituted [2.2]paracyclophanes

### 2.5.3.1     Bioreduction of 4-formyl[2.2]paracyclophane 1

Bioreductions using carbonylreductases represent one method for the kinetic resolution of racemic mixtures. In principle, isolated enzymes together with their cofactors can be used for the reduction reaction. However, in preparative bioconversions involving cofactors, reactions catalyzed by whole cells are usually favorable, because of the simple way of in vivo cofactor recycling, which can usually be accomplished in a simple way, *e.g.* by addition of glucose [12]. Because the use of chiral cyclophanes for enantioselective synthesis has been limited so far due to the limited access to optically active material, a screening for micro-organisms and commercially available alcohol dehydrogenases (ADH) able to reduce (*R,S*)-**1** stereoselectively was carried out (Scheme 2.5.1) [19].

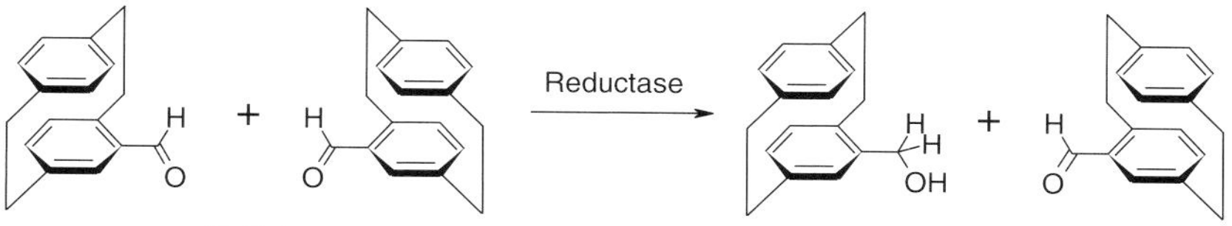

**Scheme 2.5.1.** Reaction scheme of the asymmetric bioreduction of 4-formyl[2.2]paracyclophane (**1**).

Commercially available alcohol dehydrogenases (carbonylreductases) from horse liver, yeast, and *Thermoanaerobium brockii* have been tested for activity, but were found to be completely inactive.

The micro-organisms selected for the initial screening had previously been reported to reduce planar chiral ferrocene derivatives [20, 21] or silicon-containing ketones [22] stereoselectively. The micro-organisms listed in Table 2.5.1 were cultivated, collected by centrifugation, and then assayed for activity and enantioselectivity as resting cell biocatalysts. As a result, *Saccharomyces cerevisiae* DSM 11285, originally obtained from a baker, turned out to be highly enantioselective. To allow reproducible access to this strain, single colonies were isolated and a sample was deposited at the German Collection of Micro-organisms (DSM).

In order to minimize the influence of mass transfer limitations occurring in multiphasic reaction systems such as the one used for the bioconversion of (*R,S*)-**1**, the reaction conditions had to be optimized. Eventually methanol and Tween 80 were used to suspend the solid substrate in a phosphate buffer, containing 10% glucose for in vivo cofactor recycling. Using 30 g of *Saccharomyces cerevisiae* DSM 11285, some 84 mg of (*S*)-**1** (99% *ee*, 49% yield) could be isolated after a reaction time of 3.5 h [19].

**Table 2.5.1.** Screening for micro-organisms with alcohol dehydrogenase (carbonylreductase) activity for the reduction of 4-formyl[2.2]paracyclophane (**1**).

| Micro-organism | $ee$(S) at 50% conversion | $E$-value |
|---|---|---|
| *Saccharomyces cerevisiae* DSM 11285 | 99% | >100 |
| *Yarrowia lipolytica* DSM 1345 | 87% | 39.1 |
| *Cryptococcus humiculus* DSM 70067 | 73% | 13.9 |
| *Rhodotorula rubra* IFO 889 | 56% | 6.1 |
| *Rhodotorula mucilginosa* ATCC 20129 | 47% | 4.6 |
| *Candida boidinii* DSM 70026 | 45% | 4.2 |
| *Pichia jadinii* DSM 2361 | 42% | 3.7 |
| *Hanseniaspora osmophila* DSM 2249 | 23% | 1.9 |
| *Trigonopsis variabilis* DSM 70714 | 18% | 1.7 |
| *Rhodococcus erythropolis* DSM 43066 | 63% | 8.2 |

## 2.5.3.2   Synthesis of 4-hydroxy[2.2]paracyclophane 3 using hydrolases

Since the early years of biocatalysis, and since the discovery that certain enzymes are active even in organic solvents, kinetic resolution reactions (especially those using hydrolases, *i.e.* most often lipases and esterases) have attracted much interest by organic chemists. This is mainly due to the enormous potential of these enzymes in stereoselectively hydrolyzing non-natural esters, a topic recently reviewed by Kazlaus-kas and Bornscheuer [23].

In order to synthezise optically pure 4-hydroxy[2.2]paracyclophane **3**, commercially available lipases were applied in hydrolysis, esterification, and transesterification reactions. While in esterification and transesterification reactions no activity could be found, several lipases catalyzed the kinetic resolution of (*R,S*)-**4** (Scheme 2.5.2) [24].

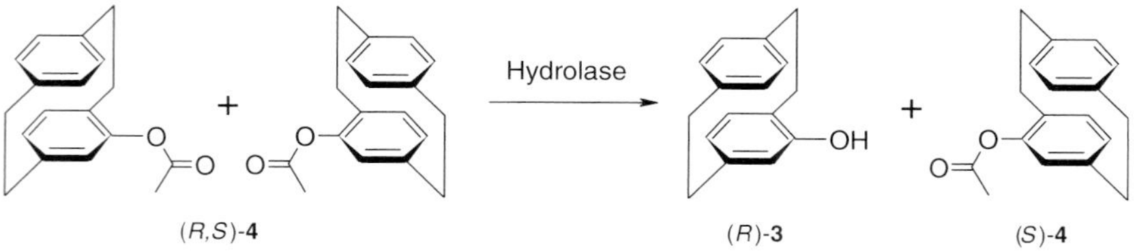

*(R,S)*-**4**          *(R)*-**3**          *(S)*-**4**

**Scheme 2.5.2.** Reaction scheme for the kinetic resolution of (*R,S*)-4-acetoxy[2.2]paracyclophane (**4**) using hydrolases.

Of the enzymes screened, lipase from *Candida rugosa* (formerly *Candida cylindracea* [25]) commercially available from the Japanese company Amano, showed the highest $E$-value ($E = 20$) [24]. Interestingly, lipase preparations from the same micro-organism (*Candida rugosa*) obtained from different suppliers, showed significant differences in enantioselectivity (see Table 2.5.4). A similar observation was reported by Barton et al [25]. This difference may be explained by the impurity of commercial lipase preparations; indeed, up to four different lipases (two in each *Candida*

**Table 2.5.2.** Influence of organic co-solvents on the activity and the enantioselectivity of *Candida rugosa* lipase, catalyzing the hydrolysis of 4-acetoxy[2.2]paracyclophane (**4**).

| Organic solvent | Log $P$ | Activity [U/mg protein $\times 10^{-5}$] | $E$-value |
|---|---|---|---|
| Acetone | −0.23 | 38.42 | 82 |
| Tetrahydrofuran | 0.49 | 0 | no conversion |
| Diethyl ether | 0.85 | 4.46 | >100 |
| Hexanone | 1.30 | 6.82 | 60 |
| Diisopropyl ether | 1.88 | 8.83 | 30 |
| Chloroform | 2.00 | conversion <10%[a] | conversion <10% |
| Toluene | 2.50 | 11.55 | 20 |
| Xylene | 3.10 | 18.86 | 75 |
| Dodecane | 6.59 | 14.81 | 4 |

[a] At this conversion, no activity could be calculated.

*cylindracea* preparation tested) have been found in commercial lipase preparations as determined recently by activity staining experiments [26]. Because of the strong dependence of the enantioselectivity on the lipase preparation used, it should be stressed that detailed information on the source of the enzyme is strictly required in order to allow reproducibility. This information is unfortunately missing in the paper of Cipiciani et al., who dealt also with the successful kinetic resolution of 4-acetoxy[2.2]paracyclophane [27].

Both, the activity and the enantioselectivity of a hydrolase-catalyzed kinetic resolution reaction may be influenced by the chemical nature of a co-solvent, the pH of the buffer solution, the reaction temperature [12], and the ratio of the co-solvent to buffer phase [28]. Some of these parameters were optimized for the *Candida rugosa* lipase-catalyzed kinetic resolution of (*R,S*)-**4** [29]. Co-solvents were chosen according to their log *P*-values [30], but as can be seen from Table 2.5.2 neither the activity nor the *E*-value seemed to be correlated with the log *P* value. With the exception of dodecane, an increased *E*-value relative to toluene (which was used in the screening experiments) was determined for all co-solvents tested, and diethyl ether gave the best result ($E > 100$).

Up to now, all attempts to link the observed enantioselectivity of enzymes to physico-chemical data of the respective solvent have been unsuccessful [12]. An interesting contribution to the present discussion was made recently by Wescott et al., who could quantitatively rationalize the energetics of desolvation of the enantiomers in the enzyme-bound transition states and could thereby explain experimental results of the solvent dependency of the enantioselectivity of a protease [31]. However, for this method the 3-D structure of an enzyme is required, making the method applicable only to enzymes whose structures are known.

Using diethyl ether as a co-solvent, the influence of the pH of the buffer was investigated. As can be seen from Table 2.5.3, both enzymatic activity and *E*-value were influenced by the reaction pH. In a relatively narrow pH range (5.5–8.0) the *E*-value varied from 14 to > 100, with an optimum at pH 7.0. In the literature, the

**Table 2.5.3.** Influence of pH on activity and enantioselectivity of *Candida rugosa* lipase-catalyzed hydrolysis of 4-acetoxy[2.2]paracyclophane (**4**).

| pH | Activity [U/mg protein $\times 10^{-5}$] | *E*-value |
|---|---|---|
| 5.5 | 1.74 | 39 |
| 6.0 | 1.97 | 62 |
| 6.5 | 2.04 | 73 |
| 7.0 | 5.68 | >100 |
| 7.5 | 2.40 | 31 |
| 8.0 | 0.86 | 14 |

effect of the pH on the enantioselectivity of hydrolytic enzymes has been investigated only scarcely. For the porcine pancreas lipase (PPL)-catalyzed hydrolysis of meso-2,3-epoxybutane-1,4-diol-diacetic acid ester an optimum of pH 6.5 has been reported [32]. For *Candida rugosa* lipase MY-catalyzed hydrolysis of *rac*-2-(4-hydroxyphenoxy)propionic acid methyl ester the optimal activity was found at pH 6.0 (phosphate buffer) and the highest enantiomeric excess at pH 4.0 (citrate buffer) [25]. The differences in pH optima of *Candida rugosa* lipase-catalyzed hydrolysis may be explained by the inhomogeneity of the commercial sources as discussed above.

In Figure 2.5.2a and 2.5.2b, typical reaction kinetics (*ee* versus conversion) are shown. For comparison, the results obtained before (dashed lines) and after (solid lines) optimization are shown. The concentrations of the product enantiomers ((*R*)-**3** and (*S*)-**3** respectively) as well as the concentration of the substrate ((*R*,*S*)-**4**) were determined by gas chromatography. From the data obtained the *E*-value according to Chen et al [18]. was calculated (Figure 2.5.2a). Using the same model, the

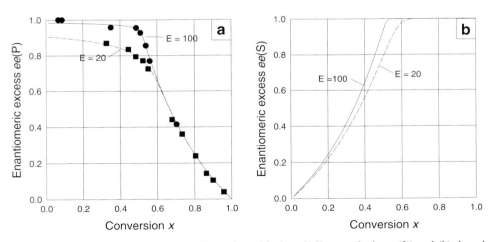

**Figure 2.5.2.** Enantiomeric excess of (a) the product (4-hydroxy[2.2]paracyclophane (**3**)) and (b) the substrate (4-acetoxy[2.2]paracyclophane (**4**)) versus conversion before (■) and after (●) optimization. Solid and dashed lines represent calculations using the model of Chen et al [18].

dependence of the enantiomeric excess of the substrate **4** on the conversion was calculated (Figure 2.5.2b) and used as a basis for the design of the preparative scale conversion. Using the mathematical model it was calculated that the enantiomeric excess of the substrate **4** is nearly 100% at a conversion of 53%.

To verify the calculation, 1.6 g of racemic **4** were subjected to the enzymatic kinetic resolution using 3 g (17 kU) *Candida rugosa* lipase under optimized reaction conditions. After 95 h the reaction was stopped at exactly 53% conversion and 0.7 g of optically pure (*S*)-**4** were isolated, showing an *ee* higher than 99% (44% yield), as expected.

## 2.5.4  Synthesis of optically active disubstituted [2.2]paracyclophanes

As discussed above, optically active disubstituted [2.2]paracyclophanes are of main interest as chiral auxilaries in asymmetric syntheses. Investigations on the direct resolution of the disubstituted racemic substrate, as well as the development of a preparative useful chemoenzymatic method for the synthesis of optically pure 5-formyl-4-hydroxy[2.2]paracyclophane (**6**), are discussed in the following.

**Table 2.5.4.** Screening for stereoselective hydrolysis of mono- and disubstituted acetoxy[2.2]paracyclophanes **4** and **5**. The *E*-value was calculated using the model of Chen et al [18].

| Enzyme | Source | Supplier | *E*-value | |
|---|---|---|---|---|
| | | | disubst. **5** | monosubst. **4** |
| Lipase | *Mucor miehei* | Biocatalysts | 4.2 | 3.0 |
| Lipase | *Candida antarctica*, fraction A | Boehringer Mannheim | 3.6 | 5.2 |
| Lipase | *Candida rugosa* | Fluka | 2.6 | 17.0 |
| Alkaline Lipase | *Stiowa* | Showa Denko | 2.3 | 3.0 |
| Lipase AY | *Candida rugosa* | Amano | 2.0 | 20.0 |
| Lipase | *Rhizopus* | Solvay-Enzymes | 1.7 | 4.0 |
| Lipase | *Candida rugosa* | Boehringer Mannheim | 1.6 | 1.8 |
| Lipase D | *Rhizopus delemar* | Amano | 1.6 | 1.5 |
| Lipase | *Pseudomonas alcaligenes* | Gist-Brocades | 1.5 | 1.5 |
| Lipase A | *Aspergillus niger* | Amano | 1.3 | 4.0 |
| Lipase | *Candida antarctica* | Amano | 1.2 | n. c. |
| Lipase M | *Mucor javanicus* | Amano | 1.2 | 1.2 |
| Lipase | *Pseudomonas fluorescens* | Biocatalysts | 1.2 | n. c. |
| Lipase | *Humicula* sp. | Boehringer Mannheim | 1.1 | Conversion <10 %[a] |
| Lipase | *Pseudomonas* sp. | Boehringer Mannheim | 1.1 | Conversion <10 %[a] |
| Lipase | *Burkholderia* sp. | Boehringer Mannheim | <1.1 | n. c. |
| Lipase | *Chromobacterium viscosum* | Asahi | <1.1 | Conversion <10 %[a] |
| Lipase | *Humicola lanuginosa* | Biocatalysts | <1.1 | 3.0 |
| Lipase G | *Penicillium camembertii* | Amano | <1.1 | n. c. |
| Lipase R | *Penicillium roquefortii* | Amano | <1.1 | 2.0 |
| Lipase | Porcine pancreas | Fluka | <1.1 | Conversion <10 %[a] |
| Lipase F | *Rhizopus javanicus* | Amano | <1.1 | 1.5 |

[a] At this conversion, no *E*-value could be calculated; n. c.: no conversion

### 2.5.4.1    Screening for bioreduction and hydrolytic activity

All micro-organisms which have previously been shown to be capable of reducing (monosubstituted) 4-formyl[2.2]paracyclophane (**1**) to 4-hydroxymethyl[2.2]paracyclophane (**2**) were applied for the bioreduction of 5-formyl-4-hydroxy[2.2]paracyclophane (**6**). While monosubstituted **1** was reduced with good activity and high enantioselectivity (see above), no reduction at all could be observed in the case of the disubstituted [2.2]paracyclophane.

For this reason all hydrolases able to perform the kinetic resolution of (monosubstituted) 4-acetoxy[2.2]paracyclophane (**4**) were applied for the resolution of 4-acetoxy-5-formyl[2.2]paracyclophane (**5**). However, transesterification and esterification reactions were not observed. Hydrolysis resulted in only low *E*-values, as shown in Table 2.5.4 [29].

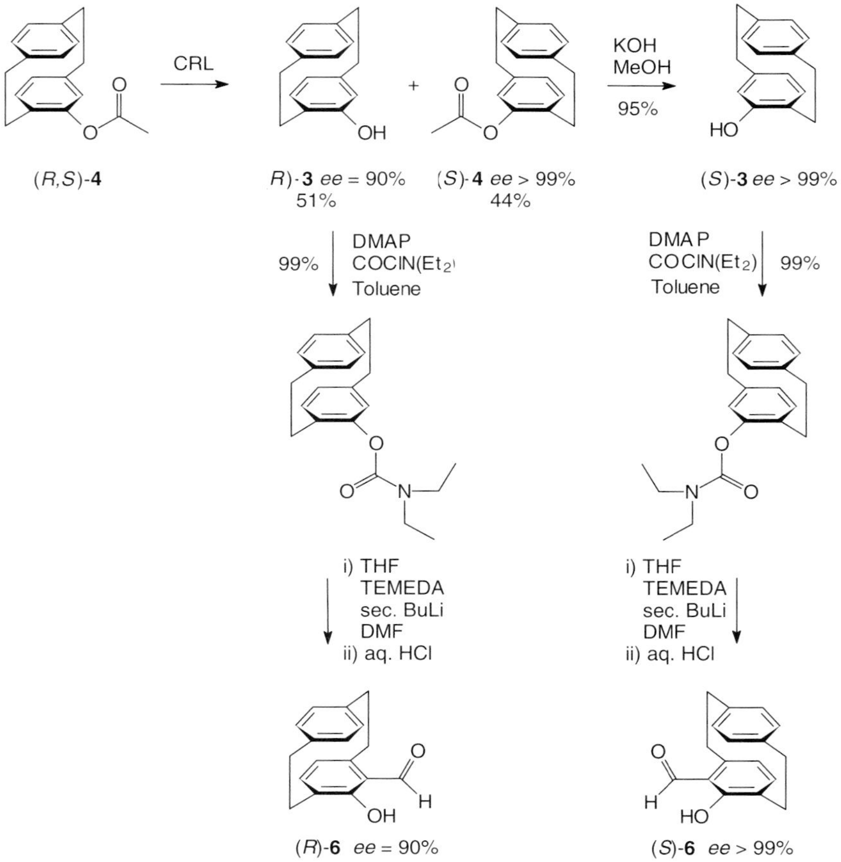

**Scheme 2.5.3.** Chemoenzymatic synthesis of 5-formyl-4-hydroxy[2.2]paracyclophane (**6**)

As can be seen from Table 2.5.4, under screening conditions the disubstituted [2.2]paracyclophane **5** was hydrolyzed by the same enzymes with significantly lower $E$-value than the monosubstituted [2.2]paracyclophane **4**.

### 2.5.4.2    Chemoenzymatic synthesis of 5-formyl-4-hydroxy[2.2]paracyclophane (6)

Because all methods published so far for the synthesis of disubstituted **6** start from monosubstituted **3**, an enantioselective step can be introduced either at the beginning (enantioselective preparation of **3**) or at the end of the synthesis (enantioselective preparation of **6**). In both cases a high enantioselectivity is required for a preparative useful overall synthesis; however, an early resolution step is favorable with respect to process economics.

The $E$-value of 20 found under screening conditions for the preparation of ($S$)-**4** [24] using *Candida rugosa* lipase (CRL) from Amano provided a more promising starting point for an optimization than the $E = 4$ found for the kinetic resolution of **5** [29]. As described above, the $E$-value for the preparation of ($S$)-**4** was enhanced to $E > 100$ after optimization. After hydrolysis, ($S$)-**3** was subjected to stereoselective *ortho*-formylation according to Hopf and Barrett [33] (Scheme 2.5.3). Some 285 mg of optically pure ($S$)-**6** ($ee > 99\%$) were derived from 430 mg ($S$)-**4** (overall yield 51%). In an analogous way ($R$)-**4**, isolated after the kinetic resolution reaction (90% $ee$) was converted to optically active ($R$)-**6** (90% $ee$) [29].

# References

[1]  F. Spindler, B. Pugin, H.-P. Jalett, H.-P. Buser, U. Pittelkow, H.-U. Blaser, in *Catalysis of Organic Reactions* (Ed.: R. E. Malz), Marcel Dekker, New York, **1996**, pp. 153–166.
[2]  A. Togni, *Angew. Chem.* **1996**, *108*, 1581–1583; *Angew. Chem. Int. Ed. Engl.* **1996**, *35*, 1475–1477.
[3]  Rosenfeld, P. Keehn, *The Cyclophanes I and II*, Academic Press, New York, **1983**.
[4]  J. Issberner, M. Böhme, S. Grimme, M. Nieger, W. Paulus, F. Vögtle, *Tetrahedron: Asymmetry* **1996**, *7*, 2223–2232.
[5]  V. Rozenberg, V. Kharitonov, D. Antonov, E. Sergeeva, A. Aleshkin, N. Ikonnikov, S. Orlova, Y. Belokoń, *Angew. Chem.* **1994**, *106*, 106–108; *Angew. Chem. Int. Ed. Engl.* **1994**, *33*, 91–92.
[6]  P. J. Pye, K. Rossen, R. A. Reamer, N. N. Tsou, R. P. Volante, P. J. Reider, *J. Am. Chem. Soc.* **1997**, *19*, 6207–6208.
[7]  H. Falk, P. Reich-Rohrwig, K. Schlögl, *Tetrahedron* **1970**, *26*, 511–527.
[8]  A. Cipiciani, F. Fringuelli, V. Mancini, O. Piermatti, F. Pizzo, R. Ruzzicioni, *J. Org. Chem.* **1997**, *62*, 3744–3747.
[9]  H. Hopf, W. Grahn, D. G. Barrett, A. Gerdes, J. Hilmer, J. Hucker, Y. Okamoto, Y. Kaida, *Chem. Ber.* **1990**, *123*, 841–845.
[10]  Faber, *Pure Appl. Chem.* **1997**, *69*, 1613–1632.
[11]  K. H. Drauz, H. Waldmann, *Enzyme Catalysis in Organic Synthesis*, VCH-Verlag, Weinheim, **1995**.
[12]  K. Faber, *Biotransformations in Organic Chemistry – A Textbook*, 3.rd ed., Springer Verlag, Berlin, **1997**.
[13]  Y. Yamazaki, K. Hosono, *Tetrahedron Lett.* **1988**, *29*, 5769–5770.

[14]  T. Izumi, S. Aratani, *J. Chem. Tech. Biotechnol.* **1993**, *57*, 33–36.
[15]  D. Lambusta, G. Nicolosi, A. Patti, M. Piatelli, *Tetrahedron Lett.* **1996**, *37*, 127–130.
[16]  M. Nakazaki, H. Chikamatsu, Y. Hirose, T. Shimizu, *J. Org. Chem.* **1979**, *44*, 1043–1048.
[17]  T. Izumi, T. Hinata, *J. Chem. Tech. Biotechnol.* **1992**, *55*, 227–231.
[18]  C.-S. Chen, Y. Fujimoto, G. Girdaukas, C. J. Sih, *Am. Chem. Soc.* **1982**, *104*, 7294–7299.
[19]  D. Pamperin, H. Hopf, C. Syldatk, M. Pietzsch, *Tetrahedron: Asymmetry* **1997**, *8*, 319–325.
[20]  S. Top, G. Jaouen, J. Gillois, C. Baldoll, S. Malorana, *J. Chem. Soc., Chem. Commun.* **1988**, 1284–1285.
[21]  Y. Yamazaki, K. Hosono, *Tetrahedron Lett.* **1989**, *30*, 5313–5314.
[22]  C. Syldatk, A. Stoffregen, F. Wuttke, R. Tacke, *Biotechnol. Lett.* **1988**, *10*, 731–736.
[23]  R. Kazlauskas, U. Bornscheuer, in *Biotechnology*, 2rd ed. (Eds.: H.-J. Rehm; G. Reed; A. Pühler; P. Stadler), VCH-Wiley, Weinheim, **1998**.
[24]  D. Pamperin, B. Ohse, H. Hopf, M. Pietzsch, *J. Molec. Catal. B: Enzymatic* **1998**, *5*, 317–319.
[25]  M. J. Barton, J. P. Hamman, K. C. Fichter, G. J. Calton, *Enzyme Microb. Technol.* **1990**, *12*, 577–583.
[26]  M. M. Soumanou, U. T. Bornscheuer, U. Menge, R. D. Schmid, *J. Am. Oil Chem. Soc.* **1997**, *74*, 427–433.
[27]  A. Cipiciani, F. Fringuelli, V. Mancini, O. Piermatti, A. M. Scappini, R. Ruzzicioni, *Tetrahedron* **1997**, *53*, 11853–11858.
[28]  E. Holmberg, K. Hult, *Biocatalysis* **1990**, *3*, 243–251.
[29]  D. Pamperin, C. Schulz, H. Hopf, C. Syldatk, M. Pietzsch, *Eur. J. Org. Chem.* **1998**, 1441–1445.
[30]  C. Laane, S. Boeren, K. Vos, C. Veeger, *Biotechnol. Bioeng.* **1987**, *30*, 81–87.
[31]  C. R. Wescott, H. Noritomi, A. M. Klibanov, *J. Am. Chem. Soc.* **1996**, *118*, 10365–10370.
[32]  D. Grandjean, P. Pale, J. Chuche, *Tetrahedron Lett.* **1990**, *32*, 3043–3046.
[33]  H. Hopf, D. G. Barrett, *Liebigs Ann.* **1995**, 449–451.

# 2.6     Multivariate analysis of biotransformations for a more effective strain selection (Intelligent screening)

*Wolf-Rainer Abraham*

For almost a century, micro-organisms were used to derivatize organic compounds. If intact cells are involved in these reactions such a procedure is called biotransformation. Biotransformations have the advantage that they proceed under mild conditions with high regio-, stereo- and enantioselectivity without producing toxic wastes. Furthermore, many strains can oxidize substrates at positions which are thermodynamically not activated and can therefore not be derivatized selectively by most organic chemical methods. However, this selectivity results in one of the main drawbacks of biotransformations, namely the need to search for the best strain for any reaction at any substrate. This means screening a large number of strains. Often only a small change in the substrate which has even not to be near the reaction center is sufficient to reduce drastically the yield of the former optimal strain. Up to now the prediction of the best-suited strain for a planned biotransformation is only vague and depends mainly on the experience of the researcher. To overcome this difficulty, a large number of strains must often be screened, and this is very time consuming. Obviously, this disadvantage is also the reason for the hesitation of organic chemists to use biotransformations.

The application of techniques developed during recent years should result in a considerable reduction in time and money for screening by identifying criteria for a preselection of strains. Such approaches are usually summarized under the term intelligent screening. Our idea was to cluster the strains of our strain collection into groups with similar potentials in biotransformations. Such a grouping of strains would lead to an immense reduction in screening efforts, because in search for a biotransformation one would first screen representatives of each group to identify the group with the most active members. With this information in hand one would then use only strains of this group for the main screen. It is perfectly obvious that such an approach would reduce the screening effort substantially.

However, how can the strains be grouped into clusters with similar biotransformation potentials? Micro-organisms possess a multitude of enzymes and it is therefore not possible to group them only by the formation of a single biotransformation product (univariate analysis). To pay attention to this enormous diversity, a grouping by a number of criteria is needed, and this can be done by multivariate statistical analysis. Therefore, we tested 13 different substrates, mainly mono- and sesquiterpenes, with 100 strains (60 fungi and 40 bacteria) and determined the $R_f$ value of the products on HPTLC plates. These data were analysed further in multivariate statistical analysis [1, 2].

A principal component analysis using the biotransformation products resulted in the identification of five groups of micro-organisms corresponding almost exactly to the large phylogenetic clades. The analysis of these groups revealed that the taxonomic position of a strain is mirrored in its ability to catalyze certain biotransformations, *i.e.*

that the phylogeny of the strains is correlated with its biotransformation potential. In these groups, fungi and bacteria can be discerned, in the course of which *Basidiomycotina*, *Ascomycotina* and *Zygomycotina* of the fungi formed discernable clusters. The *Deuteromycotina* (*Fungi imperfecti*) are only barely discernable from the *Ascomycotina* which is not surprising because it is assumed that up to 80% of this group belong to the *Ascomycotina*. Within the bacteria, Gram-positive and Gram-negative bacteria are discernable. A further grouping could not be found with statistical significance, probably due to the fact that the data-set contained only 40 bacterial strains, which is far too little with respect to the large diversity of these organisms.

At first glance the result is surprising because such a correlation is not obvious and similar attempts using secondary metabolites from micro-organisms were not as unambiguous. Therefore, why are biotransformation of substrates correlated with the phylogeny of the strains involved? The phylogeny used for comparison is based on the comparison of gene sequences. At the same time, it is assumed that the more different the gene sequences are, the more distant the organisms are related. As a gold standard for the determination of the phylogeny of organisms, the sequences of the genes of ribosomal RNAs have been established, usually comparing 16S respectively 18S rDNA. Some other genes were also used to establish the phylogeny of organisms, *e.g.* for plants the sequences of the large subunit of ribulose-1,5-biphosphate carboxylase (rbcL) are often compared. Although there are minor differences in the exact branching of the phylogenetic tree resulting from the comparison of the sequences of different genes, all in all the result is the same. An excellent review concerning the molecular view of microbial diversity was recently given by Norman Pace [3]. On the one hand, such a result leads to optimism that one can really deduce the natural relationship of life, while on the other hand it explains why a comparison of biotransformations resulted in a grouping controlled by the phylogeny of the strains.

The practical consequences of this is that there is no need to screen the whole strain collection in order to achieve such a grouping. Instead, the taxonomy can be used to assign the strains to the different groups. This helps a lot because many strains are obtained from international strain collections where they were already identified. For a small and very specialized strain collection, however, clustering the strains by their biotransformation potentials is an attractive alternative to the taxonomic grouping, because such a grouping is focused entirely on the metabolic potential of the strains.

However, at this point the limits common to both approaches should be mentioned. We are working with statistical methods and this means that we cannot be certain that a given group must show the reaction we are looking for, but merely that the probability is high that this characteristic is there. Moreover, it does not mean that this group is the only one able to perform the reaction, but merely that the ability to do the reaction is concentrated in this group [4]. A reliable prediction cannot be given if the number of strains tested is too small, or too few strains show the particular reaction, *i.e.* the reaction is too rare for the statistical approach applied. Therefore, it is always necessary to keep these limiting factors in mind and it is recommended here to calculate the significance of a prediction. Tests of this novel approach, however, produced very promising results.

## 2.6.1    Oxidation of eugenol

Eugenol **1** is available in large quantity at low prices and is used as starting material for the chemical synthesis of vanillin. However, vanillin formed by chemical methods is called a natural identical substance, in contrast to the natural vanillin. Although chemically there is no difference, the difference is in legislation, and in the price of natural vanillin. Because of this, the search for the formation of vanillin from eugenol using micro-organisms resulting in natural vanillin was in the agenda of many laboratories. In the course of our search for biotransformation products of eugenol microbial transformations were observed only under special conditions. One of these products, coniferylaldehyde **2**, displayed very interesting olfactorial characteristics and we began to look for strains forming this compound in high amounts with only minor side products (Scheme 2.6.1). We found exclusively fungi to perform the reaction, and here the *Deuteromycotina*, where the genus *Fusarium* contained the most active strains [5].

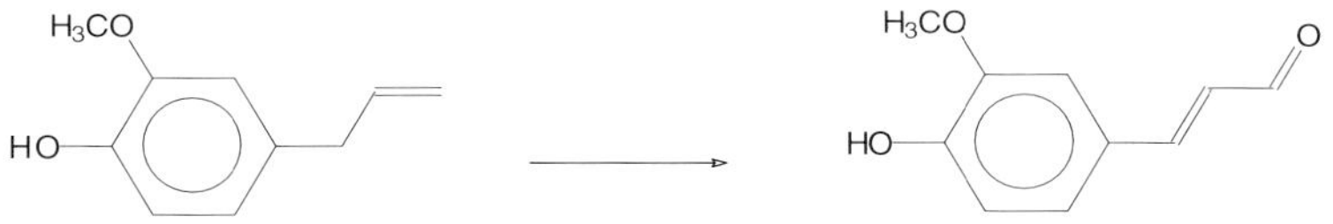

**Scheme 2.6.1.** Biotransformation of eugenol **1** to coniferylaldehyde **2**.

## 2.6.2    Enantiomer separation

To separate enantiomers of alcohols, the kinetic-controlled resolution of their esters by hydrolytic enzymes offers an attractive approach. The yield of this reaction is primarily influenced by the enzyme and the acyl group. Saponification of benzoates is sometimes an interesting variant to achieve enzymatically controlled kinetic resolution of such racemates. However, not all esterases show such an activity, therefore we were looking for other strains performing such a reaction. The occurrence of such esterases seems also to be connected to phylogenetic groups because the screening of isopinocampheol-benzoate **3** revealed that this reaction was found in 70% of the *Zygomycotina* and *Deuteromycotina*, while no strain of the *Basidiomycotina* showed such a reaction [6] (Scheme 2.6.2).

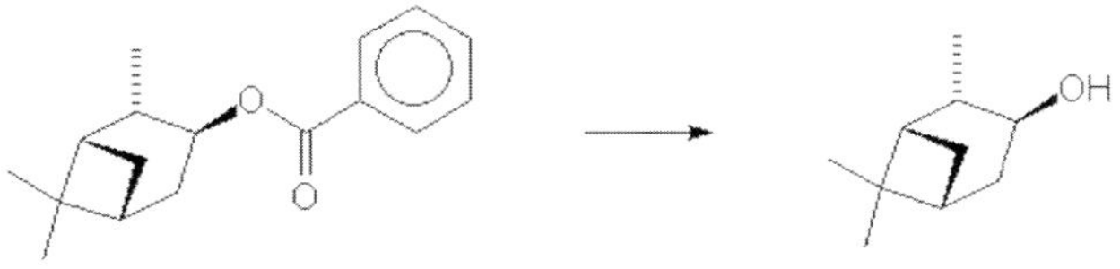

**Scheme 2.6.2.** Saponification of (1*S*)-isopinocampheol-benzoate **3** by microorganisms.

## 2.6.3  Hydration of *trans*-nerolidol

The screening for biotransformation products of *trans*-nerolidol **5** resulted in the detection of an unusual reaction where water was added to the inner 6,7-double bond of the molecule to give caparrapidiol **6** (Scheme 2.6.3). This reaction is not an epoxidation with subsequent reduction because it ran also anaerobically, producing here even higher yields. The reaction is highly specific and we found neither the corresponding reaction with *cis*-nerolidol nor any water addition to one of the other double bonds of *trans*-nerolidol was detected [7]. The water addition to 3*S*-*trans*-nerolidol gave 3*S*,7*S*-caparrapidiol 6 and no 7*R*-epimer was observed. The search for the best-suited strains identified once more the genus *Fusarium* containing the most active strains which gave yields of up to 80% of this product. It is worth mentioning that some *Fusarium* species were reported to detoxify kievitone by adding water to a terpenic double bond by means of an extracellular hydratase [8]. Although we did not prove this experimentally we speculate that similar enzymes are responsible for the water addition to the inner double bond of *trans*-nerolidol and related compounds.

**Scheme 2.6.3.** Microbial hydration of (*S*)-trans-nerolidol **5** to caparrapidiol **6**.

## 2.6.4  Epoxidation of *trans*-nerolidol

The ability to epoxidize the $\omega$-1-terminal 10,11-double bond of (*S*)-*trans*-nerolidol **5** is widespread in micro-organisms. However, a subsequent hydrolysis of the epoxide **7** to 10,11-dihydroxy *trans*-nerolidol **8** is often observed (Scheme 2.6.4). We tried to

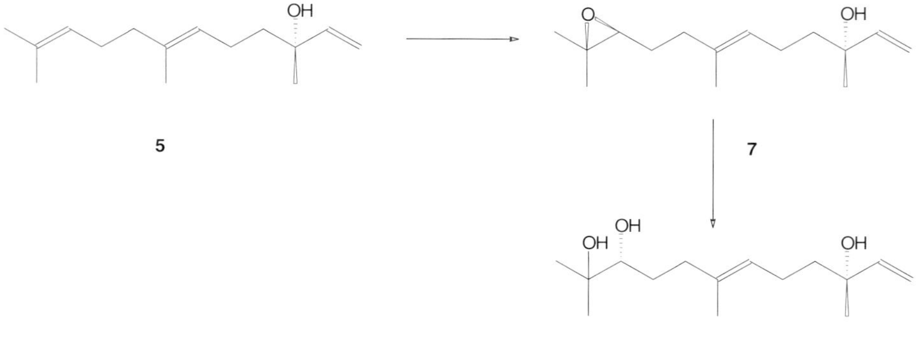

**Scheme 2.6.4.** Microbial epoxidation of *trans*-nerolidol **5** and subsequent opening of the epoxide **7** to the vicinal diol **8** catalyzed by micro-organisms.

decouple these two reactions and screened for strains carrying out the epoxidation, but not the subsequent hydrolysis. Again, in one of the phylogenetic groups this reaction was preferentially observed. This time it was the *Basidiomycotina*, where we found this reaction with all strains tested, but the hydrolysis of the formed epoxide was never observed.

## 2.6.5 Hydroxylation of aristolenepoxide

The biotransformation of the sesquiterpene aristolenepoxide **9** led to a number of hydroxylation products, some of which showed modest phytotoxicity in the cress test. Among these active metabolites, aristolenepoxide-9$\beta$-ol **10** was the most active compound, while aristolenepoxide-5-ol **11** displayed slightly less phytotoxicity [9] (Scheme 2.6.5). Surprisingly, the sum of both, aristolenepoxide-5,9$\beta$-diol which was also isolated from the reaction products of some strains, was completely inactive in our assay. The search for the most active strains for the formation of the phytotoxins revealed that the bacteria were almost completely inactive. For both hydroxylation products the most active fungi were found within the *Basidiomycotina*. The *Ascomycotina* showing a moderate activity in the formation of aristolenepoxide-9$\beta$-ol were not active at all in the formation of the 5-analog.

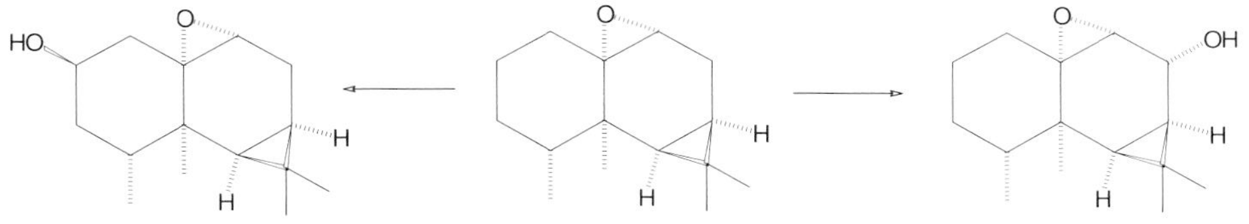

**Scheme 2.6.5.** Microbial hydroxylation of aristolenepoxide **9** to the phytotoxic alcohols **10** and **11**.

## 2.6.6 Monoterpene alcohols

There are several reports in the literature that monoterpene alcohols are poor starting materials for biotransformations, while their esters gave good yields. Biotransformations of 1$S$-(+)-isopinocampheol-benzoate **3** however led to very poor yields in terms of hydroxylation products, while the free alcohol **4** gave far better yields. This substrate is an excellent example to show the advantages of biotransformations, because it was possible to show hydroxylation reactions at all of its three tertiary carbons. Hydroxylation at the methyl group adjacent to the hydroxy group occurred under retention of the configuration at C-2, producing the *cis*-diol **12** [6]. While hydroxylations at the bridge carbons of the cyclobutane moiety are achieved chemically only with major problems, some strains produced 5-hydroxy-isopinocampheol **13** in biotransformations in more than 50% yield (Scheme 2.6.6). While the activity of the

*Basidiomycotina* to show this reaction was quite low, 60% of the *Zygomycotina* showed the 2-hydroxylation and 80% showed the 5-hydroxylation of isopino-campheol. All the strains with the highest yields within 48 h fermentation time belonged to this group again underlining the usefullness of our concept.

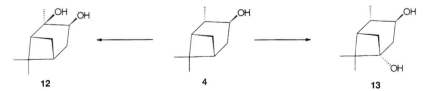

12                          4                          13

**Scheme 2.6.6.** Microbial hydroxylations of isopinocampheol 4 resulting in the tertiary alcohols **12** and **13**.

The results show that the phylogenetic position of a strain can be used to predict its performance in biotransformations. The examples given here demonstrate that the catalytic activity screened for is found preferentially in the taxonomic units, and also corroborate the approach described here. Often, the transformation rates and yields obtained under conditions used for screening are sufficient to run the fermentation on a preparative scale, thus making biotransformations an interesting tool for organic chemists.

# References

[1]  W.-R. Abraham. DECHEMA Biotechnology Conferences (*Eds.*: G. Kreysa und A. J. Driesel), Weinheim-Basel-Cambridge, 1992, Vol. *5A*, 41–44.
[2]  W.-R. Abraham. *World J. Microbiol. Biotechnol.*, 1994, *10*, 88–92.
[3]  N. R. Pace. *Science* 1997, *276*, 734–740.
[4]  W.-R Abraham, A. Riep and H.-P. Hanssen. *Bioorg. Chem.*, 1996, *24*, 19–28.
[5]  W.-R. Abraham and H.-A. Arfmann. 1992. DE, P 42 19 770.8, 17.6.1992.
[6]  W.-R. Abraham. *Zeitschr. Naturforsch.*, 1994, *49c*, 553–560.
[7]  W.-R. Abraham. *World J. Microbiol. Biotechnol.*, 1993, *9*, 319–322.
[8]  D. A. Smith, J. M. Harrer, and T. E. Cleveland. *Phytopathology* 1982, *72*, 1319–1323.
[9]  W.-R. Abraham, A. Riep, and H.-P. Hanssen. *Bioorg. Chem.*, 1996, *24*, 19–28.

# 2.7  Extending the applicability of lipases and esterases for organic synthesis

*Uwe T. Bornscheuer*

Lipases and esterases have been proven to be very suitable enzymes in the preparation of a wide variety of optically pure building blocks for organic synthesis [1]. Lipases in particular accept a broad range of substrates and exhibit high stability and activity in organic solvents. Based on a detailed examination of the stereoselectivity of lipases towards secondary alcohols, reliable empirical rules have been established which allow the prediction of the stereopreference, as well as the extent of stereoselectivity in most cases [2]. In sharp contrast, a similar size-based rule is not valid for primary alcohols, especially if an oxygen atom is attached next to the stereocenter [3]. Furthermore, the stereoselectivity of lipases towards this class of compounds is usually low. The lower stereoselectivity was related to the additional $CH_2$-group introducing a kink between the chiral center and the hydroxy function in primary alcohols (Scheme 2.7.1).

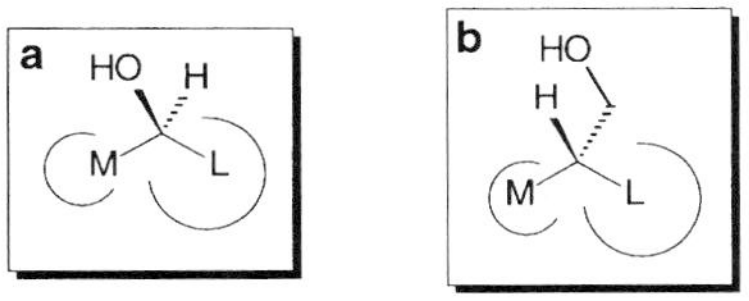

**Scheme 2.7.1.** Empiric rules showing the enantiopreference of *Pseudomonas cepacia lipase* (PCL) toward secondary (a) [3] and primary alcohols (b) [4]. (M = medium-sized substituent; L = large substituent).

   In order to extend the applicability of lipases and esterases, two different strategies have been used. To overcome the low stereoselectivity of lipases towards primary alcohols, variation of the substrate structure was used, as exemplified for the synthesis of a Bryostatin 1 precursor (see below). The other approach targets on the broadening of the substrate spectra of these enzymes, because attempts for the preparation of optically pure compounds starting from sterically demanding substrates usually failed. For instance, dihydropyridine derivatives could only be resolved by introducing a non-hindered cleavage site [4] or a tethered hydroxy function far away from bulky aromatic groups [5]. As an alternative, the mutagenesis of the biocatalyst itself by means of directed evolution was investigated as shown for the mutagenesis of an esterase, which is presented later.

## 2.7.1  Results

### 2.7.1.1  Resolution of a Bryostatin 1 building block

Bryostatins represent a class of marine macrolide natural products with antineoplastic activities. In cooperation with the Institute of Organic Chemistry, University of Hannover, the synthesis of the $C_{10}-C_{16}$ fragment of Bryostatin 1 was investigated, for which *meso* substrates represented by the general structure **1** (Scheme 2.7.2) were envisioned as suitable precursors [6].

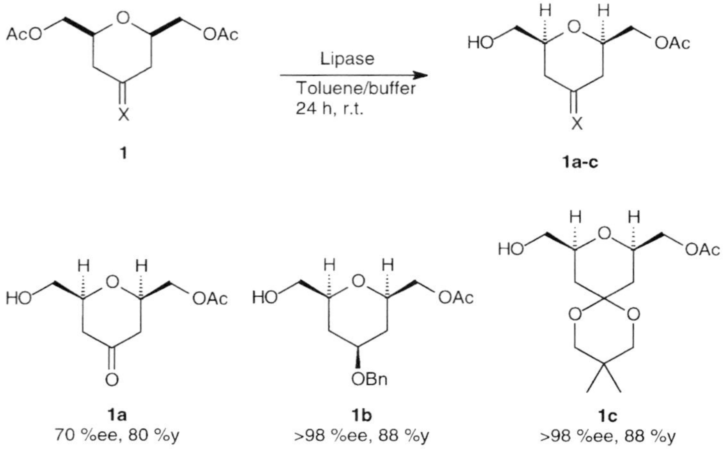

**Scheme 2.7.2.** Lipase-catalyzed desymmetrization of *meso*-diacetates afforded **1a–c** [6a].

These prostereogenic substrates are available from 8-oxabicyclo[3.2.1]oct-6-en-3-one by functionalization and oxidative cleavage of the C=C double bond. Hydrolysis of **1** using lipases should afford 2,4,6-trifunctionalized *C*-glycosides as monoacetates **1a–c**. An initial screening of several hydrolases revealed that lipase from *Pseudomonas cepacia* (PCL, Amano PS) exhibited the highest rate and stereoselectivity, but maximum enantiomeric excess was 70% *ee* for ketone **1a**. Further optimization of the reaction conditions with respect to temperature, solvent, and solvent:buffer ratio gave only slight improvements. Acylation of the corresponding *meso* diol by transesterification with vinyl acetate resulted in significant amounts of diacetates. A substantial increase in stereoselectivity was achieved by using either the diacetate of benzylether **1b** or that of spiroketal **1c** as substrates. In both cases the monoacetates obtained by lipase-catalyzed hydrolysis had an enantiomeric excess >98% *ee* and yields were improved to 88%. Thus the stereoselectivity of PCL increased, although variation of the substrate structure occurred far away from the ester group. We proposed that this change in stereoselectivity is due to a reduced flexibility in the conformation of benzylether **1b** and spiroketal **1c**, because all three substituents at the tetrahydropyrane ring adopt an equatorial position.

The same principle of 'substrate engineering' was also applied to the resolution of 1,2-*O*-protected glycerol derivatives, but the increase in stereoselectivity was only modest, *e.g.* the enantioselectivity increased from 1.1 to 8.5 *E* [7, 8].

## 2.7.1.2 Directed evolution of an esterase

In the previous section, modification of the substrate structure resulted in a substantial increase in stereoselectivity. However, this approach is not always feasible, especially when the structure is set as in the total synthesis of natural products and especially, if wild-type hydrolases do not show any activity at all. In such a case, the problem of non-reactivity might be overcome by introducing mutations into the hydrolase gene in order to alter their substrate spectra. In principle, this could be done by either a rational protein design or by random mutagenesis. The major disadvantages of a rational design by site-directed mutagenesis are the need of structural data of the enzyme and a detailed knowledge of the mechanism of catalysis. Furthermore, site-directed mutagenesis is a time-intensive method, which allows the production of relatively few enzyme variants within a reasonable time. Recently, the directed evolution of enzymes was described as a new and very elegant approach to generate and identify new enzyme variants. Principles and examples can be found in several recent reviews [9]. Prerequesites for a successful directed evolution are an effective mutation strategy for the improvement of the enzymes, the functional expression of the protein in a suitable microbial host, and a fast and reliable assay system for the identification of enzyme variants with desired properties out of a pool of $10^4$ to $>10^6$ variants. A range of methods for the generation of enzyme libraries is available, and in most cases researchers used error-prone PCR for this purpose. However, the assay system usually represents the major bottle-neck. Desired hydrolase variants were identified by use of chromogenic esters to identify more active *p*-nitrobenzyl esterases [10] or to increase the stereoselectivity of a lipase [11].

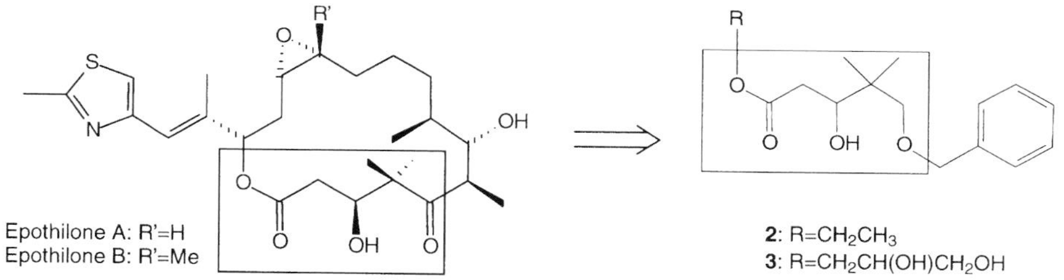

**Scheme 2.7.3.** Structures of Epothilones and retrosynthetic analysis to 3-hydroxy esters **2–3**.

Our interest in directed evolution started from the attempt to resolve 3-hydroxy ester **2** (Scheme 2.7.3), that shows close structural similarity to a key building block in Epothilones, a new class of macrolides, which show taxol-like biological activity [12]. Previously, we have shown the successful resolution of aliphatic [13] and arylaliphatic [14] 3-hydroxy esters using lipases or esterases. Unfortunately, none out of 18 lipases and two esterases showed any activity towards **2**, neither in hydrolysis in phosphate buffer nor in acylation with vinyl acetate in toluene [15]. In order to overcome this problem we decided to evolve new enzymes by directed evolution of an esterase from *Pseudomonas fluorescens* (PFE, expressed in *Escherichia coli* [16]). Esterase libraries were created by using a commercial mutator strain from *Epicurian coli* XL1-Red [17]. These were then assayed by means of an agar plate assay in the

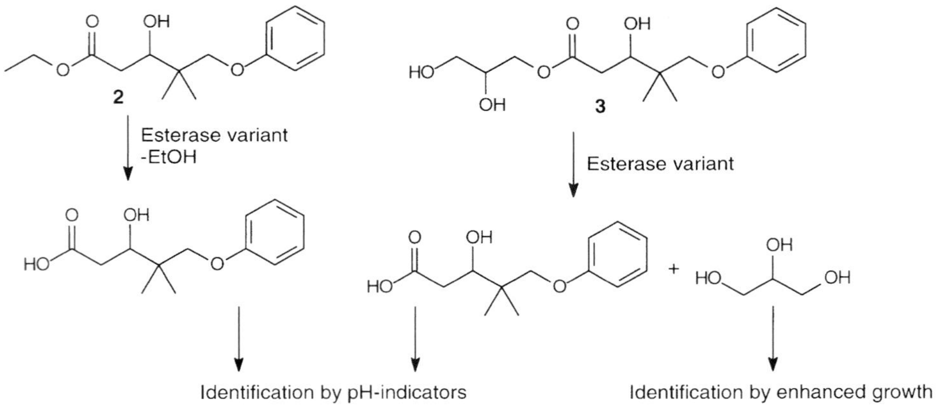

**Scheme 2.7.4.** Assay systems used for the identification of active esterase variants.

presence of pH indicators (neutral red and crystal violet) in combination with a growth assay based on the release of the carbon source glycerol from **3** facilitating growth on minimal media (Scheme 2.7.4).

After incubation for 2–6 days, two colonies were identified, which had turned red on both substrates. Moreover, the red colony on the plate containing the glycerol ester was significantly larger (diameter ~3 mM) compared to non-red colonies (diameter <1 mM) on the same plate. The plasmid was isolated using the colony from the master plate, transformed into *E. coli* and the esterase produced and isolated. This esterase variant was then subjected to a preparative hydrolysis of ethyl ester **2** and samples were analyzed by gas chromatography using a chiral column. Indeed, this esterase variant stereoselectively hydrolyzed **2**, resulting in 25% *ee* for the remaining substrate, which corresponds to an enantioselectivity of approx. $E = 5$. This is in accordance with the low enantioselectivity observed with PFE towards other chiral substrates [18]. Sequencing of this esterase variant revealed that two point mutations (A209D, L181V) were introduced.

The directed evolution of the esterase demonstrates that it is possible to alter the substrate specificity of an enzyme. Only two mutations were necessary to tip the balance from a non-substrate to a stereoselective hydrolysis of a sterically–hindered 3-hydroxy ester. However, in this specific case, further improvements in the enantioselectivity of PFE are required to allow the preparation of optically pure 3-hydroxy ester.

**Acknowledgements:** I am especially grateful to my colleagues at the Institute of Organic Chemistry, University of Hannover (Prof. H. M. R. Hoffmann, Prof. H. H. Meyer and Dr. T. Lampe) and to Dr. J. Altenbuchner, Institute for Industrial Genetics, University of Stuttgart). In addition, I would like to thank Prof. R. D. Schmid (Institute for Technical Biochemistry, University of Stuttgart) for useful discussions.

# References

[1] (a) U. T. Bornscheuer, R. J. Kazlauskas, *Hydrolases in Organic Synthesis – Regio- and Stereo-selective Biotransformations*, VCH-Wiley, Weinheim, **1999**; (b) R. D. Schmid, R. Verger, *Angew. Chem. Int. Ed. Engl.* **1998**, *37*, 1608–1633.

[2] R. J. Kazlauskas, A. N. E. Weissfloch, A. T. Rappaport, L. A. Cuccia, *J. Org. Chem.* **1991**, *56*, 2656–2665.

[3] A. N. E. Weissfloch, R. J. Kazlauskas, *J. Org. Chem.* **1995**, *60*, 6959–6969.

[4] (a) X. K. Holdgrün, C. J. Sih, *Tetrahedron Lett.* **1991**, *32*, 3465–3468; (b) H. Ebiike, Y. Terao, K. Achiwa, *Tetrahedron Lett.* **1991**, *32*, 5805–5808.

[5] H. Ebiike, K. Maruyama, K. Achiwa, *Tetrahedron: Asymmetry* **1992**, *3*, 1153–1156.

[6] (a) T. F. J. Lampe, H. M. R. Hoffmann, U. T. Bornscheuer, *Tetrahedron: Asymmetry* **1996**, *7*, 2889–2900; T. F. J. Lampe, H. M. R. Hoffmann, *Tetrahedron Lett.* **1996**, *37*, 7695–7698.

[7] C. S. Chen, Y. Fujimoto, G. Girdaukas, C. J. Sih, *J. Am. Chem. Soc.* **1982**, *104*, 7294–7299.

[8] L. Gaziola, U. Bornscheuer, R. D. Schmid, *Enantiomer* **1996**, *1*, 49–54.

[9] (a) F. H. Arnold, *Acc. Chem. Res.* **1998**, *31*, 125–131; (b) U. T. Bornscheuer, *Angew. Chem. Int. Ed. Engl.* **1998**, *37*, 3105–3108; (c) S. Harayama, *Trends Biotechnol.* **1998**, *16*, 76–82; (d) O. Kuchner, F. H. Arnold, *Trends Biotechnol.* **1997**, *15*, 523–530; (e) W. P. C. Stemmer, *Bio/Technology* **1995**, *13*, 549–553.

[10] J. C. Moore, F. H. Arnold, *Nature Biotechnol.* **1996**, *14*, 458–467.

[11] M. T. Reetz, A. Zonta, K. Schimossek, K. Liebeton, K.-E. Jaeger, *Angew. Chem. Int. Ed. Engl.* **1997**, *36*, 2830–2832.

[12] (a) K. C. Nicolaou, F. Roschangar, D. Vourloumis, *Angew. Chem. Int. Ed. Engl.* **1998**, *37*, 2014–2045; (b) L. Wessjohann, *Angew. Chem. Int. Ed. Engl.* **1997**, *36*, 715–718.

[13] U. Bornscheuer, A. Herar, L. Kreye, V. Wendel, A. Capewell, H. H. Meyer, T. Scheper, F. N. Kolisis, *Tetrahedron: Asymmetry* **1993**, *4*, 1007–1016.

[14] K. Wünsche, U. Schwaneberg, U. T. Bornscheuer, H. H. Meyer, *Tetrahedron: Asymmetry* **1996**, *7*, 2019–2022.

[15] (a) U. T. Bornscheuer, J. Altenbuchner, H. H. Meyer, *Biotechnol. Bioeng.* **1998**, *58*, 554–559; (b) U. T. Bornscheuer, J. Altenbuchner, H. H. Meyer, *Bioorg. Med. Chem.* in press.

[16] (a) N. Krebsfänger, F. Zocher, J. Altenbuchner, U. T. Bornscheuer, *Enzyme Microb. Technol.* **1998**, *22*, 641–646; (b) I. Pelletier, J. Altenbuchner, *Microbiology* **1995**, *141*, 459–468.

[17] A. Greener, M. Callahan, B. Jerpseth, *Methods Mol. Biol.* **1996**, *57*, 375–385.

[18] N. Krebsfänger, K. Schierholz, U. T. Bornscheuer, *J. Biotechnol.* **1998**, *60*, 105–111.

# 3 Carbohydrate Chemistry and Glycobiology

## 3.1 Multivalent neoglycoconjugates for the inhibition of mannose-sensitive carbohydrate-protein interactions

*Thisbe K. Lindhorst*

Cells are surrounded by carbohydrates of diverse complexity. Eukaryotic cells typically expose branched, so-called multiantennary oligosaccharides on their surfaces, which are covalently bound to lipids (glycolipids) or proteins (glycoproteins), respectively. The complex sugar portions in these glycoconjugates encode biological information by providing a three-dimensional arrangement of functionalities. This can eventually be recognized by specialized, carbohydrate-specific proteins, called lectins and selectins. Investigating carbohydrate recognition, it was observed, that simple monosaccharides are sufficient *in vitro*-probes to distinguish between carbohydrate specificities of lectins.

Interactions of carbohydrate ligands and their lectin receptors are of essential importance for the communication between cells in numerous cases, and can trigger crucial cascades of biological events [1] such as transduction of biochemical signals to the insides of cells or recruitment of leukocytes to sites of injury and inflammation [2]. In addition, microbes also often use carbohydrate-protein interactions to adhere to the surface of potential host cells. To interfere with the latter process is a promising entry to the development of anti-microbial agents, as the adhesion of microbes to their host cells is a prerequiste for infection [3].

To understand the molecular details of carbohydrate-protein interactions is a central goal in glycobiology. One of the prominent approaches is the synthesis of oligosaccharide and glycoconjugate analogs to interfere with the interactions in question, both to study their biological roles as well as to investigate their therapeutic potential.

### 3.1.1 Multivalency in carbohydrate-protein interactions

Quite characteristically, singular carbohydrate-protein interactions posess weak binding constants in the millimolar or high micromolar range, in contrast to, for example, protein-protein interactions with nanomolar dissociation constants ($K_D$s) [4]. To become effective, carbohydrate-protein interactions are designed as multivalent contacts [5]. Multivalency in carbohydrate-protein interactions is provided by the multiantennary oligosaccharide parts of cell surface glycoconjugates on one hand, and by clustered carbohydrate recognition domains (CRDs) of proteins on the

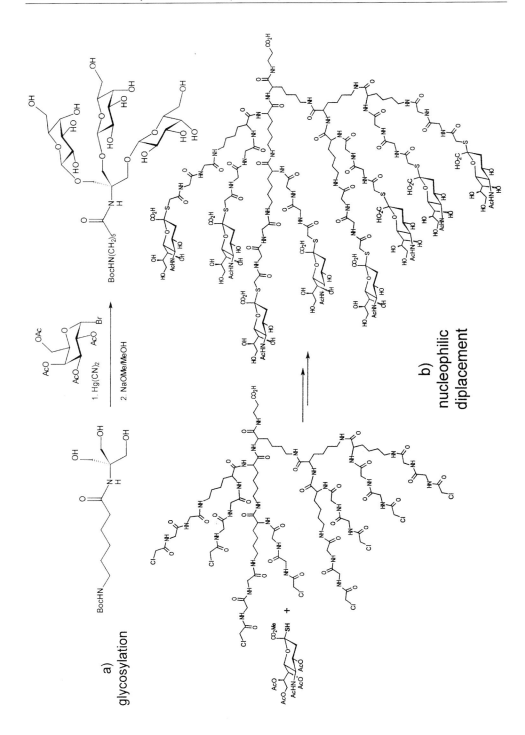

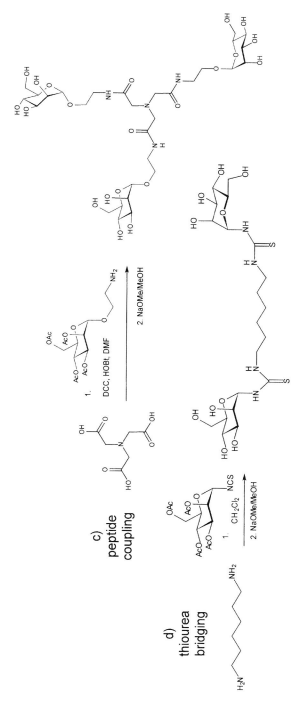

**Scheme 3.1.1.** Examples for the synthesis of glycoclusters by different synthetic strategies. (a) Synthesis of cluster galactosides by glycosylation of tris(hydroxy-methyl)methylamine (TRIS) [6]. (b) Synthesis of a polylysine-based neuraminic acid-containing glycodendrimer by nucleophilic displacement reaction [11]. Cluster glycosides obtained by peptide coupling [12]. (d) Synthesis of thiourea-bridged glycoclusters from core amines and glycosyl isothiocyanates [12, 14].

c)
peptide
coupling

d)
thiourea
bridging

other. Multivalency bears several biological advantages. It enables fine-tuning of biological response and allows a broad contact of biological surfaces, which can for instance induce their supramolecular rearrangement, followed by signal transduction. For the design of high-affinity, or high-avidity carbohydrate probes to interfere in the respective recognition systems, the role of multivalency in carbohydrate recognition has to be considered.

## 3.1.2   Synthesis of multivalent neoglycoconjugates

The importance of multivalency in carbohydrate-protein interactions was first pointed out by Y. C. Lee, in the context of the asialoglycoprotein receptor of liver cells [6]. He showed that a linear increase of monosaccharides in synthetic glycoclusters (Scheme 3.1.1a) led to a logarithmic increase in binding potency, an observation for which the term 'cluster effect' was coined [7]. It is obvious that multivalent glycoconjugates have to be designed with regard to the spacing of clustered CRDs they are thought to bind to. Different classes of multivalent neoglycoconjugates have thus been elaborated and have been shown to act as high avidity inhibitors of carbohydrate-protein interactions in *in vitro*-tests in many cases [8]. While the nomenclature of multivalent neoglycoconjugates has become increasingly difficult and confused, a clear distinction can still be made between

(i)   structurally not exactly defined glycoconjugates containing an average number of carbohydrate moieties in a random spacial arrangement; and
(ii)  monodisperse, structurally exactly defined glycoclusters, with distinct molecular weights and carbohydrate content.

The first class of multivalent neoglycoconjugates comprises classical neoglycoconjugates, where carbohydrate epitopes are linked to proteins such as bovine serum albumin (BSA), and a group of polymeric neoglycoconjugates, which have been named (neo)glycopolymers (in contrast to the natural polysaccharides). For structurally distinct multivalent carbohydrate conjugates 'glycoclusters', 'glycodendrimers', and 'carbohydrate-containing dendrimers' are common terms [9].

In contrast to polydisperse or structurally ill-defined multivalent glycoconjugates, the biological testing of multivalent neoglycoconjugates with defined structures allows us to conclude about the molecular details of the investigated interaction in the sense of quantitative structure-activity relationships (QSAR studies). This paper is focused on the synthesis and testing of structurally defined, mannose-containing glycoclusters.

### 3.1.2.1   Chemistries for glycocluster synthesis

It was shown that for binding to clustered CRDs the complementary glycoclusters have not to be designed as extremely large compounds or curiosities of any other kind [10]. Often smaller glycoclusters and glycodendrimers compared favorably to the larger ones. Rather, it seems important to choose feasible synthetic strategies, to allow the easy variation of carbohydrate moieties, linkage types and saccharide spacing, valency, and flexibility.

Typical chemistries used for the synthesis of multivalent neoglycoconjugates include the following approaches:

(i)   glycosylation of branched polyols to form cluster glycosides (Scheme 3.1.1a);
(ii)  synthesis of glycoclusters by nucleophilic displacement reaction of chloro-acetylated core molecules and, for example, 2-thio-neuraminic acid derivatives (Scheme 3.1.1b);
(iii) to use peptide chemistry and, for instance, couple amino-functionalized carbo-hydrate derivatives to oligo-carboxylic acids (Scheme 3.1.1c) or vice versa [13]; and
(iv)  the high-yielding formation of thiourea derivatives by the reaction of amines with isothiocyanates was also well suited for the synthesis of multivalent neoglyco-conjugates [14] (Scheme 3.1.1d).

The latter approach was advantageous, as branched and hyperbranched polyamines on one hand and NCS-functionalized carbohydrate derivatives on the other are easily available in great variety.

The carbohydrate-dependent adhesion of bacteria is often mannose-specific. There-fore, mannose-containing glycoclusters and glycodendrimers were designed to inhibit mannose-sensitive bacterial adhesion and this is described in the following.

### 3.1.2.2  Core molecules for glycocluster and glycodendrimer synthesis

A variety of oligofunctional molecules with terminal alcohol, amino or carboxy func-tions, respectively, are suited as cores for the synthesis of glycoclusters. Several are commercially available; others are easily derived from purchasable material (Scheme 3.1.2).

Core amines, like diamines of different chain lengths or the triamine tris(2-amino-ethyl)amine (**6**) were further branched out to PAMAM-(polyamidoamine-)type dendrimers, following standard procedures of dendrimer chemistry [15]. The synthesis of PAMAM dendrimers involves Michael-type addition of primary amines to methyl acrylate, followed by amidation of the resulting $\beta$-substituted methyl esters by an excess of ethylene diamine. Thus, stepwise growth of a molecule is achieved with doubling the number of peripheral amino functionalities in every generation. The differently spaced tetra-, hexa- and octavalent hyperbranched amines **7–11** were synthesized (Scheme 3.1.3) and eventually coated with carbohydrates by thiourea bridging.

In addition to non-carbohydrate core molecules, suitably functionalized carbo-hydrate derivatives could also be designed as scaffolds for the clustering of saccharides. Starting from fully allylated glucose (allyl 2,3,4,6-tetra-*O*-allyl-α-D-glucopyranoside; **13**) the glucose-based penta-ol **14** was obtained by hydroboration and the penta-amine **15** was synthesized by photoaddition of cysteamine hydrochloride to all five allyl functions of glucoside **13** and isolated as penta-hydrochloride [16] (Scheme 3.1.4).

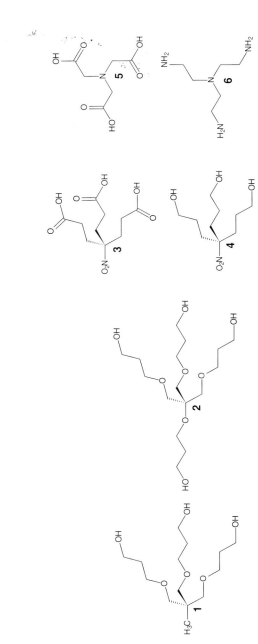

**Scheme 3.1.2.** Simple oligofunctional molecules are suitable cores for the synthesis of glycoclusters. Triol **1** was prepared from 2,2-bis-hydroxymethyl-propanol by hydroboration and tetra-ol **2** was obtained equally from pentaerythritol. Triacid **3** was derived from nitromethane and eventually was reduced to triol **24**. Titriplex I (**5**) and tris(2-aminoethyl)amine (**6**) are commercially available.

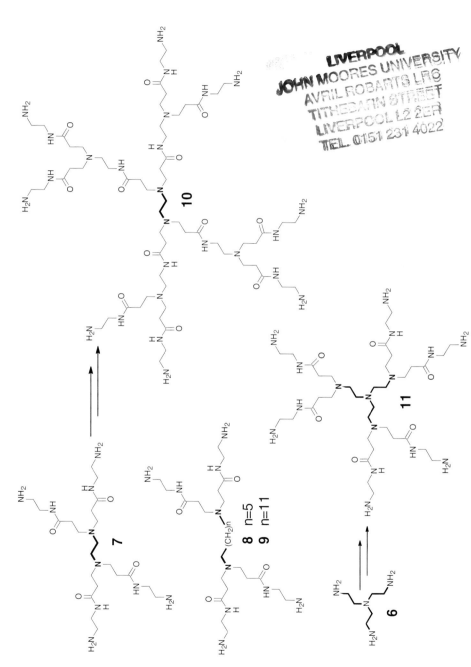

**Scheme 3.1.3.** Hyperbranched PAMAM-type polyamines were used as cores for the synthesis of thiourea-bridged glycoclusters. The octa-amine **10** was derived from **7**, the hexa-amine **11** from **6**.

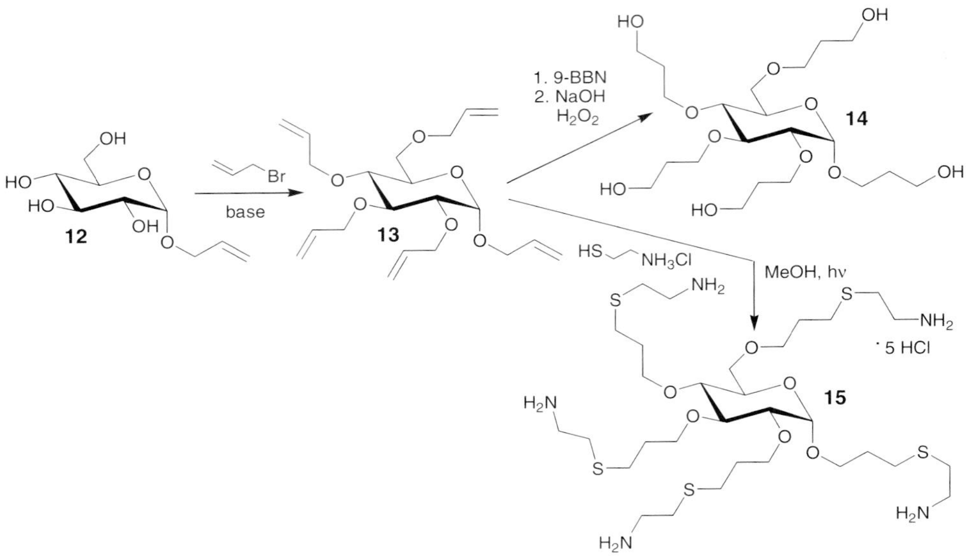

**Scheme 3.1.4.** Carbohydrate-based oligofunctional scaffolds for glycocluster synthesis were obtained from fully allylated glucose.

### 3.1.2.3    Synthesis of mannose-containing glycoclusters and glycodendrimers

*Cluster mannosides by glycosylation of oligoalcohols:* For the simultaneous manno-sylation of polyols, the glycosylation conditions have to be carefully optimized. Employing the acetylated mannosyl trichloroacetimidate **16** (Scheme 3.1.5) as the mannosyl donor, glycosyl orthoester clusters such as **17** were obtained instead of the desired cluster mannosides [17]. Trials to isomerize the tetrakis(orthoester) **17** to the respective tetrakis(glycoside) were not successful. When the benzoyl-protected tri-chloroacetimidate **18** was used for mannosylation of polyols, the reaction proceeded as desired and yielded cluster mannosides such as the trivalent cluster **19** after Zémplen deprotection of the acyl groups. Also the benzyl-protected mannosyl trichloro-acetimidate **20** was a suitable donor in oligo-mannosylation reactions as it was exemplified with the synthesis of **21** [18].

*Mannosyl clusters through the thiourea bridging strategy:* A broad variety of thiourea-bridged glycoclusters and glycodendrimers was synthesized employing a series of differently NCS-functionalized mannose derivatives. These were obtained in some variety as summarized in Scheme 3.1.6. The acetyl-protected α-mannosyl iso-thiocyanate **23** was prepared by melting KSCN with acetylated mannosyl bromide **22** [19]. *p*-Isothiocyanatophenyl α-D-mannoside (**25**) was synthesized from *p*-nitrophenyl α-D-mannoside (**24**), via reduction of the nitro group, followed by reaction with thiophosgene [20]. ω-NCS-functionalized α-mannosides such as **27–29** were obtained from the reaction of ω-azido-spacer glycosides with triphenyl phosphite and carbon disulfide. These methods make NCS-functionalized carbohydrate derivatives

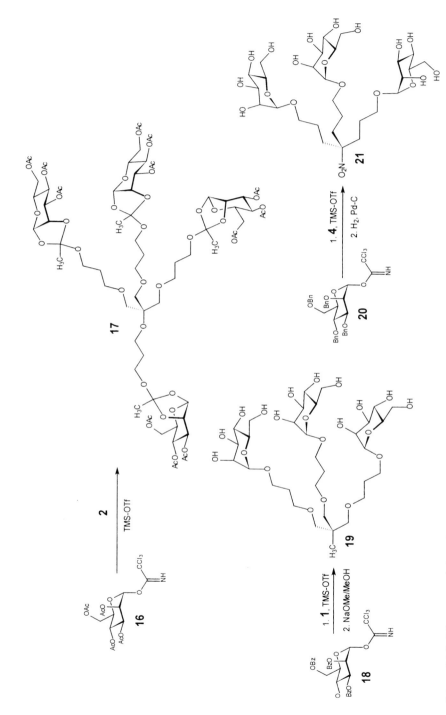

**Scheme 3.1.5.** Oligomannosylation of polyols to cluster α-D-mannosides was especially successful using trichloroacetimidates **18** and **20**. Glycosyl orthoesters were frequently formed, when the acetylated donor **16** was employed.

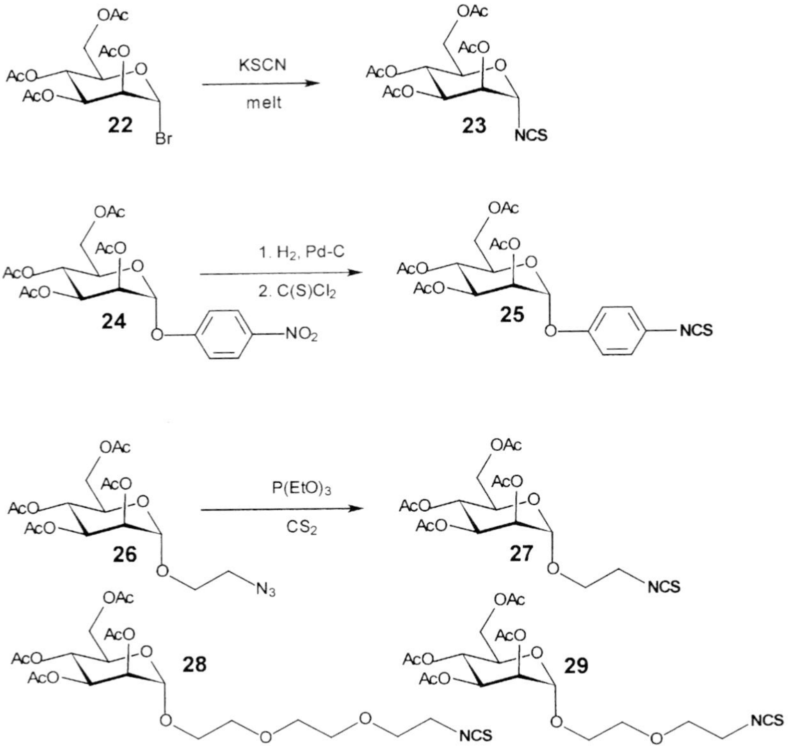

**Scheme 3.1.6.** A variety of isothiocyanato-functionalized carbohydrate derivatives were obtained using different starting materials and synthetic pathways. Mannosides **28** and **29** were prepared in analogy to **27**.

easily accessible, regardless of the sugar configuration. Even complex saccharides can equally well be turned into NCS-functionalized building blocks for glycocluster synthesis [21].

Some of the synthesized thiourea-bridged α-mannosyl clusters and dendrimers are depicted in Scheme 3.1.7. All core amines from Scheme 3.1.3 (**6–11**) were 'glycocoated' with mannosyl isothiocyanate **23** [12, 14], as drawn for the synthesis of the octavalent glycodendrimer **33**. Equally, the analogous trivalent glycocluster **30** was obtained from **6**, the tetravalent cluster **31** from **8**, and the hexavalent equivalent **32** from hexa-amine **11**. This glycocoating procedure consists of only two reaction steps, formation of the thiourea cluster and deacetylation. However, thiourea-bridged glycoclusters could even be prepared by a one-step reaction in water using unprotected *p*-nitrophenyl α-D-mannoside (**25**) and polyamines [20]. Thus, the depicted hexavalent phenyl mannoside cluster **35** was obtained directly from **11** and its trivalent analog **34** from **6**. Thiourea bridging was also used to prepare carbohydrate-centered glycoclusters as represented by **36** [22]. All synthesized multivalent glycomimetics were structurally perfect, as revealed by NMR and mass spectroscopic analysis [23].

*Non-classical cluster mannosides:* It has often been shown, that non-specific interactions added by aromatic or hydrophobic portions in glycoconjugates improve the overall binding capacity of a given ligand. This is illustrated by the approximate 100-fold higher binding potency of *p*-nitrophenyl α-D-mannoside (*p*NPMan) toward the type 1-fimbrial adhesin of *Escherichia coli* (vide supra) when compared to methyl α-D-mannoside (MeMan) [24]. Therefore, it is attractive also to include hydrophobic regions into clustered carbohydrate ligands, for instance, to introduce aromatic aglycone moieties.

To allow the variation of the aglycone portions in cluster glycosides, the carbohydrates must be linked differently to multivalent scaffolds than via their reducing ends. According to observations made with many lectins [25], these do not necessarily require the 6-hydroxy group of a sugar to recognize it as a ligand and specifically bind to it. Therefore, clustering of carbohydrate epitopes via the 6-position of the sugar ring was considered. This approach is an entry into a new class of neoglycoconjugates, realizing a linking mode of monomers which can be considered a 'reverse' one when compared to natural oligosaccharides.

Thus, 'glycoside donors' **38**, **40**, and **42** were synthesized (Scheme 3.1.8) by selective tosylation of the primary 6-OH group in starting α-D-mannosides, followed by azide substitution. The resulting protected or unprotected 6-deoxy-6-azido mannosides were either reduced to the amines or submitted to a Staudinger reaction to yield the 6-NCS-modified glycosides. Amine **38** and the isothiocyanates **40** and **42** served as mannoside donors in the peptide coupling reaction with the triacid **3** and in thiourea bridging with the triamine **6**, respectively, to yield the glycoclusters **43**, **44**, and **45** [18].

The mannoside donor **38** was also employed for the synthesis of cluster mannoside **46**, which carries biotin as a bio-label. (Scheme 3.1.9).

## 3.1.3   Testing of the antiadhesive properties of synthetic α-mannosyl clusters

As mentioned previously, adherence of microbes to their host cells is a crucial step prior to successful infection of the cell. The adhesion process is often carbohydrate-dependent and mannose-sensitive adhesion is widely distributed. The herein described mannose-containing glycoclusters and the employed synthetic strategies for their preparation have been elaborated to interfere in mannose-specific adhesion, both to investigate the details of binding as well as eventually to develop the most potent compounds toward therapeutic applications.

Bacteria bind to their host cells with the help of specific lectins, which they carry at surface organelles, called fimbriae or pili. Fimbriae are classified according to their carbohydrate specificities, type 1 fimbriae recognizing mannose-containing ligands. Type 1 fimbriae are common among enterobacteria, of which one of the most important is *Escherichia coli*. The bacterium expresses between 100 and 400 type 1 fimbriae on its cell surfaces, each of which is composed of a repeating major subunit, FimA (approximately 18 kDa), accounting for more than 98% of the fimbrial protein [26]. The FimH protein (approximately 32 kDa) is the mannose-specific adhesin [27],

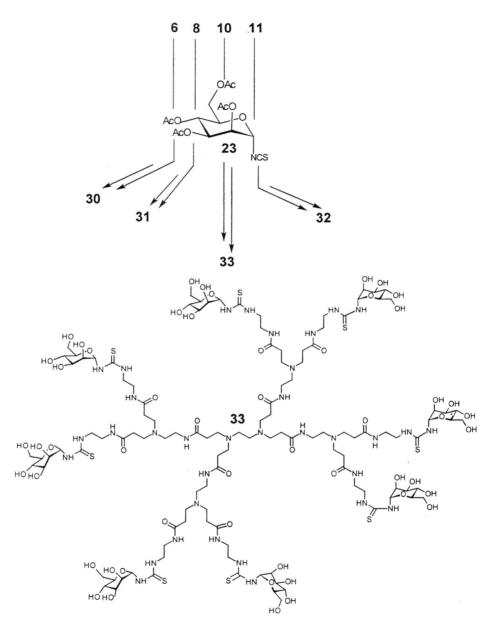

**Scheme 3.1.7.** Some of the synthesized thiourea-bridged α-mannosyl clusters and dendrimers are depicted. Glycoclusters **30**–**33** and **36** were obtained by the same procedure; the phenyl mannoside clusters **34** and **35** were prepared in water.

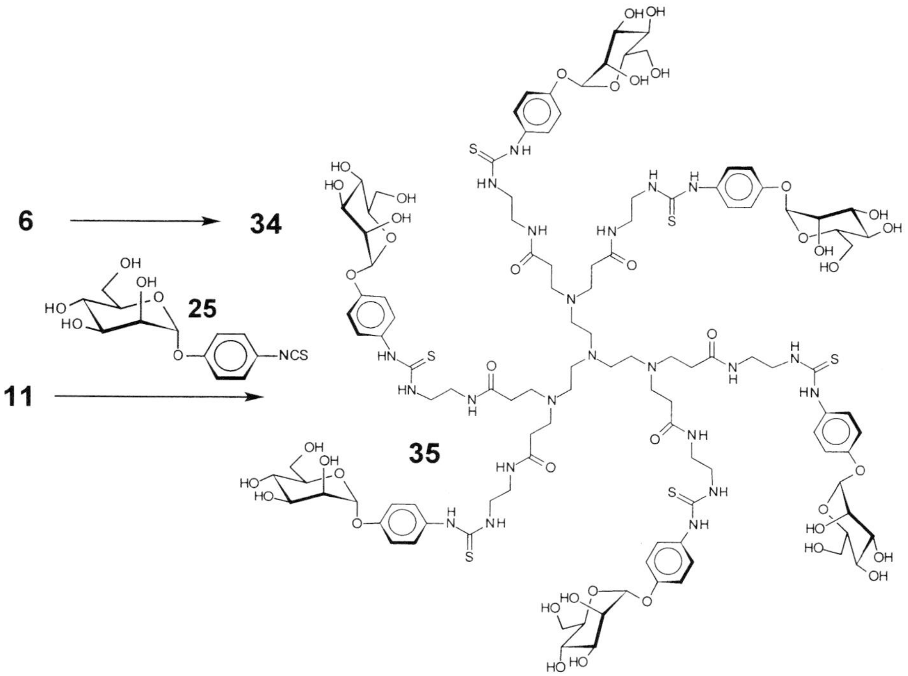
6
34
25
NCS
11
35

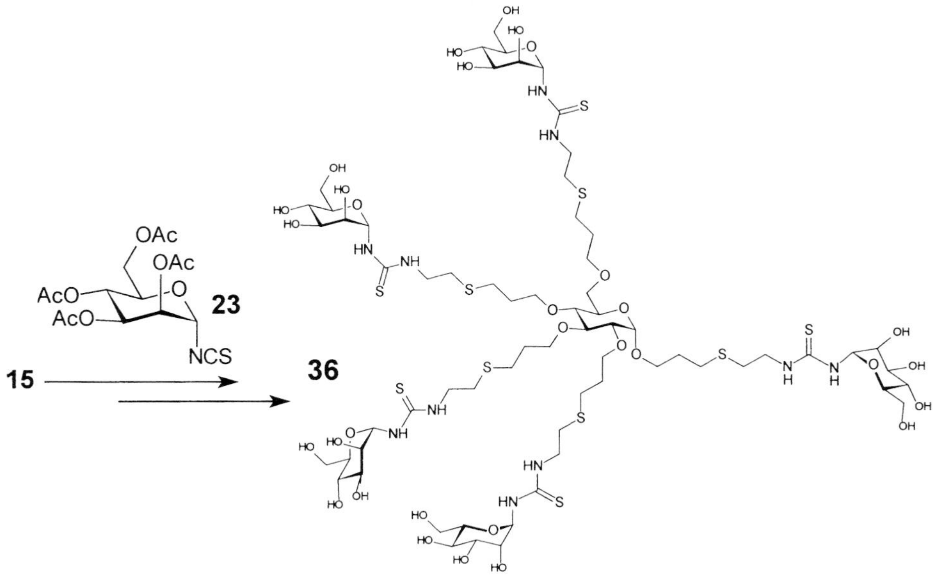
23
NCS
15
36

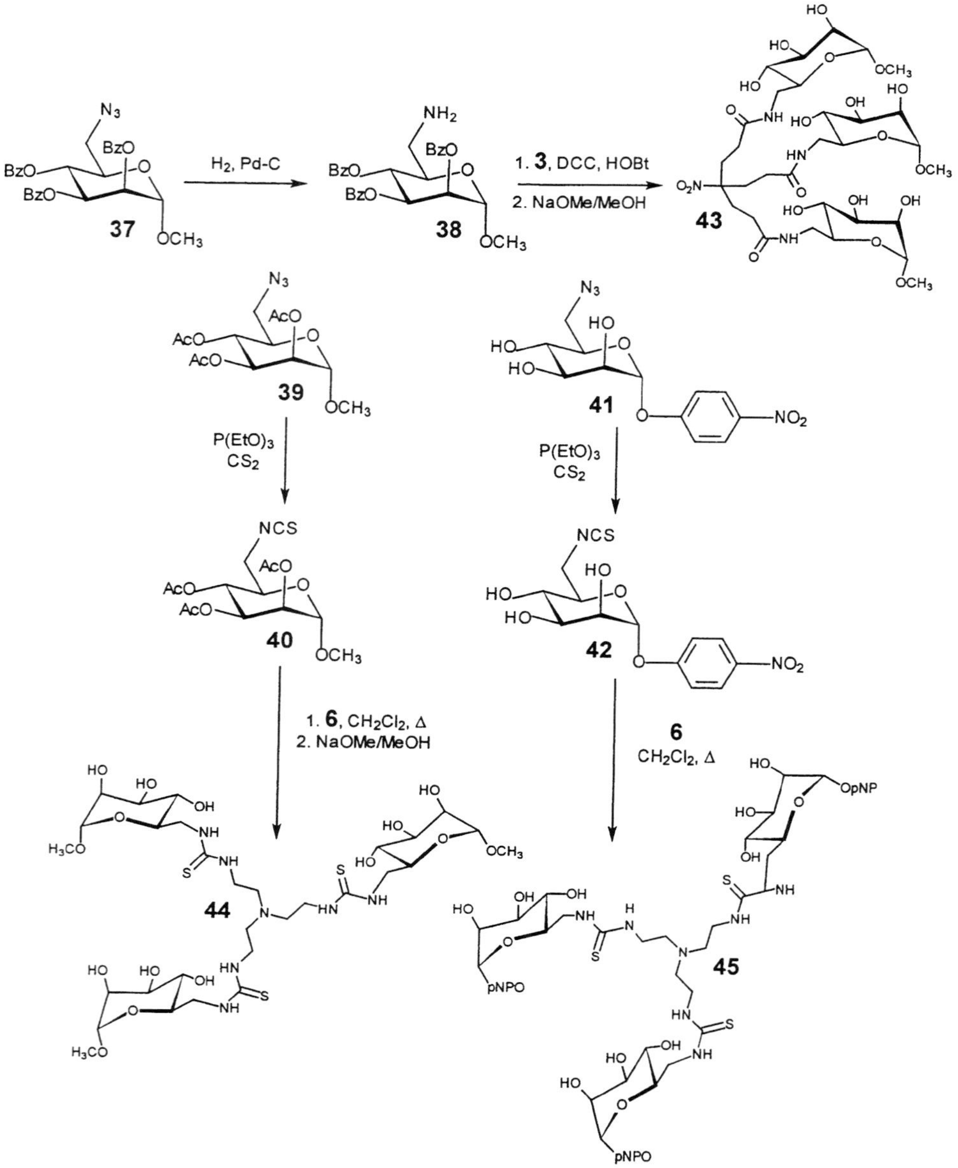

**Scheme 3.1.8.** 'α-Mannoside donors' were synthesized and clustered via the 6-position of the sugar ring. The trivalent cluster mannosides **44** and **45** differ only in their aglycone moieties.

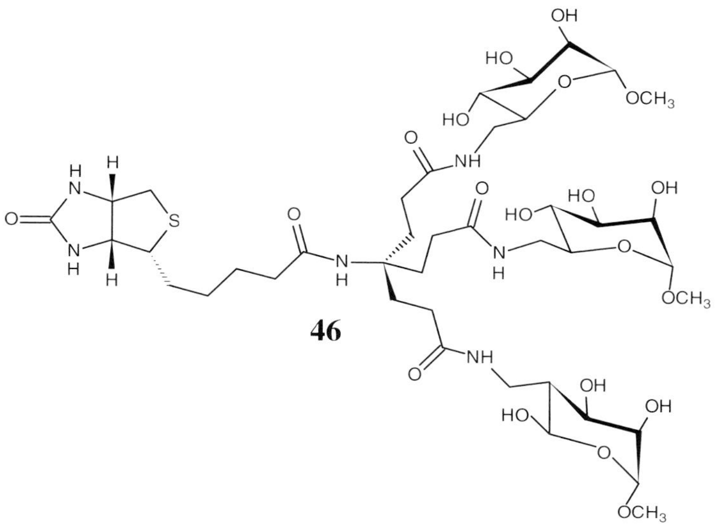

**Scheme 3.1.9.** Using molecules like **3** as cores for clustering, allows the synthesis of biolabeled glycoclusters such as the biotinylated cluster mannoside **46**.

where the carbohydrate-binding site is located. FimH is localized at the lateral ends of the fimbriae and is also distributed along their shafts [28].

*E. coli* is an abundant inhabitant of the intestinal tract of mammals and has been recognized as an important pathogen involved in a variety of intestinal and extra intestinal diseases. Different types of fimbriae are considered to be important in, for example, urinary tract infections, including P fimbriae and type 1 fimbriae; the latter also facilitate the adhesion of other fimbriated strains and thus contribute to virulence [29].

The majority of the herein described mannose clusters were tested as inhibitors of *E. coli* adhesion to erythrocytes. A recombinant type 1 fimbriated *E. coli* strain, *E. coli* HB101 (pPKl4) [30], was used for the tests.

Type 1 fimbriae-mediated adhesion of *E. coli* to erythrocytes can be inhibited by millimolar concentrations of methyl $\alpha$-D-mannoside (MeMan). This weak inhibitory potency was compared to the inhibitory effect of synthetic $\alpha$-mannosyl clusters of different valencies. The results are listed in Table 3.1.1.

Classical inhibition hemagglutination tests were used to determine the inhibition of the individual derivatives. The inhibition titer (IT) is the lowest concentration of the inhibitor at which no agglutination occurs by macroscopic inspection. Relative inhibition titers (RITs) were related to methyl $\alpha$-D-mannoside (MeMan) as standard (IT $\equiv$ 1). Clustering of $\alpha$-mannosyl residues in multivalent neoglycoconjugates such as **30–33** resulted in approximately up to 100-fold higher binding potency compared to MeMan. When the molar content of mannosyl residues in the clusters was considered, valcency-corrected RITs were obtained, revealing a maximal effect of clustering in the range of 35 in the documented testing series [12].

**Table 3.1.1.** Inhibitory potencies of α-mannosyl glycoclusters and glycodendrimers, as inhibitors of the hemagglutination of guinea pig erythrocytes by type 1 fimbriated *E. coli* HB 101 (pPK14). As controls, *p*-nitrophenyl α-D-glucopyranoside, the glucosyl analog of the trivalent mannosyl cluster **30** and non-functionalized core molecules were used. Non of these displayed any inhibitory capacity.

| Compound tested | IT [μM] | RIT | RIT based on moles mannose |
|---|---|---|---|
| MeMan | 9600 | 1 | 1 |
| *p*NPMan | 72 | 133 | 133 |
| **30** | 91 | 105.5 | 35.17 |
| **31** | 260 | 39 | 9.75 |
| **32** | 91 | 105.5 | 17.58 |
| **33** | 83 | 115.6 | 14.45 |

However, an equally large effect as received by clustering is also met by an aromatic mannoside such as *p*-nitrophenyl α-D-mannoside (*p*NPMan), as the *p*-nitrophenyl moiety adds to receptor binding by non-specific hydrophobic interactions [24]. Consequently, mannosyl clusters were further developed in order to combine favorable binding effects. A collection of selected data is listed in Table 3.2.2.

**Table 3.2.2.** Average $IC_{50}$ values of cluster α-D-mannosides as determined by ELISA.

| Cmp | MeMan | *p*NPMan | **34** | **35** | **21** | **43** | **44** | **45** |
|---|---|---|---|---|---|---|---|---|
| $IC_{50}$[μM] | 3500 | 49 | 5.1 | 9.8 | 39 | 11 | 5333 | 429 |

In this case an ELISA permitted the measurement of $IC_{50}$ values for the inhibition of *E. coli* adhesion to yeast mannan. $IC_{50}$ values reflect the inhibitor concentration which causes 50% inhibition of bacterial binding to yeast mannan. The results depicted show that clustering of phenyl α-mannoside, such as in **34** and **35**, improved the binding potency, when compared to *p*NPMan. However, when considering the content of phenyl mannoside in these clusters the effect appears rather small. Interestingly, some of the trivalent clusters (**21**, **43**) performed very well as antiadhesives against *E. coli* with $IC_{50}$ values in the low micromolar range. Furthermore, the inhibitory potencies of the latter two clusters were even better in inhibition hemagglutination tests [18]. It is important to bear such differences in mind when *in vitro* results are used as basis for the development of useful compounds for *in vivo* situations.

Although the thiourea-bridged methyl mannoside cluster **44** was an extremely weak inhibitor, the observation is promising that its analog **45** with *p*NP instead of methyl aglycon moieties is more than 100-fold more potent. This finding underlines the usefulness of the 'reverse' clustering concept.

### 3.1.4    Conclusion

It was shown that among other, more classical synthetic strategies, thiourea bridging served as an especially powerful and flexible method for the synthesis of glycoclusters. It allows the easy variation of carbohydrate epitopes, so that multivalent conjugates for the investigation of other adhesion systems are easily prepared [31]. Furthermore, it facilitates the modification of spacer properties and is compatible with a variety of concepts for the design and optimization of multivalent glycoclusters. A series of $\alpha$-mannosyl-containing glycoclusters revealed interesting effects when tested as anti-adhesives with *E. coli*. The most potent structures will be developed further in order to improve the conformational and dynamic properties of these carbohydrate ligands.

**Acknowledgements:** This work was financed by the Deutsche Forschungsgemeinschaft (SFB 470) and the Fonds der Chemischen Industrie (FCI). I would like to thank Prof. Dr. Joachim Thiem for his support and I wish to express my gratitude to my committed coworkers, Dr. Christoffer Kieburg, Sven Kötter and Michael Dubber, who have contributed substantially to the herein summarized results. I am also indebted to my collegue Dr. Ulrike Krallmann-Wenzel, who performed the assays with *E. coli*.

## References

[1]   (a) A. Varki, *Glycobiology* **1993**, *3*, 97–130; (b) M. Fukuda, *Bioorg. Med. Chem.* **1995**, *3*, 207–215.

[2]   L. A. Larsky, *Annu. Rev. Biochem.* **1995**, *64*, 113–139.

[3]   (a) E. H. Beachey, *J. Infect. Dis.* **1981**, *143*, 325–345; (b) J. Hacker, *Curr. Top. Microbiol. Immunol.* **1990**, *151*, 1–27; (c) K.-A. Karlsson, *Curr. Opin. Struct. Biol.* **1995**, *5*, 622–635.

[4]   E. Toone, *Curr. Opin. Struct. Biol.* **1994**, *4*, 719–728.

[5]   (a) L. L. Kiessling, N. L. Pohl, *Chem. Biol.* **1996**, *3*, 71–77; (b) M. Mammen, S.-K. Choi, G. M. Whitesides, *Angew. Chem. Int. Ed.* **1998**, *37*, 2754–2794.

[6]   (a) Y. C. Lee, *Carbohydr. Res.* **1978**, *67*, 509–514; (b) R. T. Lee, Y. C. Lee, *Methods Enzymol.* **1987**, *138*, 424–429.

[7]   Y. C. Lee, R. T. Lee, *Acc. Chem. Res.* **1995**, *28*, 321–327.

[8]   R. Roy, *Curr. Opin. Struct. Biol.* **1996**, *6*, 692–702.

[9]   (a) R. Roy in *Carbohydrate Chemistry* (G. van Boons, Ed.) Wiley, **1998**; (b) Th. K. Lindhorst, *Nachr. Chem. Techn. Lab.* **1996**, *44*, 1073–1079; (c) N. Jayaraman, S. A. Nepogodiev, J. F. Stoddart, *Chem. Eur. J.* **1997**, *3*, 1193–1199.

[10]  P. R. Ashton, E. F. Hounsell, N. Jayaraman, T. M. Nilsen, N. Spence, J. F. Stoddart, M. Young, *J. Org. Chem.* **1998**, *63*, 3429–3437.

[11]  R. Roy, D. Zanini, J. Meunier, A. Romanowska, *J. Chem. Soc., Chem. Commun.* **1993**, 1869–1890.

[12]  Th. K. Lindhorst, C. Kieburg, U. Krallmann-Wenzel, *Glycoconjugate J.* **1998**, *15*, 605–613.

[13]  K. Aoi, K. Itoh, M. Okada, *Macromolecules* **1995**, *28*, 5396–5393.

[14]  Th. K. Lindhorst, C. Kieburg, *Angew. Chem. Int. Ed. Engl.* **1996**, *35*, 1953–1956.

[15]  (a) E. Buhleier, W. Wehner, F. Vögtle, *Synthesis* **1978**, 155–158; (b) A. D. Meltzer, D. A. Tirrell, A. A. Jones, P. T. Inglefield, D. M. Hedstrand, D. A. Tomalia, *Macromolecules* **1992**, *25*, 4541–4548.

[16]  M. Dubber, Th. K. Lindhorst, *Carbohydr. Res.* **1998**, *310*, 35–41.

[17]  (a) Th. K. Lindhorst, to be published; (b) Th. K. Lindhorst, Carbohydrate Bioengineering Meeting, Helsingør, 1995: *En Route to New Cluster Glycosides: The Synthesis of a Tetra-antennary α-Mannosides Derived from Pentaerythritol*, abstract P6.

[18]  S. Kötter, U. Krallmann-Wenzel, S. Ehlers, Th. K. Lindhorst, *J. Chem. Soc., Perkin Trans. 1* **1998**, 2193–2200.

[19]  Th. K. Lindhorst, C. Kieburg, *Synthesis* **1995**, 1228–1230.

[20]  C. Kieburg, Th. K. Lindhorst, *Tetrahedron Lett.* **1997**, *38*, 3885–3888.

[21]  Th. K. Lindhorst, M. Ludewig, J. Thiem, *J. Carbohydr. Chem.* **1998**, *17*, 1131–1149.

[22]  C. Kieburg, M. Dubber, Th. K. Lindhorst, *Synlett* **1997**, 1447–1449

[23]  V. Havlíček, C. Kieburg, P. Novák, K. Bezouška, Th. K. Lindhorst, *J. Mass Spectrom.* **1998**, *33*, 591–598.

[24]  (a) N. Firon, I. Ofek, N. Sharon, *Carbohydr. Res.* **1983**, *120*, 235–249; (b) N. Sharon, *FEBS Lett.* **1987**, *217*, 145–157.

[25]  K. Drickamer, *Curr. Opin. Struct. Biol.* **1995**, *5*, 612–616.

[26]  (a) P. Klemm, *Eur. J. Biochem.* **1984**, *143*, 395–399; (b) P. Klemm, *Rev. Infect. Dis.* **1985**, *7*, 321–340.

[27]  E. V. Sokurenko, H. S. Courtney, D. E. Ohman, P. Klemm, D. L. Hasty, *J. Bacteriol.* **1994**, *176*, 748–755.

[28]  (a) K. A. Krogefelt, H. Bergmany, P. Klemm, *Infect. Immun.* **1990**, *58*, 1995–1998; (b) P. Klemm, K. A. Krogfelt, In *Fimbriae, Adhesion, Genetics, Biogenesis and Vaccines* (P. Klemm, Ed.), CRC Press, Boca Raton **1994**, pp 9–26.

[29]  H. Connell, W. Agace, P. Klemm, M. Schembri, S. Mårlid, C. Svanborg, *Proc. Natl. Acad. Sci. USA* **1996**, *93*, 9827–9832.

[30]  P. Klemm, B. J. Jørgensen, I. van Die, H. de Ree, H. Bergmans, *Mol. Gen. Genet.* **1985**, *199*, 410–414.

[31]  K. Bezouška, V. Křen, C. Kieburg, Th. K. Lindhorst, *FEBS Lett.* **1998**, *426*, 243–247.

## 3.2     Octyl *O*- and *S*-glycosides related to the GPI anchor of *Trypanosoma brucei* as probes for in vitro galactosylation by trypanosomal α-galactosyltransferases

*Thomas Ziegler, Ralf Dettmann, and Michael Duszenko*

Membrane attachment of eukaryotic membrane proteins occurs either by membrane-spanning hydrophobic peptide domains or by a glycosyl-phosphatidylinositol (GPI) anchor, which is covalently linked to the C-terminus of the protein. The latter was first described by Ferguson et al. and one of us (M. D.) as a membrane anchor motif of variant surface glycoprotein (VSG) [1], a surface protein, which forms a protective barrier around the bloodstream forms of the protozoan parasite *Trypanosoma brucei*. GPI-anchored proteins are found in all classes of eukaryotes [2] and VSG accounts for about 10% of the total trypanosomal proteins and is thus perfectly suited to investigate the GPI structure in detail. Indeed, VSG was the first protein of which the GPI structure was completely determined. However, a clear and defined biological function of the GPI anchor has still to be elucidated, but it is noteworthy that GPI-anchored proteins are involved in important processes of cell–cell interaction, cell adhesion, and immunoresponse [3].

Further studies of GPI structures obtained from protozoa (such as *T. brucei*-VSG [2, 4] and *T. crucei*-1G7 [4, 5]), fish (such as Torpedo-acetylcholinesterase (AchE) [6]), and mammalian sources (such as rat brain Thy-1 [7] and human AchE [8]) revealed that obviously all GPI anchors contain a highly conserved pentasaccharide core structure ABCDE (Figure 3.2.1). In the case of the VSG GPI anchor, this core structure is, depending on the trypanosomal clone investigated, differently galacto-sylated. These variable galactose side-chains may be important for the integrity of the VSG coat in order to serve as a molecular diffusion barrier. Since several hundred

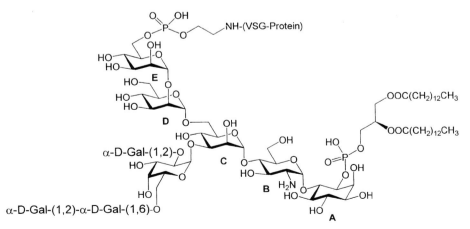

**Figure 3.2.1.** Structure of the GPI anchor of *Trypanosoma brucei* VSG.

different VSG variants are encoded within the trypanosomal genome which will be sequentially expressed during a persisting infection (antigenic variation), galactose residues could perform space-filling functions in order to compensate for differences in the three-dimensional VSG structures [9]. Thus, the respective galactosyltrans-ferases are of fundamental interest as possible targets for a new anti-trypanosomal strategy and synthetic fragments related to trypanosomal GPI anchors may serve as probes for revealing the biosynthesis of these structures.

In the case of *T.brucei* variant MITat 1.4 strain 427 [10], the core structure of the GPI anchor is modified by a diantenna $\alpha$-(1 $\rightarrow$ 3)-linked with the mannosyl residue C (Figure 3.2.1). The enzymes involved in the biosynthetic pathway are still unknown, but it was reported recently that octyl 1-thio-mannopyranosides may serve as suitable substrates for trypanosomal $\alpha$-galactosyltransferases. [11, 12] Similarly, octyl *O*-glyco-sides have been shown to exhibit sufficient hydrophobicity to act as substrates for glycosyltransferases [13].

Here, the synthesis of a series of mono- to tetrasaccharide octyl *O*- and *S*-glycosides related to the core structure CDE and its galactosylated variants (Figure 3.2.1) is described. Additionally, the suitability of these glycosides to serve as synthetic sub-strates for $\alpha$-galactosyltransferases in membrane fractions from *T. brucei* has been determined.

## 3.2.1    Results and discussion

### 3.2.1.1    Synthesis of acceptors

Several syntheses of the complete structure of the GPI anchor of *T. brucei* have already been published [14–18]. Since we intended to prepare mono- to trisaccharide fragments and galactosylated counterparts as well of the core structure of the trypanosomal GPI anchor in order to demonstrate the possible enzymatic galactosylation step (*i.e.* at site C in Figure 3.2.1), we used the glycodesilylation protocol which was recently developed in our group [19] for the construction of the desired saccharides. Therefore, octyl *O*- and *S*-mannopyranosides **1** that were shown to function as substrates for *T. brucei* $\alpha$-galactosyltransferase [11, 12] were needed. Furthermore, octyl 4,6-*O*-(1,1,3,3-tetra-isopropyl-1,3-disiloxane-1,3-diyl)-$\alpha$-D-mannopyranosides derivatives **2** and their ring-opened counterparts **3** and **4** were required for that approach (Scheme 3.2.1).

To this end, D-mannose penta-acetate was converted with 1-octanol in the presence of SnCl$_4$ [20] into octyl 2,3,4,6-tetra-*O*-acetyl-$\alpha$-D-mannopyranoside (48%), deacety-lation of which gave octyl $\alpha$-D-mannopyranoside **1a**. Silylation of **1a** with 1,3-dichloro-1,1,3,3-tetraisopropyl-1,3-disiloxane followed by selective monobenzoylation afforded 2-*O*-benzoyl-4,6-*O*-disiloxane-protected derivative **2a** in a 45% yield over three steps. Due to the steric hindrance [19] of HO-3 in disiloxane-protected **1a**, HO-2 was benzoylated preferentially. However, a small amount of the corresponding 2,3-di-*O*-benzoyl derivative was also obtained. Since the latter was originally planned to be used as the glycosyl acceptor for the elongation of the sugar chain at O-6 by a glycodesilylation reaction, the same compound was also prepared from the monobenzoate **2a** by treatment with benzoyl bromide. [19, 21] Finally, compound **2a** and its fully protected counterpart were selectively converted into the mannosyl acceptors **3a** (84%) and **4a** (70%) by protodesilylation with HF in pyridine. [12, 19] In a similar sequence, the corresponding octyl 1-thio-$\alpha$-D-mannopyranoside derivatives **1b**, **2b**, **3b**, and **4b** were prepared.

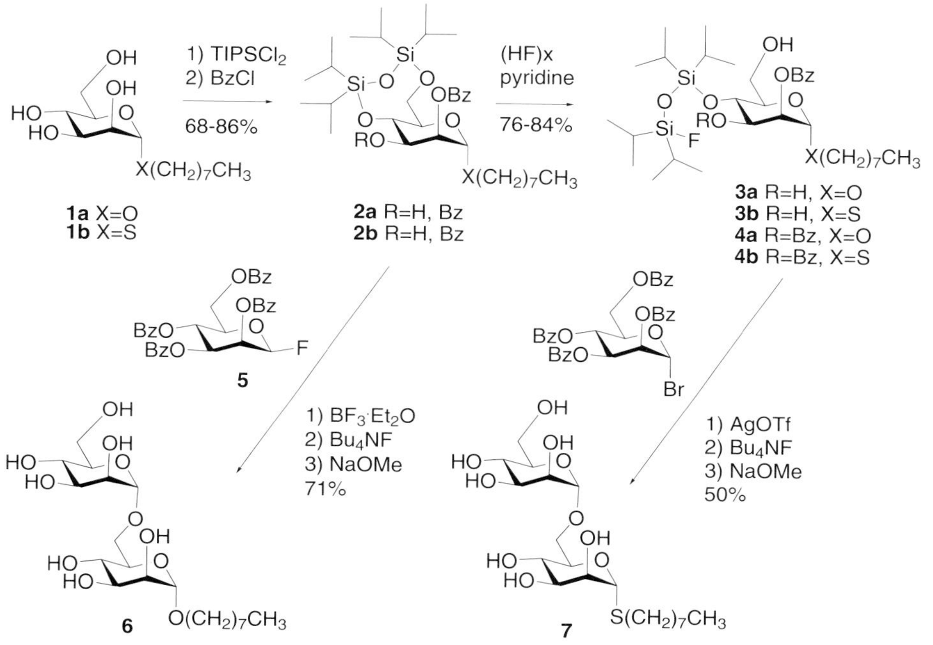

**Scheme 3.2.1.**

When compound **2a** was treated with 2,3,4,6-tetra-*O*-benzoyl-$\beta$-D-mannopyranosyl fluoride **5**, prepared in 81% yield from benzobromomannose with $KHF_2$, [22] and boron trifluoride diethyletherate as the catalyst, a clean glycodesilylation at O-6 of the acceptor occurred. The crude intermediate disaccharide was directly desilylated (tetrabutyl ammonium fluoride in THF) to give the $\alpha$-(1 → 6)-linked octyl glycoside disaccharide (72%) which was finally deacylated to afford disaccharide **6** in an almost quantitative yield. For the synthesis of the corresponding octyl 1-thio-mannoside disaccharide, compound **3b** was first condensed with benzobromomannose in the presence of silver trifluoromethanesulfonate, followed by fluoride-catalyzed desilylation and deacylation to give disaccharide **7** (50%). As was observed previously in similar cases [19] no mannosylation of O-3 in **3b** occurred.

For the construction of the corresponding trisaccharides, a blockwise synthesis was applied (Scheme 3.2.2). First, benzobromomannose was condensed with 1,3,4,6-tetra-*O*-acetyl-$\beta$-D-mannopyranose [23] to give the corresponding $\alpha$-(1 → 2)-linked disaccharide block (86%) which was converted into the disaccharide donor **8** (72%)

by sequential treatment with hydrazine acetate [24, 25], followed by trichloroaceto-
nitrile and $K_2CO_3$ [12]. Trimethylsilyl trifluoromethanesulfonate-catalyzed condensation of donor **8** with octyl *O*-mannoside **2a** proceeded in a clean reaction and afforded
trisaccharide **9a** in 72% yield after sequential removal of the fluorodisiloxane residue
at O-4 of the intermediate with tetrabutyl ammonium fluoride and deacylation.
Similarly, octyl *S*-mannoside **2b** was condensed with donor **8** and furnished the desired
trisaccharide **9b** in 55% yield [12].

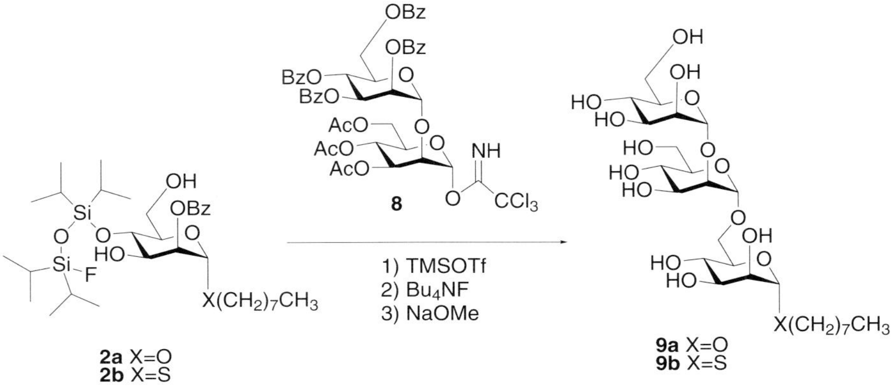

**Scheme 3.2.2.**

For the construction of the required galactosylated fragments compound **2a** was
first condensed with ethyl 2,3,4,6-tri-*O*-benzyl-1-thio-$\beta$-D-galactopyranoside (46%)
followed by HF-catalyzed ring opening of the intermediate to afford disaccharide
block **10** in 90% yield (Scheme 3.2.3). The latter served as a universal building
block for several galactosylated GPI anchor fragments as follows [12]. Desilylation
of **10** followed by deacylation and hydrogenolysis of the benzyl groups afforded disaccharide **11** (70%). AgOTf-promoted mannosylation of **10** with benzobromomannose
followed by desilylation gave first the corresponding trisaccharide (78%). Next,
deblocking of the latter afforded compound **12** in 44% yield over two steps. Finally,
in a similar sequence, disaccharide block **10** was condensed with imidate **8** in 48%
yield. Desilylation, deacylation, and debenzylation then afforded the tetrasaccharide
**13** in 69% overall yield.

   All nine saccharides **1a**, **1b**, **6**, **7**, **9a**, **9b**, **11**, **12**, and **13** were used for the enzymatic
galactosylation by a *T. brucei* $\alpha$-galactosyltransferase as described below.

### 3.2.1.2 Enzymatic $\alpha$-galactosylation

Using trypanosomal membrane fractions prepared from MITat 1.4 and radiolabeled
UDP-galactose, we have been able to show that the synthetic substrates **1**, **6**, and **7**
were all galactosylated in vitro (Figure 3.2.2) [12]. In accordance with results reported

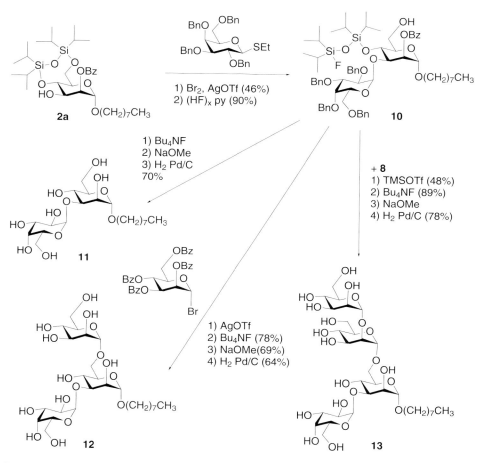

**Scheme 3.2.3.**

earlier for **1a** [11], **1a** resulted in 90% β-galactosylation and 10% α-galactosylation. The different trisaccharides used in this study gave the following results: **9a** led to at least three different tetrasaccharides, which were all sensitive to coffee bean α-galactosidase digestion, but resistant to bovine testes β-galactosidase. Digestion with α-mannosidase resulted in radioactive labeled disaccharide, trisaccharide and tetrasaccharide bands on HPTLC plates, as outlined in Figure 3.2.2. As judged from these results, the following structures are most likely. Two branched tetrasaccharide structures: (1) α-D-Man*p*-(1 → 2)-α-D-Man*p*-(1 → 6)-[α-D-Gal*p*-(1 → x)]-α-D-Man*p*-1-*O*-(CH₂)₇-CH₃, leading to a labeled disaccharide, and (2) α-D-Man*p*-(1 → 2)-[α-D-Gal*p*-(1 → x)]-α-D-Man*p*-(1 → 6)-α-D-Man*p*-1-*O*-(CH₂)₇-CH₃, leading to a labeled trisaccharide; and the terminally galactosylated structure α-D-Gal*p*-(1 → x)-α-D-Man*p*-(1 → 2)-α-D-Man*p*-(1 → 6)-α-D-Man*p*-1-*O*-(CH₂)₇-CH₃, which is not susceptible to α-mannosidase. In contrast, 1-thio-glycoside **9b** resulted in a totally different product

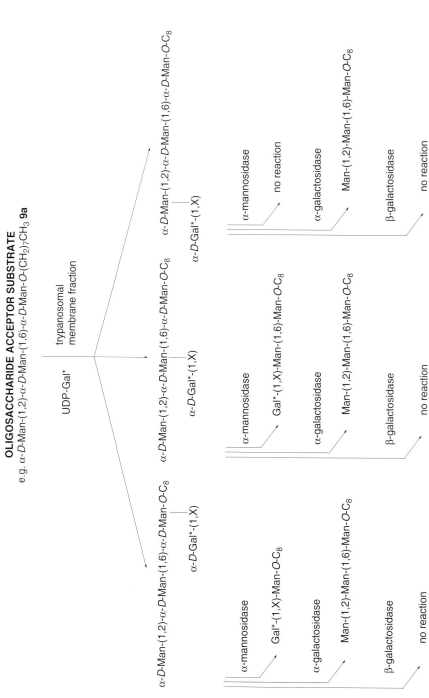

**Figure 3.2.2.** As described, the trypanosomal membrane fraction, containing galactosyltransferases, was incubated with different substrates in the presence of radiolabeled UDP-Gal. The graph shows the expected products, if the acceptor **9a** is used. The products are treated with α-mannosidase and the galactosidases and applied to TLC. Analysis of the resulting fluorography reveals which products have been formed in the initial reaction. The scheme shown was applied to all acceptors used in this study. The asterisk indicates radiolabeled products.

pattern and the total incorporation of radioactivity was about two-fold higher. The higher incorporation rate might be due to the fact, that **9b** led always to a band in the disaccharide region of the HPTLC. Considering the fact that the acceptor **9b** was pure as judged by NMR and HPTLC analysis, we assume that an $\alpha$-mannosidase activity in the membrane fraction degrades **9b** to the monosaccharide **1b** which is readily galactosylated thereafter.

In a second set of experiments we used the disaccharides **6** and **7** in concentrations ranging from 0.5 mM to 10 mM as acceptors for galactosylation. All products were sensitive to $\alpha$-galactosidase and $\alpha$-mannosidase digestion. Acceptor **6**, however, gave rise to at least two products. Using the latter, galactosylation of both, the branched and the terminal products increased with increasing acceptor concentration, whereas with the sulfur-containing disaccharide **7** galactosylation of the branched product reached a maximum at a concentration of 2 mM. Preliminary enzymatic galactosylations of compound **11**, **12**, and **13** showed that the branched GPI fragments **12** and **13** were galactosylated faster than the corresponding non-galactosylated counterparts **6** and **9a** (*i.e.* **13** is galactosylated by trypanosomal membrane fractions about 1000-fold faster than **9a**). From these results it may be concluded that the tetrasaccharide $\alpha$-D-Man$p$-(1 → 2)-$\alpha$-D-Man$p$-(1 → 6)-[$\alpha$-D-Gal$p$-(1 → x)]-$\alpha$-D-Man$p$-1-$O$-(CH$_2$)$_7$-CH$_3$obtained from galactosylation of **9a** (Figure 3.2.2) does not have structure **13**. However, further studies are required in order to solve the respective galactosylation pattern of all substrates.

All experiments described so far have been performed with membrane fractions isolated from the trypanosome clone MITat 1.4. This variant has been chosen because the GPI membrane anchor of the respective VSG variant contains a di-branched galactosyl sidechain. Galactosylation of the invariant GPI core structure is, however, variable. We have thus used membrane fractions of trypanosome variants which express a completely different galactosylation pattern of their respective VSG membrane anchors, *e.g.* MITat 1.2 and MITat 1.5 [26]. Galactosylation of **7**, however, was virtually identical in all three clones.

These results show that the responsible galactosyltransferases are always present in bloodstream forms of *T. brucei*, and that the expression of a different VSG variant does not influence the expression of the galactosyltransferases regardless of the final galactosylation pattern. This explanation would support the hypothesis that galactosyltransferases of *T. brucei* express a proof-reading function on VSG since the distinctive galactosylation pattern is responsible for the correct orientation of VSG in the membrane which allows a dense packaging and formation of the protective surface coat.

### 3.2.1.3 Preparation of trypanosome membrane fractions

Bloodstream forms of *Trypanosoma brucei* strain 427 variant MITat 1.4, MITat 1.2 and MITat 1.5 were isolated from infected rats as described previously [10]. Trypanosomes (5 10$^9$ cells/mL) were lysed for 5 min at 37°C in 10 mM PIPES buffer (pH 6.5), containing 1 mM dithiothreitol and protease inhibitors (chymostatin, leupeptin and pepstatin; 1 µM each). The lysate was centrifuged (14000 g; 10 min, 4°C) and the pellet (membrane fraction) finally resuspended in 50 mM PIPES buffer (pH 6.5), containing 15 mM $MnCl_2$, 1 mM dithiothreitol, 1 mM ATP, and protease inhibitors as before.

*Galactosyltransferase assay:* Galactosyltransferase activity was assayed in a total volume of 50 µL 50 mM PIPES buffer (see above), containing up to 40 µL membrane fraction, 0.025% Triton X-100, 50 nCi UDP-[U-$^{14}$C]galactose (1.8 µM), and a sugar acceptor in a range between 0–10 mM. The mixtures were incubated for 90 min at 35°C. The reaction was stopped by adding 50 µL of ice-cold acetone and the denatured protein was removed by centrifugation (14000 g; 5 min, 4°C). The supernatants (100 µL) were diluted with 1 mL HOAc (100 mM) and loaded onto 100 mg ISOLUTE C18 (endcapped) solid-phase extraction columns (ICT, Bad Homburg) equilibrated with 3 mL MeOH and 5 mL 100 mM HOAc. Following five washes with 1 mL 100 mM HOAc each, hydrophobic sugars were eluted with 3 250 µL MeOH. Aliquots of the combined eluates (650 µL) were dried using a Speed-vac evaporator and used for glycosidase digestions or directly applied to aluminium-backed 20 cm 20 cm silica gel 60 HPTLC plates (Merck, Darmstadt). The remaining eluate (100 µL) was assayed for radioactivity, using Ultimagold liquid scintillation cocktail (Packard, Frankfurt).

*Exoglycosidase digestion:* Galactosylated products were dried and digested as described earlier [11]. The digested samples were prepared for HPTLC analysis as described above.

*HPTLC:* Dried samples were dissolved in 4 µL 1:1 methanol:water, applied to HPTLC plates and treated as described earlier [11]. Standard sugar derivatives, 5 µg each of **6** and **9a**, were always run alongside with the samples.

**Acknowledgements:** This work was supported financially by the Deutsche Forschungs-gemeinschaft and the Fonds der Chemischen Industrie.

# References

[1]  M. A. J. Ferguson, K. Haldar, and G. A. M. Cross, *J. Biol. Chem.*, **1985**, *260*, 4963–4968; M. A. J. Ferguson, M. Duszenko, G. S. Lamont, P. Overath, and G. A. M. Cross, *J. Biol. Chem.*, **1986**, *261*, 356–362; M. A. J. Ferguson, S. W. Homans, R. A. Dwek, and T. W. Rademacher, *Science*, **1988**, *239*, 753–759.

[2]  M. J. McConville and M. A. J. Ferguson, *Biochem. J.*, **1993**, *294*, 305–324.

[3]  A. J. Turner, *Essays Biochem. (Engl.)*, **1994**, *28*, 113–127.

[4]  J. R. Thomas, R. A. Dwek, and T. W. Rademacher, *Biochemistry*, **1990**, *29*, 5413–5414.

[5]  M. L. S Günther, M. L. Cardoso de Almeida, N. Yoshida, and M. A. J. Ferguson, *J. Biol. Chem.*, **1992**, *267*, 6820–6828.

[6]  A. Mehlert, I. Silman, S. W. Homans, and M. A. J. Ferguson, *Biochem. Soc. Trans.*, **1992**, *21*, 43S.

[7]  S. W. Homans, M. A. J. Ferguson, and A. F. Williams, *Nature*, **1988**, *333*, 269–272.

[8]  M. A. Deeg, D. R. Humphrey, S. M. Yang, T. R. Ferguson, V. N. Reinhold, and T. L. Rosenberry, *J. Biol. Chem.*, **1992**, *267*, 18573–18580.

[9]  S. W. Homans, M. A. J. Ferguson, R. A. Dwek, and T. W. Rademacher, *Biochemistry*, **1989**, *28*, 2881–2887.

[10]  G. A. M. Cross, *Parasitology*, **1975**, *71*, 393–417.

[11] S. Pingel, R. A. Field, M. L. S. Günther, M. Duszenko, and M. A. J. Ferguson, *Biochem. J.*, **1995**, *309*, 877–882.

[12] T. Ziegler, R. Dettmann, V. Duszenko, and V. Kolb, *Carbohydr. Res.*, **1996**, *295*, 7–23.

[13] M. M. Palcic, L. D. Heerze, M. Pierce, and O. Hindsgaul, *Glycoconjugate J.*, **1988**, *5*, 89–98.

[14] R. Verduyn, C. J. J. Elie, C. E. Dreef, G. A. van der Marel, and J. H. van Boom, *Recl. Trav. Chim. Pays-Bas*, **1990**, *109*, 591–593.

[15] C. Murakata and T. Ogawa, *Carbohydr. Res.*, **1992**, *235*, 95–114.

[16] D. R. Mootoo, P. Konradsson, and B. Fraser-Reid, *J. Am. Chem. Soc.*, **1989**, *111*, 8540–8542; R. Madsen, U. E. Udodong, C. Roberts, D. R. Mootoo, P. Konradsson, and B. Fraser-Reid, *J. Am. Chem. Soc.*, **1995**, *117*, 1554–1565.

[17] T. G. Mayer, B. Kratzer, and R. R. Schmidt, *Angew. Chem.*, **1994**, *106*, 2289–2293; *Angew. Chem. Int. Ed. Engl.*, **1994**, *33*, 2177–2181.

[18] D. K. Baeschlin, A. R. Chaperon, V. Charbonneau, L. G. Green, S. V. Ley, U. Lücking, and E. Walther, *Angew. Chem.*, **1998**, *110*, 3609–3614; *Angew. Chem. Int. Ed. Engl.*, **1998**, *37*, 3423–3428.

[19] T. Ziegler, K. Neumann, E. Eckhardt, G. Herold, and G. Pantkowski, *Synlett*, **1991**, 699–701; T. Ziegler and E. Eckhardt, *Tetrahedron Lett.*, **1992**, *44*, 6615–6618; T. Ziegler, E. Eckhardt, and G. Pantkowski, *J. Carbohydr. Chem.*, **1993**, *13*, 81–109.

[20] R. J. Ferrier and R. H. Furneaux, *Methods Carbohydr. Chem.*, **1980**, *8*, 251–253.

[21] T. Ziegler, E. Eckhardt, K. Neumann, and V. Birault, *Synthesis*, **1992**, 1013–1017.

[22] M. Kreuzer and J. Thiem, *Carbohydr. Res.*, **1986**, *149*, 347–361; J. Thiem, M. Kreuzer, W. Fritsche-Lang, and H. M. Deger (Hoechst AG), *Ger. Offen.*, DE 3,626,028 (Cl. C07H13/06) 19 March 1987, *Chem. Abstr.* **1987**, *107*, P 176407e.

[23] P. Kovac, *Carbohydr. Res.*, **1986**, *153*, 168–170.

[24] P. L. Durette and T. Y. Shen, *Carbohydr. Res.*, **1979**, *69*, 316–322; C. M. Reichert, *Carbohydr. Res.*, **1979**, *77*, 141–147.

[25] G. Excoffier, D. Gagnaire, and J.-P. Utille, *Carbohydr. Res.*, **1975**, *39*, 368–373.

[26] A. Mehlert and M. A. J. Ferguson, personal communication.

# 3.3    Convenient synthesis of *C*-branched carbohydrates from glycals

*Thomas Sommermann, Michael Maurer, and Torsten Linker*

*C*-branched sugars are of current interest in carbohydrate chemistry. During the past 20 years, many methods have been developed for the synthesis of *C*-glycosides in which a carbon atom substitutes the glycosidic oxygen [1]. However, *C*-functionalizations at other positions of the sugar ring require many steps or the use of toxic tin and mercury compounds. Herein we present a practical and general protocol for the synthesis of *C*-analogs of carbohydrates, which is notable for easily available starting materials, high yields, and good stereoselectivities.

## 3.3.1    Manganese(III) acetate-mediated additions

During the course of our investigations on manganese(III)-mediated radical reactions [2], we revealed the first application of this methodology in carbohydrate chemistry [3]. Glycals were chosen as ideal substrates for the addition of CH-acidic compounds, since these chiral building blocks can be prepared on a multigram scale and are known to serve as precursors for a broad variety of optically active products [4]. Thus, addition of dimethyl malonate (**2a**) to tri-*O*-acetyl-D-glucal (**1a**) proceeds with high regioselectivity to afford the *C*-analogs **3a** in 66% yield (method A) (Scheme 3.3.1). However, four diastereomers are formed with only moderate selectivity. Furthermore, the unsaturated carbohydrates **4** result from an acid-catalyzed Ferrier rearrangement [5] under the drastic reaction conditions. To further extend manganese(III)-mediated reactions in carbohydrate chemistry and to suppress the undesired Ferrier rearrangement, we next investigated the addition of dimethyl malonate (**2a**) to the 5–6-unsaturated carbohydrate **1b** [6]. Indeed, the 6-*C*-analogs 5*R*-**5a** and 5*S*-**5a** were isolated in high stereoselectivities but only moderate yields, due to the drastic reaction conditions and the formation of side products (Scheme 3.3.2).

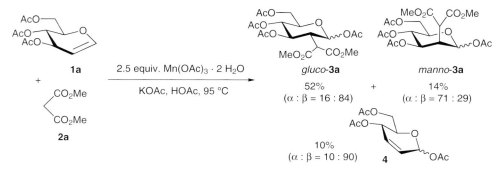

**Scheme 3.3.1.** Manganese(III)-mediated addition of dimethyl malonate (**2a**) to tri-*O*-acetyl-D-glucal (**1a**).

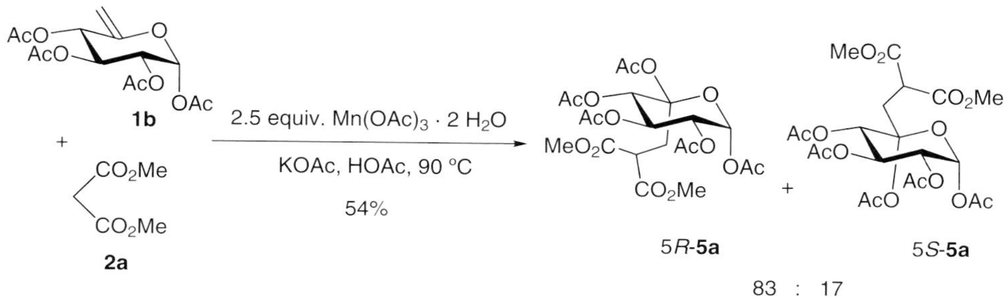

**Scheme 3.3.2.** Manganese(III)-mediated addition of dimethyl malonate (**2a**) to the 5–6-unsaturated carbohydrate **1b**.

## 3.3.2 Ceric(IV) ammonium nitrate (CAN)-mediated additions

To overcome the problems of acid-catalyzed Ferrier rearrangements and the formation of side products, milder reaction conditions were next employed. Ceric(IV) ammonium nitrate (CAN) turned out to be the reagent of choice (method B), since radical generation from CH-acidic substrates takes place even in methanol at low temperatures. Thus, the addition of malonates **2** to glycals **1** affords the C—C bond-formation products **3** and **6** highly regioselectively in 86–92% yield without competing Ferrier rearrangement (Table 3.3.1).

Furthermore, due to the lower reaction temperature, higher stereoselectivities were obtained with cerium(IV) (method B) than with manganese(III) (method A). The methodology can be applied to various glycals **1** derived from hexoses and pentoses and provides a general and convenient entry to 2-*C*-branched carbohydrates [7].

Interestingly, tri-*O*-acetyl-D-glucal (**1a**) and di-*O*-acetyl-D-xylal (**1e**) afford preferentially the *gluco*- and *xylo*-configurated products, which can be rationalized by unfavorable steric interactions with the ester group in the 3-position. To further prove this hypothesis, we investigated the addition of malonates **2** to the 3-deoxyglycal **1c**. Indeed, now the products were isolated in a ratio of approximately 1:1. On the other hand, highest stereoselectivities were observed with tri-*O*-acetyl-D-galactal (**1d**), di-*O*-acetyl-D-arabinal (**1f**), and hexa-*O*-acetyl-D-maltal (**1g**) (Table 3.3.1), since one ester group is orientated *pseudo* axial or the large carbohydrate substituent shields one face of the double bond. Therefore, severe steric interactions with the malonyl radicals are operative, which result in the exclusive formation of *galacto*-, *arabino*-, and *malto*-configurated products (Scheme 3.3.3).

## 3.3.3 Mechanistic considerations

In the first step, malonyl radicals **7** are generated from malonates **2** and manganese(III) (method A) or cerium(IV) (method B) by an inner-sphere electron transfer. Such acceptor-substituted radicals are characterized by the low energy of the singly occupied

**Table 3.3.1.** Addition of malonates **2** to various glycals **1**.

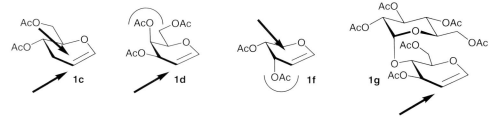

| 1 | 2 | R | method^a *dr* α : β^b | R' | 3 (%)^c | 3 (%)^c | 6 (%)^c |
|---|---|---|---|---|---|---|---|
| **1a** | 2a | Me | A    79 : 21 | Ac | *gluco*-**3a** (52)^d | *manno*-**3a** (14)^e | - |
|        | 2b | *i*-Pr | A    84 : 16 | Ac | *gluco*-**3b** (57)^f | *manno*-**3b** (11)^g | - |
| **1a** | 2a | Me | B    85 : 15 | Me | *gluco*-**3a** (62) | *manno*-**3a** (14) | *gluco*-**6a** (16) |
|        | 2b | *i*-Pr | B    91 :  9 | Me | *gluco*-**3b** (68) | *manno*-**3b** (8) | *gluco*-**6b** (16) |
| **1c** | 2a | Me | B    53 : 47 | Me | 3-H-*gluco*-**3a** (47) | 3-H-*manno*-**3a** (41) | - |
|        | 2b | *i*-Pr | B    56 : 44 | Me | 3-H-*gluco*-**3b** (49) | 3-H-*manno*-**3b** (39) | - |
| **1d** | 2a | Me | B    >98 : 2 | Me | *galacto*-**3a** (78) | - | *galacto*-**6a** (8) |
|        | 2b | *i*-Pr | B    >98 : 2 | Me | *galacto*-**3b** (73) | - | *galacto*-**6b** (17) |
| **1e** | 2a | Me | B    93 : 7 | Me | *xylo*-**3a** (81)^h | *lyxo*-**3a** (6) | - |
|        | 2b | *i*-Pr | B    87 : 13 | Me | *xylo*-**3b** (75)^i | *lyxo*-**3b** (11) | - |
| **1f** | 2a | Me | B    <2 : 98 | Me | *arabino*-**3a** (89)^j | - | - |
|        | 2b | *i*-Pr | B    <2 : 98 | Me | *arabino*-**3b** (87)^j | - | - |
| **1g** | 2a | Me | B    <2 : 98 | Me | *malto*-**3a** (89) | - | - |
|        | 2b | *i*-Pr | B    <2 : 98 | Me | *malto*-**3b** (81) | - | - |

^a Method A: 2-4 equiv. Mn(OAc)$_3$ · 2 H$_2$O, HOAc, 95 C; method B: 3-6 equiv. CAN, MeOH, 0 C. ^b Diastereomeric ratio (*dr* related to attack of malonyl radicals) determined by $^1$H NMR analysis of the crude product (600 MHz). ^c Yield of isolated product after column chromatography. ^d α : = 16 : 84. ^e α : β = 71 : 29. ^f α : β = 8 : 92. ^g α : β > 97 : 3. ^h α : β = 5 : 95. ^i α : β = 3 : 97. ^j α : β = 92 : 8.

**Scheme 3.3.3.** Preferred attack of radicals to glycals **1c**, **1d**, **1f**, and **1g**.

molecular orbital (SOMO) and exhibit electrophilic character. Thus, the interaction with the highest occupied molecular orbital (HOMO) of the double bond becomes predominant, which has the largest coefficient at the 2-position of glycals (Scheme 3.3.4). This explains the highly regioselective addition of malonates **2** to afford the adduct radicals **8**, and reveals the importance of orbital interactions in radical reactions, since for steric reasons, attack at the 1-position should be favored.

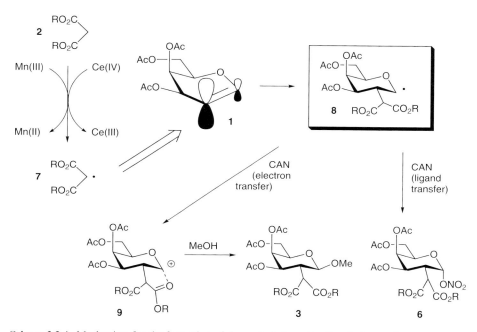

**Scheme 3.3.4.** Mechanism for the formation of the methyl glycosides **3** and nitrates **6**.

The formation of methyl glycosides **3** and nitrates **6** is interesting from the mechanistic point of view. The adduct radical **8** is readily oxidized by CAN to the cation **9**, which is trapped by the solvent to afford the methyl glycoside **3**. The exclusive formation of $\beta$-galactosides, $\beta$-glucosides, and $\alpha$-mannosides **3** can be rationalized by a neighboring group participation of the malonyl substituent (Scheme 3.3.4). On the other hand, the nitrates **6** are exclusively obtained as $\alpha$-anomers and cannot be formed *via* the intermediate **9**. A direct ligand transfer from CAN without participation of cations is more likely, which would explain the high stereoselectivity, since carbohydrate radicals like **8** are preferentially trapped from the $\alpha$-face.

## 3.3.4  Addition of malonates in the presence of nucleophiles

Another advantage of CAN-mediated radical additions consists of the possibility to perform the reactions in acetonitrile as solvent [8]. This allows the addition of nucleophiles, which trap the cationic intermediates **9** (Scheme 3.3.4). In initial experiments with *O*-nucleophiles **10**, we could demonstrate that this methodology is applicable in carbohydrate chemistry and can be used for the synthesis of derivatives **11** (Scheme 3.3.5). Even a disaccharide **11d** is directly available in one step. Of course the yields of the reactions have to be optimized, but only the major products (no *manno* isomers or nitrates) are depicted in Scheme 3.3.5.

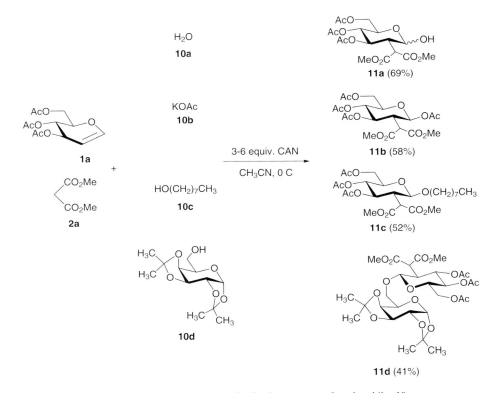

**Scheme 3.3.5.** Addition of dimethyl malonate (**2a**) in the presence of nucleophiles **10**.

## 3.3.5    Conclusion

The addition of malonates to unsaturated saccharides provides a general and convenient entry to *C*-branched carbohydrates. Our methodology is applicable to glycals derived from hexoses and pentoses and is characterized by easily available precursors. The generation of malonyl radicals by ceric(IV) ammonium nitrate (CAN) is superior to manganese(III)-mediated additions in terms of milder reaction conditions and yields. All reactions exhibit a very high degree of regioselectivity, since only 2-*C*-branched sugars were obtained. This result can be best rationalized by favorable orbital interactions between the SOMO of the malonyl radical and the HOMO of the double bond.

The substitution pattern of the glycals strongly alters the diastereomeric ratios. Thus, the addition of malonates to tri-*O*-acetyl-D-galactal and di-*O*-acetyl-D-arabinal occurs exclusively from one face of the carbohydrate. Strong evidence was found for a ligand transfer rather than electron transfer during the formation of nitrates, which sheds light on the mechanism of transition-metal-mediated radical reactions. Finally, the addition of malonates in the presence of nucleophiles in acetonitrile as solvent provides an easy entry to glycosides and disaccharides, which might serve as precursors for the synthesis of glycomimetics.

**Acknowledgments:** This work was generously supported by the Deutsche Forschungs-gemeinschaft (SFB 347 'Selektive Reaktionen Metall-aktivierter Moleküle' and a Heisenberg fellowship for T. L.) and the Fonds der Chemischen Industrie.

# References

[1] D. E. Levy, C. Tang, *The Chemistry of C-Glycosides*; Pergamon: Oxford, 1995; M. H. D. Postema, *C-Glycoside synthesis*, CRC Press: London, 1995; J.-M. Beau, T. Gallagher, *Top. Curr. Chem.* 1997, *187*, 154; F. Nicotra, *Top. Curr. Chem.* 1997, *187*, 55–83.

[2] U. Linker, B. Kersten, T. Linker, *Tetrahedron* 1995, *51*, 9917–9926; T. Linker, B. Kersten, U. Linker, K. Peters, E.-M. Peters, H. G. von Schnering, *Synlett* 1996, 468–470; T. Linker, *J. Prakt. Chem. Chem. Zt.* 1997, *339*, 488–492; Reviews: G. G. Melikyan, *Synthesis* 1993, 833–850; J. Iqbal, B. Bhatia, N. K. Nayyar, *Chem. Rev.* 1994, *94*, 519–564; P. I. Dalko, *Tetrahedron* 1995, *51*, 7579–7653; B. B. Snider, *Chem. Rev.* 1996, *96*, 339–363; G. G. Melikyan, *Org. React.* 1996, *49*, 427–675.

[3] T. Linker, K. Hartmann, T. Sommermann, D. Scheutzow, E. Ruckdeschel, *Angew. Chem., Int. Ed. Engl.* 1996, *35*, 1730–1732; T. Linker, *GIT Fachz. Lab.* 1996, 1167–1169.

[4] Reviews: R. J. Ferrier, *Adv. Carbohydr. Chem. Biochem.* 1969, *24*, 199–266; F. W. Lichtenthaler, In *Modern Synthetic Methods 1992*; Scheffold, R., Ed.; VCH: Weinheim, 1992; pp 273–376; S. J. Danishefsky, M. T. Bilodeau, *Angew. Chem., Int. Ed. Engl.* 1996, *35*, 1380–1419.

[5] R. J. Ferrier, N. Prasad, *J. Chem. Soc. C* 1969, 581–586; B. Fraser-Reid, *Acc. Chem. Res.* 1996, *29*, 57–66.

[6] R. Blattner, R. J. Ferrier, *J. Chem. Soc. Perkin Trans. 1* 1980, 1523–1527.

[7] T. Linker, T. Sommermann, F. Kahlenberg, *J. Am. Chem. Soc.* 1997, *119*, 9377–9384.

[8] E. Baciocchi, R. Ruzziconi, *J. Org. Chem.* 1991, *56*, 4772–4778; A. Citterio, R. Sebastiano, A. Marion, R. Santi, *J. Org. Chem.* 1991, *56*, 5328–5335; V. Nair, J. Mathew, J. Prabhakaran, *Chem. Soc. Rev.* 1997, 127–132.

## 3.4    Enzymatic tools for the synthesis of nucleotide (deoxy)sugars

*Lothar Elling*

During the past decade studies in glycobiology have demonstrated, that oligosaccharides in glycoconjugates such as glycoproteins, glycoplipids, hormones, and antibiotics play an active role in inter- and intracellular communication. Glycan structures are now known to mediate, *e.g.* cell-cell adhesion, cell-matrix contact, host-pathogen interaction, immunological recognition, cell and protein trafficking, proteolytic protection, and functional regulation of proteins [1–4]. The best known examples are the blood group antigens AB0(H) [5, 6], the sialyl-Lewis[x] structure as ligand of E-, P- and L-selectin [7, 8], and the Galili-epitope Gal($\beta$1–3)Gal [9] as dominant antigen in xenotransplantation. In areas of pharmaceutical research, oligosaccharide structures as well as their modifications and mimics are target molecules for drug development concerning inflammation, cancer metastasis, and infectious diseases. Efficient methods for the synthesis of naturally occurring or modified oligosaccharides are needed in order to use them as diagnostics or therapeutics. Strategies for the chemical synthesis of oligosaccharides are highly developed [10, 11]. However, the enzymatic synthesis of oligosaccharides employing glycosidases and Leloir-glycosyltransferases are now also well established [12]. These enzyme classes give typically moderate to high product yields in a one-step, highly stereoselective reaction without the need for protection and deprotection strategies. An additional characteristic feature of Leloir-glycosyltransferases is their regiospecificity. However, the relative expensive and partly labile nucleotide sugars are needed as donor substrates.

The present paper summarizes our work on the enzymatic synthesis of nucleotide (deoxy)sugars with emphasis on cheap educts and high enzyme productivities.

### 3.4.1    Enzymatic synthesis of nucleotide (deoxy)sugars from sucrose

In this context we have developed novel enzymatic tool boxes for the synthesis of nucleotide (deoxy)sugars from sucrose and nucleoside di- or monophosphates, including different techniques to realize large-scale synthesis with high enzyme productivities [13]. With sucrose synthase (EC 2.4.1.13; SuSy) we could establish a novel access to nucleotide sugars from sucrose (**1**) and nucleoside diphosphates (NDP) (Figure 3.4.1). SuSy represents a unique plant Leloir-glycosyltransferase catalyzing the cleavage of **1** with nucleoside diphosphates (with NDP: R=OH, and dNDP: R=H) *in vivo* and *in vitro* [14–18].

dTDP-ᴅ-glucose (dTDP-Glc) **3c** (B=T, R=H, Figure 3.4.1) was the first nucleotide sugar synthesized in a continuous mode in an enzyme membrane reactor with **1** and **2c** as substrates [19]. The space-time yield was $98 \, \text{g} \, \text{l}^{-1} \, \text{d}^{-1}$, with an enzyme consumption of 10 U SuSy per gram dTDP-Glc. The combination of SuSy from rice grains with a recombinant dTDP-ᴅ-glucose-4,6-dehydratase (EC 4.2.1.46) resulted in the synthesis

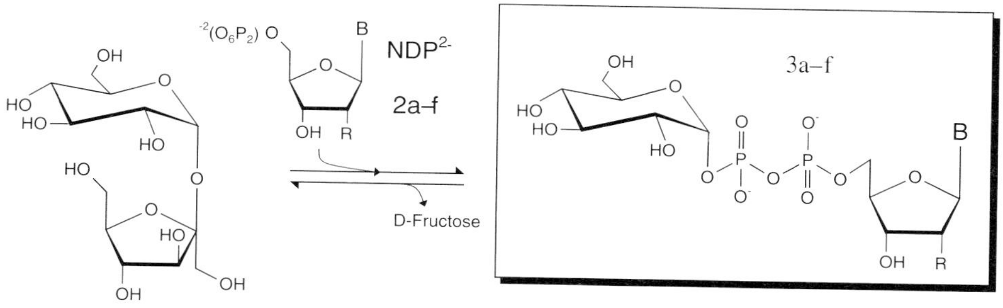

**1**      2a–f NDP: UDP>dUDP>dTDP>ADP>CDP>GDP

**Figure 3.4.1.** Substrate spectrum of sucrose synthase in the cleavage reaction of sucrose: acceptance of nucleoside diphosphates (NDP, **2–f**).

of dTDP-6-deoxy-4-ketoglucose [20], a key intermediate in the biosynthesis of dTDP-activated D- and L-deoxysugars. A gram-scale synthesis of dTDP-6-deoxy-4-ketoglucose was realized employing a fed-batch technology [21]. In the enzymatic reaction, 2.26 mmol (94% yield) activated deoxysugar was synthesized within 5 h, resulting in a space-time yield of $133\,g\,l^{-1}\,d^{-1}$. The overall yield after product isolation was 73% (1.1 g isolated product). Since nucleoside diphosphates are still expensive substrates, nucleoside monophosphates and **1** were converted to nucleotide (deoxy)sugars by a combination of kinases, SuSy and dTDP-D-glucose-4,6-dehydratase. In this way UDP-, dUDP-, ADP-and CDP-glucose, and dUDP-6-deoxy-4-ketoglucose were obtained on a 0.1 g scale [22]. Recently, we succeeded in the gram-scale synthesis of ADP-glucose (ADP-Glc) from ATP, AMP and **1**, employing myokinase and a recombinant SuSy from potato expressed in *Saccharomyces cerevisiae* [23, 24]. The repetitive use of enzymes in batch reactions (repetitive batch technique) gave high enzyme productivities for SuSy (28 mg ADP-Glc/U enzyme) and myokinase (140 mg ADP-Glc/U enzyme) with 55% yield (2.8 g ADP-Glc). After product isolation 2.2 g (44% overall yield) was obtained. In a similar way UDP-$\alpha$-D-galactosamine **(4)** was synthesized and chemically converted to UDP-*N*-acetyl-$\alpha$-D-galactosamine **(5)** (UDP-GalNAc) (Figure 3.4.2) [25].

A key step in the enzymatic reaction was the shift of the unfavorable reaction equilibrium catalyzed by galactose-1-phosphate uridyltransferase to the product side by the combination of two additional enzymatic steps (phosphoglucomutase and glucose-6-phosphate dehydrogenase) converting glucose-1-phosphate to gluconate-6-phosphate. The yield for the enzymatic synthesis of **4** was 42%. After acetylation with *N*-acetoxysuccinimide and product isolation, 82 mg of **5** was obtained, with an overall yield of 34%.

## 3.4.2 *In situ* regeneration of UDP-D-galactose from sucrose

*In situ* regeneration cycles of nucleotide sugars include the combination of Leloir-glycosyltransferases and enzymes (kinases, pyrophosporylases and others) for the synthesis of the corresponding activated sugars starting from the nucleotide (NDP

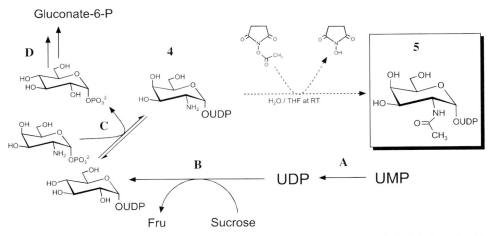

**Figure 3.4.2.** Chemoenzymatic synthesis of UDP-*N*-acetylgalactosamine (**5**) (UDP-GalNAc). **A:** nucleoside monophosphate kinase (EC 2.7.7.4), **B:** SuSy (EC 2.4.1.13), **C:** galactose-1-phosphate uridyltransferase (EC 2.7.7.12), **D:** phosphoglucomutase (EC 2.7.5.1) and glucose-6-phosphate dehydrogenase (EC 1.1.1.49).

or NMP) which is formed in the glycosyltransferase reaction [26]. These cycles are an alternative to the large-scale synthesis and avoid laborious isolation procedures of nucleotide sugars. The cycles are energetically driven by phospho(enol) pyruvate as substrate of pyruvate kinase which converts the inhibitory nucleoside di- or monophosphates to the corresponding nucleoside triphosphates.

We have developed a novel 3-enzymes-reaction-cycle for the synthesis of Gal($\beta$1–4)GlcNAc (**6**, *N*-acetyllactosamine, LacNAc) where UDP-Gal is regenerated *in situ* from UDP and sucrose by catalysis of SuSy from rice grains and UDP-galactose 4'-epimerase (Figure 3.4.3) [27]. Sucrose is the high-energy substrate and driving force of the reaction cycle [28]. The cycle was optimized to reach high enzyme productivities employing the repetitive batch technique. Besides the excellent stabilities of SuSy and

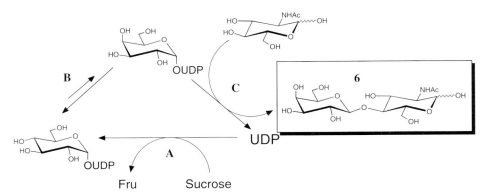

**Figure 3.4.3.** Enzymatic synthesis of *N*-acetyllactosamine (**6**) with *in situ* regeneration of UDP-galactose. **A:** SuSy (EC 2.4.1.13), **B:** UDP-galactose 4' epimerase (EC 5.1.3.2), **C:** $\beta$1–4galactosyltransferase (EC 2.4.1.38).

β1–4galactosyltransferase (β1–4GalT) in the presence of sucrose we observed, however, a rapid inactivation of UDP-galactose 4′-epimerase. In the presence of UMP and high concentrations of acceptor monosaccharide, the enzyme oxidizes the monosaccharide and is inactivated due to the presence of a tightly enzyme-bound NADH which cannot be exchanged by external NAD$^+$. We realized that the enzyme-bound cofactor can be regenerated by the addition of the transition state analogs dTDP- or dUDP-6-deoxy-4-ketoglucose (see above), thus enabling the repetitive utilization of the enzyme. In 11 batches we produced 594 mg (1.55 mmol) **6** using 10 U SuSy, 1.25 U β1–4GalT and 5 U UDP-galactose 4′-epimerase with addition of 14.3 mg (0.027 mmol) of dUDP-6-deoxy-4-ketoglucose. The novel *in situ* regeneration cycle for UDP-Gal was also employed in combination with two galactosyltransferases, β1–4GalT and α1–3galactosyltransferase, for the preparation of the Galili-epitope, Gal(α1–3)Gal(β1–4)GlcNAc [29].

### 3.4.3 Enzymatic synthesis of GDP-α-D-mannose from D-mannose

GDP-α-D-mannose (GDP-Man) is the precursor of GDP-β-L-fucose and the donor substrate of mannosyltransferases. We have synthesized GDP-Man (**10**) starting

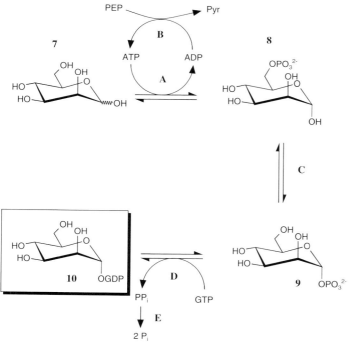

**Figure 3.4.4.** Enzymatic synthesis of GDP-α-D-mannose (**10**) from D-mannose (**7**) via mannose-6-phosphate (**8**), mannose-1-phosphate (**9**) and *in situ* regeneration of ATP. **A:** hexokinase (EC 2.7.1.1), **B:** pyruvate kinase (EC 2.7.1.40), **C:** phosphomannomutase (EC 5.4.2.8), **D:** GDP-Man pyrophosphorylase (EC 2.7.7.13), **E:** inorganic pyrophosphatase (EC 3.6.1.1).

from D-mannose **7** using hexokinase, phosphomannomutase (PMM) and GDP-Man pyrophosporylase (GDP-Man PP) (Figure 3.4.4) [30]. The latter two are recombinant enzymes from *Salmonella enterica*, group B, overexpressed in *E. coli* with specific activities of 0.1 U mg$^{-1}$ and 0.3–0.6 U mg$^{-1}$ for PMM and GDP-Man PP, respectively. With the repetitive batch technique 581 mg **10** (960 μmol, 80% yield) were synthesized in three batches after 72 h using 80 U PMM and GDP-Man PP, respectively. The overall yield after product isolation was 22.9%.

A detailed kinetic study of the GDP-Man PP reaction revealed that besides a strong product inhibition, the cofactor Mg$^{2+}$ is a key switch for effective enzymatic synthesis [31]. A very strong product inhibition was observed for **10** with a $K_i$ of 9 μM being close to the $K_m$ values of the substrates Man-1-P (15 μM) and GTP (40 μM). A more striking feature of GDP-Man PP represents the concentration ratio of the cofactor Mg$^{2+}$ and the substrate GTP in the enzymatic reaction. The maximum enzyme activity was found at a ratio of about 1, which had to be kept constant in order to obtain a constant high reaction velocity. Taking both aspects into consideration, **10** was produced continuously in a two-stage cascade of enzyme membrane reactors with a space-time yield of 28 g l$^{-1}$ day$^{-1}$ and an enzyme consumption of 0.9 U g$^{-1}$.

**Acknowledgements:** The author thanks Prof. Dr. Wandrey and PD Dr. U. Kragl (Institute of Biotechnology, Research Center Jülich), Prof. Dr. W. Piepersberg (Institut of Chemical Microbiology, University Wuppertal), Prof. Dr. W. Klaffke (Institute of Organic Chemistry, University Münster), and Prof Dr. D. H. van den Eijnden (Institute of Medical Chemistry, University Amsterdam) for their interesting and fruitful cooperation. Financial support by the Deutsche Forschungsgemeinschaft (El 135/2-1), the Hoechst Marion Roussel AG, and the EU (project: Engineering Protein O-Glycosylation for the Production of Receptor Blockers, BIO4-CT95-0138) is gratefully acknowledged.

# References

[1]  A. Varki, *Glycobiology* **1993**, *3*, 97–130.

[2]  H. Lis, N. Sharon, *Eur. J. Biochem.* **1993**, *218*, 1–27.

[3]  T. W. Rademacher, R. B. Parekh, R. A. Dwek, *Annu. Rev. Biochem.* **1988**, *57*, 785–838.

[4]  G. W. Hart, *Curr. Opin. Cell Biol.* **1992**, *4*, 1017–1023.

[5]  S.-I. Hakomori, *Am. J. Clin. Pathol.* **1984**, *82*, 635–648.

[6]  W. M. Watkins, in *Glycoproteins, Vol. 29a* (Eds.: J. Montreuil, J. F. G. Vliegenhart, H. Schachter), Elsevier Science, Amsterdam, **1995**, pp. 313–390.

[7]  A. Varki, *Proc. Natl. Acad. Sci. USA* **1994**, *91*, 7390–7397.

[8]  L. A. Lasky, *Annu. Rev. Biochem.* **1995**, *64*, 113–139.

[9]  U. Galili, K. Swanson, *Proc. Natl. Acad. Sci. USA* **1991**, *88*, 7401–7404.

[10]  H. G. Garg, K. von dem Bruch, H. Kunz, *Adv. Carbohydr. Chem. Biochem.* **1994**, *50*, 277–310.

[11]  R. R. Schmidt, W. Kinzy, *Adv. Carbohydr. Chem. Biochem.* **1994**, *50*, 21–123.

[12]  C.-H. Wong, R. L. Halcomb, Y. Ichikawa, T. Kajimoto, *Angew. Chem. Int. Ed. Engl.* **1995**, *34*, 521–546.

[13]  L. Elling, *Adv. Biochem. Eng. Biotechnol.* **1997**, *58*, 89–144.

[14] L. Elling, M.-R. Kula, *Biotechnol. Appl. Biochem.* **1991**, *14*, 306–316.
[15] L. Elling, M.-R. Kula, *J Biotechnol.* **1993**, *29*, 277–286.
[16] L. Elling, M.-R. Kula, *Enzyme Microb. Technol.* **1995**, *17*, 929–934.
[17] L. Elling, B. Güldenberg, M. Grothus, A. Zervosen, M. Péus, A. Helfer, A. Stein, H. Adrian, M.-R. Kula, *Biotechnol. Appl. Biochem.* **1995**, *21*, 29–37.
[18] L. Elling, M. Grothus, A. Zervosen, M.-R. Kula, *Ann. N. Y. Acad. Sci.* **1995**, *750*, 329–331.
[19] A. Zervosen, L. Elling, M.-R. Kula, *Angew. Chem. Int. Ed. Engl.* **1994**, *33*, 571–572.
[20] A. Stein, M.-R. Kula, L. Elling, S. Verseck, W. Klaffke, *Angew. Chem. Int. Ed. Engl.* **1995**, *34*, 1748–1749.
[21] A. Stein, M.-R. Kula, L. Elling, *Glycoconjugate J.* **1998**, *15*, 139–145.
[22] A. Zervosen, A. Stein, H. Adrian, L. Elling, *Tetrahedron* **1996**, *52*, 2395–2404.
[23] A. Zervosen, U. Römer, L. Elling, *J Mol Catalysis B: Enzymatic* **1998**, *5*, 25–28.
[24] H. Schrader, W. B. Frommer, L. Elling, unpublished.
[25] T. Bülter, C. Wandrey, L. Elling, *Carbohydr. Res.* **1997**, *305*, 469–473.
[26] Y. Ichikawa, G. C. Look, C.-H. Wong, *Anal. Biochem.* **1992**, *202*, 215–238.
[27] A. Zervosen, L. Elling, *J. Am. Chem. Soc.* **1996**, *118*, 1836–1840.
[28] L. Elling, M. Grothus, M.-R. Kula, *Glycobiology* **1993**, *3*, 349–355.
[29] C. H. Hokke, A. Zervosen, L. Elling, D. H. Joziasse, D. H. van den Eijnden, *Glycoconjugate J.* **1996**, *13*, 687–692.
[30] L. Elling, J. E. Ritter, S. Verseck, *Glycobiology* **1996**, *6*, 591–597.
[31] S. Fey, L. Elling, U. Kragl, *Carbohydr. Res.* **1997**, *305*, 475–481.

## 3.5    The enzyme dTDP-glucose-4,6-dehydratase as a tool for the synthesis of deoxy sugars

*Werner Klaffke*

Deoxy sugars are relatively widespread carbohydrates in bacteria, plants, and fungi. They are found in those polysaccharides produced by, *e.g.* lactobacilli and streptococci where they co-determine factors such as antigeneity [1, 2], or the rheology of the surrounding medium [3, 4]. As integral parts of secondary metabolites, deoxy sugar substituents in many commercially applied drugs not only regulate hydrophilicity of the aglycone, but in some cases direct the drug towards its endogenous receptor [5–12].

Therefore, a better accessibility of deoxy sugars and structures harboring them, should facilitate research in this exciting area in that it may open up new avenues in the synthesis of otherwise scarcely available material for biological and medicinal tests.

Deoxy sugar-carrying glycosides and polysaccharides are believed to be secondary gene products, which are synthesized from activated sugars by glycosyl transferases [13]. The activation is brought about in analogy to mammalian pathways by nucleoside diphosphates. It is the activated sugar which is derivatized in general before being transferred to the respective aglycone.

In the project summarized in the present account, we have focused our effort on the L-rhamnose pathway. A central characteristic of this pathway is the oxidation-reduction step on dTDP-glucose (**1**) mediated by dTDP-glucose-4,6-dehydratase, in which a formal hydride-shift from C-4 to C-6 is observed [14–16]. A hydride is abstracted by a covalently bound $NAD^+$ from the $\beta$-face of pyranoside (**1**) and re-delivered from the same side to C-6 [17, 18]. Once this central step is passed, the resulting 4-ulose (**2**) comprises a versatile chemical functionality, which is exploited in all detail by nature. Scheme 3.5.1 depicts the hitherto known routes.

The $\alpha$-positions are sufficiently reactive to allow for keto-enol tautomerism. The 3,5-epimerase therefore catalyzes the successive $\alpha$-facial abstraction of both neighboring protons H-3 and H-5, and subsequent re-protonation of this structure from the $\beta$-face. As a result of this, the configuration at C-3 and C-5 is inverted in compound **3** and the ring conformation flips from $^4C_1$ (D) to $^1C_4$ (L). Final enantioselective reduction

**Table 3.5.1.** Nucleotide sugars.

| Sugar | Activated form | Secondary metabolite |
|---|---|---|
| D-glucose (glc) | UDP-$\alpha$-D-glc | UDP-$\alpha$-D-glucuronide |
| | dTDP-$\alpha$-D-glc | dTDP-$\beta$-L-rhamnose |
| N-acetyl-D-glucosamine (GlcNAc) | UDP-$\alpha$-D-glcNAc | |
| D-galactose (Gal) | UDP-$\alpha$-D-gal | |
| N-acetyl-D-galactosamine (GalNAc) | UDP-$\alpha$-D-galNAc | |
| D-mannose (Man) | GDP-$\alpha$-D-man | GDP-$\beta$-L-fucose |
| | | GDP-L-gal |
| N-Acetylneuraminic acid (Neu5Ac) | CMP-Neu5Ac | |

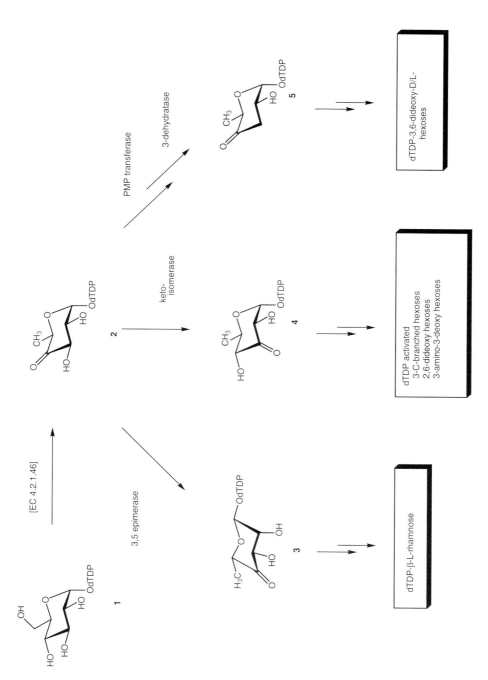

**Scheme 3.5.1.** Biological pathways towards functionalized deoxy-sugars

of C-4 then yields dTDP-$\beta$-L-rhamnose. The other routes, which are more speculative, include the keto-isomerization to yield 3-ulose (**4**), which is now perfectly set for reductive amination or deoxygenation to give either 3-amino or 2,6-dideoxy sugars. Via a pyridoxamine phosphate transferase acting on intermediate **2** and a 3-dehydratase, further reduction of C-3 renders 3,6-dideoxy sugar like compound **5** [19, 20].

As a consequence of the permanent presence of the activating group, all of the above structures are theoretically ready for glycosyl transfer at any stage of their transformation, once they are picked up by the 'correct' glycosyl transferase. Glycosylation, however, is not the only role for such intermediates in the biosynthesis of secondary metabolites: the antibiotic granaticin (**6**) has got incorporated a 2,6-dideoxy sugar into its western aliphatic ring [21–23].

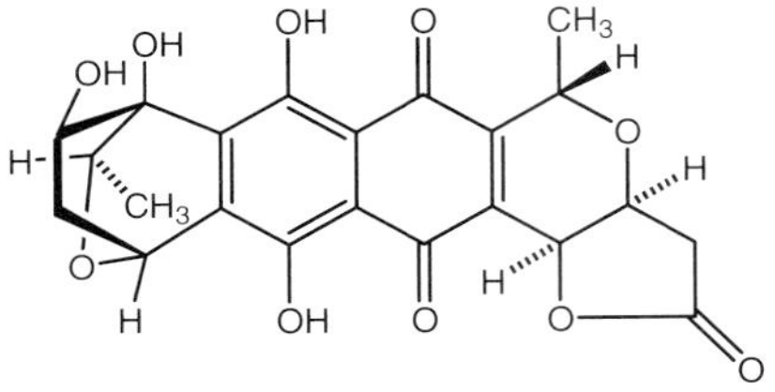

**6**

**Scheme 3.5.2.** A 2,6-dideoxy sugar is an integral part of granaticin.

## 3.5.1   Syntheses

Work performed by us and others active in the area [24–27] has been centered around four major issues:

1. Synthesis of activated sugars and derivatives thereof.
2. Enzymatic activation of deoxygenated hexoses.
3. Isolation of glucose 4,6-dehydratase from wild type and recombinant clones and enzymatic transformations to yield activated 4-keto sugars.
4. Transformations employing overexpressed proteins of the *rfb*-gene cluster from *Salmonella typhimurium*.

*Activated sugars:* The synthesis of the activated sugars has been shown to proceed smoothly via the respective hexose-1-phosphates and coupling with dTMP-morpholidate. Yields in these chemical syntheses are usually moderate in the range between 20 and 30% of isolated diphosphate [28–30] Therefore, the question arose, whether an enzymatic one-pot approach could be advantageously applied here. Scheme 3.5.3 summarizes the results of this for the three hexoses, glucose (**7**), 3-deoxy- (**8**), and 3-azido-3-deoxy glucose (**9**) which could be isolated in the form of their dilithium diphosphates [31]. The experiment shown here was carried out in 10 mM TRIS buffer at pH 7.6 containing an aliquot of dTDP-glucose pyrophosphorylase (dTDP-G-PP), previously isolated from *E. coli B* [32], commercially available yeast hexokinase, pyruvate kinase

(PK), chemically synthesized dTTP [33], phosphoenol pyruvate (PEP), phosphogluco-mutase (PGM), and glucose-1,6-bisphosphate (G-1,6-PP). After 5 days, the proteins were denatured and the entire mixture subjected to ion-exchange chromatography on Dowex 2-X8 resin. Upon elution by a LiCl-gradient, product fractions were obtained, pooled, and lyophilized. As the endogenous substrate could be isolated in 31% yield, the marked drop down to 17% and 16% for derivatives **11** and **12**, respectively, was anticipated and not surprising for artificial substrates. With amounts ranging between 20–40 mg the procedure still proved usable and sufficient for a one-pot process.

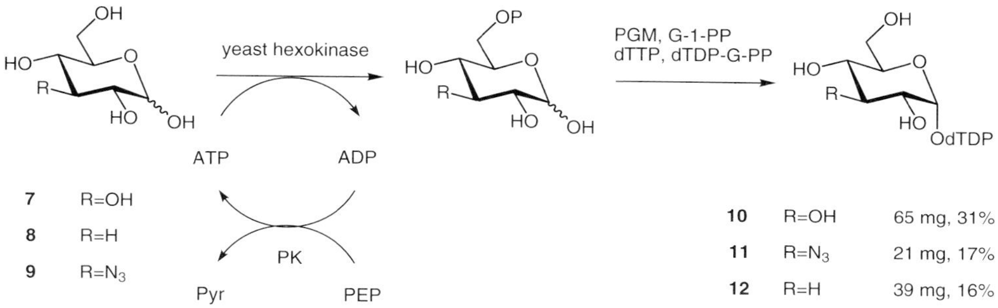

**Scheme 3.5.3.** Enzyme-mediated one-pot syntheses of activated derivatives of glucose, 3-deoxy-, and 3-azido-3-deoxyglucose. PK = phosphokinase, PGM = phosphoglucomutase, G-1,6-PP = glucose-1,6-bisphosphate, dTTP = thymidine triphosphate, dTDP-G-PP = dTDP-glucose pyrophosphorylase.

In addition, it should be noted that a derivative such as compound **12** would be extremely difficult to synthesize by classical chemical methods, since the azido-function is proven incompatible with standard phosphorylation techniques [28].

In analogy to the biosynthesis of glucose derivatives, fresh brewer's yeast extract was shown to catalyze the synthesis of GDP-mannose and also its 3-deoxy derivative **14** [34]. A previously synthesized 3-azido derivative **13** was subjected to a supplemented brewer's yeast extract containing hexokinase, chemically synthesized GTP and the GTP-regenerating enzyme system. The latter was needed for regeneration of GTP used up by the phosphorylation of the hexose to guarantee a high stationary GTP concentration required for the last step, formation of the GDP sugar, which uses an equimolar amount of that triphosphate. In a 40 ml incubation, 53% of the activated mannose derivative **14** was isolated, equalling a yield of 8%. Hence again, as for all the one-pot-processes mentioned here, a chemical synthesis would still be fighting for competition since the multi-step conventional methods have to face the facility of a single-step procedure.

*Keto-intermediates.* With the activated glucose-derived deoxy sugars in hand, the enzymatic transformation to render the 4-keto derivatives could be addressed, and it could be shown that sufficient amounts of the enzyme could be prepared in scales suitable for chemical laboratories. From a 15 l fermentation broth of *E. coli B* (ATCC 11303) [35], shaken at 37°C, 60 g of raw cell material could be isolated to

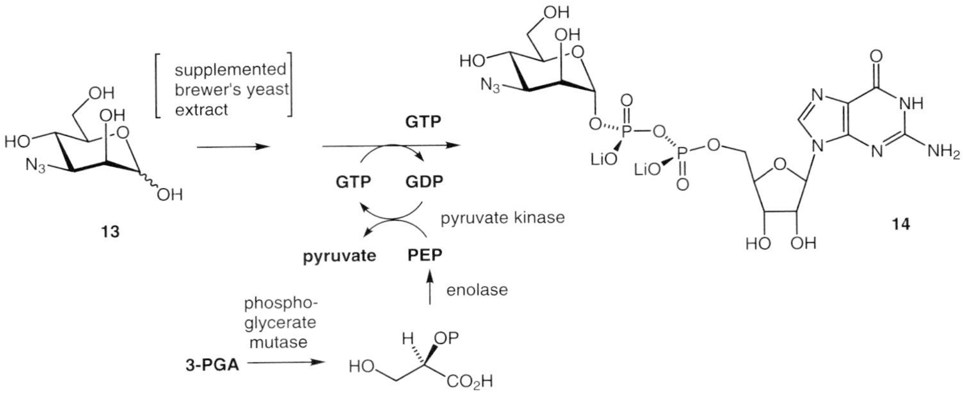

**Scheme 3.5.4.** Synthesis of GDP-3-azido-3-deoxy-ᴅ-mannose using fresh brewer's yeast extract. GTP = guanidine triphosphate, GDP = guanidine diphosphate, PEP = phoshoenol pyruvate, 3-PGA = 3-phosphoglyceric acid.

yield finally after sonication, ammonium sulfate precipitation, and DEAE-sepharose chromatography 30 U with 17 mg/ml (protein/extract).

In order to improve the purification procedure several affinity materials have been synthesized and checked against the enzyme [36]. Scheme 3.5.5 depicts a few of the synthesized structures used, of which the glycoyl derivatives **18** and **19** have been most promising. Previously described methylene diphosphonate (**15**) [37] and the new monophospho glycolate (**18**) are both stable against phosphatase activity, which makes them extremely versatile ligands for a chromatographic purification of raw extracts. Monophospho glycolate **18** has a six-fold higher affinity to the enzyme than dTDP, and a three-fold higher affinity than the substrate itself. From further studies it became clear that derivatization of the nucleobase at postion C-5 would be still accepted by the enzyme. This resulted in the design for **19** and delivered a ligand, which in the eyes of the chemist is ideally suited for affinity chromatography: it is stable towards phosphatases, and non-enzymatic phosphate/diphosphate cleavage; it exhibits a non-competitive inhibition constant of $K_i = 451 \, \mu M$. The deblocked ligand was immobilized on a Fisons-IAsys cartridge for association/dissociation studies, and in preliminary results showed a 130% higher capacity than previously reported dTDP-hexanolamine sepharose. Further investigations in down-stream processing are presently underway.

The initial experiment with crude enzyme extract and its endogenous substrate was yielding an unexpected result. After incubation of the enzyme with dTDP-glucose and work-up of the reaction mixture by chromatography on a Dowex 2X8 anionic exchange resin, the $^1$H-NMR spectrum revealed a mixture of two dTDP-activated sugars instead of the expected one [38]. Since this was not observed in the parallel experiment with recombinant enzyme from *Salmonella* [26] two questions arose:

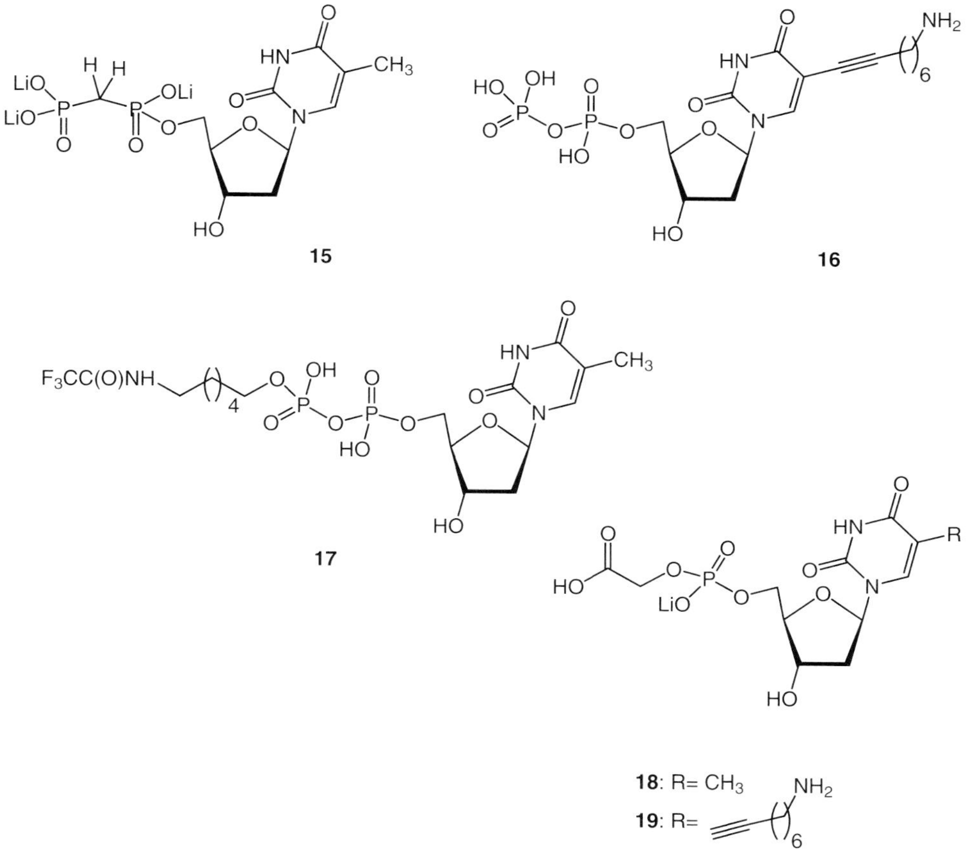

**15**    **16**

**17**

**18**: R= CH$_3$
**19**: R=

**Scheme 3.5.5.** Structures used in the inhibition experiments.

1. Was the isolation of two products instead of one due to a previously not observed side activity in the extract,
2. Was the second product a derivative of the expected 4-keto glucose?

To answer these issues, the work-up procedure was investigated first, after it had become evident that the ratio between the two inseparable products was varying depending on the time it took to perform the final chromatography on the reaction mixture. Indeed, it could be confirmed that upon storage of the product mixture over the anion exchange resin for 10 h at 4°C, the mixture was converted to a single product as judged by $^1$H-NMR. This product, however, was not the expected 4-keto glucose (**20**), but 3-ulose (**21**) instead. To confirm this hitherto uncharacterized dTDP-hexose, we embarked on a chemical synthesis of this postulated reaction product. Phenoxyacetyl-protected hexopyranose (**22**), as illustrated in Scheme 3.5.7, was phosphitylated according to Shibaev's method [39]. Subsequent hydrolysis rendered the H-phosphonate, which upon treatment with hydrazine hydrate gave alcohol **23**.

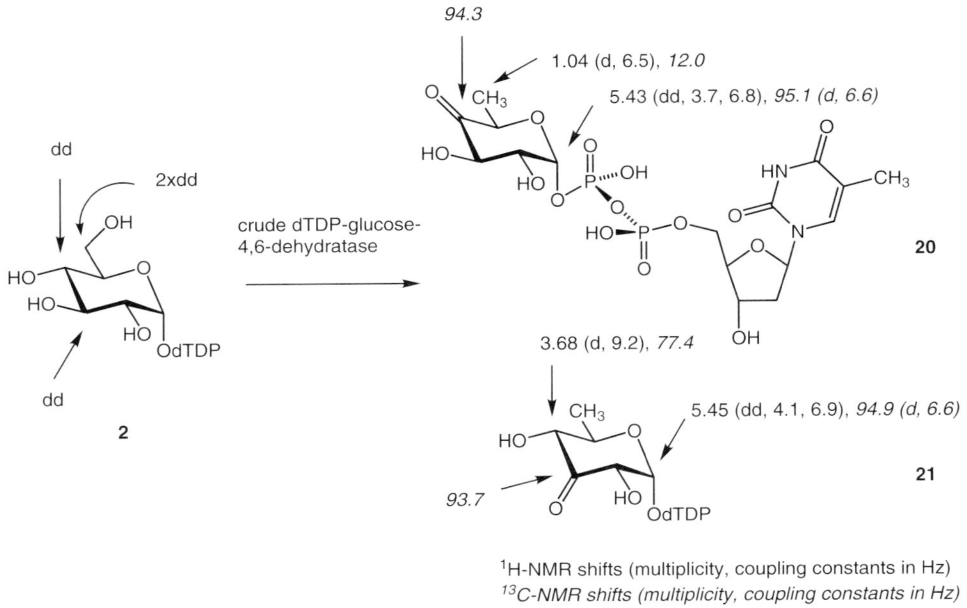

Scheme 3.5.6. Incubation experiment with crude dTDP-glucose-4,6-dehydratase.

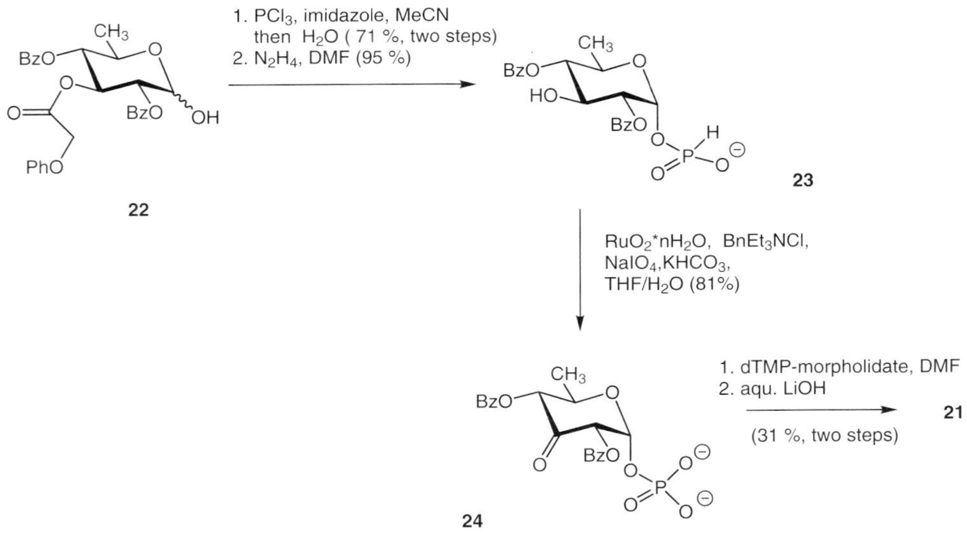

Scheme 3.5.7. Synthesis of **21**.

According to the experience of van Boom et al. [40], oxidation of phosphonate monoesters is very sluggish and yields only minor traces of the desired phosphate diesters. It was very fortunate, however, that the system described by Kiely et al. [41] simplified the procedure extensively. Not only was it possible to convert **23** to the glycosyl phosphate, but the ruthenium oxide/sodium periodate system, in a buffered THF/water mixture and in the presence of phase transfer catalyst benzyltriethyl ammonium chloride smoothly oxidized C-3 as well to yield the 3-ulosyl phosphate **24** in 81%. Standard conversion into the thymidine diphosphate then gave the compound in question, dTDP-α-D-ribo-3-ulose (**21**) [42,43]. This procedure circumvents the intermediate protection applied by Schmidt et al. [44], who reported on an own solution to this synthetic problem at the same time.

Comparison of the respective NMR data revealed that indeed, the unknown compound in the mixture was the 3-ulose (**21**), which was mixed with the expected product, dTDP-α-D-xylohexopyranosyl-4-ulose (**20**). The most plausible explanation was indeed a non-enzymatic keto-isomerization brought about by the basic anion exchange resin. This fact was underpinned by parallel experiments with recombinant enzyme and a different work-up procedure to give **20** as the only observed reaction product.

Immediately after this, the question arose whether or not structurally similar but chemically modified dTDP-hexoses would be also susceptible to enzymatic oxido-reduction of the described type. Preliminary studies were encouraging: both derivatives, 3-deoxy and 3-azido-3-deoxy-hexopyranoses **11** and **12** served as substrates for the dehydratase with a selectivity ratio of 13:3:1 (**2**:**11**:**12**). Table 3.5.2 depicts the outcome of those assaying experiments.

In the actual experiments employing both the exogenous substrates on a preparative scale, product formation was observed by tlc after 1 h. After 3 days no further formation of keto-sugar could be detected and the reaction mixture was subjected to Dowex 2-X8-based chromatography, since it was reckoned that keto-isomerization would be not interfering in these two cases. Indeed, it was the NMR experiment which showed the expected products, the assignments of which are depicted in Scheme 3.5.8.

**Table 3.5.2.** Kinetic data $K_m$ and $V_{max}$ (assay according to [35,45]) for substrates **11** and **12**. Activities and rates were measured for an extract from *E. coli* p4439.

| | 11 | 12 |
|---|---|---|
| $K_m$ {μM} | $200 \pm 35$ | $350 \pm 60$ |
| $V_{max}$ {μmol/h × mg} | $130 \pm 22$ | $90 \pm 15$ |
| $V_{max}/K_m$ {ml/s × mg} | $\approx 0.18$ | $\approx 0.07$ |

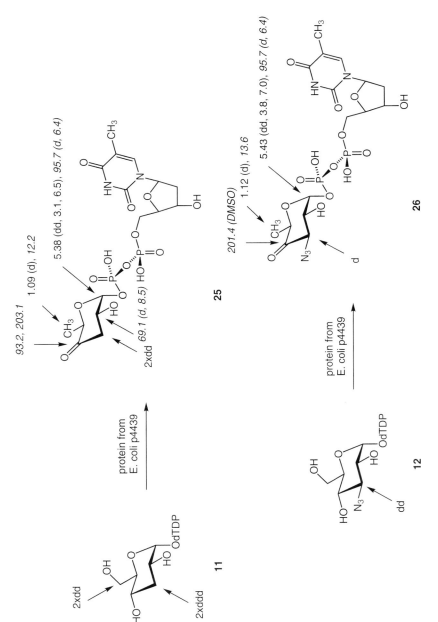

**Scheme 3.5.8.** Isolated products from incubation experiments of 11 and 12 with a crude extract from *E. coli* p4439.

The respective configurations could be unequivocally confirmed by the coupling constants, the intactness of the diphosphate bridge by ESI-MS.

## 3.5.2 Synthesis of dTDP-ascarylose

When an article by Marumo et al. [24] appeared on the synthesis of dTDP-L-rhamnose using raw extracts from overexpressing *E. coli K12*, our focus shifted immediately further from the dehydratase to the remaining enzymes of the biochemical pathway. One question was especially intriguing: would the substrate specificity of 3,5-epi-merase and 4-reductase be as equally broad as was that of the dehydratase? Both cultures, *E. coli* p4439 and p3378 were donated for these experiments to us by P. Reeves [24, 46]. The p4439 harbored the *rfb B*-gene from *Salmonella enterica* LT2 controlled by a T7-promotor, and complementarily p3378 harbored the first seven genes of the entire *rfb*-cluster.

Preincubation of **11** with the p4439 extract supplemented with NADH cofactor for 6 h yielded expected **25** as judged by tlc. A second subsequent incubation with 45 U of the p3378 extract, after denaturation of the pre-incubation proteins, was worked-up after 12 h according to our standard procedure. The experiment, after a gel-chromato-graphic desalting step and lyophilization, yielded 90 mg of a product, which was prone to decomposition in solution at room temperature. The $t_{1/2}$ in $D_2O$ was judged from the decay of the phosphorus-coupled H-1″, however, dTDP-$\beta$-L-3,6-deoxy sugar **27** (dTDP-ascarylose, cf. Scheme 3.5.7) could be assigned from typical signals for H-6″ and H-3″. It was, after all possible to keep the lyophilisate at $-18°C$ for more than two weeks.

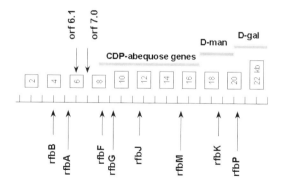

**rhamnose pathway**

| | |
|---|---|
| rfbA | glucose-1-phosphate thymidyltransferase |
| rfbB | dTDP-glucose 4,6-dehydratase |
| rfbC | orf 6.1 or orf 7.0 encode for either rfbC or D… |
| rfbD | 3,5-epimerase and reductase |

**Figure 3.5.1.** Gene cluster of *Salmonella enterica* LT 2 coding for different proteins in the biochemical pathway of deoxy sugars.

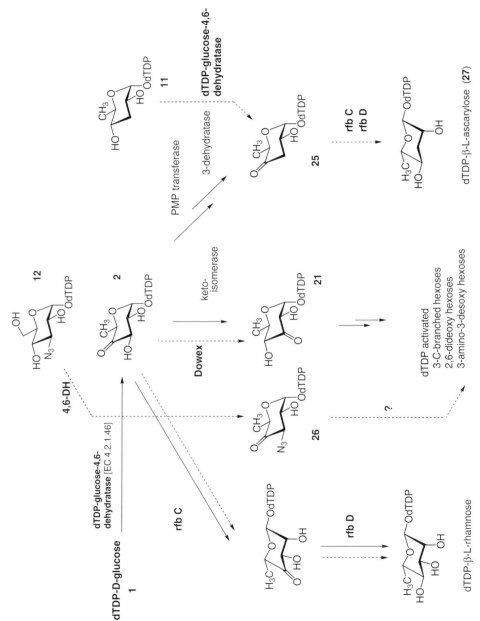

**Scheme 3.5.9.** Summarized view on synthetic possibilities along the deoxy-sugar pathway.

### 3.5.3 Conclusion

The substrate specificity of nucleoside diphosphate sugars has been shown to permit for the use of derivatives of dTDP-glucose. Not only was it possible to deoxygenate at C-6 dTDP-activated glucose on a preparative scale, but also 3-azido and 3-deoxy derivatives thereof, which have themselves been produced by phosphorylating enzymes from chemically synthesized precursors. Finally, the use of crude enzyme extracts, overexpressed by *E. coli*, made it possible to either synthesize dTDP-L-rhamnose in preparative amounts, or – by incubating with activated 3-deoxy-glucose – to have a direct access to dTDP-ascarylose (**27**). The consequences of these findings for the preparative accessibility of various deoxy sugars are summarized schematically in Scheme 3.5.9.

Dotted lines indicate the pathways chemically pre-synthesized, artificial substrates take in the biosynthetic scheme, whereas the solid lines are the paths for dTDP-glucose and its biosynthetic derivatives. By combining chemical steps with enzymatic transformation, via **26** a direct access to 3-azido and 3-amino sugars is conceivable. By using the dehydratase reaction on 3,6-deoxy derivative **11**, both enzymatic steps PMP-transferase and 3-dehydratase are circumvented and the 4-keto sugar **25** is bio-synthesized in one step. Both, rfb C and rfb D gene products take the sugar further to yield dTDP-L-ascarylose **27**.

**Acknowledgements:** I wish to express my thanks to the co-workers in this project for their enthusiasm, Brigitte Leon, Susanne Liemann, Sophie Chambon, and Andreas Naundorf. Special thanks are due to Lothar Elling (IET Jülich) for his co-operation in the dehydratase research, and Prof. P. Reeves (Sydney) for donating the overproducing *E. coli* K12 cultures. This work has enjoyed financial support by the 'Deutsche Forschungsgemeinschaft', the 'Fonds der Chemischen Industrie', and Unilever Research Laboratory Vlaardingen, The Netherlands.

### References

[1] D. R. Bundle, N. M. Young, *Curr. Opin. Struct. Biol. 2* (1992) 666–673.
[2] S.-j. Deng, C. R. MacKenzie, T. Hirama, R. Brousseau, T. L. Lowary, N. M. Young, D. R. Bundle, S. A. Narang, *Proc. Natl. Acad. Sci. U. S. A. 92* (1995) 4992–4996.
[3] G. W. Robijn, D. J. C. van-den-Berg, J. P. Kamerling, H. Haas, J. F. G. Vliegenthart, *Carbohydr. Res. 276* (1994) 117–136.
[4] G. W. Robijn, A. Imberty, D. J. C. van-den-Berg, A. M. Ledeboer, J. P. Kamerling, J. F. G. Vliegenthart, S. Perez, *Carbohydr. Res. 288* (1996) 57–74.
[5] K. C. Nicolaou, S.-C. Tsay, T. Suzuki, G. F. Joyce, *J. Am. Chem. Soc. 114* (1992) 7555–7557.
[6] T. Li, Z. Zeng, V. A. Estevez, K. U. Baldenius, K. C. Nicolaou, G. F. Joyce, *J. Am. Chem. Soc. 116* (1994) 3709–15.
[7] K. C. Nicolaou, B. M. Smith, K. Ajito, H. Komatsu, L. Gomez-Paloma, Y. Tor, *J. Am. Chem. Soc. 118* (1996) 2303–2304.
[8] K. C. Nicolaou, K. Ajito, H. Komatsu, B. M. Smith, P. Bertinato, L. Gomez-Paloma, *Chem. Commun.* (1996) 1495–1496.

[9] K. C. Nicolaou, B. M. Smith, K. Ajito, H. Komatsu, L. Gomez-Paloma, Y. Tor, *J. Am. Chem. Soc. 118* (1996) 2303–2304.

[10] C. Liu, B. M. Smith, K. Ajito, H. Komatsu, L. Gomez-Paloma, T. Li, E. A. Theodorakis, K. C. Nicolaou, P. K. Vogt, *Proc. Natl. Acad. Sci. U. S. A. 93* (1996) 940–944.

[11] X. Gao, D. J. Patel, *Biochemistry 28* (1989) 751–762.

[12] D. L. Banville, M. A. Keniry, R. H. Shafer, *Biochemistry 29* (1990) 9294–9304.

[13] R. Caputto, L. F. Leloir, C. E. Cardini, A. C. Paladini, *J. Biol. Chem. 184* (1950) 333–350.

[14] K. Herrmann, J. Lehmann, *Eur. J. Biochem. 3* (1968) 369–376.

[15] R. Okazaki, T. Okazaki, J. L. Strominger, A. M. Michelson, *J. Biol. Chem. 237* (1962) 3014–3026.

[16] M. Matsuhashi, J. M. Gilbert, S. Matsuhashi, J. G. Brown, J. L. Strominger, *Biochem. Biophys. Res. Commun. 15* (1964) 55–59.

[17] C. E. Snipes, G. U. Brillinger, L. Sellers, L. Mascaro, H. G. Floss, *J. Biol. Chem. 252* (1977) 8113–8117.

[18] Y. Yu, R. N. Russell, J. S. Thorson, L.-d. Liu, H.-w. Liu, *J. Biol. Chem. 267* (1992) 5868–5875.

[19] P. Gonzalez-Porque, J. L. Strominger, *Proc. Natl. Acad. Sci. USA 69* (1972) 1625–1628.

[20] P. A. Rubenstein, J. L. Strominger, *J. Biol. Chem. 249* (1974) 3776–3781.

[21] H. G. Floss, P. J. Keller, J. M. Beale, *J.Nat.Prod. 49* (1986) 957–970.

[22] H. G. Floss, J. H. Beale, *Angew.Chem.Int.Ed.Engl. 28* (1989) 146–177.

[23] C. E. Snipes, C. J. Chang, H. G. Floss, *J. Am. Chem. Soc. 101* (1979) 701–706.

[24] K. Marumo, L. Lindqvist, N. Verma, A. Weintraub, P. R. Reeves, A. A. Lindberg, *Eur. J. Biochem. 204* (1992) 539–545.

[25] W. Piepersberg, M. Stockmann, K. M. Taleghani, J. Distler, S. Grabley, P. Sichel, B. Bräu, *PCT WO 93/06219* (1993).

[26] A. Stein, M.-R. Kula, L. Elling, S. Verseck, W. Klaffke, *Angew. Chemie Int. Ed. Engl. 34* (1995) 1748–1749.

[27] L. Elling, *Adv. Biochem. Eng. 58* (1997) 89–144.

[28] S. Liemann, W. Klaffke, *Liebigs Ann. Chem.* (1994) 1779–1787.

[29] A. Naundorf, S. Liemann, W. Klaffke, *Eurocarb VII* (Cracow, Poland), Synthesis of Deoxy Sugar Phosphates, pp. Abstr. C004-Abstr. C004 (1993).

[30] B. Leon, S. Liemann, W. Klaffke, *J. Carbohydr. Chem. 12* (1993) 597–610.

[31] W. Klaffke, in K. Krohn, H. Kirst, H. Maas (Eds.): *Antibiotics and Antiviral Compounds*, VCH Verlagsgesellschaft mbH, Weinheim 1993, p. 389–402.

[32] R. L. Bernstein, P. W. Robbins, *J. Biol. Chem. 240* (1965) 391–397.

[33] D. E. Hoard, D. G. Ott, *J. Am. Chem. Soc. 87* (1965) 1785–1788.

[34] W. Klaffke, *Carbohydr. Res. 266* (1995) 285–292.

[35] R. D. Bevill, in H. U. Bergemeyer (Ed.), VCH, Weinheim, 1974, p. 2268–2269.

[36] W. Klaffke, A. Naundorf, *XVIII Int. Carbohydr. Symposium*, Milan, Italy, Abstr. Book BO 010 (1996).

[37] V. J. Davisson, D. R. Davis, V. M. Dixit, C. D. Poulter, *J. Org. Chem.. 52* (1987) 1794–1801.

[38] A. Naundorf, W. Klaffke, *Carbohydr. Res. 285* (1996) 141.

[39] A. V. Nikolaev, I. Ivanova, V. N. Shibaev, *Carbohydr. Res. 242* (1993) 91–107.

[40] J. van Boom, *personal communication* (1993)

[41] P. E. Morris, D. E. Kiely, G. S. Vigee, *J. Carbohydr. Chem. 9* (1990) 661–673.

[42] A. Naundorf, W. Klaffke, *Eurocarb IX*, Sevilla, Abstr. Book AO 4 (1995).

[43] A. Naundorf, PhD thesis, Universität Hamburg 1997

[44] T. Müller, R. R. Schmidt, *Angew. Chemie Int. Ed. Engl. 34* (1995) 1328–1329.

[45] R. D. Bevill, *Biochem. Biophys. Res. Commun. 30* (1968) 595–599.

[46] L. Lindqvist, K. H. Schweda, P. R. Reeves, A. A. Lindberg, *Eur. J. Biochem. 225* (1994) 863–872.

# 4 Peptide Chemistry and Applications

## 4.1 $\alpha,\alpha$-Disubstituted amino acids and bicyclic lactams: Potential building blocks for the synthesis of peptide mimics

*Bernhard Westermann, Nicole Diedrichs, Ina Gedrath and Armin Walter*

By virtue of their impressive biological profiles, coupled with their complex structural arrangements, naturally occurring peptides are of high interest. However, native peptides turn out to be tedious in pharmaceutical applications. The reasons, among others, for this behavior are that peptides are: (i) susceptible to degradation by proteases; (ii) of low oral availability; and (iii) highly flexible, which causes (i) low metabolic stability; (ii) problematic dosage; and (iii) side-effects due to alternating receptor-interactions.

The mimicry of peptide secondary structural elements offers the possibility to overcome these drawbacks. In addition, the biological active conformation can be determined more easily allowing further rational drug design [1]. In this area, much research has been devoted to the synthesis of $\beta$-turn analogs, whose native counterparts play an important role in peptide receptor recognition and antigen determination. $\beta$-Turns comprise only four amino acids, with the amino acids $i$ and $i+3$ connected via a hydrogen bridge (Figure 4.1.1). To classify certain types of $\beta$-turns, the dihedral angles $\phi$ ($N_\alpha-C_\alpha$ bond) and $\psi$ ($C_\alpha-C_{carbonyl}$ bond) must be examined [2].

The incorporation of a peptide conformation mimetic (building block which enforces a particular conformation when inserted into a peptide chain) may lead to biological active derivatives, in which the flexible peptide chain is stabilized and the dihedral angles are fixed to some extent. Conformational mimics frequently used for this purpose are bicyclic

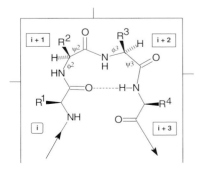

**Figure 4.1.1.** $\beta$-Turn model.

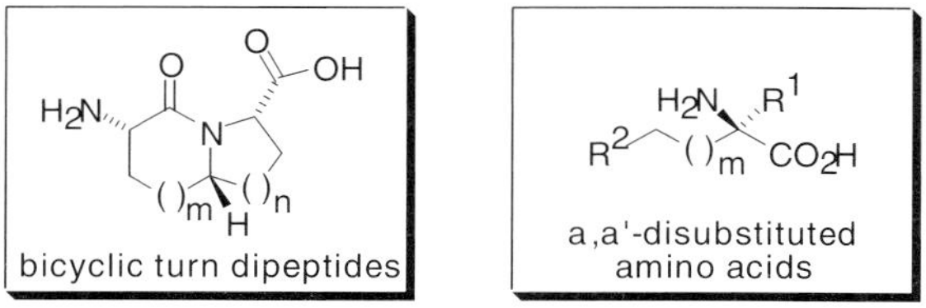

**Figure 4.1.2.** Peptide conformational mimics.

turn dipeptides (bridging amino acids $i + 1$ and $i + 2$) and $\alpha,\alpha$-disubstituted amino acids (substituting amino acid $i + 1$; Figure 4.1.1). In addition, interesting features arise when D-configured amino acids are incorporated at position $i + 1$. These turns (termed x'-turns) are strengthened due to a more stable hydrogen bridge, therefore resulting in a higher metabolic stability [2]. Thus, it is desirable to provide a synthetic route which allows the synthesis of enantiomerically pure conformational mimics (Figure 4.1.2), starting from easily available precursors via a common synthetic scheme, and preferably leading to both enantiomers.

A variety of approaches towards these products have been developed so far mainly with the aid of starting material from the 'chiral pool' [3]. Other approaches based on the utilization of chiral auxiliaries have also been successfully established. Our approach starts from easily available $\beta$-ketoesters, which can be obtained in enantiomerically pure form by enzyme-catalyzed kinetic resolution providing both enantiomers (Scheme 4.1.1).

## 4.1.1    Synthesis of enantiomerically pure $\beta$-ketoesters and derivatives via enzyme-catalyzed kinetic resolutions

In recent years we have established a protocol, by which racemic $\beta$-ketoesters $(\pm)$-**1** bearing a quaternary stereogenic center can be saponified by pig liver esterase PLE (Scheme 4.1.1) [4]. Enzymatic digestion of one enantiomer leads to a $\beta$-ketoacid, which is decarboxylated to yield **2b** during work-up. The remaining $\beta$-ketoester exhibits high enantiomeric purity (>99% ee), as revealed by chiral GLC analysis.

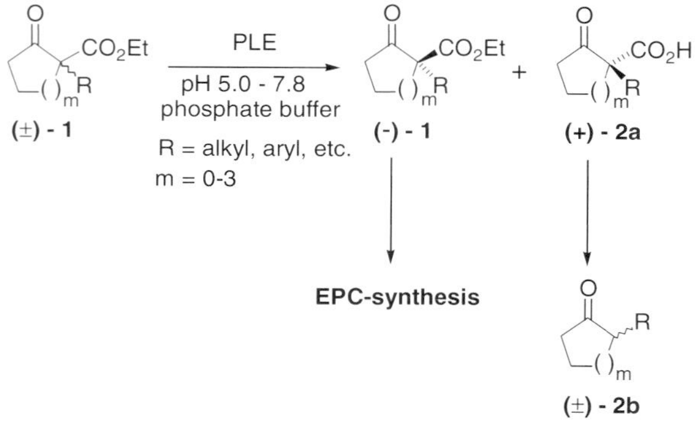

**Scheme 4.1.1.** PLE-catalyzed saponification of $\beta$-ketoesters.

Despite its great utility, this method yields only one enantiomer of the desired starting products. To overcome this drawback, we turned our attention towards the corresponding oximes. The *E*-oximes can be prepared in quantitative yield upon treatment with hydroxyl amine, the ratio *E/Z*-oxime is >50:1. It transpired that we have been able to resolve these intermediates quite efficiently utilizing Lipase PS (Amano). The transesterification of the *n*-butyrated oxime (±)-**3** in the presence of *n*-butanol is superior with respect to enantiomeric excess and reaction time in comparison to the transesterification of vinyl-*n*-butyrate with the oxime. This reaction exhibits excellent enantioselectivity (>90% ee) for both the remaining butyrated oxime (+)-**3** and for the hydrolyzed product (−)-**4**. The ester moiety is not affected under these conditions. Oxime (−)-**4** can easily be separated from (+)-**3** by crystallization and chromatography, respectively.

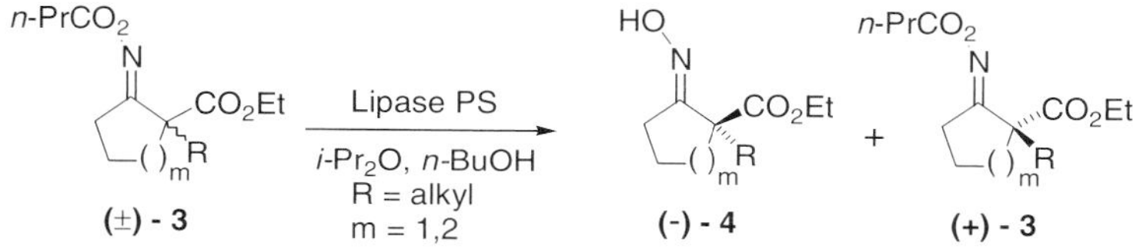

**Scheme 4.1.2.** Lipase-catalyzed kinetic resolution of *E*-oximes.

Hydrolysis of the butyrated oxime (+)-**3** can be achieved with $Na_2CO_3$/methanol in high yields. With chiral, non-racemic oximes **3** in hand, routes to non-natural amino acids and bicyclic turn dipeptides have been devised.

## 4.1.2  Synthesis of α,α′-disubstituted amino acids

Key intermediates for the synthesis of α,α′-disubstituted amino acids are N-protected lactams **6** (Scheme 4.1.3) [5]. Most obviously, the oximes (+)-**4** and (−)-**4** can easily be transformed into their corresponding lactams via Beckmann rearrangement. To initiate the rearrangement, **4** is tosylated in almost quantitative yield and stirred in the presence of silica gel. The lactams **5** can be obtained in excellent yields as the only regio- and stereoisomers. It is well known, that the Beckmann rearrangement occurs with retention of configuration at the stereogenic centre without racemization. In fact, NMR experiments employing Eu(hfc)$_3$ proved that no racemization has taken place.

The lactams **5** can be conveniently protected by treatment with Boc$_2$O. The key step in the devised synthesis towards the amino acids is the ring opening of the protected lactams **6**. This reaction can be carried out by simple treatment with a nucleophile. The nucleophiles employed here have been hydroxyl and alkoxy anions, and the products, being the corresponding esters or acids, can be obtained in high yields. Other nucleophiles such as Grignard reagents can also be used. Due to this highly chemoselective character of the nucleophilic ring opening, the synthesis of fully protected α,α′-disubstituted amino acids **7** can be achieved in three steps. Furthermore, these products exhibit a

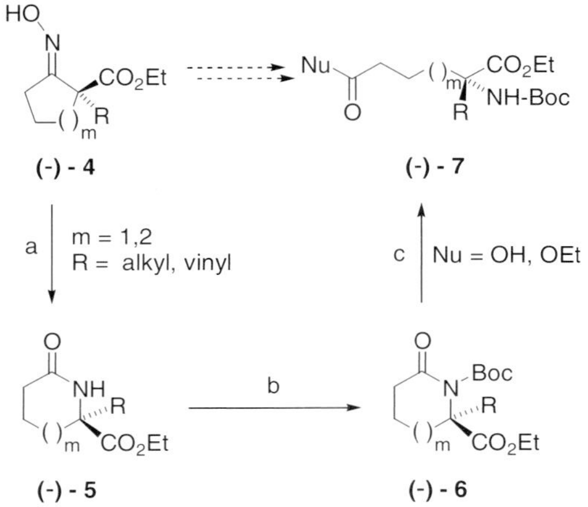

**Scheme 4.1.3.** Synthesis of $\alpha,\alpha'$-disubstituted amino acids **7**. a: (i) *p*-TsOH, py, quant.; (ii) silica gel, $CH_2Cl_2$, 73–87%; b: NaH, $Boc_2O$, 82–87%; c: aq. LiOH or cat. NaOEt, MeOH, 70–95%.

broad pattern of functionality and should be applicable for the synthesis of complex peptides.

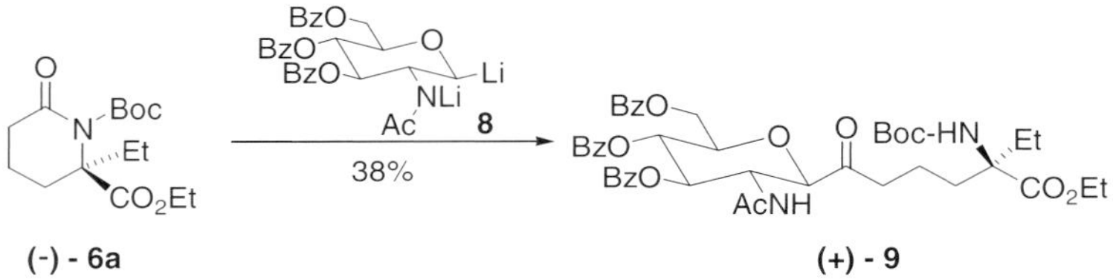

**Scheme 4.1.4.** Synthesis of *C*-glycopeptides.

This powerful method to obtain enantiomerically pure amino acids by nucleophilic ring opening of *N*-Boc-protected lactams **6a** can be demonstrated in the synthesis of *C*-glycopeptides **9** [6]. Employing the dianion of protected GlcNAc **8** as the nucleophilic source [7], the expected product **9** is formed. During this reaction no racemization occurs, yielding diastereomerically pure *C*-glycopeptides. Currently, these products are under intense investigation, because they are hydrolytically stable in comparison to their *O*-glycosylated counterparts.

## 4.1.3    Synthesis of bicyclic turn mimetics

### 4.1.3.1    Synthesis of bicyclic lactams

Bicyclic lactams can be synthesized by two principal methods. Suitable precursors for this ring anellation are monocyclic lactams **10** and **12**, which can be provided by the stereospecific Beckmann rearrangement described above.

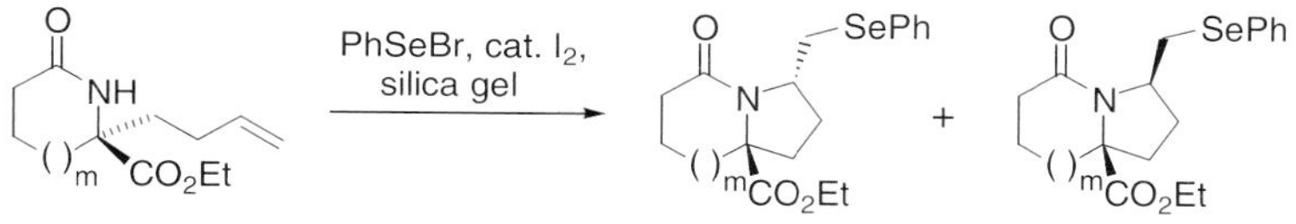

**10** (m = 1)          **11a**    6 : 1; 88%   **11b**
**12** (m = 2)          **13a**    2 : 1; 93%   **13b**

**Scheme 4.1.5.** Synthesis of bicyclic lactams via Seleno ring anellation.

The first route is based on a seleno-mediated ring anellation (Scheme 4.1.5). In the presence of PhSeBr with catalytic amounts of I$_2$, the cyclization yields smoothly the *5-exo-trig* products **11** and **13** in excellent yields. Products formed via a *6-endo-trig* cyclization could not be determined. With the δ-valerolactam (**10**) as educt, the reaction is highly diastereoselective (*trans:cis* = 6:1); in the case of the ε-caprolactam (**12**), the diastereoselctivity is only modest (*trans:cis* = 2:1). Further attempts to increase the diastereoselectivity have been unsuccessful, *e.g.* employing the Nicolaou reagent *N*-phenyl seleno phthalimide leads to a decrease in selectivity [8].

Unfortunately, any experiment to obtain homologous heterocycles by employing the ω-pentenyl-derived lactam failed. For this reason, we devised another route towards bicyclic lactams by using the olefin metathesis (Scheme 4.1.6).

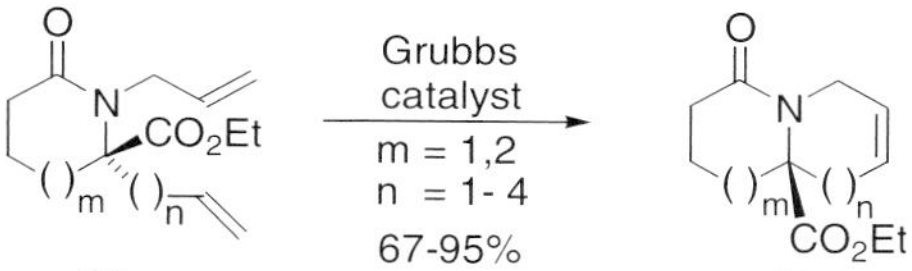

**14**                                    **15**

**Scheme 4.1.6.** Synthesis of bicyclic lactams via olefin metathesis.

Starting compounds are easily available from the monocyclic lactams by deprotonation of the amide and subsequent reaction with allyl bromide or its homologs. The ring closure can be effectively achieved by using the Grubbs catalyst [9]. In high yields the expected bicyclic lactams are formed, providing a functionalized B-ring, which can be transformed further. With these powerful methods in hand, the bicyclic heterocycles can be converted to bicyclic turn peptides.

### 4.1.3.2   Synthesis of bridged tetrapeptides

To demonstrate the synthetic utility of the bicyclic lactams **11**, **13** as building blocks for the synthesis of peptide conformational mimics, we devised routes to transform them into tetrapeptides (Scheme 4.1.7). First, the phenyl selenyl moiety in **11a** had to be converted into a carboxylic group and second, electrophilic amination at the α-position of the lactam had to carried out.

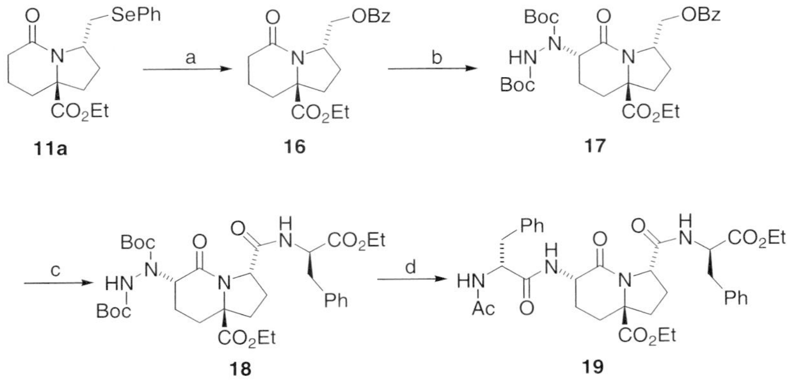

**Scheme 4.1.7.** Synthesis of tetrapeptide **19**. a: (i) m-CPBA, KOH, 82%; (ii) NaH, BzBr, 87%; b: BocN = NBoc, LiHMDS, 89%; c: (i) H$_2$, Pd/C, 72%; (ii) RuCl$_3$/NaIO$_4$, quant.; (iii) CIP, H-PheOEt, 83%; d: (i) TFA, dann H$_2$, Pd/C, 76%; (ii) CIP, AcPheOH, 64%.

The selenyl moiety in **11a** is prone to nucleophilic substitution when oxidized to the selenon. This can be achieved with a five-fold excess of *m*-CPBA. In situ treatment with KOH displaces the selenon, yielding the corresponding alcohol in high yield. Subsequent protection of the hydroxyl group with benzyl bromide leads to lactam **16**. Next, we turned our attention towards the incorporation of the amine residue. Electrophilc amination can be effectively carried out with *t*-butyl azodicarboxylate [10]. After deprotonation of **16** with LiHMDS, the desired hydrazino derivative **17** is formed. Two diastereomers are obtained (*trans*:*cis* = 4:1).

With the amino moiety protected with a group stable under basic conditions, we transformed the hydroxyl moiety. After deprotection, oxidation of the hydroxyl group is carried out with RuCl$_3$/NaIO$_4$. Under these conditions, no deprotection of the amino-protecting group occurs. Subsequent coupling with phenylalanine ethylester to yield **18** is achieved with the CIP (2-Chloro-1,3-dimethyl imidazolidinium hexafluorophosphate)-coupling reagent [11], which is superior to other known coupling reagents.

Further transformation of the hydrazino residue is done by deprotection in two steps, treatment with trifluoroacetic acid and hydrogenolysis of the remaining diimine leads to the desired product, which can be isolated as its trifluoroacetate. Finally, derivatization of the amino residue can be achieved again by coupling with *N*-acetyl phenylalanine to yield **19**. This sequence is a highly efficient process for the formation of bicyclic turn mimics. Although not shown in detail, it can also be used to synthesize tetrapeptides bearing the 7,5-bicyclic lactam (from **13**) as the bridging fragment.

The carboxylic residue at the quaternary center may also play a pivotal role. It is quite obvious, that this residue can also be used for peptide condensations. After saponification, peptide couplings can be carried out easily, leading to tetrapeptide **20** (Scheme 4.1.8).

At this stage the structural refinement of the two bridged peptides **19** and **20** should be taken into consideration. They both mimic a proline-containing dipeptide, **21** the

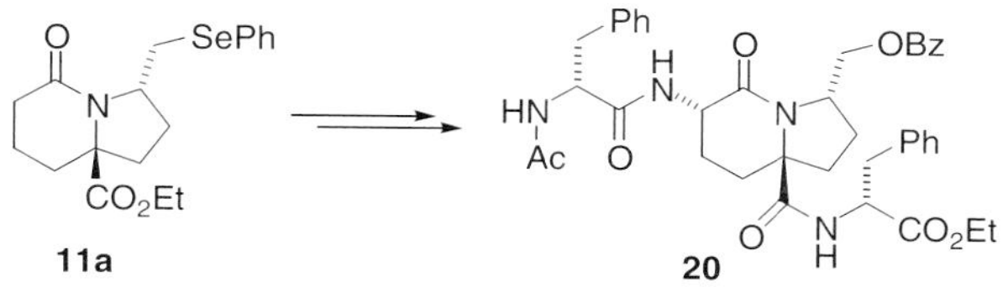

**Scheme 4.1.8.** Synthesis of tetrapeptide **20**.

*cis*-configured Xaa-proline dipeptide, **22** the *trans*-configured Xaa-proline dipeptide (Figure 4.1.3). Although derivatives exist which mimic both the proline-containing dipeptides, this synthetic strategy offers the possibility to have both conformers easily available starting from one common precursor.

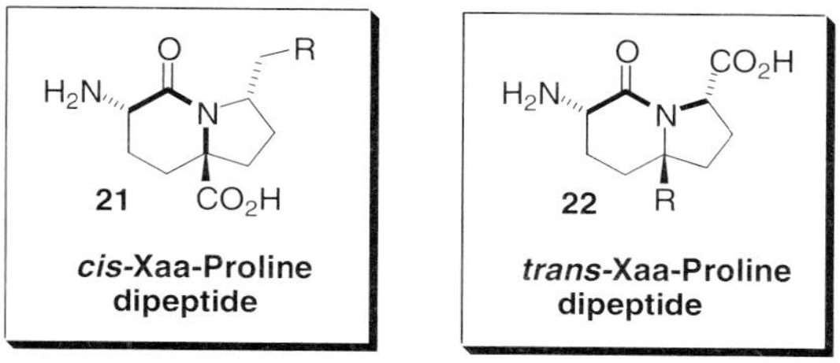

**Figure 4.1.3.** *cis/trans* dipeptide conformation mimcs.

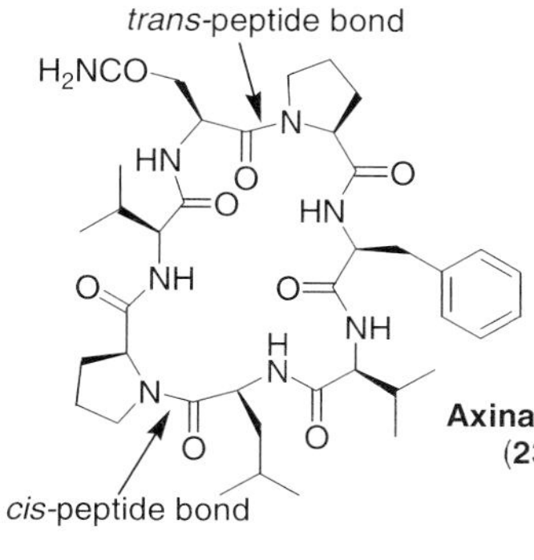

Therefore, the protocol presented in this account is of general utility for the synthesis of these products and meeting the criteria of a flexible synthetic route. It offers significant advantages over the use of 'chiral pool' synthesis, leading to homologous derivatives. Due to the enzyme-catalyzed kinetic resolution, which is often regarded as a disadvantage during a synthetic sequence, both enantiomers can be obtained. Currently, incorporation of these bicyclic lactams into cyclic peptides like axinastatin II **23** is under investigation [12]. In this heptapeptide, two proline residues are incorporated, connected via a *cis*- and *trans*-peptide bond, respectively. Therefore, it is most interesting to synthesize cyclic heptapeptides by employing bridged turn lactams, presented in this account.

**Acknowledgements:** This work was supported by the Deutsche Forschungsgemeinschaft and the Fonds der Chemischen Industrie. N. D. and I. K. like to thank the Fonds der Chemischen Industrie for a doctoral scholarship.

# References

[1]  T. Kolter, A. Giannis, *Angew. Chem.* **1993**, *105*, 1303–1326; *Angew. Chem. Int. Ed.* **1995**, *32*, 1244; J. Gante, *Angew. Chem.* **1994**, *106*, 1780–1802; *Angew. Chem. Int. Ed.* **1994**, *33*, 1699.

[2]  W. F. Degrado, *Adv. Prot. Chem.* **1988**, *39*, 51–124; G. D. Rose, L. M. Gierasch, J. A. Smith, *Adv. Prot. Chem.* **1985**, *37*, 1–109; J. B. Ball, R. A. Hughes, P. F. Alewood, P. R. Andrews, *Tetrahedron* **1993**, *49*, 3467–3478.

[3]  T. Wirth, *Angew. Chem.* **1997**, *109*, 235–237; *Angew. Chem. Int. Ed.* **1997**, *36*, 225–227; R. M. Williams, *Synthesis of Optically Active Amino Acids*, Pergamon Press, Oxford, **1988**; S. Hanessian, G. McNaughton-Smith, H.-G. Lombart, W. D. Lubell, *Tetrahedron* **1997**, *53*, 12789–12854.

[4]  B. Westermann, H. Scharmann, I. Kortmann, *Tetrahedron: Asymmetry* **1993**, *4*, 2119–2122, B. Westermann, I.Kortmann, *Biocatalysis* **1994**, *10*, 289–294; B. Westermann, S. Dubberke, *Liebigs Ann./Recueil* **1997**, 375–380.

[5]  B. Westermann, I. Gedrath, *Synlett* **1996**, 665–666.

[6]  B. Westermann, A. Walter, N. Diedrichs, *Angew. Chem.* accepted.

[7]  M. Hoffmann, H. Kessler, *Tetrahedron Lett.* **1994**, *35*, 6067–6070; V. Wittmann, H. Kessler, *Angew. Chem.* **1993**, *105*, 1138–1140; *Angew. Chem. Int. Ed.* **1993**, *32*, 1091.

[8]  J. Cossy, N. Furet, *Tetrahedron Lett.* **1993**, *34*, 7755–7756.

[9]  R. H. Grubbs, S. J. Miller, G. Fu, *Acc. Chem. Res.* **1995**, *28*, 446–452; S. Schuster, S. Blechert, *Angew. Chem.* **1997**, *109*, 2124–2145; *Angew. Chem. Int. Ed.* **1997**, *36*, 2036–2055; N. Diedrichs, B. Westermann *Synlett* **1999**, 1127–1129.

[10]  D. Gramber, C. Weber, R. Beeli, J. Inglis, C. Bruns, J. A. Robinson, *Helv. Chim. Acta* **1995**, *78*, 1588–1606.

[11]  Y. Kiso, Y. Fujiwara, T. Kimura, A. Nishitani, K. Akaji, *Int. J. Pept. Prot. Res.* **1992**, *40*, 308–314.

[12]  O. Mechnich, G. Hessler, H. Kessler, M. Bernd, B. Kutscher, *Helv. Chim. Acta* **1997**, *80*, 1338–1354.

# 4.2 β-Amino acids as building blocks for peptide modification

*Norbert Sewald*

Although being less abundant than the corresponding α-amino acids, β-amino acids [1] occur in nature both in free form and bound in peptides. They are stronger bases and weaker acids compared to the α-amino correlates. The skeleton atom pattern and, consequently, the hydrogen bond pattern of a peptide will be modified considerably upon incorporation of β-amino acids. This may result in a modulation of the physiological activity due to conformational changes in the pharmacophoric region and in increased metabolic stability [2, 3]. β-Peptides, consisting exclusively of β-amino acids, are completely stable towards proteolysis [4].

## 4.2.1 β-Amino acids in medicinal chemistry – a brief survey

In the past, mainly the simple motifs β-alanine [5] or isoaspartic acid [6] have been used for peptide modification. The modified N-terminal fragments [β-Ala$^1$]-ACTH-(1–18)-NH$_2$, [β-Ala$^1$]-ACTH-(1–23)-NH$_2$, and [β-Ala$^1$]-ACTH-(1–24)-NH$_2$ of corticotropin (ACTH, adrenocorticotropic hormone), which stimulates the synthesis of gluco- and mineralocorticoids, display higher and prolonged adrenocorticotropic and lipotropic activity [5, 7].

The nonapeptide bradykinin (Arg-Pro-Pro-Gly-Phe-Ser-Pro-Phe-Arg) plays a role in blood pressure regulation and has a serum half-life of less than 0.5 min. The metabolic cascade in vivo starts *e.g.* with a cleavage of the Pro$^7$–Phe$^8$ peptide bond by endopeptidase 24.11 and ACE (angiotensin converting enzyme), respectively, or of the Gly$^4$–Phe$^5$ or Phe$^5$–Ser$^6$ bond by endopeptidase 24.18 [8]. [β-HPro$^7$]-Bradykinin has a vasodepressant activity comparable to that of bradykinin and is stable in vitro towards dipeptidylcarboxypeptidase [2]. A *retro-inverso* analog of kelatorphan containing β-homophenylalanine efficiently inhibits neutral endopeptidase 24.11 (vide supra, EC 3.4.24.11), a zinc metallopeptidase responsible for the degradation of enkephalins (endogenous opioid peptides) in brain by cleavage of the Gly–Phe bond [9].

The nonapeptide vasopressin (ADH, antidiuretic hormone) has antidiuretic properties and also causes contractions of peripheral blood vessels, while the analog [Lys$^8$,β-Ala$^9$]-vasopressin exhibits antagonistic properties [10]. Angiotensin II (hypertensin II, Asp-Arg-Val-Tyr-Ile-His-Pro-Phe) is responsible inter alia for blood pressure control and osmoregulation. [β-Asp$^1$,Val$^5$]-Angiotensin II is more active and metabolically more stable than its parent angiotensin II [6a].

The peptide hormones gastrin and cholecystokinin (CCK) are important regulators of gastric secretion. Preferred cleavage of the Met-Asp bond occurs upon incubation of the C-terminal tetrapeptide fragment Trp-Met-Asp-Phe-NH$_2$ (Tetragastrin), which already displays gastrin activity, with a mucosal preparation. Boc-[Leu$^2$,β-HAsp$^3$]-Tetragastrin acts as a gastrin antagonist [11]. CCK is secreted from the duodenum into the blood circulation. CCK-B receptor antagonists, that may be important for the treatment of gastrointestinal or CNS targets [12], also exhibit gastrin antagonism [13a]. The dipeptide Adoc-($\alpha$-Me)Trp-$\beta$-HPhe-OH is a nanomolar CCK receptor antagonist, with the R,S-diastereomer being CCK-B selective and the S,R-diastereomer CCK-A selective.

Only recently, $\beta$-amino acids have also experienced a renaissance as aspartate mimetics in non-peptide RGD-antagonists (vide infra) [14].

## 4.2.2   Do $\beta$-amino acids induce secondary structure elements?

Predictable folding patterns of peptides, especially in the pharmacophoric regions, are desirable in the context of a structure-based drug design in order to deduce information on the relative spatial orientation of pharmacophoric groups. Certain $\alpha$-amino acids stabilize, for instance, $\beta$-turns. Proline with a *trans*-configured peptide bond has a high propensity to occur in position $i + 1$, *e.g.* of a $\beta$I- or $\beta$II-turn [15], while $i + 2$ of a $\beta$VI-turn is favored in the case of a *cis* peptide bond. D-Proline and D-amino acids in general show a high preference for position $i + 1$ of a $\beta$II$'$-turn [16], while glycine frequently is found in $i + 1$ or $i + 2$ of a $\beta$-turn. Replacement of an $\alpha$-amino acid by its $\beta$-amino analog in a peptide extends the backbone by one carbon atom. Homo-oligomers of $\beta$-amino acids ($\beta$-peptides) have recently attracted considerable attention as they form a novel class of peptide analogs that adopt predictable and reproducible folding patterns, *e.g.* helices [17]. The issue of whether peptides containing both $\alpha$- and $\beta$-amino acids display new types of secondary structures has been addressed only in a few cases [18]. Hence, as yet very little information on conformational preferences of $\beta$-amino acids is available.

The natural product cyclochlorotin (Figure 4.2.1), produced by *Penicillium* species, contains $\beta$-Phe besides the non-proteinogenic $\alpha$-amino acids $\alpha$-aminobutyric acid

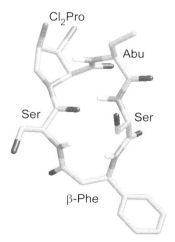

**Figure 4.2.1.** 3D structure of cyclochlorotin.

(Abu) and dichloroproline (Cl$_2$Pro) [19]. The pentapeptide conformation in the solid state comprises a combination of a β-turn around Ser-Cl$_2$Pro-Abu-Ser and a complementary γ-turn around Ser-β-Phe-Ser. This result could suggest that β-amino acids act as γ-turn mimetics. However, the strong conformational preference of a proline residue for position $i + 1$ in a β-turn will certainly dominate the secondary structure of the cyclopeptide. Therefore, it is not possible to generally deduce from this example a significant propensity of a β-amino acid for the central position ($i + 1$) of a γ-turn. However, in nearly all cyclopeptide examples examined until now, β-amino acids occur together with α-amino acids exhibiting strong conformational bias (*e.g.* proline [18c–e], D-amino acids [18d], *N*-alkyl amino acids [18f]). It has nevertheless been stated that β-amino acids display no preference for positions $i + 1$ and $i + 2$ of a β-turn [18e]. This C$_{11}$ conformation (hydrogen-bonded structure with a 'ring' consisting of 11 atoms) has been found only in a very few cases, *e.g.* Boc-Aib-Aib-β-Ala-NHMe [20], the linear nonapeptide leucinostatin A [21], and the cyclic pentapeptide *cyclo*(-Pro-Phe-Phe-β-Ala-β-Ala-) [18e]. Consequently, the question whether β-amino acids display intrinsic conformational preferences when incorporated into peptides composed of α-amino acids cannot be answered unambiguously on the basis of the literature data available.

## 4.2.3  Synthesis of enantiomerically pure β-amino acids

There are several different methods for the preparation of enantiomerically pure β-amino acids [22]. The stereoconservative chain elongation of suitably protected α-amino acids with diazomethane (Arndt–Eistert synthesis) represents the simplest strategy, but relies on the chiral pool. The usual protocol for the Wolff rearrangement of the intermediate diazoketone requires the addition of several equivalents of a base, which is detrimental for the preparation of Fmoc-protected β-amino acids. A sonochemical variant (Scheme 4.2.1) permits a base-free Ag$^+$-catalyzed Wolff rearrangement [23].

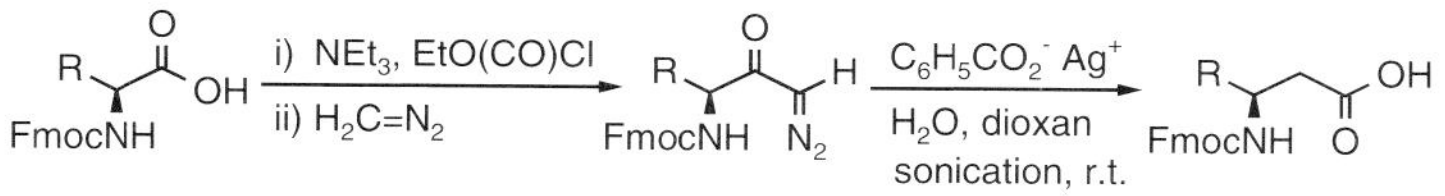

**Scheme 4.2.1.**

The 1,4-addition of chiral nitrogen nucleophiles to α,β-unsaturated esters is a much more versatile route towards β-amino acids [22]. Lithium (1-phenylethyl)trimethylsilyl amide or the corresponding lithium amidocuprate have, among others, been applied successfully as enantiomerically pure reagents (Scheme 4.2.2).

This strategy allows the synthesis of a broad variety of β-amino acids with different functional groups and substitution patterns. Trapping of the intermediate ester enolate, *e.g.* with deuterium oxide, proceeds stereoselectively to give *anti*-α-deuterio β-amino acids [24a,c].

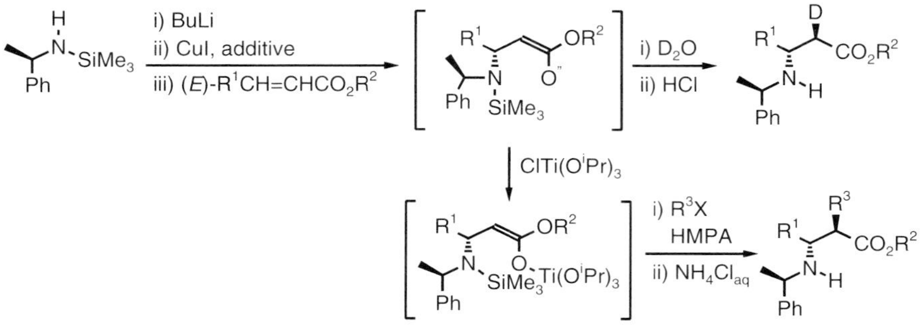

**Scheme 4.2.2.**

However, the reactivity of the intermediate lithium/copper ester enolates is not sufficient for a tandem alkylation, presumably because of the existence of sterically demanding bi- or oligonuclear copper complexes of still unknown structure. A stabilization of the intermediate via co-ordination to silicon, as it was postulated earlier, can be ruled out on the basis of $^{29}$Si NMR measurements. Transmetallation to mononuclear titanium complexes breaks up the aggregates and alkylation gives *anti*-$\alpha$-alkyl $\beta$-amino acids [24c].

Chiral recognition phenomena are observed when an additional stereogenic center is present in the substrate (Scheme 4.2.3). Addition of the chiral lithium amides results in certain cases in a purely reagent-controlled stereoselective addition. Both diastereomers of $\beta$-homothreonine (*erythro* or *threo*) and the corresponding 2,3-*anti*-2-deuterio derivatives can be obtained selectively [24b, c].

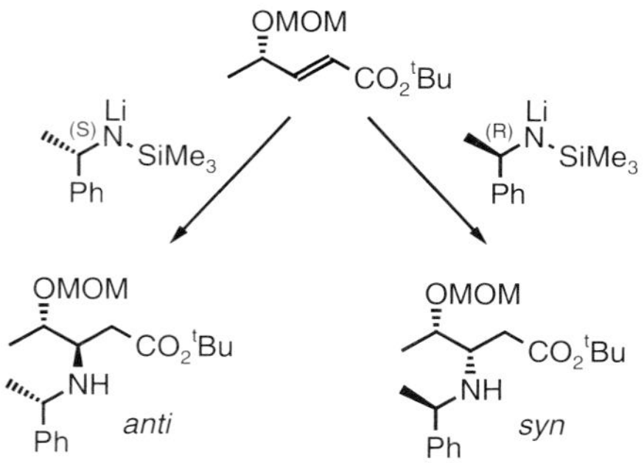

**Scheme 4.2.3.**

A new modular approach towards a catalytic asymmetric synthesis of $\beta^2$-homo-amino acids [25] and $\alpha,\beta$-disubstituted $\beta$-amino acids, respectively, is currently under investigation (Scheme 4.2.4). As the organozinc reagents used in the first step of this process tolerate a broad variety of functional groups (*e.g.* esters, acetals, etc.), multifunctional derivatives can be obtained. Both the amino and the carboxylic group of the β-amino acid are already present in the substrate in a masked form (nitro/acetal group).

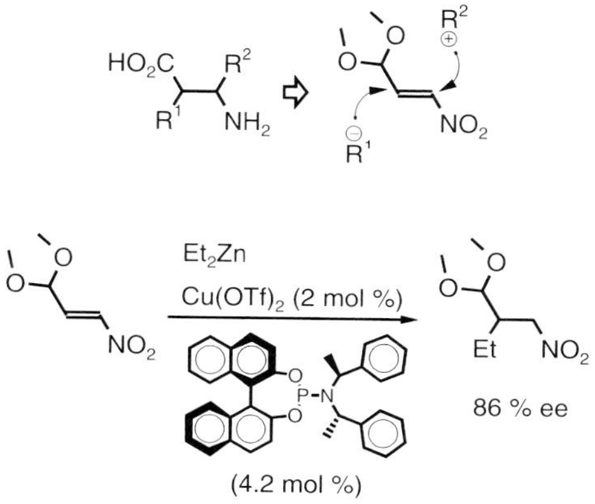

**Scheme 4.2.4.**

## 4.2.4   Cyclic RGD peptides containing β-amino acids

The amino acid sequence Arg-Gly-Asp (RGD) plays a major role in cellular binding and recognition phenomena involved in blood platelet aggregation, tumor cell adhesion, and osteoporosis. The cyclic penta- and hexapeptides c-(RGDfV) and c-(RGDfVG) containing the sequence Arg-Gly-Asp antagonize binding of proteins containing the RGD triad to the corresponding integrins (heterodimeric glycoprotein receptors) [26]. The selectivity of these peptides between the fibrinogen receptor $\alpha_{IIb}\beta_3$ (responsible for blood platelet aggregation) and the vitronectin receptor $\alpha_V\beta_3$ (responsible i.a. for tumor cell adhesion) depends strongly on the local conformation of the pharmacophoric RGD sequence, determined by the relative position of a D-amino acid as a turn-inducing element. D-Phe (f) has a strong bias for position $i + 1$ of a βII'-turn. Consequently, both cyclopeptides c-(RGDfV) and c-(RGDfVG) form βII'-turns around DfVR. This forces the pharmacophoric RGD sequence in c-(RGDfV) into a γ-turn arrangement, while in c-(RGDfVG) a second β-turn around GRGD is formed with Arg-Gly-Asp in positions $i + 1$ to $i + 3$ [26]. Hence, this class of peptides provides excellent model systems for probing the consequences of peptide modification using β-amino acids in terms of biological activity and solution structure.

Cyclic RGD peptides have been modified using β-amino acids and their activity as antagonists of blood platelet aggregation has been examined. A strategy for the synthesis of cyclic peptides was modified in order to permit on-resin cyclization of the peptide [27], allowing multiple or combinatorial syntheses of the target peptides (Figure 4.2.2).

Blood platelets are stimulated by a variety of physiologically relevant agents (*e.g.* ADP, thrombin, etc.) resulting in activation of $\alpha_{IIb}\beta_3$ receptors (GPIIb/IIIa) on the

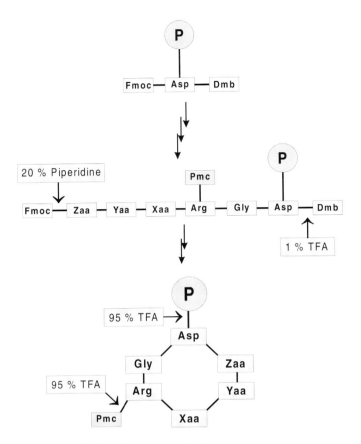

**Figure 4.2.2.** SPPS with on-resin cyclization.

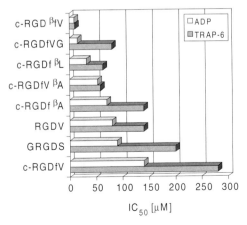

**Figure 4.2.3.** Anti-aggregatory properties (thrombocyte aggregation).

cell surface. Preliminary aggregometric studies regarding the anti-aggregatory activity of cyclic RGD peptides modified by β-amino acids reveal a dramatic change of the activity in some cases (Figure 4.2.3) [27]. The pentapeptide c-RGD$^\beta$fV is the most efficient $\alpha_{IIb}\beta_3$ antagonist among the modified analogs examined. Its potency is in the same range as of the highly active $\alpha_{IIb}\beta_3$ antagonist c-RGDfVG known from the literature. The hexapeptide c-RGD$^\beta$fVG is much less active (IC$_{50}$ > 300 µM).

These findings suggest that secondary structure and biological activity of c-(RGDfV) and c-(RGDfVG) are altered dramatically on replacement of D-Phe (f) by D-β-Phe ($^\beta$f) and that β-amino acids might exert conformational bias on the 3D structure of peptides. A detailed NMR/MD analysis of the solution conformation of relevant peptides modified by β-amino acids is currently in progress.

**Acknowledgments:** The author wishes to express his gratitude to the Deutsche Forschungsgemeinschaft (Innovationskolleg "Chemisches Signal und biologische Autwort") and the Fonds der Chemischen Industrie for continuous support, to Annett Müller, Klaus Hiller, Matthias Körner, Audrius Rimkus, Frank Schumann, and Volkmar Wendisch for their enthusiastic engagement in their projects, and to Mario Koksch, Carla Vogt, and Matthias Findeisen for cooperations.

# References

[1] The expression β-amino acids usually includes all derivatives where two carbon atoms are situated between the carboxylic group and amino group. The term β-homoamino acid describes the insertion of an additional $C_1$ unit into an α-amino acid. This class of compounds can be divided into two subclasses, $\beta^2$-homoamino acids and $\beta^3$-homoamino acids, depending on the position of the side chain, which may be attached to $C^\alpha$ or $C^\beta$.

[2] M. A. Ondetti, S. L. Engel, *J. Med. Chem.*, **1975**, *18*, 761–763.

[3] K. Stachowiak, M. C. Khosla, K. Plucinska, P. A. Khairallah, F. M. Bumpus, *J. Med. Chem.*, **1979**, *22*, 1128–1130.

[4] T. Hintermann, D. Seebach, *Chimia*, **1997**, *5*, 244–247.

[5] K. Inouye, A. Tanaka, H. Otsuka, *Bull. Chem. Soc. Japan*, **1970**, *43*, 1163–1172.

[6] (a) B. von Riniker, R. Schwyzer, *Helv. Chim. Acta*, **1964**, *47*, 2357–2374; (b) J. Seprödi, J. Érchegyi, Z. Vadász, I. Teplán, I. Mezö, B. Kanyicska, M. Kovács, S. Vigh, *Biochem. Biophys. Res. Commun.*, **1987**, *144*, 1214–1221; (c) J. Martinez, F. Winternitz, M. Bodanszky, J. D. Gardner, M. D. Walker, V. Mutt, *J. Med. Chem.*, **1982**, *25*, 589–593.

[7] M. Fujino, C. Hatanaka, O. Nishimura, *Chem. Pharm. Bull.*, **1970**, *18*, 771–778; ibid. 1288–1291.

[8] R. H. Erickson *Peptide Metabolism at Brush-Border Membranes* in *Peptide-Based Drug Design*, M. D. Taylor, G. L. Amidon, Eds. (American Chemical Society, Washington DC, **1995**) 23–45.

[9] J. F. Hernandez, J. M. Soleilhac, B. P. Roques, M. C. Fournié-Zaluski, *J. Med. Chem.*, **1988**, *31*, 1825–1831.

[10] C. M. Drey *The Chemistry and Biochemistry of β-Amino Acids* in *Chemistry and Biochemistry of Amino Acids, Peptides, and Proteins*, B. Weinstein, Ed. (Marcel Dekker, New York, **1977**) 269–299.

[11] M. Rodriguez, P. Fulcrand, J. Laur, A. Aumelas, J. P. Bali, J. Martinez, *J. Med. Chem.*, **1989**, *32*, 522–528.

[12] T. K. Sawyer *Peptidomimetic Design and Chemical Approaches to Peptide Metabolism* in *Peptide-Based Drug Design*, M. D. Taylor, G. L. Amidon, Eds. (American Chemical Society, Washington DC, **1995**) 387–422.

[13] (a) D. Hill, D. C. Horwell, J. C. Hunter, C. O. Kneen, M. C. Pritchard, N. Suman-Chauhan, *Bioorg. Med. Chem. Lett.*, **1993**, *3*, 885–888; (b) M. Higginbottom, D. C. Horwell, E. Roberts, *Bioorg. Med. Chem. Lett.*, **1993**, *3*, 881–884.

[14] (a) J. H. Hutchinson, J. J. Cook, K. M. Brashear, M. J. Breslin, J. D. Glass, R. J. Gould, W. Halczenko, M. A. Holahan, R. J. Lynch, G. R. Sitko, M. T. Stranieri, G. D. Hartman, *J. Med. Chem.*, **1996**, *39*, 4583–4591; (b) H. U. Stilz, G. Beck, B. Jablonka, M. Just, *Bull. Chem. Soc. Belg.*, **1996**, *105*, 711–719; (c) M. J. Fisher, B. P. Gunn, C. S. Harms, A. D. Kline, J. T. Mullaney, R. M. Scarborough, M. A. Skelton, S. L. Um, B. G. Utterback, J. A. Jakubowski, *Bioorg. Med. Chem. Lett.*, **1997**, *7*, 2537–2542.

[15] G. Müller, M. Gurrath, M. Kurz, H. Kessler, *PROTEINS: Structure, Function, and Genetics*, **1993**, *15*, 235–251.

[16] J. W. Bean, K. D. Kopple, C. E. Peishoff, *J. Am. Chem. Soc.*, **1992**, *114*, 5328–5334.

[17] (a) F. López-Carrasquero, M. García-Alvarez, J. J. Navas, C. Alemán, S. Muñoz-Guerra, *Macromolecules* **1996**, *29*, 8449–8459 and cited references; (b) D. H. Appella, L. A. Christianson, D. A. Klein, D. R. Powell, X. Huang, J. J. Barchi, Jr., S. H. Gellman, *Nature* **1997**, *387*, 381–384 and cited references; (c) D. Seebach, J. L. Matthews, *J. Chem. Soc., Chem. Commun.* **1997**, 2015–2022 and cited references.

[18] (a) D. F. Mierke, G. Nößner, P. W. Schiller, M. Goodman, *Int. J. Pept. Protein Res.*, **1990**, *35*, 35–45; (b) T. Yamazaki, A. Pröbstl, P. W. Schiller, M. Goodman, *Int. J. Pept. Protein Res.*, **1991**, *37*, 364–381; (c) I. L. Karle, B. K. Handa, C. H. Hassall, *Acta Cryst.*, **1975**, *B 31*, 555–560; (d) E. Graf von Roedern, E. Lohof, G. Hessler, M. Hoffmann, H. Kessler, *J. Am. Chem. Soc.*, **1996**, *118*, 10156–10167; (e) A. Lombardi, M. Saviano, F. Nastri, O. Maglio, M. Mazzeo, C. Isernia, L. Paolillo, V. Pavone, *Biopolymers*, **1996**, *38*, 693–703 and cited references; (f) K. Yoda, H. Haruyama, H. Kuwano, K. Hamano, K. Tanzawa, *Tetrahedron*, **1994**, *50*, 6537–6548.

[19] H. Morita, S. Nagashima, K. Takeya, H. Itokawa, Y. Iitaka, *Tetrahedron*, **1995**, *51*, 1121–1132.

[20] V. Pavone, B. Di Blasio, A. Lombardi, C. Isernia, C. Pedone, E. Benedetti, G. Valle, M. Crisma, C. Toniolo, R. Kishore, *J. Chem. Soc., Perkin Trans. 2*, **1992**, 1233–1237.

[21] S. Cerrini, D. Lamba, A. Scatturin, G. Ughetto, *Biopolymers*, **1989**, *28*, 409–420.

[22] (a) *Enantioselective Synthesis of β-Amino Acids*, E. Juaristi, Ed. (Wiley-VCH, New York, **1996**); (b) N. Sewald, *Amino Acids*, **1996**, *9*, 397–408.

[23] (a) A. Müller, Diploma thesis, Leipzig, **1996**; (b) A. Müller, C. Vogt, N. Sewald, *Synthesis*, **1998**, 837–841.

[24] (a) N. Sewald, K. D. Hiller, B. Helmreich, *Liebigs Ann. Chem.*, **1995**, 925–928; (b) M. Körner, M. Findeisen, N. Sewald, *Tetrahedron Lett.*, **1998**, *39*, 3463–3464; (c) N. Sewald, K. D. Hiller, M. Körner, M. Findeisen, *J. Org. Chem.*, **1998**, *63*, 7263–7274.

[25] N. Sewald, V. Wendisch, *Tetrahedron: Asymmetry*, **1998**, *9*, 1341–1344. A stoichiometric approach has been described by T. Hintermann, D. Seebach, *Synlett*, **1997**, 437–438.

[26] (a) M. Aumailley, M. Gurrath, G. Müller, J. Calvete, R. Timpl, H. Kessler, *FEBS Lett.* **1991**, *291*, 50–54; (b) M. Pfaff, K. Tangemann, B. Müller, M. Gurrath, G. Müller, H. Kessler, R. Timpl, J. Engel, *J. Biol. Chem.*, **1994**, *269*, 20233–20238; (c) R. Haubner, D. Finsinger, H. Kessler, *Angew. Chem.*, **1997**, *109*, 1440–1456; *Angew. Chem. Int. Ed. Engl.*, **1997**, *36*, 1374–1389.

[27] A. Müller, F. Schumann, M. Koksch, N. Sewald, *Lett. Pept. Sci.*, **1997**, *4*, 275–281.

# 4.3 Reactions of chelated amino acid ester enolates and their application to natural product synthesis

*Uli Kazmaier*

For some time our group has been interested in the reactions of chelated enolates such as **1** [1]. These enolates can easily be obtained from protected amino acid esters (Y = protecting group) *via* deprotonation with LDA or LHMDS and subsequent transmetallation by addition of metal salts (MX$_n$). The chelated enolates can be trapped with electrophiles such as aldehydes or ketones, giving rise to the corresponding aldol products. On the other hand, if allylic esters are used, warming the enolates to room temperature results in a Claisen rearrangement, providing $\gamma,\delta$-unsaturated amino acids (Scheme 4.3.1).

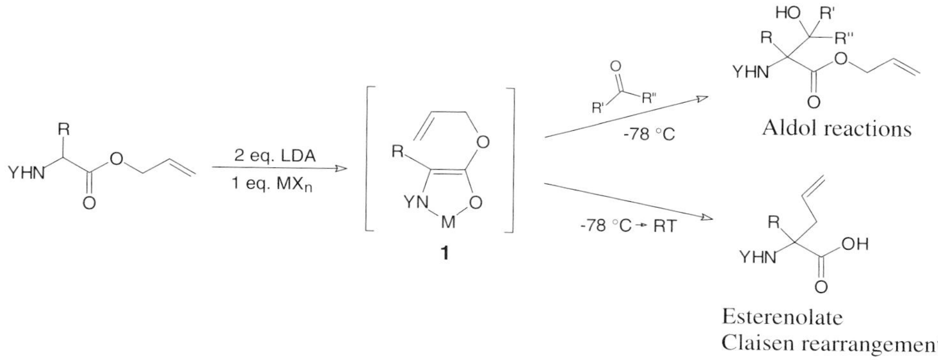

**Scheme 4.3.1.** Reactions of chelated amino acid ester enolates **1**.

These chelated enolates have several advantages in comparison to their non-chelated analogs:

1. Because of the fixation of the enolate geometry by chelation, many reactions of these enolates proceed with a high degree of diastereoselectivity. For example, aldol reactions of **1** with various aldehydes in the presence of titanium or zinc salts occur in a highly diastereoselective fashion, whereas the corresponding lithium enolates show only modest selectivity [2].
2. The chelated enolates are significantly more stable than the corresponding, probably non-chelated, lithium enolates. In contrast to these, the chelate enolates can be warmed to room temperature without decomposition. This allows the expansion of the field of enolate chemistry to reactions, such as Claisen rearrangements [3], which in general cannot be carried out with *normal*, non-stabilized enolates.
3. Many manipulations on enolates such as **1** are possible. Besides variations of the protecting group **Y**, an excessive *metal tuning* should allow modification of the

reactivity and selectivity of these enolates. Because, in most cases, the coordination sphere of the metal ion M is not saturated in the bidentate enolate complex, the additional coordination of chiral ligands on the chelated metal is possible [4].

## 4.3.1    Asymmetric aldol reactions

Aldol reactions of chelated amino acid ester enolates result in the formation of the important class of $\beta$-hydroxy amino acids. We are especially interested in the synthesis of $\alpha$-alkylated derivatives [5]. Their introduction into peptides reduces the number of peptide conformations and increases the stability towards proteases. This fact is especially interesting from a pharmaceutical point of view [6]. If derivatives of polyhydroxylated aldehydes like **3** are used, this approach allows for the synthesis of polyhydroxylated amino acids, which can serve as precursors for the synthesis of highly oxygenated piperidine alkaloids (**6**, **7**) (Scheme 4.3.2) [7].

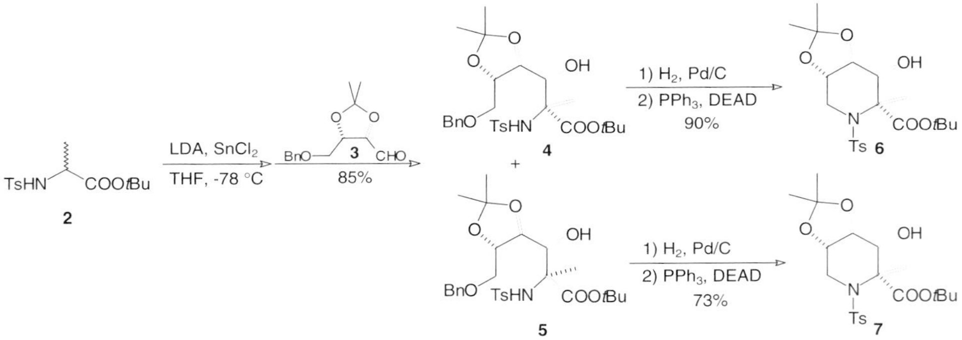

**Scheme 4.3.2.** Synthesis of $\alpha$-methylated pipecolinic acid derivatives **6,7** *via* aldol reaction.

Deprotonation of alanine ester **2** with an excess of LDA and subsequent transmetalation with tin chloride, probably results in the formation of a chelated enolate. Addition of chiral aldehyde **3** gives rise to the aldol products **4** and **5** in a 4:1 ratio and excellent yield and induced diastereoselectivity (95% ds). Diastereomerically pure **4** can be obtained by a single crystallization step (55% yield from **2**). The pure minor diastereomer **5** can be obtained by flash chromatography. After cleavage of the benzyl ether and subsequent cyclization under Mitsunobu conditions [8], the $\alpha$-methylated pipecolinic acid derivatives **6** and **7** are formed in high yields [9].

This sequence can also be applied to the synthesis of azasugars and piperidine alkaloids like 1-deoxyaltronojirimycin (**11**) (Scheme 4.3.3) [10]. In this case, aldol reaction of the glycine ester **8** provided amino acid ester **9** as an 1:1 epimeric mixture. This mixture can be used directly for the subsequent piperidine synthesis, because in the Mitsunobu cyclization an epimerization takes place and *only one* diastereomer of the pipecolinic acid derivative **10** is obtained. Addition of an excess Red-Al results in a simultaneous cleavage of the N-tosyl protecting group and reduction of the ester moiety. Removal of the acid-labile groups with Dowex 50Wx8 gives rise to 1-deoxyal-tronojirimycin (**11**).

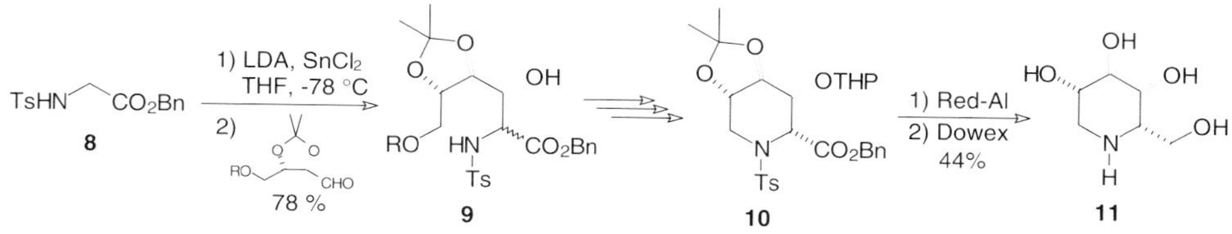

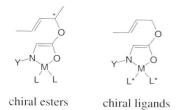

**Scheme 4.3.3.** Synthesis of 1-deoxyaltronojirimycin (**11**).

## 4.3.2   Asymmetric chelate Claisen rearrangements

Recently, we have developed a new variation of the ester enolate Claisen rearrange-ment proceeding *via* chelated allylic ester enolates (Scheme 4.3.1). This method is suitable for the synthesis of various types of amino acids and peptides [1]. Due to the fixed enolate geometry, as a result of chelate formation, the rearrangement pro-ceeds with a high degree of diastereoselectivity and is independent of the substitution pattern and the protecting groups (**Y**) used. The formation of the *syn* product can be explained by a preferential rearrangement via the chair-like transition state. For an application to amino acid synthesis, it is important not only to control the relative, but also the absolute configuration of the stereogenic centers. Therefore, we investigated two different approaches to solve this problem (Figure 4.3.1).

chiral esters        chiral ligands

**Figure 4.3.1.** Asymmetric chelate Claisen rearrangements.

### 4.3.2.1   Chelate Claisen rearrangements of chiral esters

The classical approach for the introduction of chirality is the rearrangement of chiral allylic esters. Therefore, this approach is used very often in the asymmetric synthesis of various types of natural products. Because of our interest in the synthesis of poly-hydroxylated amino acids and alkaloids, we investigated the rearrangement of chiral allylic esters [11], easily obtained from *chiral pool* materials [12]. As expected, the rearrangement occurs with a high degree of chirality transfer, giving rise to optically active amino acids. Based on this strategy we developed another approach towards the synthesis of piperidine alkaloids (Scheme 4.3.4) [13].

Starting from the chiral allylic ester **12**, chelate Claisen rearrangement using LDA as a base and zinc chloride as chelating agent, gave rise to the $\gamma,\delta$-unsaturated amino acid **13**. For the following decarboxylation step, the Barton procedure, developed for amino acids, was used [14]. Subsequent Sharpless dihydroxylation [15] of the double bond provided the corresponding diol, which was converted to the isopropylidene derivative **14**. The diastereoselectivity in the hydroxylation step was excellent (>97% ds, *matched case*). Removal of the protecting group (R) gave the corresponding alcohol, which was converted into the triflate. The triflate was used directly in the cyclization step without further purification. Subsequent deprotection and ion-exchange chromatography provided 5-*epi* isofagomine (**15**).

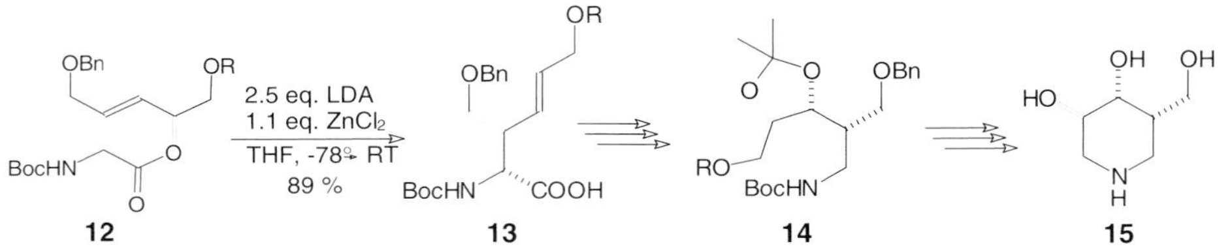

**Scheme 4.3.4.** Synthesis of 5-*epi* isofagomine (**15**) via asymmetric chelate Claisen rearrangement.

### 4.3.2.2    Chelate Claisen rearrangements in the presence of chiral ligands

Besides the described substrate controlled asymmetric rearrangements, we also focused on rearrangements in the presence of chiral ligands. This approach is of special interest because not many examples of asymmetric rearrangements of this type have been described in the literature to date [16]. In view of the many different metal salts which can be used in the chelate enolate rearrangement, and the even higher number of chiral ligands which are commonly used in asymmetric synthesis, we undertook an excessive metal and ligand screening. As a model reaction we investigated the rearrangement of TFA-protected glycine crotyl ester **16** (Scheme 4.3.5). The crotyl ester was chosen because the rearrangement provides the corresponding dehydroisoleucine derivative or the *allo*-isomer **17**, respectively, as the major product. The TFA-protecting group allows the GC-analytical determination of the absolute configuration of the rearrangement product after hydrogenation and by comparison with the naturally occurring isoleucine derivatives.

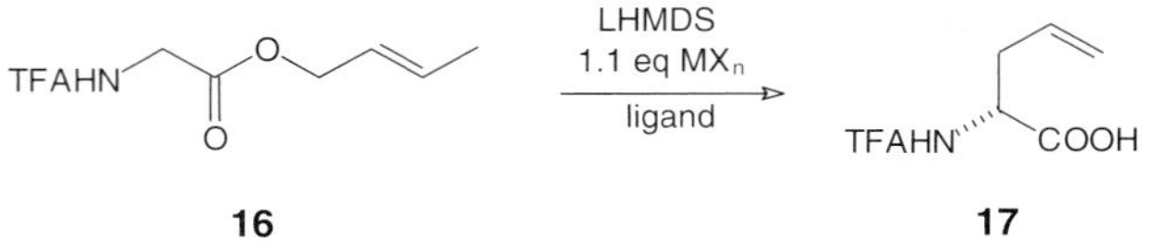

**Scheme 4.3.5.** Chelate Claisen rearrangement in the presence of chiral ligands.

The best results by far are obtained with Al(O*i*Pr)₃ and Mg(OEt)₂ as chelating metal salts and quinine as a chiral ligand. This rearrangement allows for the synthesis of various unsaturated amino acids including those with quaternary centers and allenic side chains (Scheme 4.3.6) [4].

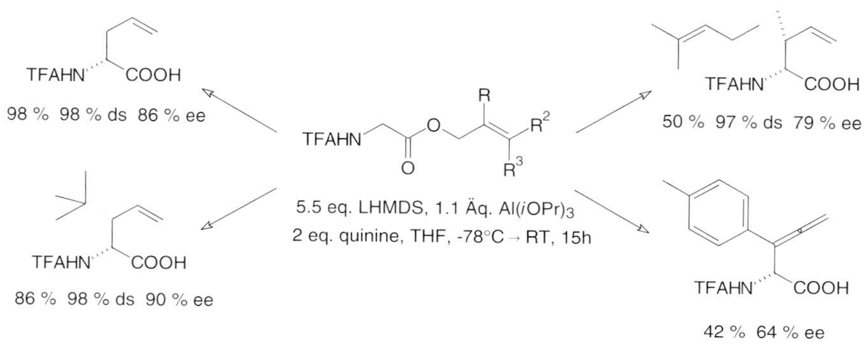

**Scheme 4.3.6.** Applications of the asymmetric chelate Claisen rearrangement.

While (*R*)-amino acids are obtained in the presence of quinine, the isomeric quinidine gives rise to the corresponding (*S*)-amino acids. Based on this asymmetric chelate Claisen rearrangement, we developed a short and very convenient synthesis of suitable protected isostatine (Scheme 4.3.7), with *de novo* generation of all stereogenic centers [17]. Isostatine is an essential amino acid of the didemnines, a group of cyclic peptides which show strong antitumor and antiviral, as well as immunosupressive, activity [18].

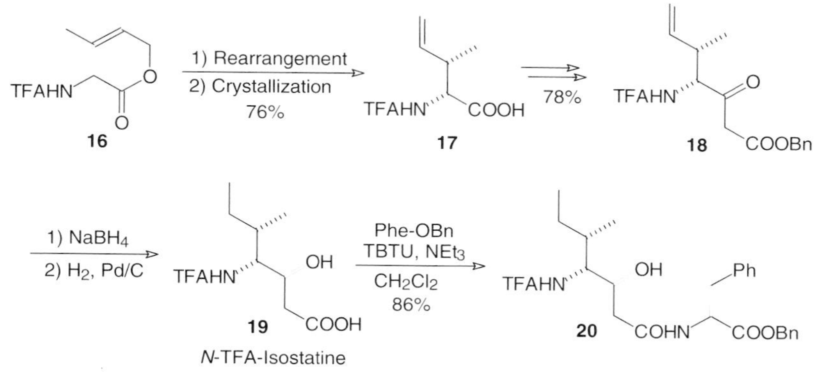

**Scheme 4.3.7.** Synthesis of protected isostatine *via* asymmetric chelate Claisen rearrangement.

Starting from achiral glycine crotyl ester **16**, the corresponding amino acid **17** was obtained in nearly quantitative yield by deprotonation with LHMDS in the presence of Mg(OEt)₂ and quinine. The enantiomerical purity was increased by single crystallization of the crude amino acid with (*S*)-phenethylamine. For the prolongation of the carbon chain, a Claisen condensation *via* the corresponding imidazolide was chosen, giving rise to $\beta$-ketoester **18**. The enantiomerically pure amino acid **19** was obtained after reduction, catalytic hydrogenation and subsequent crystallization. The free amino acid should be obtained by saponification or reduction of the protecting group. However, even more interesting is the direct introduction of **19** into peptides by coupling with other amino acids. For example, reaction with phenylalanine benzyl ester under assistance of TBTU [19] gives rise to dipeptide **20**, which can selectively be prolonged on both sides, because of the orthogonality of the protecting groups.

**Acknowledgments:**  I want to thank my coworkers for their tremendous help and encouragement during the past years. I also thank Prof. Dr. G. Helmchen for his generous support of our work. Financial support by the Graduiertenkolleg, the Deutsche Forschungsgemeinschaft and the Fonds der Chemischen Industrie is gratefully acknowledged.

# References

[1]   Review: U. Kazmaier, *Liebigs Ann. Chem./Recueil* **1997**, 285–295.
[2]   (a) U. Kazmaier, R. Grandel, *Synlett* **1995**, 945–946. (b) R. Grandel, U. Kazmaier, B. Nuber, *Liebigs Ann. Chem./Recueil* **1996**, 1143–1150.
[3]   U. Kazmaier, *Angew. Chem.* **1994**, *106*, 1096–1097; *Angew. Chem., Int. Ed. Engl.* **1994**, *33*, 998–999.
[4]   (a) U. Kazmaier, A. Krebs, *Angew. Chem.* **1995**, 2213–2214. (b) A. Krebs, U. Kazmaier, *Tetrahedron Lett.* **1996**, *37*, 7945–7946.
[5]   A. B. Hughes, A. J. Rudge, *Nat. Prod. Rep.* **1994**, *11*, 135–162, references cited therein.
[6]   C. Tomiolo, M. Crisma, S. Pegoraro, E. L. Becker, S. Polinelli, W. H. J. Boesten, H. E. Schoemaker, E. M. Meijer, J. Kamphuis, R. Freer, *Peptide Res.* **1991**, *4*, 66.
[7]   R. Grandel, U. Kazmaier, *Tetrahedron Lett.* **1997**, *38*, 8009–8012.
[8]   O. Mitsunobu, *Synthesis* **1981**, 1–28.
[9]   R. Grandel, U. Kazmaier, *J. Org. Chem.* **1998**, *63*, 4524–4528.
[10]  U. Kazmaier, R. Grandel, *Eur. J. Org. Chem.* **1998**, 1833–1840.
[11]  (a) U. Kazmaier, C. Schneider, *Synlett*, **1996**, 975–977. (b) U. Kazmaier, C. Schneider, *Synthesis* **1998**, 1321–1326.
[12]  C. Schneider, U. Kazmaier, *Synthesis* **1998**, 1314–1320.
[13]  (a) U. Kazmaier, C. Schneider, *Tetrahedron Lett.* **1998**, *39*, 817–818. (b) C. Schneider, U. Kazmaier, *Eur. J. Org. Chem.* **1998**, 1155–1159.
[14]  D. H. R. Barton, D. Crich, W. B. Motherwell, *Tetrahedron* **1995**, *41*, 3901–3924.
[15]  H. C. Kolb, M. S. vanNieuwenzhe, K. B. Sharpless, *Chem. Rev.* **1994**, *94*, 2483–2547.
[16]  Review: D. Enders, M. Knopp, R. Schiffers, *Tetrahedron Asym.* **1996**, *7*, 1847–1882.
[17]  U. Kazmaier, A. Krebs, *Tetrahedron Lett.,* **1999**, *40*, 479–482.
[18]  R. Sakai, K. L. Rinehart, V. Kishore, B. Kundu, G. Faircloth, J. B. Gloer, J. R. Carney, M. Namikoshi, F. Sun, R. G. Hughes, Jr., D. G. Gravalos, T. G. de Quasada, G. R. Wilson, R. M. Heid, *J. Med. Chem.* **1996**, *39*, 2819–2834, and reference cited therein.
[19]  R. Knorr, A. Trzeciak, W. Bannwarth, D. Gillessen, *Tetrahedron Lett.* **1989**, *30*, 1927–1930.

# 4.4    Cyclodepsipeptides: From natural product to anthelmintically active synthetic enniatins

*Peter Jeschke, Winfried Etzel, Achim Harder, Michael Schindler, Axel Göhrt, Ulrich Pleiß, Horst Kleinkauf, Rainer Zocher, Gerhard Thielking, Wolfgang Gau, and Gerhard Bonse*

The 24-membered ring cyclic octadepsipeptide PF 1022A (**1**), which was first isolated from the fungus imperfectus *Mycelia sterilia* and structurally characterized by Sasaki et al. [1], is a very active member of a novel class of anthelmintic compounds [2]. Due to this result, the structurally related, naturally occurring enniatins were screened in order to identify further, possibly more potent anthelmintics.

A number of 18-membered cyclic hexadepsipeptides has been isolated from natural sources such as several strains of the genus *Fusarium* and *Beauveria bassiana*, or has been obtained through synthetic efforts [3]. Compared to PF 1022A, the enniatin ring system is reduced by one didepsipeptide unit. Naturally occurring enniatins are composed of three variable residues, each consisting of an *N*-methylated branched-chain (*S*)-amino acid (MeXaa), in the one, three and five positions, and three residues of an (*R*)-2-hydroxy-carboxylic acid (HyCar), particularly (*R*)-2-hydroxy-isovaleric acid (HyIv) [4], in the positions two, four and six of the molecule, *i.e.* arranged in an alternating fashion.

$$\textit{Cyclo}(\text{-MeLeu}^1\text{-D-PhLac}^2\text{-MeLeu}^3\text{-D-Lac}^4\text{-MeLeu}^5\text{-D-PhLac}^6\text{-MeLeu}^7\text{-D-Lac}^8\text{-})$$

PF 1022A

**1**

$$\textit{Cyclo}(\textbf{-MeXaa}^1\text{-D-HyIv}^2\textbf{-MeXaa}^3\text{-D-HyIv}^4\textbf{-MeXaa}^5\text{-D-HyIv}^6\text{-})$$

**2a MeXaa**$^{1,3,5}$: **MeIle** *Enniatin A*

**2b MeXaa**$^{1,3,5}$: **MePhe** *Beauvericin*

Comparative screening of the enniatins ($A_1$, B, $B_1$, C) revealed that enniatin A (**2a**) showed the best anthelmintic activity in vitro. On the other hand, beauvericin (**2b**) exerted only weak activity against nematodes in vivo [5].

In order to optimize the anthelmintic activity against gastrointestinal nematodes in livestock animals, numerous enniatins were synthesized by a route similar to that for PF 1022A (**1**) and its derivatives [6].

## 4.4.1    Total synthesis of enniatins

The method for preparing the cyclic depsipeptides involved formation of linear precursor depsipeptide hexamers from three dimeric fragments in a convergent strategy as reported [6].

The macrocyclization was accomplished by two different strategies in a highly dilute solution, namely by cyclization of the N-terminal Z-protected linear hexadepsipeptide pentafluorophenyl ester in 31%–63% yield [7], and by ring closure of deprotected hexadepsipeptides **I** to form **II** (Scheme 4.4.1). The latter methodology was a useful simplification and permitted preparation of enniatins on scales of up to 10 g [8].

H-**MeXaa**[1]-D-HyCar[2]-MeXaa[3]-D-HyCar[4]-MeXaa[5]-**D-HyCar**[6]-OH

**I**

BOP-Cl, DIEA, 25 °C, 2 d

*Cyclo*(-**MeXaa**[1]-D-HyCar[2]-MeXaa[3]-D-HyCar[4]-MeXaa[5]-**D-HyCar**[6]-)

**II**

**Scheme 4.4.1.** Macrocyclization of deprotected hexadepsipeptides.

A comparison of 40 macrocyclization yields revealed for the ring closure of MeXaa[1], *e.g.* with (*R*)-lactic acid (Lac) as well as with HyCar in position six the ring closure tendency in the following order: MePhe[1] ≪ MeGly[1] < MeVal[1] < MeAla[1], MeAbu[1], MeLeu[1] < MeNva[1], MeIle[1], MeNle[1]. Schmidt et al. [7] described a similar ring closure dependence on yield and linking position.

## 4.4.2    Optimization strategy of enniatins

The stepwise optimization of the naturally occurring enniatins gave an insight into the structure-activity relationships at each position for the various MeXaa and HyCar residues.

Replacement of MeIle[3] in enniatin A (**2a**) with MePhe[3] as a structural unit of the in vivo active beauvericin (**2b**) improved the activity significantly. Catalytic hydrogenation of the benzyl residue in the MePhe[3]-containing enniatin **3a** by the usual method ($H_2$/$PtO_2$, 50 psi) [9] gave the *N*-methyl-cyclohexylalanine (MeCha) containing **3b** in 38% yield without racemization. This exhibited significant higher anthelmintic activity against nematodes in livestock animals.

*Cyclo*(-MeIle-D-Hylv-**MeXaa**[3]-D-Hylv-MeIle-D-Hylv-)

**MeXaa**[3]: MeIle (**2a**) < MePhe (**3a**) ≪ **MeCha** (**3b**)

In order to identify the essential MeXaa which is necessary for anthelmintic activity in **2**, simple (*R*)-Lac-containing enniatins with three identical MeXaa units were examined. In these derivatives **4a–4f** the anthelmintic activity was clearly dependent on the length of the alkyl side chain of the moderately hydrophobic MeXaa residues.

*Cyclo*(-**MeXaa**[1]-D-Lac-**MeXaa**[3]-D-Lac-**MeXaa**[5]-D-Lac-)

**MeXaa**[1,3,5]: MeAla (**4a**), MeNva (**4b**), MeVal (**4c**) ≪

MeLeu (**4d**) < MeNle (**4e**), **MeIle** (**4f**)

Among the analogs with MeXaa that have unbranched chain alkyl side chains, the MeNle analog **4e** was the most potent. However, maximum activity was obtained with the MeIle analog **4f** as first lead structure containing branched butyl side chains (Figure 4.4.1).

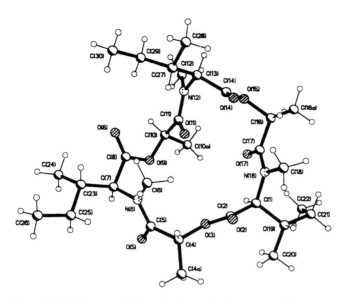

**Figure 4.4.1.** X-ray crystallography determined structure of the MeIle analog **4f**.

Based on the depsipeptide sequence of PF 1022A (**1**), two enniatins **5** and **6a** containing partial sequences of this natural product were formed by deletion of one depsipeptide unit (Scheme 4.4.2).

*Cyclo*(-MeLeu³-D-Lac⁴-MeLeu⁵-D-PhLac⁶-MeLeu⁷-D-Lac⁸)

**5**

⇑

*Cyclo*(-MeLeu¹-D-PhLac²-MeLeu³-D-Lac⁴-MeLeu⁵-D-PhLac⁶-MeLeu⁷-D-Lac⁸-)

*PF 1022A*

**1**

⇓

*Cyclo*(-**MeXaa**¹-D-PhLac²-**MeXaa**³-D-Lac⁴- **MeXaa**⁵-D-PhLac⁶-)

**MeXaa**¹,³,⁵: MeLeu (**6a**) << **MeIle** (**6b**)

**Scheme 4.4.2.** Enniatins containing partial sequences of the natural product PF 1022A (**1**).

Despite the structural similarity to PF 1022A (**1**), both derivatives exhibited relatively weak anthelmintic activity in vivo. With regard to the enniatin **4f**, it is interesting that replacement of all three MeLeu with MeIle in the one, three and five positions of the most active derivative **6a** leads to **6b**, which exhibited higher anthelmintic activity. These results suggest that the presence of MeIle is beneficial for the anthelmintic activity of enniatins.

The 18-membered ring cyclodepsipeptide **4f**, a suitable model for structural variation, was optimized via a [4 + 2]-fragment condensation reaction. Compounds **7a-7c** represent modifications at the HyCar$^2$ position of **4f**.

*Cyclo*[-MeIle-**D-HyCar**$^2$-(MeIle-D-Lac)$_2$-]

**D-HyCar**$^2$: D-HyCa (**7a**) ≪ D-Hylv (**7b**) < **D-Lac** (**4f**), **HyAc** (**7c**)

Surprisingly, introduction of the achiral 2-hydroxyacetic acid (HyAc$^2$) in **4f**, led to no essential loss of in vivo activity (**7c**) (Figure 4.4.2).

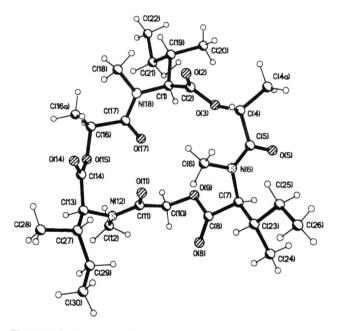

**Figure 4.4.2.** X-ray crystallography determined structure of the HyAc analog **7c**.

The subsequent replacement of only one MeXaa (in 1 position) led to the enniatins **8a–8i**, whose in vivo anthelmintic activities correlated with the onset of activity in the following order:

*Cyclo*[(-**MeXaa**$^1$-D-Lac-(MeIle-D-Lac)$_2$-]

**MeXaa**$^1$: MeGly (**8a**), MeLeu (**8b**), MeAbu (**8c**) < MeNle (**8d**), MeIle (**4f**),

MePhe (**8e**) ≪ MeVal (**8f**), MeNva (**8g**), MeCha (**8h**) < **MeAla** (**8i**)

Among all synthesized analogs, **8i** (MeAla at pos. 1) exhibited the most potent activity. These findings imply that a short residue is preferable at this position. Further structural optimization of **8i** was carried out in view of its unsymmetrically folded major conformer (see next section).

Modifications of **8i** with retention of the unsymmetrically folded conformation led to structure-activity relationships that clarified the influence of substituents on the enniatin backbone conformation and provided a useful basis for molecular modeling studies. With regard to the unsymmetrically folded major conformer of **8i**, replacement of (*R*)-Lac with HyAc in the two or six positions, as exemplified by **9** and **10**, or the simultaneous replacement of two (*R*)-lactic acids with (*R*)-HyIv in the positions four and six is possible.

*Cyclo*(-MeAla-**D-HyCar**²-MeIle-D-Lac-MeIle-**D-HyCar**⁶-)

**D-HyCar**²: HyAc (**9**) < **D-Lac** (**8i**)

**D-HyCar**⁶: HyAc (**10**) < **D-Lac** (**8i**)

*Cyclo*(-MeAla-D-Lac-MeIle–**D-HyCar**⁴-MeIle–**D-HyCar**⁶-)

**D-HyCar**⁴ᐟ⁶: D-HyIv (**11**) < **D-Lac** (**8i**)

However, the resulting derivatives are anthelmintically less active in vivo than **8i**. Singular replacement of MeAla¹, as well as the simultaneous replacement of both MeIle residues in the three and five positions of 8i by further MeXaa, resulted in less potent analogs (**12a–12d** and **13a–13b**).

*Cyclo*(-**MeXaa**¹-D-Lac-MeIle-D-Lac-MeIle-D-Lac-)

**MeXaa**¹: MePhe (**12a**) < MeCha (**12b**) < MeNva (**12c**),

MeVal (**12d**) ≪ **MeAla** (**8i**)

*Cyclo*(-MeAla-D-Lac-**MeXaa**³-D-Lac-**MeXaa**⁵-D-Lac-)

**MeXaa**³ᐟ⁵: MeLeu (**13a**) < MeVal (**13b**) < **MeIle** (**8i**)

Nonetheless, replacement of one MeIle⁵ in 5 position gave the MeAla⁵ and MeAbu⁵ analogs **14e** and **14f**, which were almost equally active as **8i**.

*Cyclo*(-MeAla-D-Lac-MeIle-D-Lac-**MeXaa**⁵-D-Lac-)

**MeXaa**⁵: Me*allo*Ile (**14a**) < MeNva (**14b**) ≪ MeVal (**14c**), MeLeu (**14d**) <

**MeIle** (**8i**), **MeAla** (**14e**), **MeAbu** (**14f**)

These results suggest that the branched butyl side chain is not very important for binding affinity. The Me*allo*Ile⁵ analog **14a** exhibited relatively weak anthelmintic activity in vivo compared with **8i**, a fact, which implies that the receptor or binding site strictly discriminates the absolute stereochemistry at the $\beta$-carbon in the side chain.

### 4.4.3   NMR spectroscopic investigations of enniatins

The structures and sequences of 35 enniatins were confirmed using COSY-, TOCSY-, HMQC-, and HMBC-NMR spectra. The conformations were examined by NOESY-NMR-spectra. The molecules are in the positive NOE regime in CDCl$_3$ solution, in spite of their size. We have identified four different types of main conformations. Each enniatin, dissolved in CDCl$_3$, exists in 2, 3 or 4 conformations that are in dynamic exchange with another one, as can be deduced from the corresponding exchange cross peaks in their NOESY spectra. Most enniatins exist in one main conformation in solution. Based on coalescence measurements, the energy $\Delta G^{\#}$ for conversion between the conformers was estimated about 75 kJ/mol. The main conformer types are: (i) molecules with C-3 symmetry (**4a–4f**); (ii) molecules with C-2 symmetry (**7a, 8d**); (iii) molecules with one *cis*-amide bond; and (iv) molecules with an unsymmetrical folded conformation (**8i, 9–11, 12a–12d, 13a–13b, 14a–14f**). In the NOESY-spectra, molecules with an unsymmetrical folded conformation show a medium intensity NOE cross peak between a MeXaa-α-H and the *N*-methyl group of the same residue. A correlation between the conformation of the main conformer of a molecule and its anthelmintic activity could be observed. Molecules with strong in vivo activity exhibit a main conformer that is unsymmetrically folded or has a *cis*-amide bond. Molecules with C-2/C-3 symmetry show only weak activity against nematodes.

### 4.4.4   Synthesis of a tritium labeled enniatin

For receptor binding studies and the elucidation of the mode of action of the most potent anthelmintic enniatin **8i**, a tritium-labeled compound with very high specific activity was required. Tritium was introduced by *N*-methylation of the *N*-demethyl precursor **15** with an excess of [methyl-$^3$H] methyl iodide in presence of silver(I)-oxide in 43% yield (Scheme 4.4.3) [5].

*Cyclo*(-MeAla-D-Lac-**Ile**-D-Lac-MeIle-D-Lac-)    $\xrightarrow[\text{DMF}]{\text{[}^3\text{H] MeI, AgO}}$

**15**

*Cyclo*(-MeAla-D-Lac-[***N*-methyl-$^3$H]Ile**-D-Lac-MeIle-D-Lac-)

**[*N*-methyl-$^3$H] 8i**

**Scheme 4.4.3.** Synthesis of the tritium-labeled enniatin **8i**.

After work-up and purification by HPLC, [*N*-methyl-$^3$H] **8i** was available with a specific radioactivity of 84 Ci/mmol (3.11 TBq/mmol), as determined by mass spectrometry. The radiolabeled enniatin **8i** showed an efficient and specific binding to a membrane fraction from the pig intestinal nematode *Ascaris suum*. Displacement by unlabeled **8i** was half-maximal at about $0.72 \pm 0.06\ \mu$M [5].

## 4.4.5 Enzymatic biosynthesis of (*R*)-lactic acid containing enniatins

New enniatin-like compounds with anthelmintic properties were produced enzymatically using two different strategies, namely by in vitro synthesis with the multienzyme enniatin synthetase (ESyn), isolated from *Fusarium* species (*F. sambucinum, F. scirpi*), and by in vivo precursor feeding of these enniatin-producing *Fusaria* strains. Apart from some considerable variations in the MeXaa portion of the enniatin molecule, it was possible to replace the (*R*)-2-hydroxy-isovaleric acid (HyIv) moiety, which is exclusively present in all naturally occurring enniatins, by a number of homologous substrates and even by (*R*)-lactic acid (Scheme 4.4.4).

main culture of *F. sambucinum*

or

ESyn, (S)-Ile, Mg$^{2+}$-salt,

S-[methyl-$^{14}$C]-adenosyl-(*R*)-methionine (SAM),

adenosine triphosphate (ATP)

$\xrightarrow{(R)\text{-lactic acid}}$

*Cyclo*(-MeIle-D-HyIv-MeIle-**D-HyCar**$^4$-MeIle-**D-HyCar**$^6$-)

**D-HyCar**$^{4,6}$: D-HyIv  (**2a**)  *Enniatin A*
**D-HyCar**$^{4,6}$: D-Lac  (**7b**)

**Scheme 4.4.4.** Synthesis of the enniatin **7b** by in vivo- and in vitro incorporation of two (*R*)-Lac in enniatin A (**2a**) via *F. sambucinum* or ESyn isolated therefrom.

The spectroscopic data for the authentic (fermented) enniatin were in agreement with those of its synthetic analog **7b** [10].

## 4.4.6 Conclusions and outlook

Systematic modifications of the naturally occurring enniatins clarified the structure-activity relationships and led to a significant enhancement of their anthelmintic activity. It may be speculated that the unsymmetrical folded major conformation might mimic the active conformation of enniatins. On the other hand, by knowing the conformation of the most active enniatins, the rational drug optimization and design of enniatin mimics may be possible. Radiolabeled enniatins are useful agents for receptor binding studies and the elucidation of the mode of action of these novel class of cyclodepsipeptide anthelmintics. The enzymatic biosynthesis could lead to practical applications of the fermentation process to produce enniatins containing (*R*)-lactic acid.

## References

[1]  T. Sasaki, M. Takagi, T. Yaguchi, S. Miyadoh, T. Okada, M. Koyama, *J. Antibiotics,* **1992**, *45*, 692–697.

[2]  G. Geßner, S. Meder, T. Rink, G. Boheim, A. Harder, P. Jeschke, J. Scherkenbeck, M. Londer-shausen, *Pestic. Sci.* **1996**, *48*, 399–407.

[3]  P. A. Plattner, K. Vogler, R. O. Studer, P. Quitt, W. Keller-Schierlein, *Helv. Chim. Acta*, **1963**, *46*, 927–935.

[4]  Abbrevations follow the recommendations of the IUPAC-IUB Joint Commission on Biochemical Nomenclatur for amino acids and peptides (*Eur. J. Biochem.* **1984**, *138*, 9–37). Additional abbrevations are defined in the text or as follows: D-HyCa, D-2-hydroxy-caproic acid; D-PhLac, D-phenyllactic acid; MeAbu, *N*-methyl-L-aminobutyric acid; MeAla, *N*-methyl-L-alanine; MeGly, *N*-methyl-glycine; MeIle, *N*-methyl-L-isoleucine; MeLeu, *N*-methyl-L-leucine; MeNle, *N*-methyl-L-norleucine; MeNva, *N*-methyl-L-norvaline; MePhe, *N*-methyl-L-phenylalanine; MeVal, *N*-methyl-L-valine; DIEA, diisopropylethylamin; HPLC, high-performance liquid chromatography.

[5]  U. Pleiss, A. Turberg, A. Harder, M. Londershausen, P. Jeschke, G. Boheim, *J. Lab. Comp. Radiopharm.*, **1996**, *38*, 651–659.

[6]  J. Scherkenbeck, A. Plant, A. Harder, N. Mencke, *Tetrahedron*, **1995**, *51*, 8459–8470.

[7]  U. Schmidt, A. Lieberknecht, H. Griesser, R. Utz, T. Beuttler, F. Bartkowiak, *Synthesis*, **1986**, 361–366.

[8]  P. Jeschke, J. Scherkenbeck, G. Bonse, N. Mencke, A. Harder, M. Londershausen, E. Bischoff, H. Müller, *WO 93/25 543*, (**1993**) Bayer AG.

[9]  W. J. Hoekstra, S. S. Sunder, R. J. Cregge, L. A. Ashton, K. T. Stewart, C.-H. R. King, *Tetrahedron* **1992**, *48*, 307–318.

[10]  P. Jeschke, A. Harder, N. Mencke, H. Kleinkauf, R. Zocher, *Eur. Pat. Appl. EP 669,543* (**1995**) Bayer AG.

# 4.5    Artificial biomimetic receptor molecules

*Thomas Schrader, Michael Herm, and Christian Kirsten*

## 4.5.1    Synthetic adrenaline receptors

### 4.5.1.1 Recognition of Amino Alcohols

The $\beta$-adrenergic receptor is located in the membrane of adrenal gland cells [1]. Seven transmembrane $\alpha$-helices form a binding pocket, which recognizes inter alia adrenaline derivatives. The hormone signal is then transduced from the extracellular space into the cytoplasm where, according to a still unknown mechanism, it subsequently activates the G-protein. This triggers an enzyme cascade, ultimately resulting in the instantaneous excretion of glucose – the adrenaline surge. Many pharmaceuticals act as agonists or antagonists for the $\beta$-adrenergic receptor, *e.g.* the well known $\beta$-blockers [2]. However, the membrane-bound receptor has never been isolated in pure form and has hence proven elusive to X-ray structure analysis [3]. It would be of great value for biochemists, as well as pharmacologists, to gain access to a small defined model of the adrenergic receptor that imitates the natural receptor–ligand interactions (Figure 4.5.1) [4].

Electrostatic and hydrogen bond interactions with an aspartate carboxylate bind the ammonium functionality; in addition, the $NH_3^+$-protons experience $\pi$-cation stabilization because they are surrounded by three electron-rich aromatic residues. Each of the aliphatic and catechole hydroxyls is hydrogen-bonded to a serine OH, while the aromatic ring is buried in a deep cleft flanked by two phenylalanine aromatic rings that are involved in double $\pi$-stacking.

Our simplest model for the binding site is based on a *p*-xylylene bisphosphonate, which not only imitates the ammonium–aspartate interaction but also amplifies it

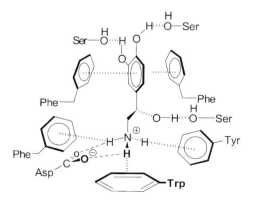

**Figure 4.5.1.** Non-covalent interactions between noradrenaline and the $\beta$-adrenergic receptor.

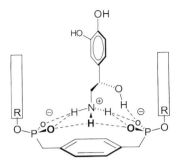

**Figure 4.5.2.** Energy-minimized complex of *p*-xylylene bisphosphonates with noradrenaline.

due to its bidentate character (Figure 4.5.2) [5]. The phosphonate ester functionalities can at a later stage serve to introduce substituents for lateral and chiral recognition of the substrate [6]. Force-field calculations show, that when an alkylammonium ion approaches the *p*-xylylene bisphosphonate, not only are strong electrostatic attractions produced, but also an almost ideal array of short linear hydrogen bonds is formed [7].

Job's method of continuous variations confirms the postulated 1:1 stoichiometry. Maxima in the range of $M = 0.5–0.55$ are observed for guest as well as for host signals [8].

We performed NMR titrations of alkyl ammonium chlorides with the simple phosphonate receptors **1–3** in [D$_6$] DMSO (Figure 4.5.3). We measured the CIS (chemically induced shift) of the $\alpha$-C*H*-N$^+$ and analyzed the binding curves by non-linear regression methods [9]. The calculated association constants for representative molecules are summarized in Table 1. The drastic increase in $K_a$-values from monophosphonate (**1**) to *m*-and *p*-diphosphonate (**2** and **3**) is strongly indicative for the chelate effect resulting in double electrostatic attraction for the alkylammonium cation.

We were surprised when we titrated various related amino alcohols (*e.g.* **4**) and found that the binding constants increase again by one order of magnitude – we had indeed found a basic structural motif for a receptor of amino alcohols. Very few synthetic receptors are known that selectively bind to this biologically and pharmaceutically important structure [10]. Another feature of these hosts is remarkable: Secondary amines are complexed at least as strongly as their primary counterparts.

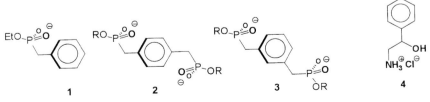

**Figure 4.5.3.** Simple phosphonate receptor molecules **1–3** and adrenaline model compound **4**.

**Table 4.5.1.** Association constants $(K_{1:1})[M^{-1}]$ from NMR titrations in DMSO at 20°C.

| Receptor | Benzylamine ∗ HCL $10^3$ $[M^{-1}]$ | (±)-**4** $10^3$ $[M^{-1}]$ |
|---|---|---|
| **1** | 0.2 | – |
| **2** | 2.8 | 15.5 |
| **3** | 7.4 | 55.0 |

Contrary to all artificial ammonium receptors known to date, bisphosphonates **2** and **3** can be expected to bind especially well to adrenaline itself, as well as to the respective $\beta$-blockers. A systematic study of the binding behavior of 1,n amino alcohols revealed that *m*-xylylene bisphosphonate **3** is a selective host for 1,2- and 1,3-amino alcohols.

### 4.5.1.2 Recognition of the catechole

The simplest modification of a bisphosphonate binding site which imitates the biological recognition of the catechole ring via $\pi$-stacking interactions consists of the synthesis of bisaryl esters. With the triphenylmethyl ester **5** we hoped to mimic the sandwich-type structure of the natural example [11]. Unfortunately, this attempt failed with adrenaline derivatives, presumably due to too many rotatory degrees of freedom of the open chain receptor molecules. However, when we compared the association constants for **5** and adrenaline derivatives with those obtained for **5** and related amino alcohols with electron–poor arenes, we observed an up to five–fold increase in $K_a$ as well as distinct upfield shifts for the aromatic protons, a clear indication of the expected $\pi$-stacking interactions (Figure 4.5.4).

If the phosphonate groups are attached to the periphery of a macrocycle, the catechole ring may be inserted into the interior of the macrocycle; in addition to $\pi$-stacking interactions the stability of the developing complexes should be markedly enhanced by hydrophobic interactions, especially in water [12]. Figure 4.5.5 shows our first example of such a host molecule: in a convergent synthesis the macrocycle

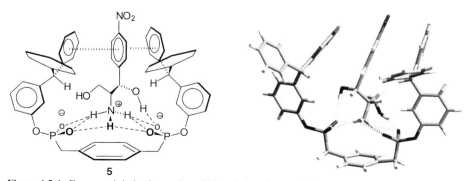

**5**

**Figure 4.5.4.** Energy-minimized complex of bisarylphosphonate **5** with an electron–poor amino alcohol.

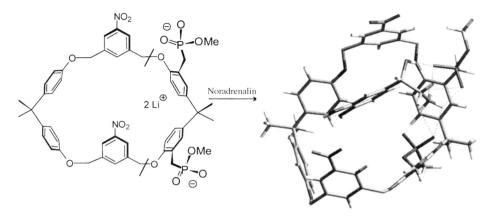

**Figure 4.5.5.** Macrocyclic host **6** with peripheral phosphonates – calculated inclusion complex with nor-adrenaline; strategic bonds for the macrocyclization step are marked with /.

is formed in the penultimate step by a template-assisted reaction of the U-type predecessor with the phosphonate-functionalized bisphenol A.

In the $^1$H-NMR spectra of 1,2-amino alcohol complexes of **6** in DMSO the small ethanolamine broadens only the aromatic host signals of the bisphosphonate (*A*), the medium size agonist noradrenaline interacts with those up to the nitroarene (*B*), whereas the larger antagonist propranolol reaches far to the other bisphenol A endpiece of the host (*C*) (Figure 4.5.6). The signal broadening thus represents a molecular imprint of the guest molecule which hinders the free rotation of the aromatic rings in the host.

In methanol, adrenaline derivatives and **6** gave the highest association constants ever measured by us in this solvent ($K_a \geq 1000\,\mathrm{M}^{-1}$). The assistance of hydrophobic interactions in these cases becomes visible in comparison to the small ethanolamine, which is bound much less tightly ($K_a = 520\,\mathrm{M}^{-1}$). In water, the macrocyclic bis-phosphonate **6** undergoes strong self-association. During the NMR titration in a buffered solution (pH = 7) both the aliphatic as well as the aromatic NMR signals

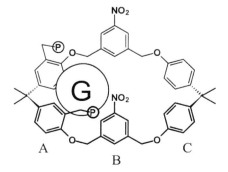

**Figure 4.5.6.** Molecular imprint of different guest molecules G inside the cavity of macrocyclic receptor **6**.

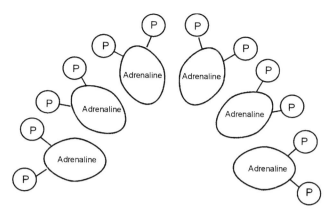

**Sketch 1.**

of noradrenaline shift considerably and lead to an unexpectedly high binding constant of $\sim 3000 \, M^{-1}$; the binding curve, however, differs from the ideal $1:1$-binding isotherm. A good explanation seems to be formation of micelles, for the amphiphilic structure of the receptor molecule remarkably resembles that of phospholipids (see above illustration) [13]. Moreover, concentrated receptor solutions in $D_2O$ show a Tyndall effect and foam on shaking. To check this hypothesis we are currently examining the aggregates in water by electron microscopy.

### 4.5.1.3  Biomimetic and chiral recognition of adrenaline

Even closer to the natural example is the structure of our second macrocycle **7** (Figure 4.5.7): According to molecular modeling, it completely embraces the adrenaline

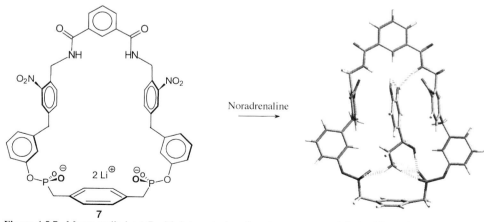

**Figure 4.5.7.** Macrocyclic host **7** with internal phosphonates – energy minimized inclusion complex with noradrenaline.

**Figure 4.5.8.** Key step in the synthesis of macrocyclic host **7** – the construction of the diphenylmethane unit.

molecule. In addition, it is able to exert with its nitroarenes the biomimetic double π-stacking with the catechole ring. Finally the isophthalamide bridge is designed to form two hydrogen bonds with the phenolic OH groups, similar to the effect used in Hunter's catenane syntheses [14].

The introduction of an OH-group at the diphenylmethane methylene carbon should facilitate at a later stage the biomimetic chiral recognition of adrenaline's aliphatic hydroxyl functionality. The key step in our synthesis of **7** consists in the construction of the highly functionalized diphenylmethane centerpiece. It is realized by means of a Pd-0-catalyzed *Negishi*-coupling of a benzylzinc reagent with the respective arylbromide (Figure 4.5.8) [15].

Preliminary binding studies of the new adrenaline receptor molecule **7** with various biologically important amino alcohols show very promising results, and indicate that this host is indeed capable of forming the postulated multiple non-covalent interactions with adrenaline derivatives.

### 4.5.1.4  Conclusions and perspectives

We intend to incorporate our adrenaline receptor models into liposomes, after which it may be possible to transmit a signal into the interior of the vesicle when adrenaline

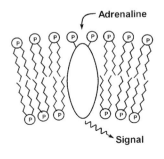

**Sketch 2.**                    **Sketch 3.**

docks on to the binding site in the artificial membrane. Furthermore we want to build adrenaline sensors. A highly sensitive direct detection can be achieved if a charge-transfer band develops upon formation of the sandwich arrangement (*e.g.* in the complex of adrenaline with **7**). The above-discussed chiral host molecules may serve as new stationary phases for the chromatographic separation of enantiomers of adrenaline derivatives and related pharmaceuticals. Finally, and with great caution, a potential medical application is suggested. Phaeochromocytoma patients suffer from uncontrolled overproduction of adrenaline, leading to severe heart damage [16]. A certain amount of an adrenaline-binder may lower the adrenaline level back to normal. For medical applications, however, many transport- and toxicology investigations have yet to be carried out.

## 4.5.2   Synthetic peptide receptors

### 4.5.2.1   Intermolecular β-sheet stabilization

The highly important β-sheet secondary structure in peptides and proteins is still poorly understood, especially with respect to its spontaneous formation, stabilization, and biological function. Lack of control is often fatal: thus β-amyloid deposition leads to Alzheimer's desease [17] and the conversion of α-helices to larger β-sheet aggregates causes BSE/Creutzfeldt–Jakob and other prion deseases [18]. In-vitro experiments are likewise hampered by the fact that larger peptides precipitate as soon as they form a β-sheet due to formation of multiple hydrogen bonds. Therefore, the synthesis of small, soluble model compounds is of high interest and represents an area of intense research [19]. The gain in knowledge about protein structure and folding may serve as a basis for future drug design. Most approaches use intramolecular interactions, which are facilitated by a covalent β-turn link between the peptide and its template. These β-sheet models have been used to examine the cause for spontaneous self-organization by formation of a hydrophobic cluster; [20] recently even the relative propensities of proteinogenic amino acids to form a parallel β-sheet have been measured [21].

To the best of our knowledge, we recently published the first example for a purely intermolecular stabilization of a β-sheet model [22]. In a three-point binding mode, cooperative hydrogen bonds clamp together a dipeptide and its rigid template. Force-field calculations demonstrate that five hydrogen bonds can be formed when two molecules of a 3-aminopyrazole approach an *N*-acylated dipeptide ester, involving every available hydrogen bond donor and acceptor of the peptide (Figures 4.5.9 and 4.5.10). This is, however, only possible, when the peptide exists in the β-sheet conformation. Viewed from above (right structure), the aminopyrazoles appear as linear bars lying exactly perpendicular above and below the peptide backbone.

Binding to the top face of the peptide is strongly favored because it forms three cooperative hydrogen bonds simultaneously to the receptor molecule, whereas the bottom face has only two. Thus, with Job's method of continuous variations we found a 1:1 stoichiometry for the top face (NH (1), see Figure 4.5.11) but a 2:1 ratio for the bottom face (NH (2)). With ¹H NMR titrations, binding constants for the 1:1

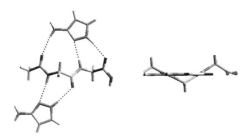

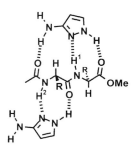

**Figure 4.5.9.** 2:1-complex between 3-aminopyrazole and Ac-Gly-Gly-OMe. Left: Side view; Right: Top view. The geometry was energy minimized using CERIUS² with the DREIDING 2.21 force field.

**Figure 4.5.10.** 2:1-complex between 3-aminopyrazole and a *N/C*-protected dipeptide. The top face of the peptide is involved in three hydrogen bonds (ADA), the bottom face only in two (DA).

complex of up to $880 \, M^{-1}$ have been determined in chloroform (Figure 4.5.12). The association constants are strongly influenced by the electronic character of the aminopyrazole derivative, as well as by the steric demand of the peptide residues.

Variable temperature studies prove that the complex is formed by dynamic hydrogen bonds and confirmed the preferential binding of the receptor molecules at the top face [23]. By detailed Karplus analysis of the *NH-α-CH* coupling constants in the complex, a remarkable correlation between the dihedral angle $\theta$ and the degree of complexation was found, which shows that several amidopyrazoles are capable of forcing the dipeptide into an almost ideal $\beta$-sheet conformation [24].

In glycine-containing dipeptides the third hydrogen bond slows down the free rotation around the C−C/C−N-bond to almost zero (Figure 4.5.13). Intramolecular nuclear Overhauser enhancements (NOE) provide additional evidence for the peptide's extended conformation, while strong reciprocal *inter*molecular NOEs give the

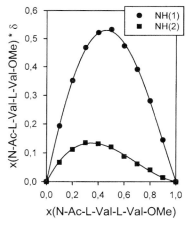

**Figure 4.5.11.** ¹H NMR Job-plot: 3-amino-5-methylpyrazole **8**/Ac-L-Val-L-Val-OMe.

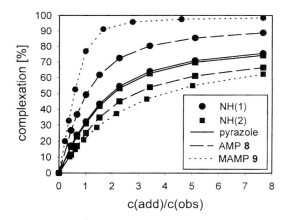

**Figure 4.5.12.** ¹H NMR titration curves for the divaline-complex with pyrazole, 3-amino-5-methylpyrazole **8** and 3-methacryloylamino-5-methylpyrazole **9** (MAMP).

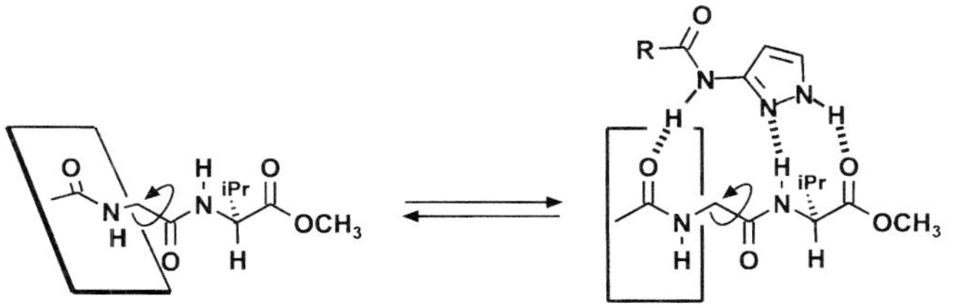

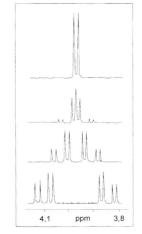

**Figure 4.5.13.** Left: Conformational lock of a rotationally free glycine-containing dipeptide by complexation with one equivalent of amidopyrazole. Right: $^1$H NMR signal of the methylene protons in Ac-Gly-L-Val-OMe. From top to bottom: without, and with 0.4, 2.3 and 6 equivalents of MAMP (**9**).

final proof of the existence of the critical third hydrogen bond and the postulated mutual orientation of the complexation partners. Preliminary $^1$H NMR titrations with tripeptides show very promising results concerning the application of this concept to oligopeptides.

### 4.5.2.2 Conclusions and perspectives

The association constants should rise markedly when aminopyrazole molecules are lined up on both sides of a larger peptide, *e.g.* in the form of aminopyrazole oligomers (Figure 4.5.14). Ultimately this could be used in a new approach for solubilizing large $\beta$-strands, with possible implications for the treatment of the above-mentioned deseases. Polymerizable 3-amino- and 3-amidopyrazoles (*e.g.* **10**) have been made accessible in excellent yields by a general route starting from *p*-toluic acid; with these compounds we introduce for molecular imprinting a new class of functional monomers, which fulfils all requirements for an efficient binding site. At present, we

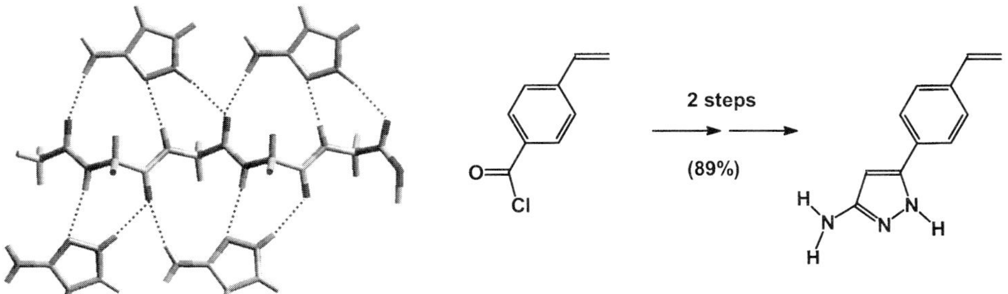

**2 steps**

**(89%)**

**Figure 4.5.14.** (a) Lining up of aminopyrazoles on both sides of a tetrapeptide; (b) Polymerizable amino-pyrazoles (*e.g.* **10**) from *p*-vinylbenzoyl chloride.

examine the racemic resolution of dipeptides with imprinted polymers using the new functional monomers. In batch experiments, separation factors $\alpha$ of up to 2.7 have been achieved with some of these polymers. Experiments to use aminopyrazoles for catalysis with imprinted polymers are underway in our laboratory.

# References

[1]  C. D. Strader, T. M. Fong, M. R. Tota, D. Underwood, *Annu. Rev. Biochem.* **1994**, *63*, 101–132.
[2]  C. R. Craig, R. E. Stitzel, *Modern Pharmacology*, Little, Brown & Co, Boston, **1990**.
[3]  Recently, electron crystallography has been carried out on two-dimensional rhodopsin crystals: G. F. X. Schertler, C. Villa, R. Henderson, *Nature* **1993**, *362*, 770–772.
[4]  (a) J. Ostrowski, M. A. Kjelsberg, M. C. Caron, R. J. Lefkowitz, *Annu. Rev. Pharmacol. Toxicol.* **1992**, *32*, 167–183; (b) S. Trumpp-Kallmeyer, J. Hoflack, A. Bruinvels, M. Hibert, *J. Med. Chem.* **1992**, *35*, 3448–3462.
[5]  T. Schrader *Angew. Chem., Int. Ed. Engl.* **1996**, *35*, 2649–2651.
[6]  A bisphosphonate and a tetraphosphate have recently been used for strong glucoside binding in acetonitrile: (a) G. Das, A. D. Hamilton, *Tetrahedron Lett.* **1997**, *38*, 3675–3678; (b) U. Neidlein, F. Diederich, *J. Chem. Soc., Chem. Commun.* **1996**, 1493–1494.
[7]  Molecular Modelling Program: CERIUS$^{2}$ TM from Molecular Simulations Inc., Force field: Dreiding 2.21.
[8]  (a) Job, P. *Compt. Rend.* **1925**, *180*, 928; (b) M. T. Blanda, J. H. Horner, M. Newcomb *J. Org. Chem.* **1989**, *54*, 4626.
[9]  (a) H. J. Schneider, R. Simova, U. Schneider *J. Am. Chem. Soc.* **1988**, *110*, 6442; (b) C. S. Wilcox, in *Frontiers in Supramolecular Chemistry and Photochemistry*; H. J. Schneider, H. Dürr, Eds.; VCH, Weinheim, **1991**.
[10]  P. B. Savage, S. H. Gellman, *J. Am. Chem. Soc.* **1993**, *115*, 10448–10449.
[11]  T. Schrader, *J. Org. Chem.* **1998**, *63*, 264.
[12]  H. J. Schneider, *Angew. Chem. Int. Ed. Engl.* **1991**, *30*, 1419.
[13]  *Liposomes – a practical approach*, R. R. C. New, Ed.; Oxford: Oxford University Press, **1992**.
[14]  C. A. Hunter, *Chem. Soc. Rev.* **1994**, 1101–109.
[15]  E. Negishi, A. O. King, N. Okukado, *J. Org. Chem.* **1977**, *42*, 1821–1823.
[16]  Klinisches Wörterbuch, W. Pschyrembel, Ed.; Berlin: de Gruyter **1988**, 1232.
[17]  (a) Prusiner, S. B. *Sci.Am.* **1995**, 48; (b) Taubes, G. *Science* **1996**, *271*, 1493.
[18]  (a) Mestel, R. *Science* **1996**, *273*, 184; (b) Prusiner, S. B. *Curr. Top. Microbiol. Immunol.* **1996**, 207.
[19]  (a) D. W. Choo, J. P. Schneider, N. R. Graciani, J. W. Kelly, *Macromolecules*, **1996**, *29*, 355; (b) D. S. Kemp, Z. Q. Li, *Tetrahedron Lett.*, **1995**, *36*, 4175; (c) J. S. Nowick, D. L. Holmes, G. Makkin, G. Noronha, A. J. Shaka, E. M. Smith, *J. Am. Chem. Soc.*, **1996**, *118*, 2764; (d) V. Brandmeier, W. H. B. Sauer, M. Feigel, *Helv. Chim. Acta*, **1994**, *77*, 70; (e) J. D. Hartgerink, J. R. Granja, R. A. Milligan, M. R. Ghadiri, *J. Am. Chem. Soc.*, **1996**, *118*, 43.
[20]  K. Y. Tsang, H. Diaz, N. Graciani, J. W. Kelly, *J. Am. Chem. Soc.* **1994**, *116*, 3988.
[21]  J. S. Nowick, S. Insaf, *J. Am. Chem. Soc.* **1997**, *119*, 10903.
[22]  (a) T. Schrader, C. Kirsten, *J. Chem. Soc., Chem. Commun.*, **1996**, 2089; (b) T. Schrader, C. Kirsten, *J. Am. Chem. Soc.* **1997**, *119*, 12061.
[23]  (a) A. A. Ribeiro, M. Goodman, F. Naider *Int. J. Peptide Protein Res.* **1979**, *14*, 414; (b) S. H. Gellman, B. R. Adams *Tetrahedron Lett.* **1989**, *30*, 3381.
[24]  (a) M. Delepierre, C. M. Dobson, F. M. Poulson *Biochemistry* **1982**, *21*, 4756; (b) V. F. Bystrov *Prog. Nucl. Magn. Reson. Spectrosc.* **1979**, *10*, 41.

# 4.6 Design and synthesis of modulators of sphingolipid biosynthesis

*Christoph Arenz and Athanassios Giannis*

Sphingolipids (SLs) are found in all eukaryotic cells where they are primarily components of the plasma membrane. Sphingolipids contain a hydrophobic ceramide backbone, which anchors them in the outer leaflet of the lipid bilayer. Ceramide itself consists of a long-chain aminoalkohol, D-erythrosphingosine, which is acylated with a fatty acid. The ceramide backbone can be modified by attachment of phosphorylcholine, to form sphingomyelin (SM), a structural component of the plasma membrane, or by attachment of one or more sugar residues, to form glycosphingolipids (GSLs), which build up cell-type-specific patterns at the outer surface of the cell membrane. These glycosphingolipids interact with toxins, viruses and bacteria as well as membrane-bound receptors and enzymes. Since their biosynthesis and degradation proceed within cellular organelles, glycosphingolipids and their precursors are also found on intracellular membranes. The enzymes involved in sphingolipid biosynthesis are membrane-bound proteins.

Sphingolipid biosynthesis starts with the condensation of L-serine and usually palmitoyl-CoA to 3-ketosphinganine. This reaction is catalyzed by the PLP-dependent serine palmitoyl transferase (SPT), the rate-limiting enzyme of sphingolipid synthesis. In the following step, 3-ketosphinganine is reduced to D-erythrosphinganine by the NADPH-dependent 3-ketosphinganine reductase, after which sphinganine is acylated to dihydroceramide (DH-Cer) by sphinganine *N*-acyl transferase. Subsequently, DH-Cer is desaturated to ceramide by dihydroceramide desaturase. The attachment of ceramide with various sugar residues leads to GSLs such as cerebrosides and sialic acid-containing gangliosides.

## 4.6.1 The need for synthetic modulators of sphingolipid biosynthesis

In the past few years some of the lower sphingolipids such as ceramide, sphingosine and sphingosine-1-phosphate (SPP) have become the focus of intense research, since it is known that they act as second messengers and influence a number of fundamental biologic processes in living cells [1]. For example, ceramide is believed to take part in the inhibition of cell growth, arrest of meiotic cell cycle [2], senescence of cells [3], programmed cell death (apoptosis) [4] and to be part of a signal transduction pathway which leads to the outspread of various inflammatoric processes [5].

The main route on which the ceramide concentration in various cell types is regulated is still under discussion. The predominant opinion is that the so-called sphingomyelin cycle [6], which produces ceramide from sphingomyelin through action of a membrane-bound neutral sphingomyelinase (N-SMase), is most important. Moreover, the lysosomal acid sphingomyelinase (A-SMase) is also thought to take part in signal pathways through cleavage of sphingomyeline to ceramide [7]. Recent studies discuss in addition to the sphingomyelinases the participation of sphingomyelin synthase (SMS) in signal transduction pathways [8].

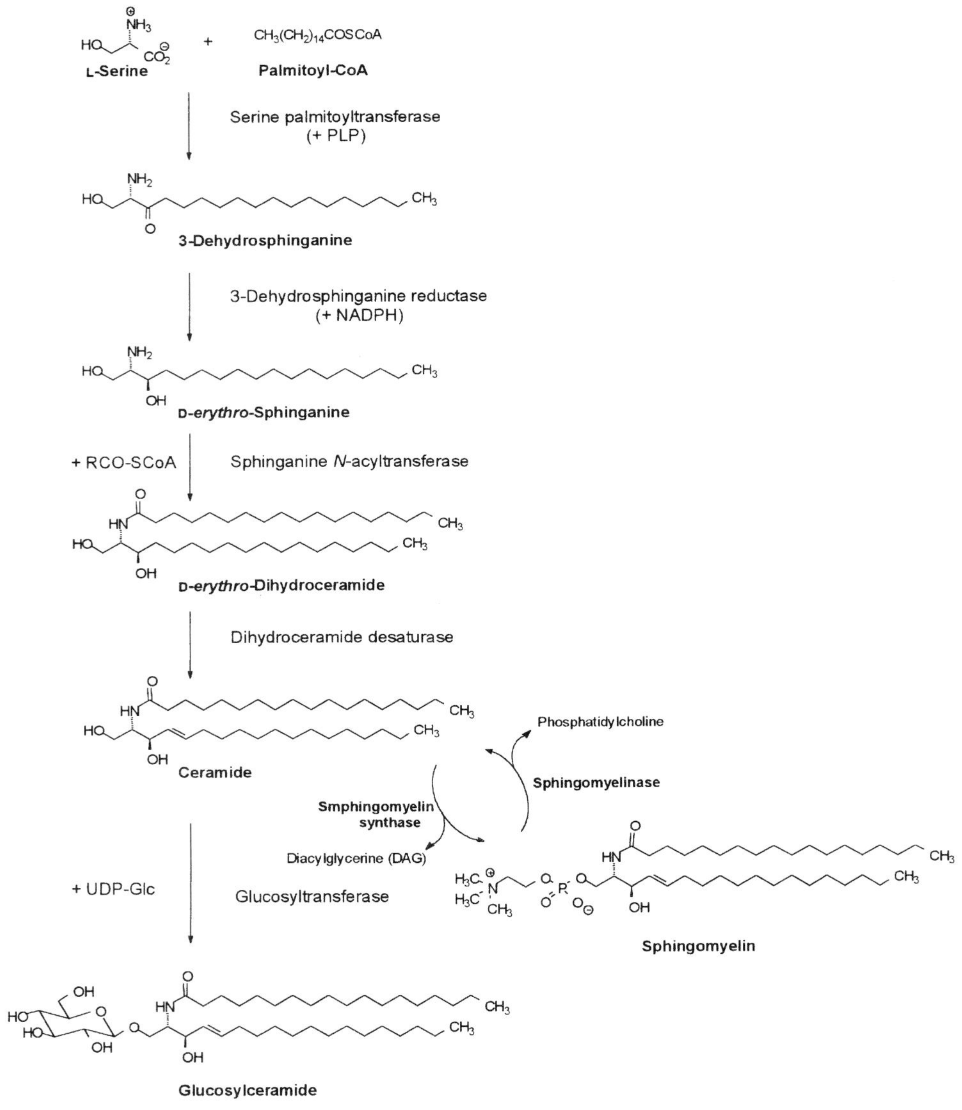

**Scheme 4.6.1.** Sphingolipid biosynthesis.

Potent specific inhibitors of the enzymes involved in sphingolipid metabolism should help to clarify physiological mechanisms leading to the above-mentioned biological effects. Last, but not least, potent inhibitors of SMS or the sphingo-myelinases may have relevance as drugs against malaria [9], multiple sclerosis [10], or overwhelming inflammatory processes [11].

The importance of the glycosphingolipids for the living cell is well documented in the evolutionary invariance of these compounds and in the fact that no inherited defects of glyco-sphingolipid biosynthesis are known [12]. However, only limited knowledge is available about the precise in vivo function of glycosphingolipids, although a number of observations indicate that they can participate in different biological events. The importance of glycosphingolipids in neuritogenesis and possibly synaptogenesis, for example, has been studied intensively [13]. In these studies, inhibitors of glycosphingo-lipid biosynthesis [12] have been employed. Fumonisin B1 (FB 1), a potent inhibitor of sphinganine *N*-acyltransferase [14] reduces axonal growth in cultured hippocampal neurons [15]. However, the effects of FB 1 may occur as a result of inhibition of glycosphingolipid synthesis or be due to an inhibition of ceramide production, which, as mentioned above, is a signaling molecule. Another problem with FB 1 is that the compound is relatively toxic, as it leads to an accumulation of toxic intermediates such as sphinganine. The development of non-toxic and specific inhibitors of the glycosyltransferases, leading to different kinds of glycosphingolipids, is important to clarify their role in the living cell. Recently, *N*-butyldeoxynojirimycine, a potent but low toxicity inhibitor of ceramideglucosyltransferase has been used to treat Tay-Sachs disease in a mouse model [16].

## 4.6.2 The concept

The goal of our work was to find a powerful inhibitor for ceramideglucosyltransferase [17], the first glycosylating step in GSL biosynthesis. We thought, perhaps that the sulfonium salt **7** may act as suicide inhibitor on this enzyme [18], because of the alkylating properties of the sulfonium group [19] instead of the original hydroxyl group at this position.

The synthesis started with N-Boc-S-methyl-L-cysteine **1** which was converted with N,O-dimethylhydroxylamine hydrochloride in the presence of 1-hydroxybenzo-triazole (HOBT), dicyclohexylcarbodiimide (DCC) and ethyldiisopropylamine, leading to amide **2**. After alkylation with hexylmagnesium bromide, the resulting ketone **3** was reduced with NaBH$_4$, resulting in the two diastereomeric alcohols **4**, which were easily separated by column chromatography on silica gel. After removal of the Boc group with HCl gas in diethyl ether, subsequent acylation of the free amino group afforded the ceramide analogs **6a**, **b**. These were converted to the corresponding sulfonium salts by treatment with methyl iodide.

## 4.6.3 Results and discussion

### 4.6.3.1 Results

Biochemical investigation revealed that the sulfonium salt **7** does not inhibit ceramide-glucosyltransferase. However, 1-methylthiodihydroceramide (1-MSDH-Cer) **6** reduced de novo ceramide production in primary cultured cerebellar neurons by about 90% at a

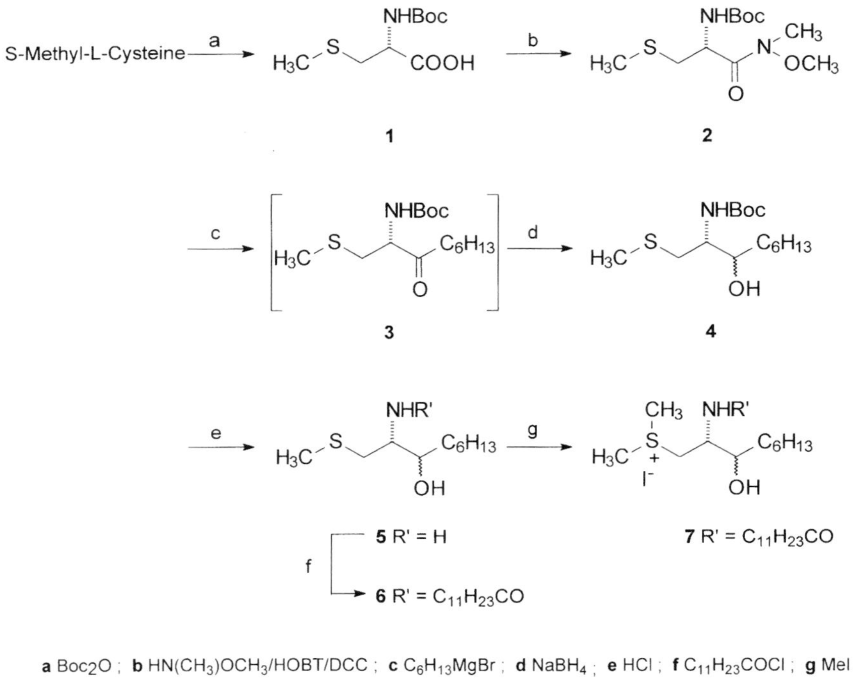
S-Methyl-L-Cysteine
a
b
c
d
e
f
g
NHBoc
COOH
H3C
S
CH3
N
OCH3
O
C6H13
OH
NHR'
I-
1
2
3
4
5 R' = H
6 R' = C11H23CO
7 R' = C11H23CO
a Boc2O ; b HN(CH3)OCH3/HOBT/DCC ; c C6H13MgBr ; d NaBH4 ; e HCl ; f C11H23COCl ; g MeI

**Scheme 4.6.2.** Synthesis of MSDH-ceramide.

concentration of 10 μM [20]. Accordingly *de novo* formation of sphingomyelin and of GSL was reduced in a time- and dose-dependent manner by up to 80%. The effects were studied by measuring the incorporation of L-[3-$^{14}$C]serine into SLs of primary cultured cerebellar neurons after 48 h preincubation with 10 μM of each compound and an additional 24 h of radioactive labeling.

The changes in sphingolipid formation could be obtained with both D-*erythro*-1-MSDH-Cer **6a** and L-*threo*-1-MSDH-Cer **6b** and were antagonized upon addition of DH-Cer or ceramide in micromolar concentrations to the culture medium; this suggests that none of the glycosyltransferases involved in GSL biosynthesis is inhibited. The corresponding free base 1-methylthiodihydrosphingosine **5** was inactive. Assays of the enzymes upstream of ceramide did not show any inhibition. For the investigation of structure-activity relationships, derivatives **8** and **9** were synthesized analogously to **7** [21]. Both compounds showed no effects on ceramide formation. These results suggest that, the thioether group is essential for biological activity.

1-MSDH-Cer does not inhibit biosynthetic enzymes, but depletes cells of free sphinganine by stimulating its catabolism. The in vitro assay with 1-MSDH-Cer (100 μM) showed only slight activation of sphinganine kinase (crude enzyme), whereas incubation of primary cultured neurons with 10 μM of 1-MSDH for 24 h and measuring the levels of free sphinganine showed a clear activation of sphinganine kinase. Reduction of cellular ceramide levels could also be obtained when intracellular

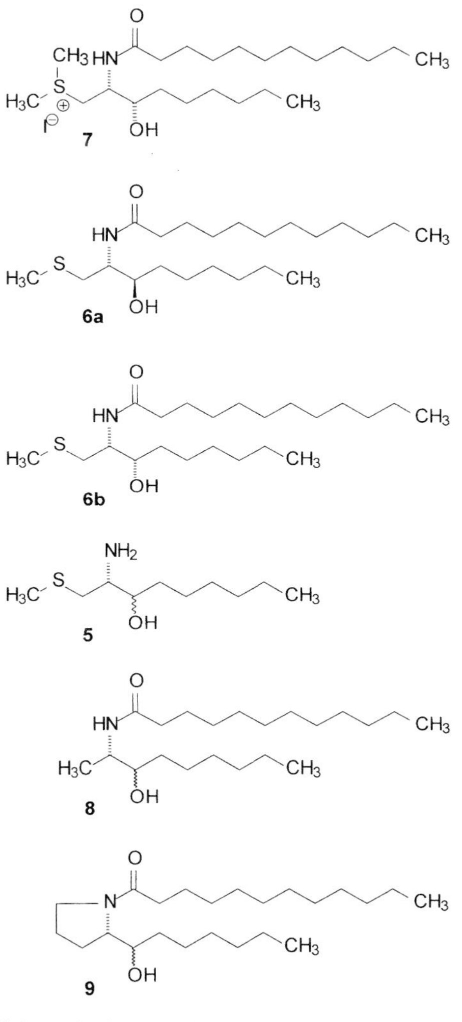

O
CH3 HN
H3C S⊕
⊖ 7 OH
CH3
CH3

O
HN
H3C S
6a OH
CH3
CH3

O
HN
H3C S
6b OH
CH3
CH3

NH2
H3C S
5 OH
CH3

O
HN
H3C
8 OH
CH3
CH3

O
N
9 OH
CH3
CH3

**Scheme 4.6.3.** Potential modulators of SL metabolism.

sphinganine levels were increased by preincubation with fuminosin B$_1$. In subsequent studies we could show that the decrease of de novo ceramide synthesis is a result of a 2.5-fold increase in sphinganinekinase activity.

The fact 1-MSDH up-regulates sphinganine kinase only after preincubation in cultured neurons and not in the *in vitro* assay suggests that the effect of 1-MSDH-Cer is not a direct interaction with the enzyme on the molecular level, but more likely seems to be a complex process which requires cell integrity and longer incubation.

Moreover, MSDH-ceramide disrupted axonal growth in cultured hippocampal neurons in a manner similar to the above-mentioned inhibitor of sphinganine-*N*-acyl-transferase fumonisin B$_1$. Those two effects can be fully reversed by the addition of

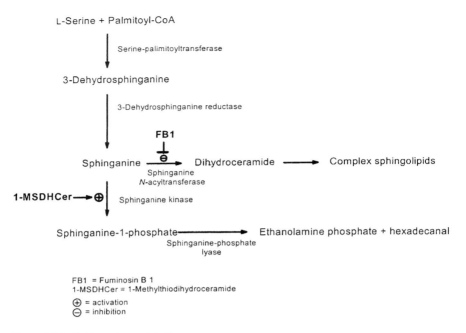

**Figure 4.6.1.** Sphinganine metabolism.

ceramide to the culture medium, supporting the idea that ceramide is involved in neuronal growth.

### 4.6.3.2   Discussion

The sulfonium salt **7** was synthesized as a potential inhibitor of ceramide glucosyltransferase. Although this compound did not affect sphingolipid biosynthesis, we could show that its precursor 1-methylthiodihydroceramide **6** stimulates sphinganine degradation.

Thus, 1-methylthiodihydroceramide is a powerful new tool for manipulating sphingolipid biosynthesis. In contrast to known inhibitors, as for example $FB_1$, no accumulation of toxic intermediates such as sphinganine and sphingosine occurs.

Therefore, 1-methylthiodihydroceramide may offer a new approach in the treatment of inherited disorders of sphingolipid degradation such as Tay-Sachs, Fabry or Nieman-Pick diseases [22], which often cause death in early childhood, as a result of a fatal accumulation of sphingolipids in the lysosomes [23].

## References

[1]  For review, see Y. A. Hannun, *sphingolids in signal transduction* (Y.A. Hannun ed.), Springer **1997**.

[2]   (a) C. S. Rani, A. Abe, Y. Chang, N. Rosenzweig, A. R. Saltiel, N. S. Radin, J. A. Shayman, *J. Biol. Chem.* **1995**, *270*, 2859–2867; (b) S. Jayadev, B. Liu, A. Bielawska, J. Y. Lee, F. Nazaire, M. Y. Pushkareva, L. M. Obeid, Y. A. Hannun, *J. Biol. Chem.* **1995**, *270*, 2047–2052.

[3]   M. E. Venable, J. Y. Lee, M. J. Smyth, A. Bielawaska, L. M. Obeid, *J. Biol. Chem.* **1995**, *270*, 30701–30708.

[4]   Hannun, Y. A. Obeid, L. M. *Trends Biochem. Sci.* **1995**, *20*, 73–77.

[5]   For reviews, see (a) C.A. Smith, T. Farrah, R.G. Goodwin, *Cell* **1994**, *76*, 959–962; (b) B. Beutler, C. Van Huffel, *Science* **1994**, *264*, 667–668.

[6]   Y. A. Hannun, *J. Biol. Chem.* **1994**, *269*, 3125–3128.

[7]   R. Schwandner, K. Wiegmann, K. Bernardo, D. Kreder, M., Krönke, *J. Biol. Chem.* **1998**, *273*, 5916–5922.

[8]   C. Luberto, Y.A., Hannun, *J. Biol. Chem.* **1998**, *273*, 14550–14559.

[9]   S. A. Lauer, N. Ghori, K. Haldar, *Proc. Natl. Acad. Sci. USA* **1995**, *92*, 9181–9185.

[10]   R. W. Ledeen, G. Chakraborty, *Neurochem Res.* **1998**, *23*, 277–289.

[11]   S. Mathias, A.Younes, C. C. Kan, I. Orlow, C. Joseph, R. N Kolesnick., *Science* **1993**, *259*, 519–522.

[12]   T. Kolter, K. Sandhoff, *Chem. Soc. Rev.* **1996**, *25*, 371–381.

[13]   K. Hirschberg, R. Zisling, G. van Echten-Deckert, A. H. Futerman, *J. Biol. Chem.* **1996**, *271*, 14876–14882.

[14]   E. Wang, W. P. Norred, C. W. Bacon, T. T. Riley, A. H. Merrill, *J. Biol. Chem.* **1991**, *266*, 14486–14490.

[15]   (a) R. Harel, A. H. Futerman, *J. Biol. Chem.* **1993**, *268*, 14476–14481. (b) A. Schwarz, E. Rapaport, K. Hirschberg, A. H. Futerman, *J. Biol. Chem.* **1995**, *270*, 10990–10998.

[16]   (a) F. M. Platt, G. R. Neises, G. Reinkensmeier, M. J. Townsend, V. H. Perry, R. L. Proia, B. Winchester, R. A. Dwek, T. D. Butters, *Science* **1997**, *276*, 428–431. (b) for review see T. Kolter, *Angew. Chem.* **1997**, *109*, 2044–2048; *Angew. Chem Int. Ed.*, **1997**, *36*, 1955–1959.

[17]   For review, see S. Ichikawa, Y. Hirabayashi, *Trends Cell Biol.* **1998**, *8*, 198–202.

[18]   G. Brenner-Weiß, A. Giannis, K. Sandhoff, *Tetrahedron* **1992**, *48*, 5855–5860.

[19]   H. Heydt, E. Vielsmaier, *in Houben Weyls Methoden der Organischen Chemie* (D. Klamann ed.), G. Thieme Verlag, **1995**, *E11*, 405–413.

[20]   G. van Echten-Deckert, A. Giannis, A. Schwarz, A. H. Futerman, K. Sandhoff, *J. Biol. Chem.* **1998**, *273*, 1184–1191.

[21]   G. Brenner-Weiß, *Ph.D. thesis*, University of Bonn, **1993**.

[22]   For review, see T. Kolter, K. Sandhoff, *Brain Path.* **1998**, *8*, 79–100.

[23]   For review, see K. Sandhoff, T. Kolter, *Trends Cell Biol.* **1996**, *6*, 98–103.

# 5 Nucleic Acid Chemistry: Mechanisms and Mimetics

## 5.1 Structural alterations of the isopolar phosphonate bond in nucleotide and oligonucleotide analogs

*Magdalena Endová, Šárka Králíková, Radek Liboska, Miroslav Otmar, Dominik Rejman, Zdeněk Točík, and Ivan Rosenberg*

More than four decades ago, a tremendous effort was initiated by chemists in the area of synthesis and biochemical evaluation of nucleotide analogs. Initially, the group of enzymatically stable, isopolar, phosphonate-based nucleotides, containing the bridging P-C linkage instead of the P-O one, played an important role as an efficient tool for the study of the mechanism of enzymes catalyzing the cleavage or transfer of the phosphorus moiety. Since that time, a number of phosphonate nucleotide analogs, mostly their acyclic congeners, have been synthesized and examined biologically with respect to their potential antiviral properties [1]. On the other hand, although the chemistry of phosphonate nucleotide analogs has provided a pool of compounds which are potentially usable as the building blocks of oligonucleotides, few results in the area of isopolar phosphonate oligonucleotides have been reported so far [2, 3].

This paper deals with the findings we have acquired at the search for novel isopolar, phosphonate-based, structurally diverse nucleotides as potential antivirals, and for new oligonucleotides as the candidates for the regulation of gene expression.

### 5.1.1 5-(Adenin-9-yl)pentofuranosylphosphonates and 2-(adenin-9-yl)cycloalkyloxymethanephosphonates

The discovery of the remarkable antiviral properties of *O*-phosphonomethyl derivatives of the acyclic nucleosides **1a, 1b,** and **2** (HPMPA, HPMPC, and PMEA) [1] at the end of the 1980's has led to the preparation of two types of their conformationally restricted, cyclic congeners differing in the manner of fixation of acyclic chain into the sugar-representing ring.

The first type compounds, various $\alpha$-**3a** and $\beta$-5-(adenin-9-yl)pentofuranosyl-phosphonates **3b,** were designed to secure the same configuration of the chiral center as in *(S)*-HPMPA **1**, while the configurations of hydroxyls were varied[4]. Thus, the synthesis of the anomeric phosphonates **3** with L-*arabino* ($R^1=R^4=OH$, $R^2=R^3=H$), 2-deoxy-L-*erythro* ($R^1=R^2=R^3=H$, $R^4=OH$), 2-deoxy-L-*threo* ($R^1=R^2=R^4=H$, $R^3=OH$), L-*ribo* ($R^1=R^3=H$, $R^2=R^4=OH$) and L-*xylo* ($R^1=R^4=H$, $R^2=R^3=OH$) configuration started from the corresponding L-pentofuranosylphosphonates.

These synthons were prepared from L-sugars, except for the synthons for L-*ribo*- and L-*xylo*-configured compounds which were obtained by intramolecular rearrangement of geminal methanesulfonyloxy phosphonate derivative of D-pentofuranoses under acidic conditions [4].

The conformationally fixed compounds of the second type **4** are distinguished by the presence of the cycloalkane **4a, 4c** or tetrahydrofuran **4b, 4d** ring mimicking the sugar moiety of nucleotides. This ring closure gives rise to the existence of enantiomeric *trans* **4a, 4b** and *cis* **4c, 4d** isomers [5]. Each of the types of cyclic compounds, *i.e.* **3a, 3b** or **4a, 4c** or **4b, 4d** represents structures resembling two 'limit' conformations of parent comounds **1, 2** given by mutual position of the phosphorus moiety and the nucleobase; the flexibility of carbocyclic compounds **4**, however, increases dramatically with the enlargement of the cycloalkane ring. The chemical synthesis of the carbocyclic nucleotides **4a, 4b** containing *trans*-configured cycloalkane ring is based on the nucleophilic displacement of the tosyl group in diisopropyl tosyloxymethanephosphonate by the hydroxyl group of the corresponding carbocyclic *trans*-nucleosides which were easily available by oxirane ring opening of the cycloalkene oxides with nucleobases. The carbocyclic *trans*-nucleosides also served as the starting compounds for the synthesis of *cis*-configured nucleosides *via* inversion of configuration on the carbon atom bearing the secondary hydroxyl group, and these derivatives were then phosphonylated to *cis*-nucleotides **4c, 4d**. While none of these cyclic compounds was found to be active against DNA viruses and retroviruses, their further evaluation in oligonucleotide chemistry is envisaged. Whereas compounds **3** could be used immediately for this purpose, the structural modification of compounds **4** by the introduction of a hydroxy group at the cycloalkane ring will be needed.

## 5.1.2 *O*-Phosphonomethylnucleosides

The investigation performed on 5'-, 3'- or 2'-*O*-phosphonomethylribonucleosides **5a–c** as the isopolar non-isosteric nucleotide analogs containing the P–CH$_2$–O moiety [6] revealed that all the prepared regioisomeric *O*-phosphonomethylribonucleosides were resistant against the phosphomonoesterase and nucleotidase cleavage, and some of them were even found to inhibit the latter enzyme. The synthesis of **5a** was accomplished by nucleophilic displacement of the tosyl group in dialkyl tosyloxy-methanephosphonate by the 5'-hydroxyl group of a nucleoside. On the contrary, the phosphonomethyl ether bond in the 2'- (**5b**) or 3'-isomer (**5c**) was predominantly generated by the intramolecular cyclization of the 3'- or 2'-chloromethanephosphonyl esters of ribonucleosides under alkaline conditions [6].

The *O*-phosphonomethyl derivatives of nucleosides in the 2'-deoxyribo series have also been prepared [7]. In this case, the synthesis of building blocks **6a**, **6b** suitable for the phosphotriester method of oligonucleotide assembly has been elaborated.

The biochemical investigation performed at the level of ribonucleoside 5'-triphosphate analogs **7a** revealed the ability of the non-isosteric P–CH$_2$–O bond to mimic the phosphomonoester one. Thus, all four triphosphate analogs **7a** were competitive selective inhibitors of incorporation of the natural ribonucleoside 5′-triphosphates into RNA in the reaction catalyzed by DNA-dependent RNA polymerase; [8] no incorporation of these compounds into RNA was found. Similar results were obtained with the adenine diphosphate analog **7b** in the reaction catalyzed by polynucleotide phosphorylase; no incorporation was found, but this analog stimulated the formation of polyA with a significantly higher molecular weight [9]. Besides, both cytosine and uracil derivatives **7a** are efficient allosteric effectors of uridine kinase from L1210 cells, as natural triphosphates themselves, and in addition, the adenine or guanine triphosphate analogs **7a** are capable of replacing natural ATP or GTP as a donor of the phosphate group in the reaction catalyzed by the same enzyme [10].

### 5.1.3    *O*-Phosphonoalkylidenenucleosides

The physico-chemical properties of the *O*-phosphonomethyl derivatives of ribonucleosides **5** stimulated the preparation of their conformationally restricted congeners **8** and **9**. These novel compounds [11], the *O*-phosphonoalkylidene ribonucleosides **8** or **9**, exist in epimeric pairs due to the chirality of the carbon atom bearing phosphoryl group and represent two limit conformations of their flexible 2′- and 3′- or 5′-*O*-phosphonomethyl congeners **5**. A number of 2′,3′-*O*- and 3′,5′-*O*-phosphonoalkylidenenucleosides and their derivatives differing in both nucleobase and alkylidene moiety have been prepared and their properties investigated [12]. Their syntheses started from the *ribo*-2′,3′- or *xylo*-3′,5′-orthoesters of nucleosides which smoothly afforded, in a redox reaction with dialkyl chlorophosphites, the appropriate phosphonates. The epimeric 2′,3′-*O*-phosphonoalkylidene derivatives (**8**) offer a possibility to construct oligonucleotides with a non-isosteric, conformationally restricted internucleotide linkage; the dinucleoside monophosphate analogs **10** and **11** were prepared [13] and their physico-chemical properties evaluated (see below). As for the biological activity of *O*-alkylidene derivatives **8**, they did not exhibit any significant antiviral properties; their 5′-*O*-*tert*-butyldiphenylsilyl derivatives, however, caused a dramatic and yet unclear effect on L 1210, L 929, and HeLa S3 cell growth. The release of the L 1210 cells from the surface was observed in the presence of the tested compounds and, in the case of L 929 or HeLa S3 cells which grew in the suspension, the cell lysis took place. A possible explanation for this phenomenon could be seen in the formation of some kind of miceles

B = A,G,C,T,U
R$^1$ = H, CH$_3$, CH$_2$OH, CH$_2$X, CH$_2$NH$_2$, CH$_2$(CH$_2$)$_n$X
**8**     (X=halogen, N$_3$)     **9**

**a** R = H
**10**    **b** R = C$_6$H$_5$          **11**

with a high affinity to the cell surface. This hypothesis is supported by the finding that a simultaneous addition of the active compound and γ-cyclodextrin (*i.e.*, a compound known as a solubilizer and/or a carrier for lipophilic compounds) disturbed the cell release and lysis completely. This effect will be further studied.

## 5.1.4   5′-*C* and 3′-*C*-phosphononucleosides

Further investigation of the isopolar non-isosteric nucleotide analogs has led to compounds which could hypothetically be derived from the nucleosides by attachment of the phosphoryl group to the 5′- or 3′-carbon atom of the sugar moiety. Although a synthetic route to these shortened types, the epimeric 5′- **12** and single epimer of 3′-geminal hydroxy phosphonates **13**, was elaborated and improved in our laboratory [14], the priority in the synthesis of the 3′-type **13** belongs to Wiemer's group [15]. Thus, the synthesis of compounds **12** and **13** started from the nucleoside 5′-aldehydes and 3′-ketonucleosides, respectively. The nucleophilic addition of *tris*-trimethylsilyl phosphite to the carbonyl group of 3′-ketonucleosides afforded single epimer **13**, whereas in the case of nucleoside 5′-aldehydes, both epimeric hydroxy phosphonates **12** were formed. The absolute configuration of the 5′-epimers **12** could be successfuly resolved by 2-D ROESY NMR experiments because one of the epimers was found, surprisingly, to display a strongly restricted rotation around the 4′-5′ bond. For the study of the physico-chemical properties of the shortened internucleotide linkage, both epimers of thymidine dimers **14** were prepared. These compounds exhibited excellent chemical and biological stability; no degradation was observed after treatment with concentrated aqueous ammonia or L 1210 cell free extract. Further evaluation of these compounds for oligonucleotide assembly is underway.

## 5.1.5   5′-Deoxy-4′-phoshonomethoxynucleosides

The family of structural variants of isosteric phosphonate nucleotides and oligonucleotides was further enlarged by the preparation of epimeric α-L-*threo*- and β-D-*erythro*-configured 4′-phosphonomethoxy derivatives **15a** of nucleosides [16]. Their synthesis was accomplished by means of an electrophile-promoted addition of hydroxyl group of dialkyl hydroxymethanephosphonate to the double bond of the $4',5'$-didehydronucleosides. The configuration in $4'$-position was determined on the basis of 2D-ROESY NMR spectroscopy. Both compounds were utilized in their protected form **15b**, as P-components in the preparation of appropriate dinucleoside phosphonates **16a** which exhibited resistance against cleavage by L 1210 cell-free extract. The facts obtained with this isosteric modification of internucleotide bond point to its utilization in oligomeric chains. The means of assembling such oligonucleotides **16b** has already been elaborated [17].

## 5.1.6   *O*-Phosphonomethyl group in oligonucleotides

The presence of an extra methylene group in the molecule of regioisomeric $2'$-, $3'$- and $5'$-*O*-phosphonomethylribonucleosides **5** provided the possibility to construct a set of four non-isosteric diribonucleoside monophosphate analogs in both $3',5'$ (**17a**, **17b**) and $2',5'$ (**18a**, **18b**) series which differed in the position of the methylene group in the internucleotide linkage. All these types of dimers with various combinations of nucleobases were synthesized [18]. The NMR spectroscopy study on these compounds revealed close conformational similarities with the natural counterparts, especially with regard to the sugar conformation and base stacking [19]. The resistance of the dimers against enzymatic cleavage was excellent; neither ribonucleases nor exonucleases

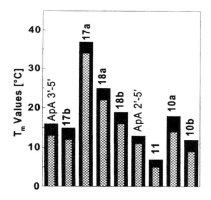

**Figure 5.1.1.** Melting points of ApA complexes with polyU.

cleaved any of our new types of internucleotide linkages [18]. The chemical stability of these compounds under strongly alkaline conditions depended on the type of the dimer. Whereas the dimers **17a** and **18a** were quite stable, the dimers **17b** and **18b** underwent elimination of the nucleoside by a mechanism known for the cleavage of RNA (participation of the 2'- or 3'-hydroxyl group). In this case, the presence of an appropriate six-membered 2',3'-cyclic phosphonate was detected. On the other hand, under acidic conditions, no cleavage was found but the $2' \leftrightarrow 3'$ isomerization took place in the case of the dimers **17a** and **18a**.

**19**
**a** n = 0; m = 1
**b** n = 1; m = 0

In order to study the conformational differences among several types of the non-isosteric internucleotide linkages, in comparison with the natural phosphodiester one, the thermal stabilities of the complexes of appropriate phosphonate ApA analogs

**17, 18** and **10, 11** with polyU were measured [20]. Most of these non-isosteric ApA dimers exhibited, somewhat surprisingly, excellent thermal stabilities that matched or even surpassed that of natural ApA itself (Figure 5.1.1). Considerable differences were found, indeed, suggesting the key role of particular arrangements of the individual phosphonate internucleotide linkage. Because these findings suggest that dinucleoside monophosphate analogs with non-isosteric, ester ether type of internucleotide linkage display two main attributes of oligonucleotides, the base stacking and base pairing, longer oligonucleotides in thymidine series **19a, 19b** containing variable numbers of the individual phosphonate internucleotide linkages in the chain were consequently prepared [21], and some of them were studied with respect to their ability to form duplexes/triplexes with natural counterparts. In spite of the fact that this study is still being continued and final conclusions have not been drawn, the partial results have shown excellent nuclease stability of modified oligonucleotides and good hybridization properties which were further enhanced if an alternating mode of phosphodiester and phosphonate ester ether linkage was employed.

## 5.1.7   Conclusions

A number of new isopolar phosphonate nucleotide analogs were prepared. Synthetic modifications featured several kinds of conformational restrictions as well as novel, both isosteric and non-isosteric, variants of the phosphonate moiety. While none of the tested compounds displayed any remarkable antiviral activities, the syntheses of new phosphonate oligonucleotides were successfully elaborated, and a large pool of structurally diverse linkages, capable of modifying the properties of oligomer chains, created. Current promising results of hybridization capabilities, along with excellent nucleolytic stability, seem to qualify the isopolar phosphonate oligonucleotides for testing in antisense strategy and related techniques of gene therapy.

**Acknowledgements:**   This work was supported by grant No. 4055605 of the Grant Agency of the Academy of Sciences of the Czech Republic, by IGA grant No. 3313–3 of the Ministry of Public Health of the Czech Republic, and by NIH grant # R03 TW00673–01. The authors are also indebted to Prof. E. DeClercq (Rega Institute, Leuven, Belgium) for antiviral screening, and to Dr. I. Votruba (this Institute) for the evaluation of cytostatic properties of prepared compounds.

## References

[1] E. De Clercq, A. Holý, I. Rosenberg, T. Sakuma, J. Balzarini, P. C. Maudgal, *Nature*, **1986**, *323*, 464; R. Pauwels, J. Balzarini, D. Schols, M. Baba, J. Desmyter, I. Rosenberg, A. Holý, E. De Clercq, *Antimicrob. Agents Chemother.*, **1988**, *32*, 1025.
[2] T. Szabo, A. Kers, J. Stawinski, *Nucleosides Nucleotides* **1995**, *23*, 893.
[3] V. A. Efimov, M. V. Choob, A. A. Buryakova, A. L. Kalinkina, O. G. Chakhmakhcheva, *Nucleic Acids Research* **1998**, *26*, 566.

[4]  M. Otmar, I. Rosenberg, M. Masojídková, A. Holý, *Collect. Czech. Chem. Commun.*, **1993**, *58*, 2159; *ibid*, **1993** *58*, 2180.

[5]  R. Liboska, M. Masojídková, I. Rosenberg, *ibid*, **1996**, *61*, 313; *ibid*, **1996**, *61*, 778; R. Liboska, *Collect. Czech. Chem. Commun. (Special Issue)*, **1996**, *61*, S72.

[6]  A. Holý, I. Rosenberg, *ibid*, **1982**, *47*, 3447; I. Rosenberg, A. Holý, *ibid*, **1983**, *48*, 778; *ibid*, **1985**, *50*, 1507.

[7]  D. Rejman, Z. Točík, R. Liboska, I. Rosenberg, unpublished results.

[8]  K. Horská, I. Rosenberg, A. Holý, K. Šebesta, *Collect. Czech. Chem. Commun.*, **1983**, *48*, 1352.

[9]  A. Holý, M. Nishizawa, I. Rosenberg, I. Votruba, *ibid*, **1987**, *52*, 3042.

[10]  J. Vesel, I. Rosenberg, A. Holý, *ibid*, **1982**, *47*, 3464; *ibid*, **1983**, *48*, 1783.

[11]  M. Endová, M. Masojídková, M. Buděšínský, I. Rosenberg, *Tetrahedron Lett.*, **1996**, *37*, 3497.

[12]  M. Endová, M. Masojídková, M. Buděšínský, I. Rosenberg, *Tetrahedron*, **1998**, *54*, 11151; *ibid*, **1998**, *54*, 11187.

[13]  M. Endová, I. Rosenberg, *Book of Abstracts, 5. Nachwuchswissenschaftler – Symposium Bioorganische Chemie, Paderborn, Germany, September 16–18*, **1996**.

[14]  Š. Králíková, I. Rosenberg, *Collect. Czech. Chem. Commun. (Special Issue)*, **1996**, *61*, S124.

[15]  W. L. McEldoon, K. Lee, D. F. Wiemer, *Tetrahedron Lett.*, **1993**, *34*, 5843.

[16]  Z. Točík, I. Kavenová, I. Rosenberg, *Collect. Czech. Chem. Commun. (Special Issue)*, **1996**, *61*, S76.

[17]  Z. Točík, D. Rejman, I. Rosenberg, unpublished results.

[18]  I. Rosenberg, A. Hol, *Collect. Czech. Chem. Commun.*, **1987**, *52*, 2572.

[19]  J. Zajíček, I. Rosenberg, R. Liboska , Z. Točík, G. V. Fazakerley, W. Guschlbauer, A. Holý, *ibid*, **1997**, *62*, 83.

[20]  M. Pressová, M. Endová, Z. Točík, R. Liboska, I. Rosenberg, *Bioorg. Med. Chem. Lett.*, **1998**, *8*, 1225.

[21]  D. Rejman, I. Rosenberg, *Book of Proceedings, 4th Cambridge Symposium on Oligonucleotide Chemistry and Biology, Queens College, Cambridge, UK, Aug. 31–Sept. 3*, **1997**, p. 84; R. Liboska, Z. Točík, I. Rosenberg, *ibid*, p. 58.

# 5.2    DNA Repair: From model compounds to artificial enzymes

*Thomas Carell, Lars Burgdorf, Jens Butenandt, Robert Epple, and Anja Schwögler*

## 5.2.1    The degradation and repair of the genetic information

The genetic information – the construction plan of any organism – is stored as a base sequence in the form of single- or double-stranded oligonucleotides within every cell [1]. For the survival and the reproduction of organisms it is vital that the genetic information is not modified or damaged during the life span of the cell [2]. Exogenous events like UV and $\gamma$-irradiation and endogenous reactive species like OH-radicals, however, cause the continuous degradation of the genetic material [3]. In addition, spontaneous alterations due to the deamination of cytidine to uridine, or the cleavage of glycosidic bonds, question the integrity of the genome. Organisms have developed efficient DNA repair systems that are able to detect and to remove these genome lesions in order to counteract the information loss [3]. It is now clear that DNA repair is one of the main cellular defense systems required for the survival of organisms.

To date, a large variety of genetic repair systems has been identified. The complicated nucleotide excision repair system [4, 5], targets bulky DNA adducts that arise if carcinogens like benzo[a]pyrene or aflatoxine [6] react with DNA bases. Such lesions are involved in the development of lung [7, 8] and liver cancers. The base excision repair systems [9] detect specific base degradation products like uridine, formamido-pyrimidine and 8-hydroxyguanosine that are formed as base deamination or oxidation products. Some lesions occur so frequently that their repair is secured by more than one repair system. The main environmentally induced lesions, namely the UV DNA photo products, belong to this category. UV-irradiation, which is ubiquitous in sun-light-exposed habitats, induces the formation of cyclobutane pyrimidine dimers as depicted in Figure 5.2.1 [10]. These lesions are formed in a photo induced $[2\pi + 2\pi]$ cycloaddition reaction of two pyrimidine bases located above each other in the DNA stack.

*In vitro* and *in vivo* data currently support that UV lesions cause cell death and are (in humans) responsible for the development of various kinds of skin tumors [11, 12]. Repair of these environmentally induced lesions is consequently essential for all organisms [12–14].

## 5.2.2    DNA Photolyase Repair Enzymes

Some organisms like *M. luteus* and T4-infected *E. coli* contain a photo dimer-specific DNA-glycosylase [15, 16]. Archaebacteria, amphibians, marsupials, and plants

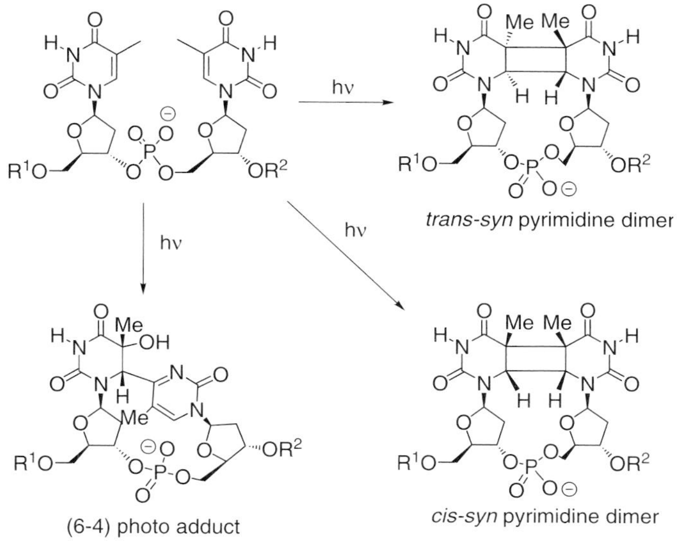

**Figure 5.2.1.** Structures of the most abundant DNA photo lesions formed upon irradiation of cells with UV light. $R^1$, $R^2$ = Continuation of the DNA strand.

possess a DNA-photolyase repair enzyme [17, 18]. A special feature of DNA photo-lyases is the fact that the enzyme utilizes long-wavelength sunlight (300 nm–500 nm) to remove the most abundant *cis-syn* dimers in a light-dependent repair reaction. The overall repair process is depicted in Figure 5.2.2.

DNA-photolyases are monomeric flavin-dependent repair enzymes with a molecular weight between 55 and 65 kDa [19,20]. The enzyme binds tightly to *cis-syn* cyclobutane uracil or thymine dimers in single- and double-stranded DNA with a $K_d = 10^{-8}$ M (thymine dimer) [17]. Photons possessing the required wavelengths are initially absorbed by either a methenyl-tetrahydrofolate (MTHF) [21] cofactor in type I photolyases or a 8-hydroxy-5-deazaflavin (8-HDF) cofactor in type II photolyases

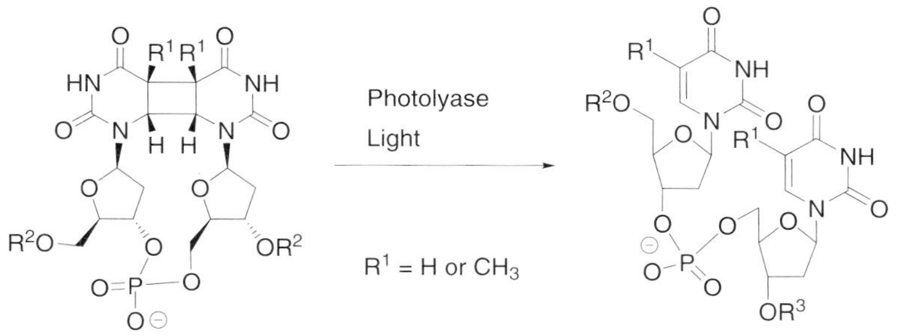

**Figure 5.2.2.** Depiction of the repair reaction initiated by DNA photolyases. $R^2$, $R^3$ = Continuation of the DNA strand.

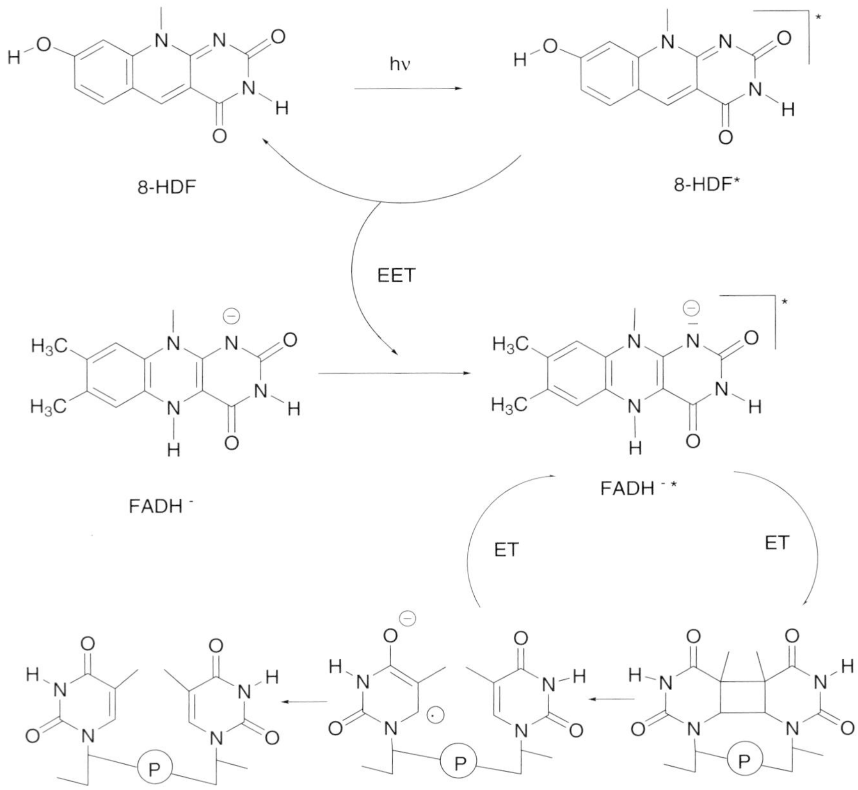

**Figure 5.2.3.** Repair mechanism of the flavin and deazaflavin containing type II DNA photolyase from, e.g. *Anacystis nidulans* [25]. EET = Excitation Energy Transfer; ET = Electron Transfer.

(Figure 5.2.3). This second cofactor transfers the light energy to a crucial FAD-cofactor, present in all known photolyases, by utilizing Förster dipole–dipole energy transfer [22] that is also operational in the photo antenna systems of the photosynthetic apparatus [23, 24].

The $FADH^-$-transfers after excitation one electron to the cyclobutane pyrimidine dimer unit. The formed dimer radical anion splits spontaneously into the monomers, and the surplus electron is ultimately transferred back to the flavin radical intermediate (Figure 5.2.3). Although the general light-driven splitting scenario depicted schematically in Figure 5.2.3 is well understood, several crucial aspects of the repair process remain ambiguous.

First, it is unknown how repair enzymes – including DNA-photolyases – recognize single DNA-lesions with high precision in a structurally heterogeneous prokaryotic or eukaryotic genome. Recent co-crystal structures of lesioned DNA in complex with repair enzymes indicate that the recognition requires the flipping of the DNA lesion out of the DNA double helix [26–28].

Second, it is currently unclear how photolyases mediate the energy and electron transfer processes. It is not known if and how the enzyme interacts with the dimer radical anion during the cleavage reaction in order to achieve repair with almost maximal efficiency (quantum yield $\phi = 0.7$–$0.9$). It is possible that the enzyme has to stabilize potentially short-lived intermediates, in order to increase the lifetime of the charge separated state. This may also include rapid deprotonation and protonation reactions required temporarily to neutralize appearing charges [29].

To gain further insight into the photolyase mechanism, it is crucial to understand the principle cleavage reaction, its dependencies, and the dimer recognition step. This can be achieved with model compounds that allow one to simulate the cleavage reaction outside the protein environment. For the understanding of the lesion recognition step, crystal structures of dimer-containing DNA double strands alone and in complex with photolyases are essential.

## 5.2.3   Mechanistic investigations with model compounds

### 5.2.3.1   General aspects and cleavage assay

In order to enable model compound investigations, we designed chemical systems (Scheme 5.2.1) [30], which mimic the enzymatic cleavage process. The prepared compounds contain a bis(carboxymethyl) functionalized cyclobutane uracil

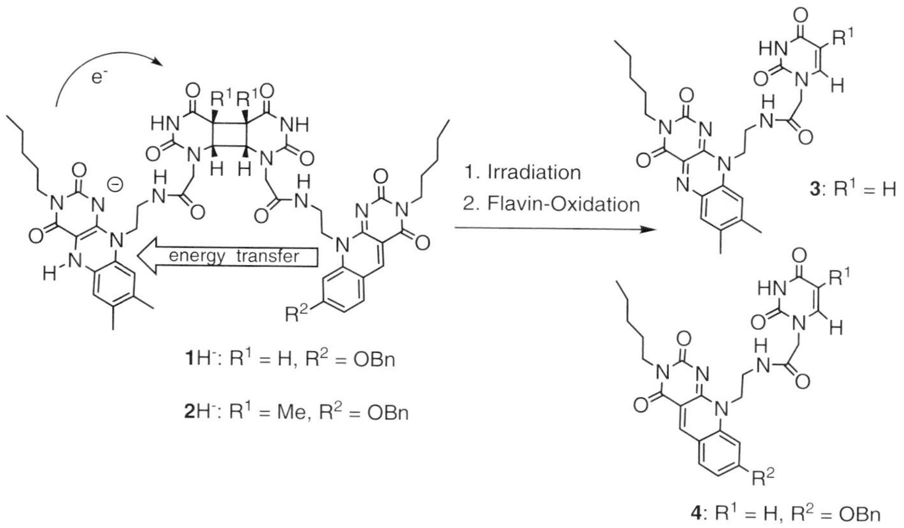

**Scheme 5.2.1.** Depiction of the two model compounds **1H⁻** and **2H⁻** (after reduction of the flavin chromophore) and of the investigated intramolecular light-induced splitting reaction of **1H⁻** to **3** and **4**.

($R^1 = H$) or thymine ($R^2 = Me$) dimer as the lesion mimic [31]. Depending on the question to be investigated, the model compounds may possess a unique set of flavin and deazaflavin cofactors like in **1** and **2** that are covalently linked via amide bonds to the dimer unit. In this way, a series of model compounds were prepared, allowing the investigation of the cleavage process, *e.g.* of **1H$^-$** and **2H$^-$** to **3** and **4** (after reoxidation) depending on external parameters such as the pH value and the medium polarity, or on internal parameters like the cofactor composition, the cofactor arrangement, and the dimer structure [32]. The fact that all data were obtained with a structurally closely related set of model compounds ensures maximal comparability of the results.

A splitting assay was developed in order to quantify the photo-induced cleavage. The model compounds were dissolved and the solutions were deoxygenated. Following reduction of the flavin (*e.g.* **1** to **1H$^-$**), the samples were stirred and irradiated with monochromatic light. After certain time intervals, small aliquots were removed from the solution, immediately reoxidized and analyzed by reversed-phase HPLC. The series of HPL-chromatograms (cleavage of **5H$_2^{2-}$** to **6**) depicted in Figure 5.2.4, underlines that the reaction of **5H$_2^{2-}$** to **6** (after reoxidation) is a clean conversion. The

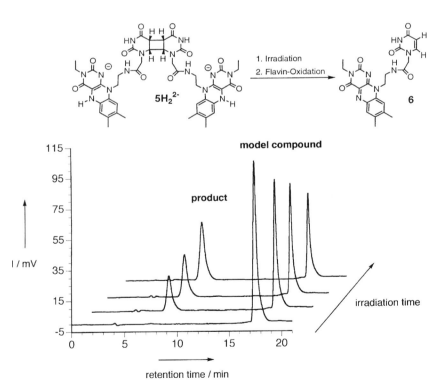

**Figure 5.2.4.** Reverse-phase HPLC diagrams taken during an irradiation experiment show the clean conversion of the model compound **5H$_2^{2-}$** to the photo-splitted product **6**.

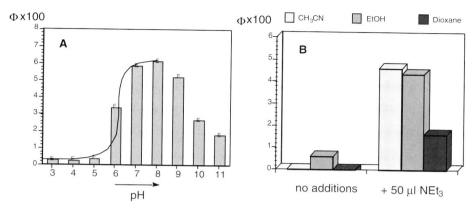

**Figure 5.2.5.** (a) pH-dependent measurement of the splitting reaction. (b) Splitting reaction in organic media in the absence and in the presence of triethylamine ($NEt_3$) as the base.

obtained peaks of the model compound at 17 min and of the reaction product at 8 min were integrated. These values were used to calculate the reaction yield at any given irradiation time. For the calculation of the quantum yields $\phi$ [33] ($\phi$ = number of reacted molecules/number of absorbed photons), the intensity of the light beam was measured using ferrioxalate actinometry [34]. Control experiments were performed in order to unequivocally establish the strict requirement for a reduced flavin, covalently attached to the dimer unit [32].

### 5.2.3.2  pH-Dependency of the cleavage reaction

Utilizing this cleavage assay, the pH-dependence of the cleavage reaction was investigated in order to clarify how the deprotonation of the reduced flavin chromophore ($FlH_2 \rightarrow FlH^-$) affects the DNA-repair process. Figure 5.2.5a shows the quantum yields obtained for the photo-cleavage of model compounds (e.g. **5**) at $\lambda = 366$ nm in water, buffered at various pH values. Very low cleavage activity was observed at pH < 6 and maximal splitting rates were measured at pH > 7. Intermediate rates were obtained around pH = 6. These data are in full agreement with the $pK_a$-value of the reduced flavin ($pK_a = 6.5$) and therefore show that the deprotonation of the reduced flavin is absolutely required for the efficient photo-induced splitting. Further measurements in organic solvents (acetonitrile, ethanol and dioxane) supported this result. As depicted in Figure 5.2.5b, no cleavage was observed in the absence of base. Addition of triethylamine into the reaction mixture caused, upon irradiation the immediate cleavage of the model compounds which prove that the addition of base is strictly necessary for the photo reaction. UV-spectroscopic investigations in the absence and presence of triethylamine support the presence of the neutral reduced flavin species ($FlH_2$) directly after catalytic hydrogenation and of the deprotonated reduced flavin-unit ($FlH^-$) in the presence of base.

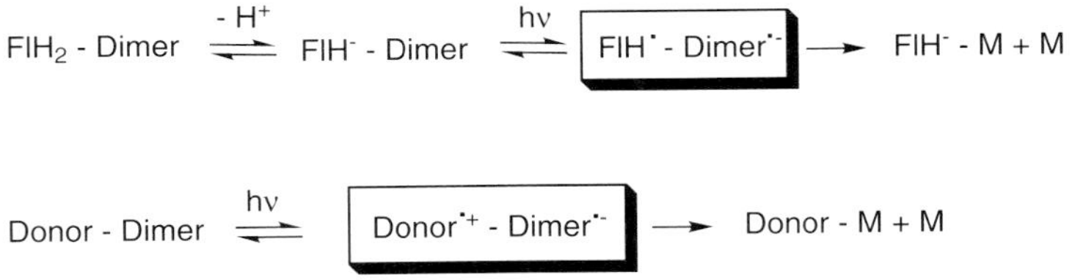

$$\text{FlH}_2\text{-Dimer} \;\underset{}{\overset{-\,H^+}{\rightleftharpoons}}\; \text{FlH}^-\text{-Dimer} \;\underset{}{\overset{h\nu}{\rightleftharpoons}}\; \boxed{\text{FlH}^\bullet\text{-Dimer}^{\bullet-}} \longrightarrow \text{FlH}^-\text{-M} + \text{M}$$

$$\text{Donor-Dimer} \;\underset{}{\overset{h\nu}{\rightleftharpoons}}\; \boxed{\text{Donor}^{\bullet+}\text{-Dimer}^{\bullet-}} \longrightarrow \text{Donor-M} + \text{M}$$

**Figure 5.2.6.** Splitting mechanism in a Donor-Dimer and in a Flavin (Fl)-Dimer model system. M = Monomer.

The data clearly show that it is the $FlH^-$ species that is required for the efficient repair. The deprotonation increases the electron richness of the flavin electron donor, and therefore facilitates the electron transfer reaction. Another factor that must be considered is the lifetime of the post-electron transfer intermediate. Model flash photolysis investigations with dimethylaniline as the electron donor showed that the dimer cleavage process proceeds on the microsecond time scale ($k = 10^6\,\text{s}^{-1}$) [35]. Electron transfer from a neutral electron donor, such as a reduced flavin ($FlH_2$) to the dimer unit gives the zwitterionic intermediate $Donor^{\bullet+}$-$Dimer^{\bullet-}$ (Figure 5.2.6), which should possess a high driving force for charge recombination. Electron transfer from a negatively charged electron donor, like a reduced and deprotonated flavin ($FlH^-$), however, yields a non-zwitterionic intermediate $FlH^\bullet$–$Dimer^{\bullet-}$ in a charge-shift reaction (Figure 5.2.6). Such a negatively charged intermediate could possess a much longer lifetime, which might allow a more efficient splitting. The neutral flavin radical intermediate $FlH^\bullet$ formed is significantly stable, particularly if the flavin radical is bound to the photolyase apoprotein [17].

### 5.2.3.3   Solvent dependency of the cleavage reaction

In order to gain support for a non-zwitterionic intermediate $FlH^\bullet$-$Dimer^{\bullet-}$, solvent dependency measurements were performed. The solvent dependency of a reaction allows us to distinguish between reactions with zwitterionic and non-zwitterionic intermediates. Polar solvents stabilize dipoles and accelerate the reaction [36–39]. In order to clarify the solvent dependency with flavin-containing model compounds, two sets of experiments were performed. In the first, the splitting rates were measured in water/ethanol and water/ethylene glycol mixtures (Table 5.2.1). These solvents possess similar H-donor/acceptor properties, but different dielectric constants. The most efficient cleavage was observed in pure water ($\phi = 0.06$). Addition of the more apolar ethanol or ethylene glycol (etgl) reduced the cleavage efficiencies by a factor of approximately 2 to $\phi = 0.03$. The second set of experiments included measurements in various organic solvents (Table 5.2.2). Here, the best cleavage efficiencies were again obtained in the most polar solvents like methanol, ethanol, and acetonitrile ($\phi = 0.04$). The cleavage rate is approximately twice as slow in apolar solvents like dioxane or ethyl acetate ($\phi = 0.02$).

**Table 5.2.1.** Solvent-dependence of the quantum yield, $\Phi$.

| Solvent mixture (% etgl) | $\Phi$ (splitting) |
|---|---|
| water/etgl (0%) | 0.062 |
| water/etgl (33%) | 0.052 |
| water/etgl (50%) | 0.041 |
| water/etgl (66) | 0.033 |
| water/etgl (100%) | 0.051 |
| DMF | 0.051 |

DMF: Dimethyl formamide, etgl: Ethylene glycol

**Table 5.2.2.** Solvent-dependence of the quantum yield, $\Phi$.

| Solvent | $\varepsilon_r$ [a] | $\Phi$ [b] |
|---|---|---|
| 1,4-dioxane | 2.21 | 0.016 |
| benzene | 2.27 | _[c] |
| ethyl acetate | 6.02 | 0.024 |
| *tert*-butanol | 12.47 | 0.037 |
| ethanol | 24.55 | 0.044 |
| methanol | 32.66 | 0.040 |
| acetonitrile | 35.96 | 0.046 |

[a] Relative permittivity (dielectric constant) for the pure liquid at 25°C. [b] Quantum yields of the model compounds. [c] No quantum yield detectable, even after prolonged irradiation.

Both experiments confirmed increased splitting efficiencies in polar solvents. The total solvent dependency, however, is very low. Both results strengthen the presence of the charge shift mechanism and exclude the formation of a zwitterion in the course of the reaction. The data therefore support the postulated non-zwitterionic $FlH^{\cdot}$–$Dimer^{\cdot-}$ intermediate.

## 5.2.4 The role of the second cofactor

### 5.2.4.1 Bis-deazaflavin model systems

Today, all well-studied DNA-photolyases contain, in addition to the crucial FAD, either a methenyltetrahydrofolate (MTHF) or a 8-hydroxy-5-deazaflavin (8-HDF) as a second cofactor. Kinetic investigations of photolyases lacking this second cofactor revealed that they still possess repair activity but (i) that the second cofactor increases the repair efficiency at a given irradiation intensity [25]; and (ii) the second cofactor shifts the wavelength at which maximal repair occurs from 370 nm in "flavin-only" photolyases to 400 nm in MTHF and even to 420 nm in 8-HDF containing enzymes. These data confirmed that the second cofactor has an energy transfer function closely related to antenna pigments in the photosynthetic apparatus [40, 41]. Such an antenna function requires an intensive interaction of the MTHF or the 8-HDF with the corresponding FAD, suggesting a close arrangement of both cofactors within the enzyme. The center-to-center distance between both cofactor-units was, however, found to be surprisingly large with 16.8 Å in the *E. coli* (MTHF, FAD) enzyme [19] and 17.5 Å in the *A. nidulans* (8-HDF, FAD) protein [20]. Time-resolved fluorescence investigations of the *A. nidulans* enzyme showed, despite the large cofactor separation, a rapid energy transfer process ($k_{EET} = 1.9 \times 10^{10} \, s^{-1}$) and a high transfer efficiency (98%) [42]. These observations are explained by a favorable orientation of the transition dipole moments. In the *E. coli* enzyme, however, the orientation of both cofactors is very unfavorable, which yields an energy transfer that is far from optimal with a transfer rate of $k_{EET} = 4.6 \times 10^9 \, s^{-1}$ and a low efficiency of only 62% [43].

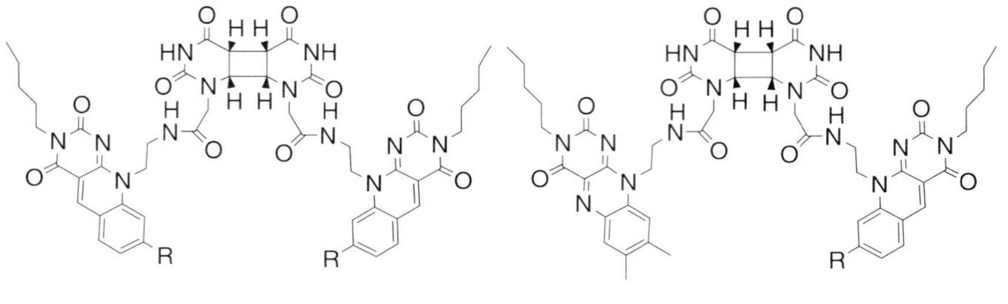

**8**, R = OBn
**9²⁻**, R = O⁻

**1**, R = OBn
**7⁻**, R = O⁻

**Figure 5.2.7.** Depiction of the deazaflavin model compounds **8** and **9²⁻** and of the mixed flavin and deaza-flavin model compounds **1** and **7⁻**, which mimic the active cofactor status in type II DNA photolyases.

In order to perform a systematic study of how deazaflavins influence the dimer cleavage depending on their oxidation and protonation states, a series of deazaflavin-containing model compounds **1**, **7⁻**, **8**, **9²⁻** was prepared (Figure 5.2.7) [44]. The first (bis-deazaflavin)-model compounds **8** and **9²⁻** contain a *cis-syn* uracil dimer and two deazaflavin units covalently attached to the dimer. The deazaflavins possess either a benzylated 8-OH-group (OH-form, in **8**) or a debenzylated and deprotonated 8-OH-group (O⁻-form, in **9²⁻**) [45]. Irradiation of both model compounds **8** and **9²⁻** at various wavelengths and subsequent HPLC analysis of the assay solutions showed that both compounds were completely stable upon irradiation. No photo-splitting into the expected photo-cleavage products was detected, which shows that both deazaflavins are unable to perform the cleavage reaction under those conditions.

### 5.2.4.2  Mixed flavin and deazaflavin model systems

The two mixed, flavin- and deazaflavin-containing model compounds **1** and **7⁻** were prepared in order to study how the presence of the deazaflavins affects the flavin-driven cleavage process. Steady-state fluorescence studies show that the deazaflavin fluorescence in the mixed model compounds **1** and **7⁻** is strongly quenched by the adjacent flavin chromophore, indicating an efficient energy transfer from the deazaflavins to the flavin units. Selective reduction of the flavin unit in both model systems was performed in order to simulate the active cofactor status in type II DNA-photolyases. Prior to each irradiation experiment, the cofactor status in the model compounds was unequivocally confirmed by UV/Visible and fluorescence spectroscopy.

The absorption spectra of these compounds in their various oxidation and protona-tion states are depicted in Figure 5.2.8a. The UV spectra show the absorption of the oxidized flavins in the flavin and deazaflavin model compounds **1** and **7⁻** at 340–370 nm and 440–~500 nm and the absorption of the oxidized deazaflavin units at 390–410 nm (**1**) and 430 nm (**7⁻**). The typical flavin absorptions dissapear after

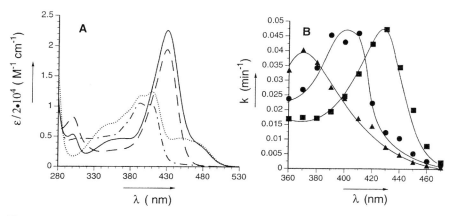

**Figure 5.2.8.** UV/Visible-spectra and action spectra of the mixed flavin and deazaflavin model compounds. (a) —— $7^-$, ------ $7H^{2-}$, ········ $1$, -·-··-·· $1H^-$. (b) ▲ = $5H_2^{2-}$, ● = $1H^-$, ■ = $7H^{2-}$.

selective reduction of the flavin units in the model compounds **1** and $7^-$ to $1H^-$ and $7H^{2-}$. The action spectra depicted in Figure 5.2.8B show the splitting rates depending on the irradiation wavelength for $1H^-$ and $7H^{2-}$ in comparison to the "flavin-only" model compound $5H_2^{2-}$. From the maxima in the action spectra, it is evident that both types of deazaflavins are able to transfer excitation energy to the reduced and deprotonated flavin (FlH$^-$) species. In the model compound $5H_2^{2-}$, maximal repair occurs at 370 nm as expected for a "flavin-only"-model compound. This maximum is shifted to 400 nm in the presence of a deazaflavin in its OH-form (model compound $1H^-$) and to 430 nm in $7H^{2-}$, which contains a deprotonated deazaflavin (O$^-$-form). The maxima in the action spectra are in excellent agreement with the absorption maxima of the corresponding deazaflavin species, which proves that the energy for the splitting process is initially absorbed by the oxidized deazaflavin chromophores and subsequently transferred to the reduced flavin. Since the action spectrum of $7H^{2-}$ is in full agreement with the *A. nidulans* spectrum, the experiments support the presence and the light gathering function of the deprotonated 8-HDF in *A. nidulans* photolyases and show that the active cofactor status of the enzyme contains a reduced and deprotonated flavin and an oxidized and also deprotonated deazaflavin (FlH$^-$ and dFl$^-$) [44].

## 5.2.5   Future directions

### 5.2.5.1   The third generation model compounds

In order to investigate how the flavin-deazaflavin distance influences the repair rate, a second series of model compound, depicted in Figure 5.2.9, is currently under investigation. These compounds feature a constant flavin-thymine dimer distance and consequently an identical electron transfer rate. The flavin-deazaflavin distance,

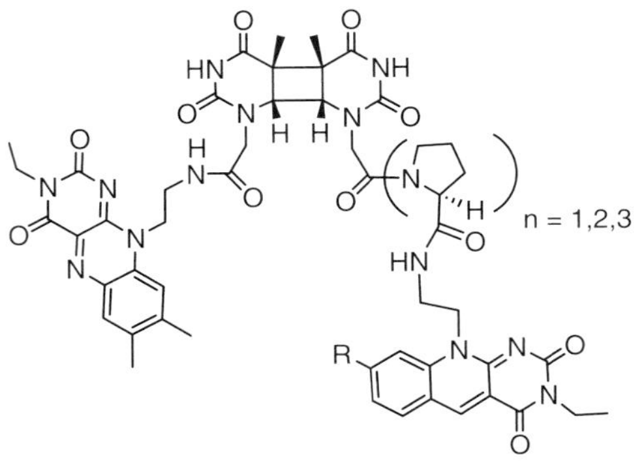

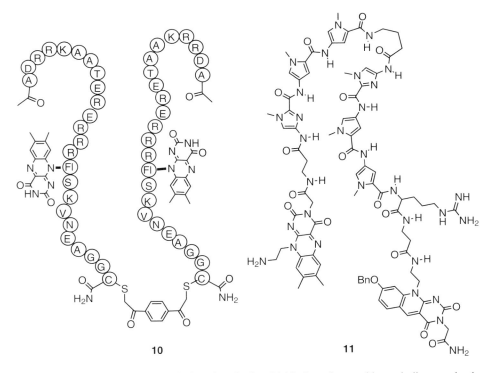

**Figure 5.2.9.** Proline spacer model compounds for the investigation of how the splitting reaction depends on the cofactor–cofactor distance. R = OBn.

however, is systematically increased [46]. This is achieved by the incorporation of one to three proline-units [47] as "rigid" spacers between the dimer and the deazaflavin. So far, fluorescence measurements showed that the energy transfer from the deazaflavin to the flavin still occurs but decreases in efficiency as expected with

**10**

**11**

**Figure 5.2.10.** Flavin and deazaflavin functionalized DNA binding oligopeptides and oligopyrrolcarboxamides. Both prepared cofactor-peptide conjugates possess a primitive DNA repair function.

increasing cofactor–cofactor distance. Splitting measurements with these compounds will further clarify the dependence of the splitting reaction on the cofactor–cofactor distance.

#### 5.2.5.2    Towards artificial DNA photolyases

Future model studies are also directed at the preparation of catalytic systems that are able to repair dimer lesions within the DNA double helix (Figure 5.2.10). To this end, we are using common DNA-binding elements as vehicles to deliver the cofactors required for the repair, into the major and the minor grooves of the DNA duplex. For the functionalization of the DNA-binding elements, cofactor amino acids were recently developed. These can be incorporated into any peptidic material by using solid-phase synthesis methods [48].

These building blocks even allow the preparation of combinatorial libraries from which DNA-repair enzymes can be potentially selected. Two of the recently investigated cofactor-peptide compounds **10** and **11** are depicted in Figure 5.2.10 [49]. Both compounds contain as a DNA-binding element either a polypeptide with the sequence of a transcription factor (**10**), which binds into the major groove of DNA or a minor groove binding distamycin derivative (**11**) [50]. The cofactors were incorporated in both units as cofactor amino acids using Fmoc-based solid phase synthesis technology. The obtained cofactor-peptide chimeras depicted in Figure 5.2.10 were recently shown to repair pyrimidine dimers cleanly in single- and double-stranded DNA, which makes us currently optimistic that the goal to develop artificial DNA-photolyases can be reached in the near future [51, 52].

# References

[1]   B. Lewin, *Gene, Lehrbuch der Molekularen Genetik*, VCH Verlagsgesellschaft, Weinheim **1991**.
[2]   T. Lindahl, *Nature* **1993**, *362*, 709–715.
[3]   E. C. Friedberg, G. C. Walker, W. Siede, *DNA repair and mutagenesis*, ASM Press, Washington, D.C. **1995**.
[4]   A. Sancar, *J. Biol. Chem.* **1995**, *270*, 15915–15918.
[5]   A. Sancar, *Science* **1994**, *266*, 1954–1956.
[6]   E. A. Bailey, R. S. Iyer, M. P. Stone, T. M. Harris, J. M. Essigman, *Proc. Natl. Acad. Sci. U.S.A.* **1996**, *93*, 1535–1539.
[7]   P. Rademacher, *Chem. unserer Zeit* **1975**, *9*, 79–84.
[8]   R. G. Harvey, *Acc. Chem. Res.* **1981**, *14*, 218–226.
[9]   H. E. Krokan, R. Standahl, G. Slupphaug, *Biochem. J.* **1997**, *325*, 1–16.
[10]  P. F. Heelis, R. F. Hartman, S. D. Rose, *Chem. Soc. Rev.* **1995**, 289–297.
[11]  J.-S. Taylor, *J. Chem. Ed.* **1990**, *67*, 835–841.
[12]  J.-S. Taylor, *Acc. Chem. Res.* **1994**, *27*, 76–82.
[13]  T. Carell, *Chimia* **1995**, *49*, 365–373.
[14]  T. P. Begley, *Acc. Chem. Res.* **1994**, *27*, 394–401.
[15]  K. Morikawa, O. Matsumoto, M. Tsujimoto, K. Katayanagi, M. Ariyoshi, T. Doi, M. Ikehara, T. Inaoka, E. Ohtsuka, *Science* **1992**, *256*, 523–526.

[16] D. G. Vassylyev, T. Kashiwagi, Y. Mikami, M. Ariyoshi, S. Iwai, E. Ohtsuka, K. Morikawa, *Cell* **1995**, *83*, 773–782.

[17] A. Sancar, *Biochemistry* **1994**, *33*, 2–9.

[18] T. Carell, *Angew. Chem.* **1995**, *107*, 2697–2700. *Angew. Chem. Int. Ed. Engl.* **1995**, *34*, 2491–2494.

[19] H.-W. Park, S.-T. Kim, A. Sancar, J. Deisenhofer, *Science* **1995**, *268*, 1866–1872.

[20] T. Tamada, K. Kitadokoro, Y. Higuchi, K. Inaka, A. Yasui, P. E. de Ruiter, A. P. M. Eker, K. Miki, *Nature Struct. Biol.* **1997**, *11*, 887–891.

[21] T. Bugg, *An Introduction to Enzyme and Coenzyme Chemistry*, Blackwell Science, Oxford **1997**.

[22] T. Förster, *Discuss. Faraday Soc.* **1959**, *27*, 7–17.

[23] N. Kraus, W.-D. Schubert, O. Klukas, P. Fromme, H. T. Witt, W. Saenger, *Nature Struct. Biol.* **1996**, *3*, 969–973.

[24] R. va Grondelle, J. P. Dekker, T. Gillbro, V. Sundstrom, *Biochim. Biophys. Acta* **1994**, *1187*, 1–65.

[25] A. P. M. Eker, J. K. C. Hessel, R. H. Dekker, *Photochem. Photobiol.* **1986**, *44*, 197–205.

[26] R. S. Lloyd, X. Cheng, *Curr. Opin. Chem. Biol.* **1998**, *4*, 139–151.

[27] R. J. Roberts, *Cell* **1995**, *82*, 9–12.

[28] G. L. Verdine, S. D. Bruner, *Chem. Biol.* **1997**, *4*, 329–334.

[29] T. Carell, R. Epple, *Eur. J. Org. Chem.* **1998**, 1245–1258.

[30] T. Carell, R. Epple, V. Gramlich, *Angew. Chem.* **1996**, *108*, 676–623. *Angew. Chem. Int. Ed. Engl.* **1996**, *35*, 620–623.

[31] T. Carell, R. Epple, V. Gramlich, *Helv. Chim. Acta* **1997**, *80*, 2191–2203.

[32] R. Epple, E.-U. Wallenborn, T. Carell, *J. Am. Chem. Soc.* **1997**, *119*, 7440–7451.

[33] H. G. O. Becker, *Einführung in die Photochemie*, Deutscher Verlag der Wissenschaften, Berlin **1991**.

[34] C. G. Hatchard, C. A. Parker, *Proc. R. Soc. A.* **1956**, *235*, 518–536.

[35] S.-R. Yeh, D. E. Falvey, *J. Am. Chem. Soc.* **1991**, *113*, 8557–8558.

[36] C. Reichardt, *Solvents and Solvent Effects in Organic Chemistry*, VCH Verlagsgesellschaft, Weinheim **1988**.

[37] D. G. Hartzfeld, S. D. Rose, *J. Am. Chem. Soc.* **1993**, *115*, 850–854.

[38] S.-T. Kim, R. F. Hartman, S. D. Rose, *Photochem. Photobiol.* **1990**, *52*, 789–794.

[39] S.-T. Kim, S. Rose, *J. Photochem. Photobiol., B:* **1992**, *12*, 179–191.

[40] G. R. Fleming, *Chimia* **1997**, *51*, 365.

[41] H. Kurreck, M. Huber, *Angew. Chem.* **1995**, *107*, 929–947. *Angew. Chem. Int. Ed. Engl.* **1995**, *34*, 849–866.

[42] S.-T. Kim, P. F. Heelis, A. Sancar, *Biochemistry* **1992**, *31*, 11244–11248.

[43] S.-T. Kim, P. F. Heelis, T. Okamura, Y. Hirata, N. Mataga, A. Sancar, *Biochemistry* **1991**, *30*, 11262–11270.

[44] R. Epple, T. Carell, *Angew. Chem.* **1998**, *110*, 986–989. *Angew. Chem. Int. Ed. Engl.* **1998**, *37*, 938–941.

[45] C. Walsh, *Acc. Chem. Res.* **1986**, *19*, 216–221.

[46] R. Epple, T. Carell, *unpublished results*.

[47] C. A. Slate, D. R. Striplin, J. A. Moss, P. Chen, B. W. Erickson, T. J. Meyer, *J. Am. Chem. Soc.* **1998**, *120*, 4885–4886.

[48] T. Carell, H. Schmid, M. Reinhard, *J. Org. Chem.* **1998**, *63*, 8741–8747.

[49] T. Carell, J. Butenandt, *Angew. Chem.* **1997**, *109*, 1590–1593. *Angew. Chem. Int. Ed. Engl.* **1997**, *36* 1461–1464.

[50] D. E. Wemmer, P. B. Dervan, *Curr. Opin. Struct. Biol.* **1997**, *7*, 355–361.

[51] P. J. Dandlicker, R. E. Holmlin, J. K. Barton, *Science* **1997**, *275*, 1465–1468.

[52] C. Hélène, M. Charlier, *Photochem. Photobiol.* **1977**, *25*, 429–434.

# 5.3    Alanyl-PNA: A model system for DNA using a linear peptide backbone as scaffold

*Ulf Diederichsen*

The backbone of nucleic acids is composed of uniformly connected deoxyribosyl or ribosylphosphodiester units. While recognition and information is typically encoded by the type and sequence of nucleobases connected to the sugar moiety, the backbone contributes not only by lining up the functional units, but it also plays a crucial role by defining specificity and selectivity of base pairing. For this purpose, the backbone of a double-stranded nucleic acid needs a repetitive secondary structure which defines pairing mode and strand orientation. Base pairing is driven by the formation of hydrogen bonds, the stacking interactions of adjacent nucleobase pairs separated by 3.4 Å, and the solvent effect gained by removing the apolar nucleobases from water inside the double helix [1]. Nevertheless, double-strand formation would not be possible unless the backbone adopts the required helical secondary structure. The conformation of a nucleic acid backbone is rather sensitive to small changes, as can be shown by the difference between canonical DNA and RNA structures due to the sugar pucker, as well as by the polymorphism of DNA structures [2]. The DNA backbone conformation is defined by six torsion angles, together with an additional four angles in the sugar moiety and one for the nucleobase connection. This complex conformational arrangement limits DNA double-strand formation to G-C and A-T Watson Crick base pairs. There is, however, a certain amount of conformational flexibility that allows DNA to adopt to interactions like intercalation (unwinding the helix) or coordination of cationic molecules to the backbone (kink of the double strand). Despite the complexity of DNA and RNA, research is focused on these naturally occurring biooligomers because of their key role in biology.

## 5.3.1    Alanyl pepeptide nucleic acid (PNA): constitution and model for duplex formation

By introducing alanyl-PNA (Figure 5.3.1), an oligopeptide with nucleobases substituted at the $\beta$-position of an alanyl unit, we intend to gain further insight into structure, base pairing, and interactions of oligonucleotides. Alanyl-PNA as a model system has a number of advantages: it is simple, linear, uncharged and has a low degree of conformational freedom [3]. Consecutive side chains in an extended peptide can be oriented towards one side by alternating the chirality of the amino acids. Furthermore, the spacing of the side chains in a $\beta$-sheet-like conformation is with 3.6 Å very close to the distance that favors stacking interactions in DNA. Since systems where base pairing occurs tend to adopt the favored stacking distance, the alanyl-PNA duplex with an extended backbone should be linear by design. In contrast to a double helix, linear pairing does not restrict the base pair in size or pairing mode. Without the topological restriction by a double helix, additional purine purine, purine

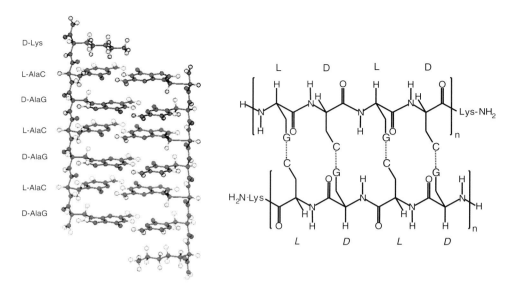

**Figure 5.3.1.** Alanyl-PNA double strand based on G-C Watson Crick pairing. Alternating chirality of the nucleo amino acids is required. The linearity of alanyl-PNA is structurally inherent, as long as the distance of two consecutive side chains is very close to the DNA stacking distance of 3.4 Å.

pyrimidine and pyrimidine pyrimidine combinations and different pairing modes were expected in addition to the standard Watson Crick pairs. We also introduce alanyl-PNA as a model system for the study of interactions of intercalators or amino acid side chains with the stack of nucleobases. In addition, we demonstrate their usefulness as a DNA motif analog.

## 5.3.2  Pairing properties of alanyl-PNA

In order to study the pairing properties of alanyl-PNA, hexamers were synthesized using standard solid-phase peptide synthesis. A lysine residue was added to the C-terminus to increase the solubility in aqueous solution (0.1 M NaCl, 0.01 M $Na_2HPO_4/H_3PO_4$, pH 7) [4]. The oligomers were purified by HPLC and characterized by ESI-MS, CD spectroscopy and, in selected cases, by $^1$H-NMR spectrometry. Homo-oligomers of the sequence XXXXXX-Lys-NH$_2$ can be used to investigate X–X selfpairing, while oligomers with the sequences XXYXYY-Lys-NH$_2$ (X and Y representing nucleo amino acids) are suited to form antiparallel double strands with X-Y base pairs. The pairing stability was examined by UV melting curves. A sigmoidal increase of absorption indicates the cooperative destacking of the pairing complex and the melting temperature (T$_m$) gives a value for the double-strand stability.

Both of the two enantiomers X̲X̲Y̲X̲Y̲Y̲-L̲y̲s̲-NH$_2$ and XXYXYY-Lys-NH$_2$ (amino acids with D-configuration are underlined) have to be considered for double-stranded alanyl-PNA with alternating configuration. As shown for the G-C pairing

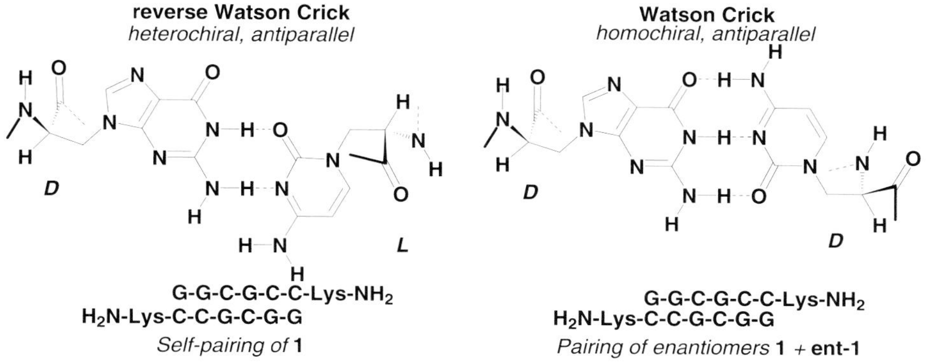

**reverse Watson Crick**
*heterochiral, antiparallel*

**Watson Crick**
*homochiral, antiparallel*

G-G-C-G-C-C-Lys-NH$_2$
H$_2$N-Lys-C-C-G-C-G-G
*Self-pairing of* **1**

G-G-C-G-C-C-Lys-NH$_2$
H$_2$N-Lys-C-C-G-C-G-G
*Pairing of enantiomers* **1** + **ent-1**

**Figure 5.3.2.** G-C Watson Crick pairing modes are dependent on the configuration of the nucleo amino acids.

in Figure 5.3.2, antiparallel selfpairing requires heterochiral nucleo amino acids for the formation of base pairs, while pairing of enantiomeric oligomers is possible with homochiral base pairs. The base pair geometry is a function of the base pair chiralities involved. That is, the Watson Crick mode requires homochiral nucleo amino acids (pairing of enantiomers) while the reverse Watson Crick mode (selfpairing) is only possible with heterochiral nucleo amino acids. According to our model, six additional hydrogen bonds for the pairing of enantiomers H-(AlaG-AlaG-AlaC-AlaG-AlaC-AlaC)-Lys-NH$_2$ (**1**) and H-(AlaG-AlaG-AlaC-AlaG-AlaC-AlaC)-Lys-NH$_2$ (**ent-1**) result in a higher stability (T$_m$ = 58°C, each 6 μM) compared to the respective self-pairing complex of oligomer **1** (T$_m$ = 40°C, 12 μM) [5].

Double strand formation was also demonstrated by CD spectroscopy, gel filtration HPLC, and ultracentrifugation. In 2D-NMR experiments (500 MHz, D$_2$O) only one set of peaks was observed, corresponding to the spin system of one strand, as it is expected for a symmetrical double strand. The ROESY spectrum of the self-pairing complex of oligomer **1** gives insight into the conformation of the side chain: cross peaks between the H-C6 of cytosine with their respective H-Cα indicate the side chain conformation suggested for the double strand.

Analogous to G-C pairing, we investigated all canonical base pair combinations. The stabilities and preferred pairing modes are summarized in Table 5.3.1. Similar

**Table 5.3.1.** Thermal melting temperatures (T$_m$) of alanyl-PNA hexamers determined with sequences of the type XXYXYY or XXXXXX, respectively (6 μM, 0.1 M NaCl, 0.01 M Na$_2$HPO$_4$/H$_3$PO$_4$, pH 7). WC = Watson Crick, rev WC = reverse Watson Crick, rev H = reverse Hoogsteen.

|  | Adenine | Thymine | Guanine | Cytosine |
|---|---|---|---|---|
| A | 21°C (rev WC/rev H) | | | |
| T | 25°C (rev H) | – | | |
| G | 32°C (rev H) | 28°C (WC) | 41°C (rev WC/rev H) | |
| C | 22°C (rev WC) | – | 58°C (WC) | – |

to DNA, the most stable pairing complex in alanyl-PNA also was found for Watson Crick G-C. In contrast to DNA, however, the alanyl-PNA A-T pairing is not isosteric to the G-C pairing since it favors the reverse Hoogsteen mode. Furthermore, base pairing was observed for all possible purine purine and purine pyrimidine combinations. As a consequence of the linear pairing complex that does not restrict the base pair size and geometry, there is no specific complementarity of the nucleobases. Therefore, the ambiguity of alanyl-PNA base pairing underlines the topological importance of the DNA double helix for the integrity of genetic code.

### 5.3.3    Alanyl-PNA: A DNA i-motif analog

Pyrimidine pyrimidine pairing was not observed, probably due to the limited stacking contribution of pyrimidines. One exception occured at pH 4.5 for the C—C pairing, where cytosine becomes semiprotonated. This makes a tridentate reverse Watson Crick pairing possible. C—C$^+$-pairing is known from DNA in the telomeric and centromeric regions of chromosomes as the intercalating motif (i-motif): [6] Two C—C$^+$ pairing double strands interpenetrate to form a tetramer. This base pair intercalation between all consecutive C—C$^+$ pairs requires unwinding of the DNA helix close to linearity (rise of 12.4–16°). The extended pairing complex of alanyl-PNA is therefore expected to function as a structural analogue of the DNA i-motif [7].

The PNA oligomer H-(AlaC-AlaC)$_4$-Lys-NH$_2$ (**2**) does not form a double strand or tetrad even at pH 4.5. Since unwinding of the alanyl-PNA to generate the required spacing for intercalation is not possible, stabilization by interpenetration is suppressed. Introducing a glycine at alternating positions leads to a cytidinyl nucleo amino acid that

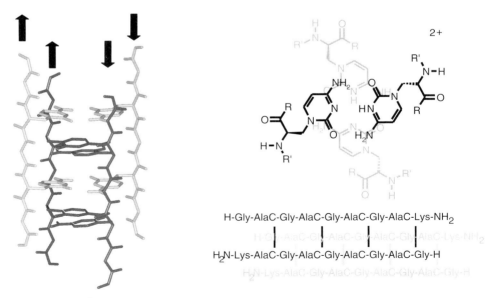

H-Gly-AlaC-Gly-AlaC-Gly-AlaC-Gly-AlaC-Lys-NH$_2$

H$_2$N-Lys-AlaC-Gly-AlaC-Gly-AlaC-Gly-AlaC-Gly-H

**Figure 5.3.3.**  C—C$^+$ pairing alanyl-PNA tetrad: a DNA i-motif analog.

has an optimal spacing for interpenetration. Using the enantiomers H-(Gly-AlaC)$_4$-Lys-NH$_2$ (**3**) and H-(Gly-AlaC)$_4$-Lys-NH$_2$ (**ent-3**) provides the heterochiral base pairing which is required for the antiparallel reverse Watson Crick pairing mode. For the equimolar mixture of enantiomers **3** + **ent-3**, stable pairing was observed at pH 4.5 (T$_m$ = 34°C, each 14 μM). This was the only C–C pairing complex that could form at acidic or neutral pH. In case of the DNA i-motif, stacking of the exocyclic amine and carbonyl groups with opposite dipoles has been demonstrated in the crystal structure and by NMR analysis [6]. Since C–C$^+$ pairing in alanyl-PNA is possible only in a tetrad, the importance of exocyclic stacking for the stabilization of the i-motif tetrad is emphasized. The alanyl-PNA C–C$^+$ tetrad might gain significance in the studies of the biological function of the i-motif, either as an analog in alanyl-PNA/DNA chimeras, or as an overall positively charged model.

## 5.3.4   Intercalation to alanyl-PNA

An advantage of alanyl-PNA double strands is the well-defined and rigid linear structure with base pairs stacking in an orientation similar to DNA. Because of the linearity and the restricted conformational freedom (two backbone torsion angles per residue) it is possible to modify the oligomers, e.g. by introduction of abasic positions, without significant changes in the conformation of the double strand. In contrast, DNA (six backbone torsion angles per nucleotide) reacts to modifications with helicalization or unwinding. DNA-intercalation is statistical and the stoichiometry cannot be controlled. In the case of the alanyl-PNA, double strand **1** + **ent-1** intercalation does not occur since the spacing between nucleobasepairs cannot be increased. However, intercalation was shown to be effective in the case of the abasic double strand formed by oligomers H-(AlaG-AlaG-AlaC-Gly-AlaC-AlaC)-Lys-NH$_2$ (**4**) and H-(AlaG-AlaG-Gly-AlaG-AlaC-AlaC)-Lys-NH$_2$ (**5**) (Figure 5.3.4) [8].

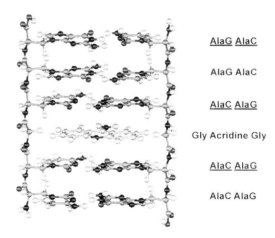

AlaG AlaC

AlaG AlaC

AlaC AlaG

Gly Acridine Gly

AlaC AlaG

AlaC AlaG

**Figure 5.3.4.** Alanyl-PNA with an abasic site is a suitable model system for intercalation.

This pairing complex with one base pair missing still has a stability of $T_m = 42°C$ (6 µM each). Upon intercalation of 9-aminoacridine, a stabilization of 10°C was observed. The binding of the intercalator within the base stack was also shown by CD-spectroscopy. Other intercalators like ethidium bromide and a bipyridylplatin complex were also bound to alanyl-PNA **4 + 5**, leading to a stabilization by 20 and 11°C, respectively. The mechanism of intercalation to alanyl-PNA differs from DNA: Since the backbone is uncharged, adhesion of the intercalator to the backbone prior to insertion into the base stack has to be ruled out. Conformational reorientation by unwinding is not necessary. Therefore, intercalation to alanyl-PNA **4 + 5** does not affect the base stacking of existing base pairs, and is specific to the abasic position. Currently, we are investigating the possibility of designing specificity for different intercalators into the abasic site by effecting its size and polarity. Finally, alanyl-PNA has the potential to become a valuable tool in the study of base stack-mediated transport phenomena.

### 5.3.5    Amino acid side chain-nucleobase interactions

Analogous to abasic positions, all kinds of mismatch interactions can be introduced in an alanyl-PNA double strand, without major distortion of the overall conformation. Therefore, alanyl-PNA can also serve as a model system for interactions between amino acid side chains and nucleobases. The natural amino acids can be incorporated into an alanyl-PNA sequence, and their influence detected by changes in the stability of the double strand. Possible interactions are hydrogen bonds and ionic interactions between acidic or basic side chains and the opposed nucleobase, stacking interactions of aromatic amino acids, and van der Waals contacts of aliphatic groups. Further, steric effects and solvation might be important. As a preliminary result the self-pairing of oligomers H-(AlaG-AlaG-AlaC-Xaa-AlaC-AlaC)-Lys-NH₂ (Xaa = glycine (**6a**), lysine (**6b**), arginine (**6c**), glutamine (**6d**), norleucine (**6e**), trypto- phan (**6f**) and phenylalanine (**6g**, Figure 5.3.5)) were compared to the self-pairing of

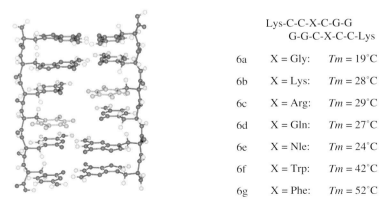

|  |  |  |
|---|---|---|
| | Lys-C-C-X-C-G-G | |
| | G-G-C-X-C-C-Lys | |
| 6a | X = Gly: | *Tm* = 19°C |
| 6b | X = Lys: | *Tm* = 28°C |
| 6c | X = Arg: | *Tm* = 29°C |
| 6d | X = Gln: | *Tm* = 27°C |
| 6e | X = Nle: | *Tm* = 24°C |
| 6f | X = Trp: | *Tm* = 42°C |
| 6g | X = Phe: | *Tm* = 52°C |

**Figure 5.3.5.** Alanyl-PNA: Model system for amino acid side chain/nucleobase intercactions.

H-(AlaG-AlaG-AlaC-AlaG-AlaC-AlaC)-Lys-NH$_2$  (**1**, T$_m$ = 40°C). The double strand with two glycines formed by self-pairing of **6a** showed a stability similar to the respective G-C pairing alanyl-PNA tetramer (T$_m$ = 19°C), whereas self-pairing of **6b**, **6c** and **6d** were more stable (T$_m$ = 28°C), probably due to hydrogen bonds between the side chains and cytosine and van der Waals contacts. The importance of van der Waals interactions might be highlighted by the norleucine derivative (**6e**), which lacks the possibility for hydrogen bonding. Alanyl-PNAs with aromatic amino acids are slightly (**6f**: T$_m$ = 42°C) or even distinctly more stable (**6g**: T$_m$ = 52°C) than the G-C hexamer. The high stability most likely is due to stacking of the aromatic amino acid side chains with the nucleobases, and an additional contribution from moving the aromatic rings into an apolar environment within the base stack of the double strand.

In conclusion, the analysis of alanyl-PNA pairing complexes provides insight into the variety of pairing possibilities formed by the canonical nucleobases without the topological restriction of the DNA backbone. Furthermore, the conformationally highly restricted and rigid alanyl-PNA double strands are a simpler system than DNA, and therefore a valuable model to study nucleic acid motifs as well as interactions with DNA.

**Acknowledgments:** This work was supported by the Deutsche Forschungsgemeinschaft and the Fonds der Chemischen Industrie. I am grateful to Professor Horst Kessler for generous support. I would also like to thank Harald Schmitt, Daniel Weicherding and Elke Vockelmann for their contributions and valuable discussions.

# References

[1]  W. Saenger, 'Principles of Nucleic Acid Structure' Springer Verlag, New York **1984**.

[2]  G. M. Blackburn, M. Gait, 'Nucleic Acids in Chemistry and Biology' Oxford University Press, New York **1996**.

[3]  (a) U. Diederichsen, *Angew. Chem.* **1996**, *108*, 458; *Angew. Chem. Int. Ed. Engl.* **1996**, *35*, 445; (b) U. Diederichsen, H. W. Schmitt, *Tetrahedron Lett.* **1996**, *37*, 475; (c) U. Diederichsen, *Bioorg. Med. Chem. Lett.* **1998**, *8*, 145.

[4]  The monomers Boc-AlaC(Z)-OH and Boc-AlaG-OH were prepared by ring opening of *N*-Boc-serine lactone with *N*4-benzyloxycarbonylcytosine or 2-amino-6-chlorpurine. The guanine derivative was obtained by further conversion with TFA/H$_2$O followed by Boc$_2$O; see P. Lohse, B. Oberhauser, B. Oberhauser-Hofbauer, G. Baschang, A. Eschenmoser, *Croatica Chim. Acta* **1996**, *69*, 535. The oligomers were synthesized by solid phase peptide synthesis on a 4-methylbenzhydrylamine polystyrene support loaded with *N*-Boc-L- or D-lysine(Z)-OH. Activation with HATU provided coupling yields of >95%.

[5]  U. Diederichsen, *Angew. Chem.* **1997**, *109*, 1966; *Angew. Chem. Int. Ed. Engl.* **1997**, *36*, 1886.

[6]  (a) K. Gehring, J.-L. Leroy, M. Guéron, *Nature* **1993**, *363*, 561. (b) Chen, L. Cai, X. Zhang, A. Rich, *Biochemistry* **1994**, *33*, 13540.

[7]  U. Diederichsen, *Angew. Chem.* **1998**, *110*, 2395; *Angew. Chem. Int. Ed.* **1998**, *37*, 2273.

[8]  U. Diederichsen, *Bioorg. Med. Chem. Lett.* **1997**, *7*, 1743.

## 5.4    New PNA building blocks for antisense research

*Stephan Jordan and Christoph Schwemler*

Antisense oligonucleotides are of great interest as potential molecular biological tools and therapeutic agents [1]. They have the ability to interact with a complementary sequence on mRNA and specifically to inhibit gene expression and protein biosynthesis. In contrast to conventional (nowadays, protein targeting) drugs, less or no undesired side effects are anticipated.

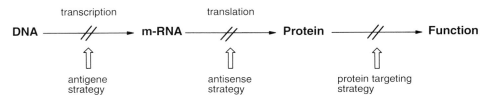

**Scheme 5.4.1.** Mode of action for different drug strategies.

Although natural antisense oligonucleotides have been employed successfully for the inhibition of viral replication, their low stability in biological media has limited their use as new therapeutic agents. This problem can be solved by the use of modified non-natural analogs and derivatives with improved nuclease resistance. Some of these new compounds have been approved for clinical trials within the past years [2].

Since their discovery, peptide nucleic acids (PNAs) [3] have become an important class of DNA-analogs for antisense and molecular biological purposes. They bind with high affinity to complementary DNA or RNA by Watson–Crick base pairing. Due to their structure they are not substrates for enzymatic degradation. They hybridize selectively to their target mRNAs without non-specific interactions which can lead to unfavourable side effects, as reported for the phosphorothioates [4]. In PNAs, the complete sugar-phosphate backbone is replaced by a polyamide backbone containing *N*-(2-aminoethyl) glycine (aeg) units. The nucleobase is attached to the nitrogen of the glycine through an acetic acid linker, the number of atoms between the ends of two monomeric units is the same as in DNA.

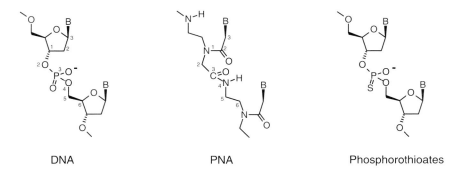

**Scheme 5.4.2.** Structure of DNA and analogs.

## 5.4.1 Synthesis of new PNAs

Formally, the primary structure of a PNA is very similar to that of an oligonucleotide, although PNA binds more strongly to complementary oligonucleotides than DNA itself. If a duplex of DNA being formed of two monomers with eight subunits in length shows a melting temperature ($T_m$) of only 16°C, the corresponding PNA/DNA duplex shows a $T_m$ value of about 42°C. This drastic increase in $T_m$ values is attributed to the lack of repulsion by the use of the uncharged PNAs and makes PNAs very attractive for antisense research. Based on the similarity model of the structures and in addition to the literature-known [3] building block **1**, we have synthesized a number of new PNA-building blocks, *e.g.* the glycylglycine-type **2**, the ornithine-type **3**, the aminobutyryl-type **4**, the aminoproline-type **5** and **6** and the pyrrolidine-type **7** and **8** [5].

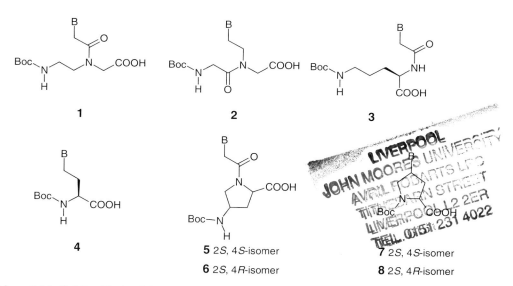

**Scheme 5.4.3.** Building blocks of the new PNA analogs.

The preparation of Boc-protected aminoproline monomers of type **6** is outlined in Scheme 5.4.4. Starting from commercially available L-4-*trans*-hydroxyproline, protection of the nitrogen followed by esterification affords compound **9**. The configuration at C-4 was inverted using the Mitsunobu reaction with *p*-nitrobenzoic acid, followed by hydrolysis of the corresponding ester. The hydroxy compound was converted into the azide **12** via the mesylate. After reduction to the amine and protection with di-*tert*-butyl dicarbonate the Cbz-group was removed by hydrogenation. Thymin-1-yl and $N^4$-Cbz-cytosin-1-yl acetic acid (synthesized as described in the literature) were activated with EDC (1-(3-dimethylaminopropyl)-3-ethylcarbodiimide) and HOBt (1-hydroxybenzotriazole) and coupled with the amine to furnish fully protected

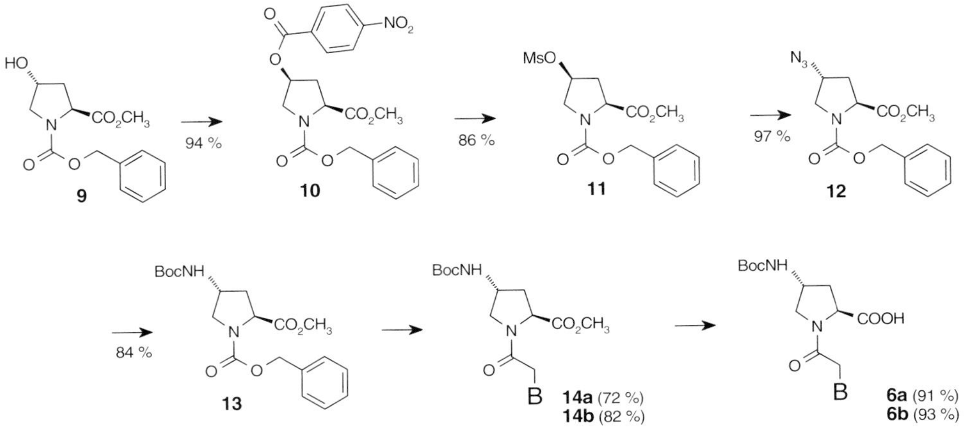

**Scheme 5.4.4.** Synthesis of the L-4-*trans*-aminoproline monomers **6a** (B = T) and **6b** (B = C$^{Cbz}$).

derivatives. Finally, the methyl ester was hydrolyzed with LiOH to compounds **6a** or **6b** [6].

Starting from L-4-*trans*-hydroxyproline **9**, mesylation followed by substitution and reduction of the azido-group affords the L-4-*cis*-aminoproline analog **5** with overall retention of the configuration at C-4.

The synthesis of another new class of PNA building blocks based on pyrrolidine-2-carboxylic acids is outlined in Scheme 5.4.5. The key step in this synthesis is the introduction of the nucleobase by Mitsunobu reaction at the 4-position of diprotected hydroxyproline **22**. Removal of the *N*-benzoyl group of the heterobase to **23** and

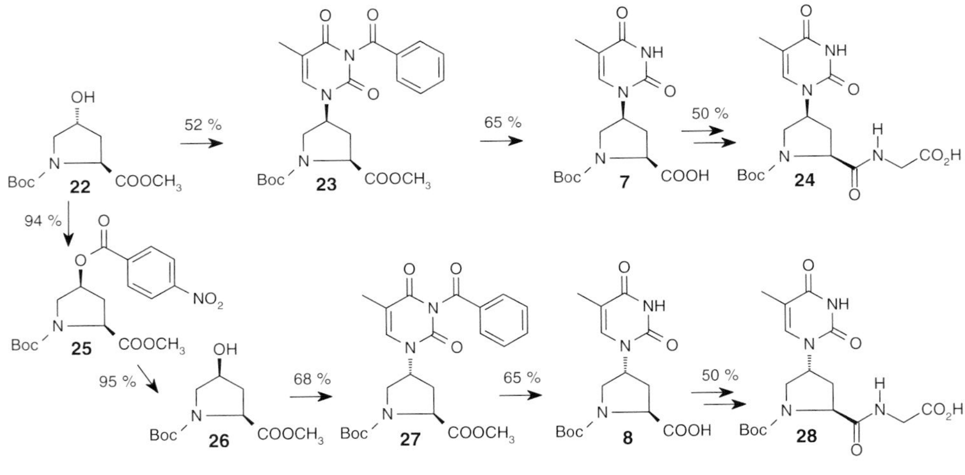

**Scheme 5.4.5.** Synthesis of the 2*S*,4*S*- and 2*S*,4*R*-N-Boc-4-(thymin-1-yl)-pyrrolidine-2-carboxylic acid.

cleavage of the ester leads to compound **7**, the 2*S*,4*S*-isomer. Successive Mitsunobu reactions of **22** results in the 2*S*,4*R*-isomer **8** via a double inversion (via **25**, **26** and **27**). To introduce cytosine as nucleobase we used $N^4$-Cbz-cytosine instead of $N^3$-Benzoyl-thymine during the Mitsunobu reaction.

Since the pyrrolidine-2-carboxylic acids lack three carbon atoms in their backbone in comparision to aeg, the new compounds are coupled to derivatives of glycine, *e.g.* compound **7** was coupled with glycine benzyl ester in the presences of HOBt and a carbodiimide. The last step in the synthesis is the hydrogenation of the ester to the carboxylic acid **24**. Using amino acids other than glycine allows backbone modification.

The oligomeric compounds were synthesized using a peptide synthesizer and standard Boc-chemistry on PAM (phenylacetamidomethyl)- or MBHA (4-methyl-benzhydrylamine)-resins. In order to suppress the tendency for aggregation in thymine PNA oligomers and to avoid base migration, firstly Boc-Lys(2-chloro-Z)-OH was coupled to the resin by activation with HOBt and DCC (*N*,*N'*-dicyclohexylcarbodiimide) in NMP (1-methyl-2-pyrrolidinone) according to Buchardt [3]. The Boc-group was removed by TFA (trifluoroacetic acid) and the resin washed to neutral pH with *N*-ethyldiisopropylamine. At the end of the synthesis the compounds were cleaved from the resin by HF and anisole at 0°C. Remaining anisole was removed by washing with ether. The oligomer was extracted with 30% acetic acid and the polymer support filtered off. The resulting filtrate was lyophilized and purified by reverse-phase HPLC using a gradient of TFA in water and TFA in water/acetonitrile. The oligomers were characterized by mass spectroscopy using LD (laser desorption)-methods.

## 5.4.2 Properties of the new homo-oligomers

The resulting oligomers were tested for stability to enzymatic degradation with nucleases and proteinase K by known procedures [6]. No products of degradation were found following reaction mixture analysis by HPLC, demonstrating that the oligomers are not substrates for these enzymes. We have studied the hybridizing properties of the modified oligomers with their complementary DNAs by measuring UV-absorption to determine $T_m$ values. Unfortunately, no hybridization was found for our homo-oligomers.

Due to the impressive binding properties of the homo-oligomers containing aeg, we have sought explanations for these unexpected results. In the case of the ornithines **3** (Scheme 5.4.2), the lengthening of the side chain from the backbone to the nucleobase may be responsible for the loss in binding capacity. For the other compounds, we have carried out molecular modeling experiments, allowing one to conclude that there is a pre-orientation of the oligomers containing an aeg backbone, leading to a pre-formed structure fitting very well to complementary DNA. In contrast, oligomers of the type described above show only random three-dimensional structures. Therefore, we planned to test this hypothesis by creating hetero-oligomers of our new building blocks and aeg.

**Table 5.4.1.** Properties of some of the modified oligonucleotides

| Sequence | Yield (%) | Purity (HPLC) | Stability against nucleases and proteases |
|---|---|---|---|
| | 42 | >95% | + |
| | 20 | >95% | + |
| | 74 | >95% | + |
| | 30 | >95% | + |

## 5.4.3   Synthesis and properties of hetero-oligomers

We synthesized hetero-oligomers of our new PNA building blocks **2–8** in combination with various amounts of the known *N*-(2-aminoethyl) glycine derivative (aeg) **1**. The same procedure was used as described for the homo-oligomeric compounds. To study the hybridizing properties of the modified oligomers with their complementary DNA, $T_m$ values were determined by UV measurement. The compounds show well-defined, single-phased melting profiles. The $T_m$ values and the sequences are listed in Table 5.4.2. In our hands, the combination of aeg **1** with L-4-*trans*-amino-proline **6** shows stronger binding properties to complementary DNA than homo-oligomers of aeg.

Best results were obtained with a 1:1 mixture of building blocks **1** and **6a,** as shown by $T_m$ values for the hybridization of compounds **16** or **19** with their complementary DNA. Compared with the original PNAs, they showed a 6–7°C increase in $T_m$. These combinations appear as promising materials for the generation of new antisense oligonucleotides. Oligomers containing the D-*trans* isomer **5a** resulted in a large decrease in hybridization compared with oligomers containing the analog **6a**. The incorporation of the other sub-units (**2, 7, 8**) also dramatically reduced the binding to $A_8$ or $A_{12}$ oligomers. The duplexes formed between these compounds and DNA were significantly destabilized.

The L-4-*trans*-amino-proline containing PNAs hybridize in a sequence-selective manner to double-stranded DNA. Hybridization and strand displacement in double-stranded plasmid DNA is shown for these oligomers (compounds **16, 19, 20, 21**) by incubation in samples containing plasmid DNA and single-strand DNA recognizing S1 nuclease. For example, compound **16** binds sequence-selectively to complementary target sequences within coiled plasmid DNA. Specific single-strand displacement of double-stranded plasmid DNA results in the generation of characteristic DNA fragments by S1 nuclease digestion. Analysis of formed DNA fragments is conducted by agarose gel electrophoresis, as shown in Figure 5.4.1. The size of all detectable DNA fragments (*i.e.* 4880, 3820, 2670, 2210, 1150, and 1060 base pairs) is consistent with the expected linearization of the plasmid (4880 base pairs) and partial cleavage of plasmid DNA at the binding sites of compound **16**.

The same pattern of S1 nuclease-generated fragmentation of plasmid DNA can be obtained using compounds **20** or **21**. Stoicheiometric amounts of plasmid DNA and octameric PNA carrying alternating amino-proline and *N*-(2-aminoethyl) glycyl building blocks resulted in detectable and specific S1 nuclease-mediated DNA fragmentation under the assay conditions used. Addition of non-complementary amino-proline containing PNA as a negative control did not result in detectable DNA fragmentation [7].

We also checked hybridization of compound **16** in a Tris-buffer in a different ratio (0.2:1; 0.25:1; 0.33:1; 0.5:1; 1:1; 2:1 and 3:1 DNA to PNA) to the complementary DNA sequence. Probes (45 $\lambda$M) in Tris–HCl (5 mM, 0.1 ml) were heated up to 93°C for 5 min and slowly cooled to room temperature. The resulting hybrids were characterized by dynamic gel capillary electrophoresis [8]. The yield of the hybridization products as a function of the concentration of complementary $d(A)_8$ shows a rapid increase up to a ratio of 0.5:1. Although by increasing the ratio of $d(A)_8$ to compound **16** the hybridization product remains relatively constant, an increasing amount of free DNA is detected. Best results were obtained using a ratio of 1 to 0.5, suggesting a 2:1 hybridization. In further experiments we successively increased the amounts of monomer **6a** in ratio to **1** from 1:1 up to 1:2 and 1:3. Again, a loss in hybridization properties was observed. This suggests that aeg seems to form a structure pre-formed for recognition of DNA. Molecular modeling shows the possibility of forming a helical structure in aeg-polymers stabilized by intramolecular hydrogen bonds between the carbonyl function in the side chain with the nucleobase and an amide proton of the backbone. Incorporation of L-4-*trans*-amino proline can increase binding by forming structures which are better fitting to DNA-helices.

**Table 5.4.2.** $T_m$ (°C) of modified oligonucleotides with complementary DNA[a].

| No. Sequence | No. of monomers | Yield (%)[b] | $T_m$ value (°C) |
|---|---|---|---|
| 15 | 8 | [c] | 40.0–42.0 |
| 16 | 8 | 61.2 | 47.0 |
| 17 | 8 | 34.9 | 27.0 |
| 18 | 8 | 67.0 | 33.0 |
| 19 | 8 | 76.6 | 46.0 |

**Table 5.4.2.** Continued.

| No. | Sequence | No. of monomers | Yield (%)[b] | T_m value (°C) |
|---|---|---|---|---|
| 20 | | 12 | 13.7 | 48.0 |
| 21 | | 12 | 18.4 | 34.4 |

[a] DNA synthesized using standard phosphoramidite technology on an Applied Biosystem ABI 380B[R] synthesizer; [b] after purification by HPLC (purity >95%); [c] not determined.

## 5.4.4  Future improvements

In conclusion, we have prepared modified PNAs for antisense research. These modified compounds contain various amounts of *N*-(2-aminoethyl) glycine and new thymidine or cytidine monomers. They form duplexes with the complementary strands of DNA and have improved base-pairing properties compared to the previously known PNAs. The new compounds also show excellent stability against enzymatic degradation.

So far, no carrier system for the membrane transport of PNAs is known, although the site of action for the antisense oligonucleotides is within the cell. This difficulty must first be overcome before PNAs can be used as therapeutic agents, though much work is needed in this area to solve this problem. However, the use of the above-mentioned modified oligonucleotides for diagnostic applications is currently possible.

Despite these ongoing problems, the potential of the antisense approach is impressive. In the future, the specific modulation of gene expression without undesired side effects may lead to a new generation of drugs. Possible targets are for example the sequences of

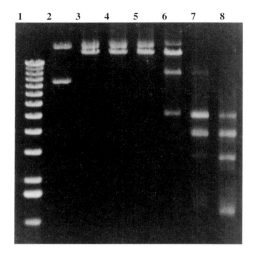

**Figure 5.4.1.** Agarose gel electrophoretic analysis of compound **16** mediated strand displacement in double-stranded plasmid DNA. Specific plasmid DNA fragmentation by S1 nuclease digestion of single-strand DNA cleavage is detected in samples 6–8. Composition of samples in lanes 1–8 is as follows. 1: molecular weight marker; 2: 1 µg plasmid DNA; 3: 1 µg plasmid DNA plus 10 U S1 nuclease; lanes 4–8, as for lane 3 with increasing amounts of **16** containing 0.0001 µg, 0.001 µg, 0.01 µg, 0.1 µg, and 1.0 µg, respectively.

oncogenes or growth factors. It is quite possible that inhibition of virus replication (*e.g.* HBV or HIV) can also be achieved with antisense oligonucleotides.

**Acknowledgements:** Today's modern drug research can only be successful within a group of highly qualified experts. The results contained in this article summarise our work with colleagues Dr. Kretschmar, Dr. Stropp, Dr. Mielke, Dr. Kosch and Dr. Schwenner. We wish to thank them, and all others involved in this project.

# References

[1]  Uhlmann, E., Peyman, A. *Chem. Rev.* **1990**, *90*, 543.
[2]  For reviews see: Szymkowski, D.E. *Drug Discovery Today*, **1996**, *1*, 415 and Altmann, K.-H., Dean, N. M., Fabbro, D., Freier, S. M., Geiger, T., Häner, R., Hüsken, D., Martin, P., Monia, B. P., Müller, M., Natt, F., Nicklin, P., Phillips, J., Pieles, U., Sasmor, H., Moser, H. E. *Chimica*, **1996**, *50*, 168.
[3]  Egholm, M., Buchardt, O., Nielsen, P. E., Berg, R. H. *J. Am Chem. Soc.* **1992**, *114*, 1895; Egholm, M., Nielsen, P. E., Buchardt, O.; Berg, R.H. *ibid.* **1992**, *114*, 9677.
[4]  Sarmieto, U. M., Perez, J. R., Becker, J. M., Narayanan, R. *Antisense Res. Dev.* **1994**, *4*, 99.
[5]  For example: Almarsson, Ö., Bruice, T. C., Kerr, J., Zuckermann, R. N. *Proc. Natl. Acad. Sci.*, **1993**, *90*, 7518; Löbberding, A., Mielke, B., Schwemler, C., Schwenner, E., Stropp, U., Springer, W., Kretschmer, A., Pötter, T. Europian Patent EP 0646596 A1 **1995** and EP 06465956 A1 **1995**; Lioy, E., Kessler, H. *Liebigs Ann.* **1996**, 201; Ceulemans, G., Khan, K., Van Schepdeal, A., Herdewijn, P. *Nucleosides & Nucleotides*, **1995**, *14*, 813; Petersen, K. H., Buchhardt, O.,

Nielsen, P. E. *Bioorg. Med. Chem. Lett.*, **1996**, *6*, 793; Lenzi, A., Reginato, G., Taddei, M. *Tetrahedron Lett.*, **1995**, *36*, 1713.

[6] Jordan, S., Schwemler, C., Kosch, W., Kretschmer, A., Stropp, U., Schwenner, E., Mielke, B. *Bioorg. Med. Chem. Lett.*, **1997**, *7*, 681, 687 and references therein.

[7] The plasmid used for tests was composed of pSP64 backbone (Promega Corp., Madison, WI, USA) carrying a globin polyadenylation sequence and a coding sequence for delta-subunit of *Torpedo californica* nicotinic receptor. Test samples contained 1.0 µg plasmid, 14 µl water, and varying amounts of **16**. For further details, see [6].

[8] Rose, D. J. *Anal. Chem.*, **1993**, *65*, 3545.

## 5.5    RNAse active site model systems

*Markus Kalesse and Thorsten Oost*

Bis(guanidinium) moieties have been used in the synthesis of artificial receptors in order to demonstrate the ability of these groups to bind to phosphates and to facilitate the cleavage of phosphodiesters. Various approaches have been published in which the ability of basic polyamines [1] like poly(Arg-Leu) to coordinate to phosphates has been utilized to cleave RNA. The work published by Anslyn [2], Göbel [3], Hamilton [4] and Schmidtchen [5] clearly demonstrates the ability of various guanidinium receptors to bind to phosphates, and by doing so accelerate the rate of transesterification/cleavage of RNA and RNA model compounds. Anslyn described the cleavage of RNA with a bis(guanidinium) receptor in water [2b]. In this case imidazole was required for substantial cleavage. Hamilton synthesized a bis(guanidinium) receptor with an internal base, but the cleavage reactions were performed in acetonitrile as the solvent [4c].

The use of guanidinium groups in the design of artificial ribonucleases has been stimulated by the fact that the active site of enzymes such as staphylococcal nuclease (SNase) [6] exhibit two essential arginine residues (Arg35, Arg87) (Scheme 5.5.1). On the basis of X-ray structural analysis, kinetic studies, and active site mutant investigation, the mechanism of SNase has been unraveled [7]. The mode of action involves augmenting the hydrogen bonding of the substrate by Arg87 in the transition state. The nucleophile necessary for the hydrolysis of DNA is generated by water coordination to $Ca^{2+}$ and deprotonation of the coordinated water by the carboxylate group at Glu43.

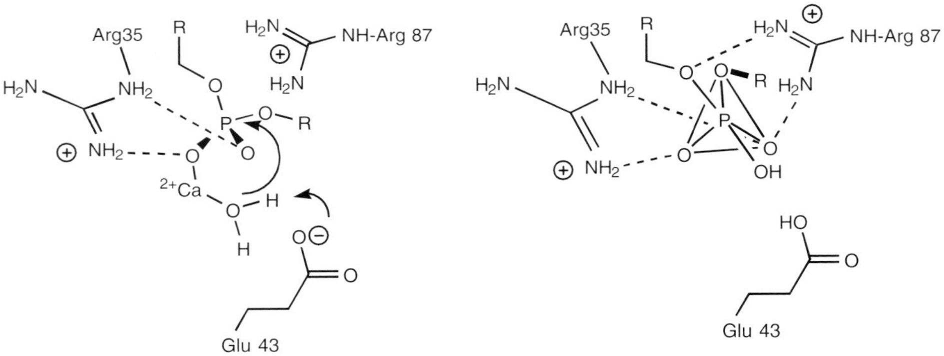

**Scheme 5.5.1.** Transition state of the staphylococcal nuclease-catalyzed phosphodiesterase reaction.

In the case of SNase, as in the case of all other enzymes, the tremendous rate acceleration in the reaction to be catalyzed is due to the distinct arrangement of the functional groups in the enzyme's active site, which in turn is generated by the three-dimensional structure of the peptide backbone. Therefore, our approach in synthesizing an active site model system that is capable of cleaving RNA in water is based on the assumption that preorganization of functional groups with some degree of conformational flexibility can generate active compounds. We envisioned that a peptide backbone mimetic at which functional groups can be assembled should lead to a substantial rate enhancement compared to unstructured analogs. Additionally, a fine-tuning of the artificial active site can be accomplished by varying the conformational freedom and configuration at the junctions between the rigid backbone and the functional groups. The great challenge was to generate a ribonuclease model system that is capable of cleaving RNA in aqueous solution, a more relevant solvent than acetonitrile or DMF.

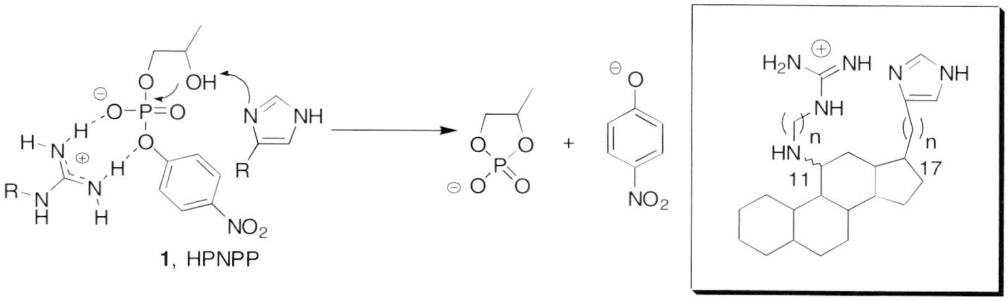

**Scheme 5.5.2.** Cleavage of the RNA analog HPNPP (**1**) [8] by steroid derivatives.

In our search for a system that allows us the synthesis of rigid bis(guanidinium) compounds with some configurational flexibility, corticosterone (**2**) was chosen as the template on which the bis(guanidinium) subunit could be varied. It was our goal to bring mono- or bis(guanidinium) moieties and the imidazole ring into close proximity. The task of the steroid system was to provide a readily accessible backbone on which the reactive functionalities can be assembled and the configuration changed in order to find an active compound. Additionally, the steric bulk provided by the steroid restricts the conformational flexibility of the guanidinium groups. Since the phosphodiester to be cleaved is RNA, our model system did not require an external nucleophile but still an internal base in order to deliver the 2'-OH to the phosphodiester linkage (Scheme 5.5.2). Therefore we synthesized active site analogs employing both the rigidity of the steroid backbone, the configurational flexibility of substituents at the C11 position, and an internal base at C17. The synthesis was achieved by utilizing a procedure published by Guthrie [9] (Scheme 5.5.3) that yielded oxime **5** after transformation of the C17 side chain into an imidazole moiety by a Weidenhagen reaction [10]. The oxime can then be transformed to the 11α-amine (**7**) or to the 11β-amine (**6**) by reduction either with Na/n-PrOH [11] or by high-pressure hydrogenation with Pt/H$_2$ under acidic conditions (Scheme 5.5.3). Both amines were obtained in good diastereomeric purity. The β-isomer (**6**) was the only detectable diastereomer during the reaction and the α-isomer (**7**) was produced in a 10:1 ratio upon reduction with sodium in PrOH, but both diastereomers could be separated via flash chromatography.

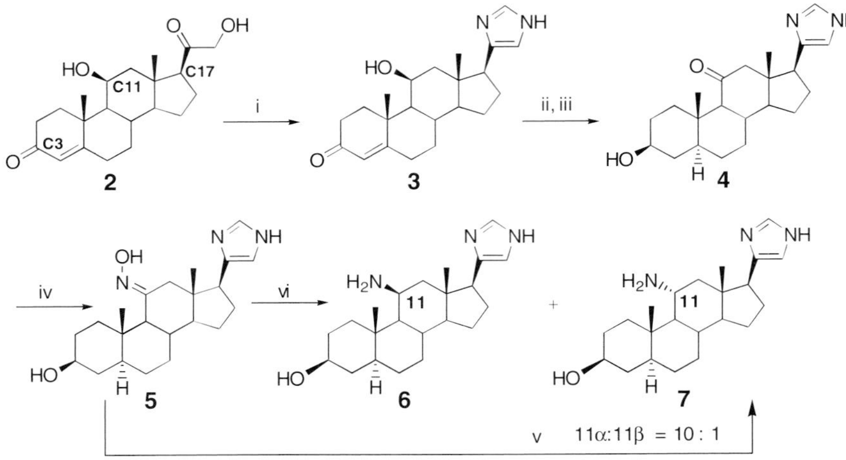

**Scheme**  5.5.3. (i) Cu(OAc)$_2$, NH$_3$, CH$_2$O, EtOH, reflux, 83%; (ii) Li, NH$_3$ (l)/MeOH; (iii) Jones Ox. (CrO$_3$, HOAc, H$_2$SO$_4$); (iv) NH$_2$OHHCl, pyridine reflux, 72% (three steps); (v) Na, *n*-PrOH reflux, 78%; (vi) H$_2$ (100 bar), Pt, HOAc/HCl, 76%.

Different side chains were attached to these two stereoisomers, yielding five different guanidinium compounds (Schemes 5.5.4, 5.5.5, 5.5.6).

From comparison of the cleavage rate constants of the steroids (Figure 5.5.1) it became clear that the steroids with the α-configuration at C11 and a 1,3-distance between the guanidinium groups are superior RNAse model systems compared to their C11-β-configured diastereomers.

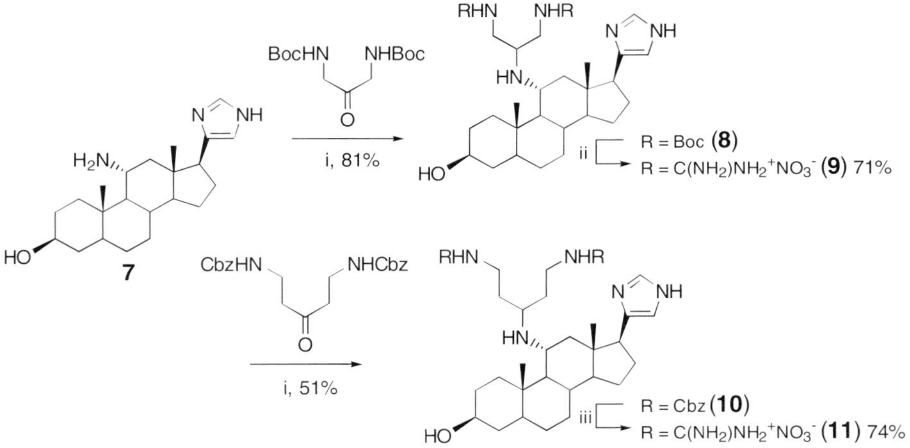

**Scheme 5.5.4.** (i) NaCNBH$_3$, MeOH, pH = 7; (ii) HCl (4 M), basic ion-exchange resin, 3,5-dimethylpyrazole-1-carboxamidine nitrate, Hünig's base, DMF; (iii) H$_2$, Pd/C, HCl, MeOH, basic ion exchange resin, 3,5-dimethylpyrazole-1-carboxamidine nitrate, Hünig's base, DMF.

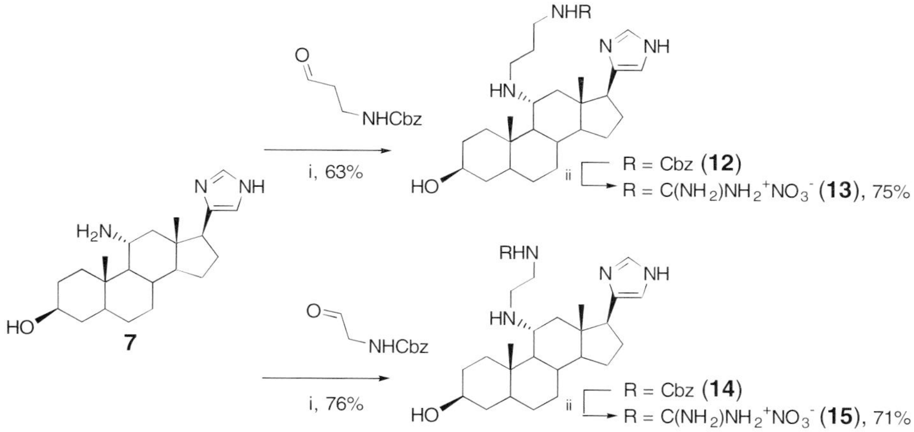

**Scheme 5.5.5.** (i) NaCNBH₃, MeOH, pH = 7; (ii) H₂, Pd/C, HCl, MeOH, basic ion-exchange resin, 3,5-dimethylpyrazole-1-carboxamidine nitrate, Hünig's base, DMF.

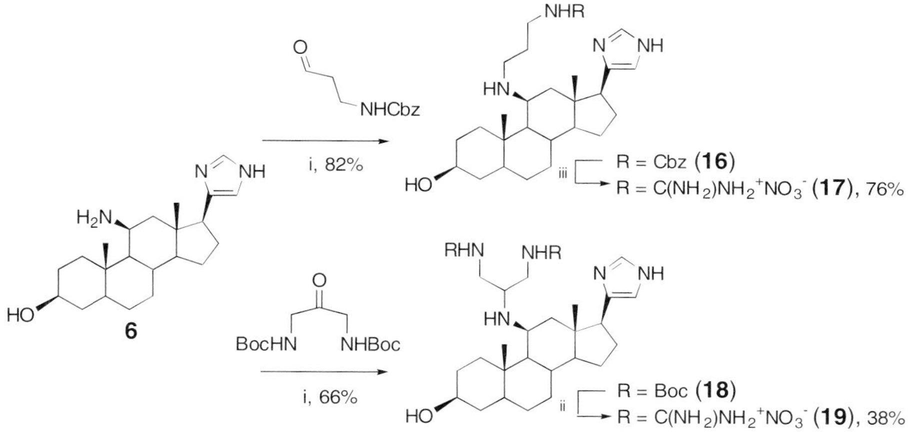

**Scheme 5.5.6.** (i) NaCNBH₃, MeOH, pH = 7; (ii) HCl (4 M), basic ion-exchange resin, 3,5-dimethylpyra-zole-1-carboxamidine nitrate, Hünig's base, DMF; (iii) H₂, Pd/C, HCl, MeOH, basic ion exchange resin, 3,5-dimethylpyrazole-1-carboxamidine nitrate, Hünig's base, DMF.

Taking this into account, we decided to synthesize the above-mentioned steroid derivatives in its C11-α-configured series with a more flexible imidazole side chain at C17. The synthesis of steroid **27** begins with a tri-ketone which can be obtained in quantitative yield from the bis-acetal **20** [12] by acid hydrolysis with SiO₂ and HCl. Formation of the thio ketal (**22**) can be achieved in methanol with 1,2 ethane dithiol and BF₃-etherate [13].

After protecting the C3 keto group, the steroid exhibits two unprotected carbonyl groups of which the C11 carbonyl group is notoriously unreactive towards

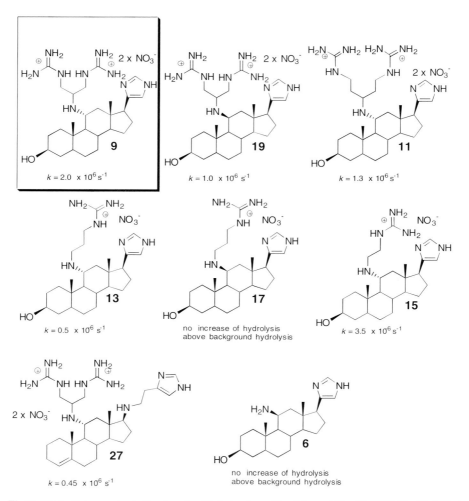

**Figure 5.5.1.** Rate constants for the steroid derivatives. Conditions: 100 mM HEPES; pH 8.00; $I = 0.5$ NaCl; 10 mM steroid; 2 mM **1**. The pseudo first-order rate constants are the rate constants for the increase of hydrolysis above background hydrolysis.

nucleophilic attack. Therefore reductive amination with histamine occurs only at the C17 carbonyl to yield the desired steroid (**23**) [14]. Analogous to our previously employed strategy [15], the C11-carbonyl was transformed into oxime **24** and reduced under dissolving metal conditions. This transformation not only generates diastereoselectively the C11-$\alpha$-amine, but also removes the undesired functionality at C3. Reductive amination with the Boc-protected bis-amine spacer followed by deprotection and bis(guanidation) generated steroid **27**.

The cleavage experiments with steroid **27** gave a pseudo first-order rate constant of only $0.45 \times 10^{-6}$ [s$^{-1}$], which is significantly lower than for steroid **9** and is of the

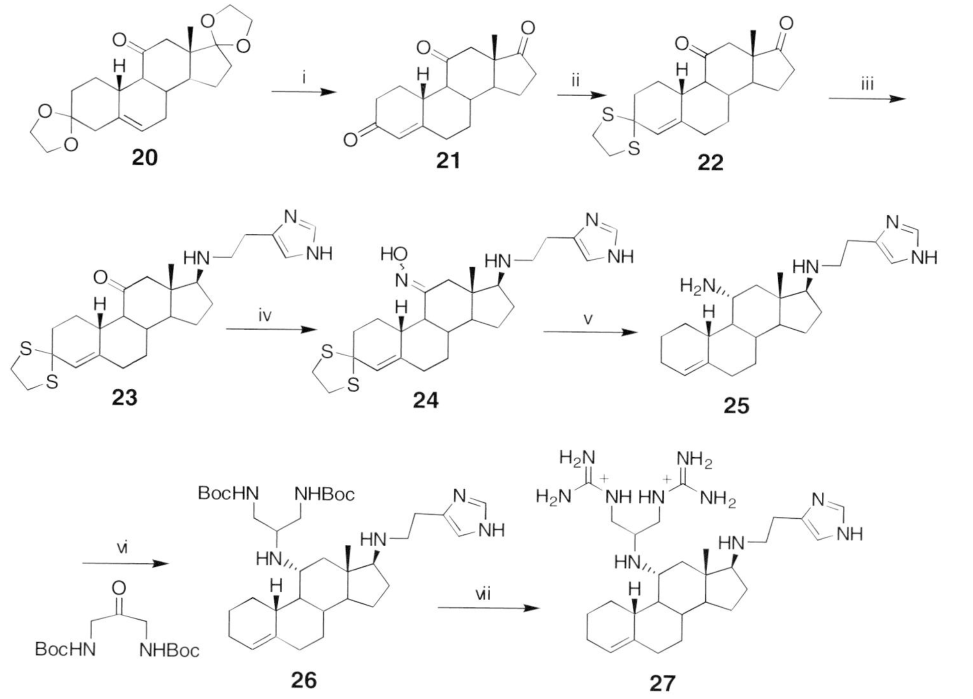

**Scheme 5.5.7.** Synthesis of steroid **27**: (i) 3 M HCl/SiO$_2$, CH$_2$Cl$_2$; (ii) ethane dithiol, BF$_3$-etherate, MeOH, 93% over two steps; (iii) histamine, toluene, reflux, then NaCNBH$_3$, MeOH, pH = 7, 0°C, 75%; (iv) NH$_2$OH·HCl, pyridine, reflux, 94%; (v) Na, n-PrOH, reflux, 75%; (vi) NaCNBH$_3$, MeOH, pH = 7, 57%; (vii) CF$_3$COOH, basic ion-exchange resin, then 3,5-dimethylpyrazole-1-carboxamidine nitrate, Hünig's base, DMF, 61%.

order of the rate constants for the mono(guanidinium) compound **15** [16] $(0.35 \times 10^{-6} \, [\mathrm{s}^{-1}])$.

Even though the various steroids exhibit different rate constants, we did not know whether these rate constants are significant changes in the hydrolytic behavior, or if similar results can be obtained with simple, unstructured bis(guanidinium) compounds. Since no data regarding the cleavage of RNA or RNA-model compounds with guanidinium compounds in water were available at that time, we synthesized various unstructured bis(guanidinium) compounds with and without internal base as a comparison to our steroid derived model systems (Figure 5.5.2).

The same transesterification/cleavage of **1** (Scheme 5.5.2) as in the cases with the steroid systems was utilized in order to determine the rate constants of the various enzyme models. All experiments were recorded against background hydrolysis under our conditions, and the reported values for the pseudo first-order rate constants are the acceleration above background hydrolysis. Figure 5.5.3 visualizes the different rate constants for the various unstructured bis(guanidinium) compounds and the steroid model systems. It is remarkable that even the bis(guanidinium) compounds

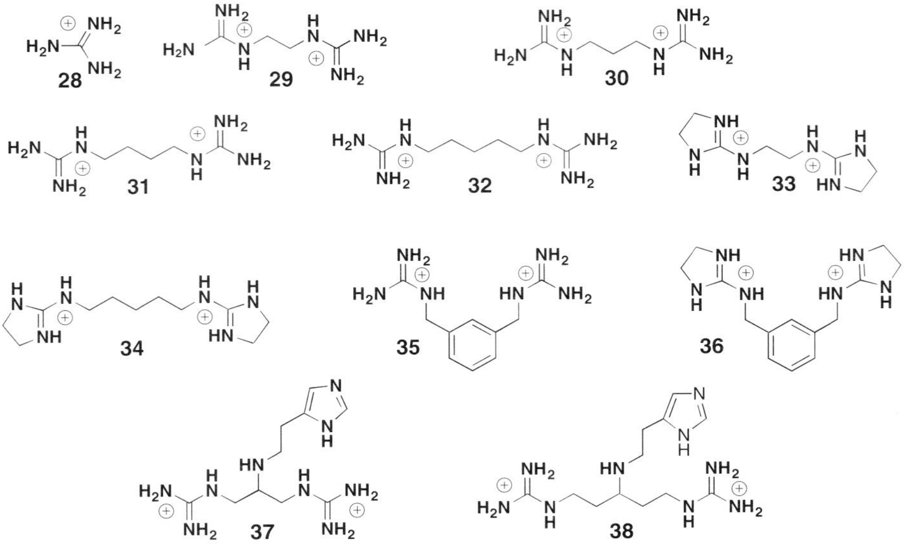

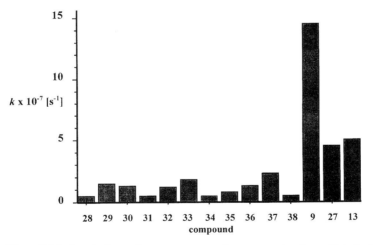

**Figure 5.5.2.** Unstructured bis(guanidinium) compounds.

**Figure 5.5.3.** Pseudo first-order rate constants of unstructured and steroid-derived guanidinium compounds. The hydrolysis experiments were performed at 55°C in HEPES buffer with pH = 7.55 (100 mM HEPES, $I = 0.5$ NaCl, 10 mM guanidinium compound, 2 mM **1**). The fact that the observed cleavage is actually a transesterification reaction as pointed out in Scheme 5.5.2 and not a hydrolysis reaction, was proven by [$^{31}$P] NMR experiments. The only detectable [$^{31}$P] NMR signals were observed at $\delta = -4.2$ for the substrate and at $\delta = 18.8$ for the cyclic phosphate.

**37** and **38** with an internal imidazole group do not exhibit cleavage rates comparable to those of the steroid derivatives. Also, the cleavage rate for steroid **27** with the same spacer and configuration as the most active steroid **9**, but with a more flexible imidazole side chain drops significantly compared to steroid **9**.

We also used compound **9** in hydrolysis experiments with UpU as the crucial probe to determine whether such bis(guanidinium) compounds can be used for cleaving of non-activated phosphodiesters as in RNA. The hydrolysis experiments were performed in the presence of HEPES buffer (100 mM) to assure a pH of 7.5. Steroid **9** shows remarkable hydrolysis of UpU at 60°C. The pseudo first-order rate constant was $2.4 \times 10^{-4}\,h^{-1}$ at a concentration of 10 mM steroid and 0.1 mM UpU. The $2',3'$ cyclic phosphate could be identified by HPLC analysis as well as the $2'$- and the $3'$-phosphate that were generated from hydrolysis of the $2',3'$ cyclic phosphate.

## Discussion

With every steroid exhibiting a different rate constant, we observed what we initially expected: Changes in conformation can provide changes in kinetics. Nevertheless, we did not expect steroid **11** to be less effective than **9**. The reason for the change in the reactivity can be seen in the increasing distance between the guanidinium groups and the steroid backbone by changing the 1,3 spacer into the 1,5 spacer. The effect of an internalized base is demonstrated by the fact that the rate constant increased from pH 7.3 to 7.65, and leveled out at pH 8.0, indicating that the imidazole moiety (pH = 7.6) actually acts as base in the transesterification reaction [14b]. The decrease of the rate of the hydrolytic cleavage with the steroid (**27**) exhibiting a more flexible imidazole side indicates, that preorganization is necessary for efficient hydrolysis.

An obstacle in assessing the potential of model systems is the fact that some systems have been tested in aqueous media whereas other systems have been used in aprotic solvents or aprotic solvents with traces of water. However, comparison of the steroid derivatives with simple bis(guanidinium) compounds clearly shows that preorganization is a crucial requirement, even for active site model systems.

The rate constant for the cleavage of the RNA model compound HPNPP strongly depends on the three-dimensional structure of the core of the RNAse model system. Within the corticosterone-derived steroid series, the C11-$\alpha$-configurated compounds were superior to their $\beta$-diastereomeres. The 1,3 distance between the two guanidinium groups gave higher rates than the 1,5 distance. This seems to reflect the fact that a position with a relatively high positive charge can coordinate phosphate anions much more efficiently. The fact that a more flexible imidazole group at position C17 leads to a substantial decrease in the hydrolytic behavior of the steroid derivatives further validates our strategy of a preorientated phosphate receptor that is able to adapt to the geometric changes in the transition state of the hydrolysis reaction. The more flexible steroid **27** does not exhibit the preorientation of steroid **9,** and is therefore not as efficient for coordinating and hydrolyzing phosphodiesters. Taking all this together, we can say that preorientation can lead to a substantial increase in the efficiency of enzyme model systems. These results not only contribute to a better understanding of enzyme mechanism, but may also lead to the development of drugs for the treatment of various diseases.

# References

[1]  (a) B. Barbier, A: Brack, *J. Am. Chem. Soc.* **1988**, *110*, 6880–6882; (b) B. Barbier, A: Brack, *J. Am. Chem. Soc.* **1992**, *114*, 3511–3515.

[2]  (a) K. Ariga, E. V. Anslyn, *J. Org. Chem.* **1992**, *57*, 417–419; (b) J. Smith, K. Ariga, E. V. Anslyn, *J. Am. Chem. Soc.* **1993**, *115*, 362–364; (c) D. M. Kneeland, K. Ariga, V. M. Lynch, C.-Y. Huang, E. V. Anslyn, *J. Am. Chem. Soc.* **1993**, *115*, 10042–10055.

[3]  (a) M. W. Göbel, J. W. Bats, G. Dürner, *Angew. Chem.* **1992**, *104*, 217–218; *Angew. Chem. Int. Ed. Engl.* **1992**, *31*, 207–209; (b) R. Groß, G. Dürner, M. W. Göbel, *Liebigs Ann. Chem.* **1994**, 49–58; (c) R. Groß, J. W. Bats, M. W. Göbel, *Liebigs Ann. Chem.* **1994**, 205–210; (d) G. Müller, G. Dürner, J. W. Bats, M. W. Göbel, *Liebigs Ann. Chem.* **1994**, 1075–1092.

[4]  (a) R. P. Dixon, S. J. Geib, A. D. Hamilton, *J. Am. Chem. Soc.* **1992**, *114*, 365–366; (b) V. Jubian, R. P. Dixon, A. D. Hamilton, *J. Am. Chem. Soc.* **1992**, *114*, 1120–1121; (c) V. Jubian, A. Veronese, R. P. Dixon, A. D. Hamilton, *Angew. Chem.* **1995**, *107*, 1343–1345; *Angew. Chem. Int. Ed. Engl.* **1995**, *34*, 1237–1239.

[5]  (a) F. P. Schmidtchen, *Tetrahedron Lett.* **1989**, *30*, 4493–4496; (b) P. Schiessl, F. P. Schmidtchen, *J. Org. Chem.* **1994**, *59*, 509–511.

[6]  (a) F. A. Cotton, E. E. Hazen, Jr., M. J. Legg, *Proc. Natal. Acad. Sci. USA* **1979**, *76*, 2551–2555; (b) E. H. Serpersu, D. Shortle, A. S. Mildvan, *Biochemistry* **1987**, *26*, 1289–1399; (c) P. J. Loll, E. E. Lattman, *Proteins* **1989**, *5*, 183; (d) J. Aqvist, A. Warshel, *Biochemistry* **1989**, *28*, 4680–4689; (e) E. H. Serpersu, D. W. Hilber, J. A. Gerlt, A. S. Mildvan, *Biochemistry* **1989**, *28*, 1539–1548; (f) D. J. Weber, E. H. Serpersu, D. Shoetle, A. S. Mildvan, *Biochemistry* **1990**, *29*, 8632–8642; (g) D. J. Weber, A. K. Meeker, A. S. Mildvan, *Biochemistry* **1991**, *30*, 6103–6114; (h) J. K. Judice, T. R. Gamble, E. C. Murphy, A. M. de Vos, P. G. Schultz, *Science* **1993**, *261*, 1578–1581.

[7]  (a) D. W. Hibler, J. N. Stolowich, M. A. Reynolds, J. A. Gerlt, J. A. Wilde, P. H. Bolton, *Biochemistry* **1987**, *26*, 6278–6286; (b) E. H. Serpersu, D. Shortle, A. S. Mildvan, *Biochemistry* **1986**, *25*, 68–77; (c) E. H. Serpersu, J. McCracken, J. Peisach, A. S. Mildvan, *Biochemistry* **1988**, *27*, 8034–8044.

[8]  D. M. Brown, D. A. Usher, *J. Chem. Soc.* **1965**, 6558–6564.

[9]  J. P. Guthrie, *Can. J. Chem.* **1972**, *50*, 3993–3997.

[10]  (a) R. Weidenhagen, R. Herrmann, *Ber. Dtsch. Chem. Ges.* **1935**, *68*, 1953–1965; (b) R. Weidenhagen, R. Herrmann, H. Wegner, *Ber. Dtsch. Chem. Ges.* **1937**, *70*, 570–583.

[11]  G. H. Phillips, G. B. Ewan, (Glaxo, Glaxo Laboratories Ltd), DT 2715078 A,. **1997**, [*Chem. Abstr.* **1978**, *88*, 38078].

[12]  A generous donation of compound **20** by the Schering company is acknowledged.

[13]  (a) J.R. Williams, G.M. Sarkisian, *Synthesis* **1974**, 32–33. (b) M.P. Bosch, F.Camps, J. Coll, A. Guerrero, T. Tatsuoka, J. Meinwald, *J. Org. Chem.* **1986**, *51*, 773–784.

[14]  Y. Langlois, C. Poupat, H.-P. Husson, P. Potier, *Tetrahedron* **1979**, *26*, 1967–1979.

[15]  (a) T. Oost, A. Filippazzi, M. Kalesse, *Liebigs Ann/Recueil* **1997**, 1005–1011; (b) T. Oost, M. Kalesse, *Tetrahedron* **1997**, *53*, 8421–8438.

[16]  T. Oost, **1997**, Doktorarbeit, Universität Hannover.

# 5.6     Studies on hairpin ribozymes

*Sabine Müller*

A ribozyme is an RNA molecule that can break and/or form covalent bonds. It greatly accelerates the rate of reaction and is highly specific with respect to the substrates it works on and the reaction products it produces. Since the first description of a catalytically active RNA molecule [1], many examples of RNA which can carry out functions previously ascribed to proteins, and are thus capable of acting as enzymes, have been described. Among the reactions observed, cleavage and ligation of phosphodiester bonds are most common. Other new biochemical activities are being developed using *in vitro* selection techniques. RNA catalysis has been demonstrated in members of group I and II introns, the genomes of viroids, virusoids, and satellite RNAs of a number of viruses, and in the prokaryotic *pre*-t-RNA processing machinery.

The hairpin ribozyme belongs to the group of small plant pathogenic RNAs (viroids, virusoids, and stellite RNAs) [2a–c] and is a region of the negative strand of the satellite RNA from *Tobacco Ringspot Virus* that undergoes self-cleavage (Figure 5.6.1) [3a–c].

It is thought that the self-cleavage reaction is an integral part of the *in vivo* rolling circle mechanism of replication used by such RNAs. The self-cleavage of the virus

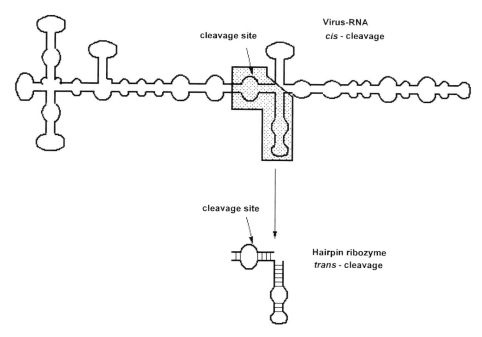

**Figure 5.6.1.** Schematic view of the secondary structure of the negative strand of *Tobacco Ringspot Virus* satellite RNA [3c]. The minimum self-cleaving domain is identified (stippled area) and separately shown as the *trans*-cleaving hairpin ribozyme.

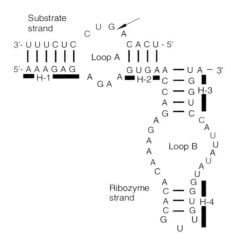

**Figure 5.6.2.** Secondary structure of the hairpin ribozyme shown in *trans*-cleaving mode. The arrow denotes the site of cleavage. The four helices (H-1 through H-4) are indicated by bars. The overhanging five nucleotides GGG AG at the 5′-end of the ribozyme strand resulting from *in vitro* transcription (see section 5.6.2) are not shown.

RNA is an intramolecular reaction (*cis*-cleavage); the substrate and ribozyme are part of one and the same molecule. However, the hairpin ribozyme is one of several small RNAs, which include the Hammerhead and Hepatitis Delta Virus RNA, that can be engineered for cleavage in *trans* [4a, b] (Figure 5.6.1). In this case, a substrate RNA strand is cleaved following incubation with a catalytic RNA (ribozyme) strand in the presence of divalent metal ions such as magnesium.

The power of such ribozymes is being harnessed for gene therapy applications through intracellular, vector-driven RNA production (for example see [5]) as well as for exogenous delivery as potential therapeutics [6a, b].

Figure 5.6.2 shows the secondary structure of a *trans*-cleaving hairpin ribozyme. Mutagenesis and *in vitro* selection studies have revealed that the hairpin ribozyme consists of four regions of Watson–Crick duplex RNA and two internal loops (A and B) [7a–f]. All of the essential nucleotides, those of which are required for catalytic activity, are located in the two internal loop regions [7d–f]. Site-specific reversible cleavage occurs 5′ to G within the sequence NGUC (N = each of the four bases) in the substrate strand [7a, f] generating 5′-hydroxyl and 2′,3′ cyclic phosphate termini [3a–c].

Functional group alteration data [8a, b] strongly support previous suggestions that the two loop regions A and B are involved in inter-domain interactions during catalysis. Linker insertion results [9a–c] indicated that in the catalytically active structure, helix 2 is not coaxially stacked upon helix 3 but that the ribozyme–substrate complex bends at the junction between helices 2 and 3. The two domains, the first containing the substrate strand (internal loop A and helices 1 and 2) and the second consisting of internal loop B and helices 3 and 4, can be completely separated and ribozyme activity reconstituted by mixing solutions of the two domains in the presence of magnesium ions [10]. Studies involving preparation of synthetic ribozymes fixed in a paperclip structure by an inter-strand disulfide cross-link between the substrate RNA and loop B of the ribozyme strand revealed further evidence of a close proximity of loop A and B in the catalytically active complex [11a, b].

In order to obtain further information about the active conformation of the ribozyme-substrate complex, we have synthesized a hairpin ribozyme containing an additional helix 5 (Figure 5.6.3) and have studied the kinetic properties in comparison to those of the wild-type (WT) ribozyme. Lastly, in order to avoid the use of radio-labeled samples, we have evaluated a quantitative assay for sequence-specific ribozyme activity utilizing fluorescently labeled RNA substrates in conjunction with an automated DNA sequencer.

## 5.6.1  Design of the HpH5 ribozyme and its substrate

For our studies, in addition to the WT ribozyme, we have synthesized a new type of hairpin ribozyme, which has an extra sequence at the 3′-side of the hairpin ribozyme to allow it to hybridize with the 5′-side sequence of the substrate. Thus, in comparison to the WT ribozyme, the new HpH5 ribozyme consists of an additional helix 5 between the ribozyme and part of the substrate strand (Figure 5.6.3).

There exists a linker of five unpaired cytidine residues at the hinge between helices 3 and 5, which was introduced in order to preserve the necessary hinge flexibility to allow the ribozyme to bend and thus, to bring loop A and loop B in close proximity to each other. If the linker length of five cytidines proves to be sufficient for the forma-tion of the active bent structure, the amount of cytidines in the linker can be reduced stepwise until a rigid structure without any unpaired bases at the hinge is formed and

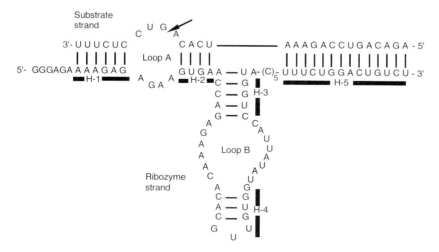

**Figure 5.6.3.** Secondary structure of the HpH5 ribozyme shown in *trans*-cleaving mode. The arrow denotes the site of cleavage. The five helices (H-1 through H-5) are indicated by bars.

the catalytic activity of such ribozymes can be investigated. This is of interest not only for ribozyme activity but also for structural investigation of RNA.

## 5.6.2    Preparation of ribozymes

The hairpin ribozyme, which is 50 nucleotides in length, is a relatively small catalytic RNA. However, we found it still too large to be produced through chemical solid-phase synthesis. Thus, both ribozymes (WT = 50 nucleotides, HpH5 = 69 nucleotides) have been prepared by *in vitro* transcription. Our strategy involved the synthesis of DNA templates containing the ribozyme sequence linked to the sequence of the T7-RNA-Polymerase promoter for ribozyme transcription. The templates have been generated from two synthetic DNA strands, overlapping by nine complementary bases. After annealing, the DNA templates have been completed enzymatically using T7-DNA-sequenase and the four deoxynucleoside triphosphates (Figure 5.6.4).

After purification of the DNA double strands by polyacrylamide electrophoresis and ethanol precipitation, ribozymes have been synthesized by transcribing the duplex DNA templates with T7-RNA-polymerase. The overhanging five nucleotides GGG AG at the 5′- end of both ribozymes (Figures 5.6.2 and 5.6.3) were necessary since the T7-RNA-polymerase needs an optimal start sequence for transcription [12]. DNA templates missing this specific sequence could not be transcribed successfully (data not shown).

Final purification of the synthesized ribozymes was achieved by electrophoresis on denaturing polyacrylamide gels and precipitation from ethanol.

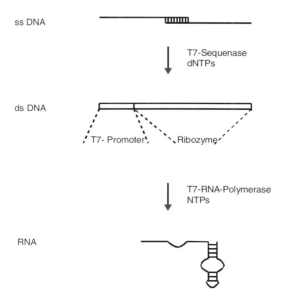

**Figure 5.6.4.**  Preparation of ribozymes by *in vitro* transcription.

### 5.6.3  Synthesis of substrate RNAs

Substrates of the WT and the HpH5 ribozyme (14 and 28 nucleotides, respectively) have been synthesized by solid-phase chemistry. Over the past few years, synthesis methodologies for assembly of oligoribonucleotide chains have become reasonably well established. Solid-phase phosphoramidite procedures have proved most popular, and comprehensive reviews of synthetic methods have appeared [13a, b].

Our own procedure is based on 2′-*O*-*t*-butyldimethylsilyl (TBDMS) nucleoside protection [14a, b] and involves sequential couplings of β-cyanoethyl-(*N*,*N*-diisopropyl)-phosphoramidites of 5′-*O*-dimethoxytrityl-2′-*O*-TBDMS-nucleosides (Figure 5.6.5), essentially by the method of Scaringe et al. [15]. As solid phase, we predominantly have used controlled pore glass (CPG), to which the first ribonucleoside derivative is attached via its 3′-position as a succinate linkage [16]. The standard synthesis was on the 1 μmole scale, and for coupling of phosphoramidites we used primarily standard tetrazole activation and coupling times of 12 minutes.

According to the assay used for quantitation of ribozyme activity, both the substrate strands had to be fluorescently labeled. This was achieved by using commercially available fluorescein phosphoramidite or support-linked fluorescein, dependent on labeling the substrate at the 5′-end or the 3′-end.

The synthesized oligoribonucleotides were deprotected by overnight suspension of the CPG in methanolic ammonia, decanting and evaporation of the resulting solution to dryness. Treatment with triethylamine trihydrofluoride/DMF (3 : 1) at 55°C for 1.5 h [17] was carried out to remove silyl groups, followed by precipitation with *n*-butanol. Finally, oligoribonucleotides were purified by strong anion-exchange chromatography on a Mono Q column (Amersham Pharmacia Biotech) using buffer A (20 mM sodium acetate, 20% acetonitrile) with a gradient of buffer B (20 mM sodium acetate, 1 M sodium chloride, 20% acetonitrile): 0–60% over 50 min. Desalting was achieved by extensive dialysis against water.

For analysis of chemically or enzymatically synthesized oligoribonucleotides, we routinely used total enzymatic digestion with snake venom phosphodiesterase and alkaline phosphatase to give the corresponding nucleosides, followed by separation

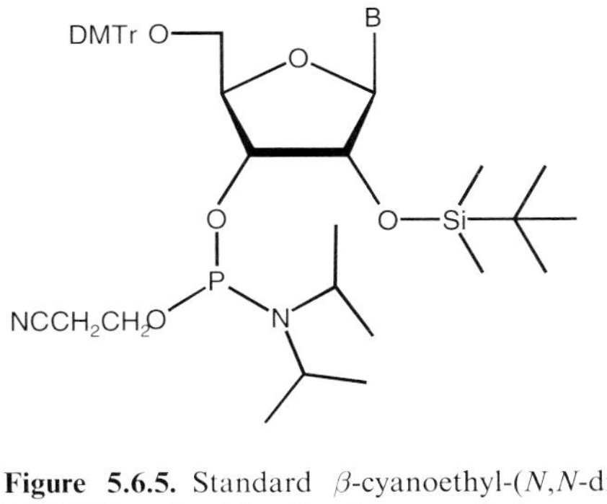

**Figure 5.6.5.** Standard β-cyanoethyl-(*N*,*N*-diisopropyl)-phosphoramidite of 5′-*O*-dimethoxytrityl-2′-*O*-TBDMS-nucleosides. B = heterocyclic base.

by reversed-phase HPLC. The four nucleosides could be identified and quantitated by comparison with standards. Finally, the purity of the oligomers was analyzed by subjection of the samples to electrophoresis on denaturing polyacrylamide gels and reversed-phase HPLC.

### 5.6.4    Quantitative assay of ribozyme activity

As yet the application of substrate RNAs labeled with fluorescein for quantitation of hairpin ribozyme activity has not been reported. Most assays are based on [$^{32}$P]-end-labeled substrates. Thus, the amount of reactants and products after reaction can be analyzed by subjecting the reaction mixture to denaturing polyacrylamide gels followed by autoradiography. In analogy to our previous work on sequence-specific endonuclease activity [18], we have evaluated a quantitative assay of ribozyme activity based on DNA sequencer technology. The procedure depends on fluorescently labeled substrates and an automated DNA sequencer to determine amounts of both reactant and product of the reaction, each individual measurement being standardized internally.

To utilize this assay for quantitation of ribozyme activity, it was necessary first to investigate whether the rather bulky fluorescein molecule had any influence on the formation of the ribozyme–substrate complex. Substrate RNAs were fluorescently labeled at either the 5′- or 3′-end. Hairpin ribozymes were assembled from the catalytic strand and the 5′- or 3′-fluorescently labeled substrate strand. The ribozyme cleaves the substrate next to G (Figure 5.6.1), reducing the length of the RNA strand carrying the fluorescent label at the 5′-end from 14 to five nucleotides (and from 14 to nine nucleotides if the fluorescein is attached to the 3′-end). Thus, for both cases, reactants and products of the reaction could be separated by polyacrylamide gel electrophoresis and detected by virtue of their fluorescence. Electrophoretic analysis was carried out with an Automated Laser Fluorescence DNA Sequencer (A.L.F., Amersham Pharmacia Biotech) which has a laser beam which passes horizontally through the entire width of the gel, 19 cm down from the sample loading line. The gel has 40 lanes and, correspondingly, 40 photodiodes are placed behind the gel in a linear array along the path of the laser beam, each one consisting of an individual channel of parallel and continous fluorescence measurement. Ribozyme cleavage was visualized as a conversion of the slow-moving into the faster-moving species. Output data of the DNA sequencer were analyzed by using A.L.F.-Manager (Amersham Pharmacia Biotech); peaks were integrated using this software package. Progress of the reaction was monitored by the change of the relative peak areas. Since both reactant and product were measured, combined areas could be standardized to 100% (linear response provided, see [18]) and the measurement becomes independent of pipetting errors when loading the gel, and also of the quite pronounced sensitivity differences between individual channels. The assay achieves high sample throughput by parallel measurement of multiple samples. Because of its capacity to produce large sets of experimental data, the system is particularly well suited for the determination of reaction kinetics. The substrates are non-radioactive and can be stored for long periods of time.

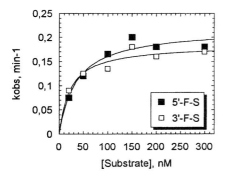

**Figure 5.6.6.** Steady-state analysis of the hairpin ribozyme cleavage reaction at different substrate concentrations. Cleavage assays were carried out under multiple-turnover conditions using the WT hairpin ribozyme and the 5'-fluorescein labeled substrate (■) or the 3'-fluorescein labeled substrate (□). Non-linear least-square fits to the Michaelis–Menten equation (continous lines) were used to calculate the steady-state parameters ($K_M$ and $k_{cat}$) for the cleavage reaction.

Steady-state parameters of the ribozyme reactions were obtained from determination of the initial rate of reaction under multiple turnover conditions using the described quantitative assay. In both cases (5'-or 3'-fluorescein labeling) the reaction velocities, once normalized to a fixed ribozyme concentration, followed a conventional Michaelis–Menten curve and the obtained catalytic parameters $K_M$ and $k_{cat}$ are in the same range (Figure 5.6.6 and Table 5.6.1).

Thus, fluorescently labeled RNAs are suitable substrates of the hairpin ribozyme, whether the fluorescein label is attached at the 5'- or 3'-end of the substrate RNA. The obtained kinetic data of the ribozyme cleavage are in good agreement with $K_M$ and $k_{cat}$ values found for the hairpin ribozyme using [$^{32}$P]-end-labeled substrates. Therefore, our results strongly support previous suggestions that either the active-site chemistry or a conformational change in the hairpin ribozyme-substrate complex is rate limiting, whereas substrate binding and product dissociation occur rather rapidly.

## 5.6.5    Kinetic analysis of HpH5 ribozyme cleavage

For kinetic analysis, both the WT ribozyme and the HpH5 ribozyme were assembled from the catalytic strand and the respective 3'-fluorescein-labeled substrate RNA. Kinetic parameters were obtained from determination of the initial rates of reaction at concentrations of substrate around $K_M$. Comparison of the kinetic behavior reveals a difference between the cleavage rates of the WT and the HpH5 ribozyme, the rate constant $k_{cat}$ of HpH5 ribozyme is about twice as high as $k_{cat}$ of WT (Table 5.6.1).

This result is in good agreement with the outcome of recent kinetic studies of Esteban et al. [19]. The authors have studied the kinetic mechanism of the hairpin ribozyme in detail and have observed a suboptimal cleavage activity of the WT hairpin ribozyme. Their studies strongly support a model in which the ribozyme partitions

**Table 5.6.1.** Kinetic data for cleavage of the WT hairpin ribozyme and the HpH5 ribozyme.

|  | $k_{cat}$ [min$^{-1}$] | $K_M$ [µM] | $k_{cat}/K_M$ [min$^{-1}$ µM$^{-1}$] |
|---|---|---|---|
| 5'-F-HpWT | $0.218 \pm 0.016$ | $0.036 \pm 0.01$ | 6.06 |
| 3'-F-HpWT | $0.184 \pm 0.012$ | $0.023 \pm 0.07$ | 8 |
| 3'-F-HpH5 | $1.461 \pm 0.081$ | $0.067 \pm 0.037$ | 6.88 |

between two major conformations leading to active and inactive ribozyme–substrate complexes. Although the results of Esteban et al. do not provide information about the specific structure of the inactive conformation, the authors discuss the concept of an inactive conformation of the ribozyme–substrate complex in which coaxial stacking of helices 2 and 3 results in an extended conformation that prevents interaction of the two domains, one containing loop A and one containing loop B, of the ribozyme (Figure 5.6.7). It is widely accepted that the folded structure of RNA is critical for its catalytic activity. The additional helix in the HpH5 ribozyme possibly assists proper folding of the ribozyme–substrate complex and thus, increases the proportion of substrate molecules bound in the active conformation. As discussed earlier, the constant $k_{cat}$ reflects a rate-limiting conformational transition or slow-reaction chemistry in the WT ribozyme. If this is also true for the HpH5 ribozyme, the higher cleavage rate can be explained by a stronger tendency of the ribozyme–substrate complex to fold into the active conformation. The actual population of substrate bound to the ribozyme in the active complex is larger, resulting in a higher *turnover* of substrate molecules.

Our studies demonstrate that the introduction of five additional cytidine residues at the hinge between helices 3 and 5 guarantees sufficient flexibility to allow bending into the active conformation. The structure of the HpH5 ribozyme may be described as a three-way junction.

For DNA, in addition to three-way junctions with unpaired bases, rigid three-way junctions without additional bases have been described [20a, b]. To our knowledge, similar studies with RNA do not exist so far. However, the structure of the hammerhead

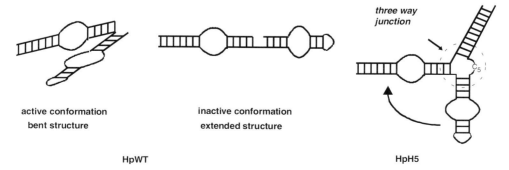

active conformation
bent structure

inactive conformation
extended structure

three way
junction

HpWT

HpH5

**Figure 5.6.7.** Topology of hairpin ribozymes.

ribozyme may be described as such a three-way junction too. Since the HpH5 ribozyme possesses high catalytic activity, it is very likely that hairpin ribozymes of this type without additional bases at the hinge will also form stable three-way structures. We are currently investigating the ability of such structures to catalyze cleavage of substrate RNAs. Preliminary results suggest that hairpin ribozymes of the HpH5 type having less then five cytidine residues at the hinge are also catalytically active. Detailed kinetic studies are in progress.

**Acknowledgements:** I thank Beate Scholz for carrying out the kinetic analyses and Christina Schönherr for technical assistance. I am grateful to Prof. Ulrich Koert for supporting my work and to Dr. Derek Levison for critical comments on the manuscript.

# References

[1] T. R. Cech, A. J. Zaug, P. J. Grabowski, *Cell* **1981**, *27*, 487–496.
[2] (a) R. H. Symons, *Annu. Rev. Biochem.* **1992**, *61*, 641–671; (b) D. M. Long, O. C. Uhlenbeck, *FASEB J.* **1993**, *7*, 25–30; (c) A. M. Pyle, *Science* **1993**, *261*, 709–714.
[3] (a) J. Haselhoff, W. L. Gerlach, *Gene* **1989**, *82*, 43–52; (b) P. A. Feldstein, J. M. Buzayan, G. Bruening, *ibid.*, 53–61; (c) A. Hampel, R. Tritz, *Biochemistry* **1989**, *28*, 4929–4933.
[4] (a) R. H. Symons, *Curr. Opin. Struct. Biol.* **1994**, *4*, 322–330; (b) T. Tuschl, J. B. Thomson, F. Eckstein, *Curr. Opin. Struct. Biol.* **1995**, *5*, 296–302.
[5] M. Yu, M. C. Leavitt, M. Maruyama, D. Young, A. D. Ho, F. Wong-Staal, *Proc. Natl. Acad. Sci. USA* **1995**, *92*, 699–703.
[6] (a) P. Marschall, J. B. Thomson, F. Eckstein, *Cell. Mol. Neurobiol.* **1994**, *14*, 523–538; (b) R. E. Christoffersen, J. J. Marr, *J. Med. Chem.* **1995**, *38*, 2023–2037.
[7] (a) A. Hampel, R. Tritz, M. Hicks, P. Cruz, *Nucleic Acids Res.* **1990**, *18*, 299–304; (b) B. M. Chowrira, J. M. Burke, *Biochemistry* **1991**, *30*, 8518–8522; (c) A. Berzal-Herranz, S. Joseph, J. M. Burke, *Genes Dev.* **1992**, *6*, 129–134; (d) A. Berzal-Herranz, S. Joseph, B. M. Chowrira, S. E. Butcher, J. M. Burke, *EMBO J.* **1993**, *12*, 2567–2574; (e) S. Joseph, J. M. Burke, *J. Biol. Chem.* **1993**, *268*, 24515–24518; (f) P. Anderson, J. Monforte, R. Tritz, S. Nesbitt, J. Hearst, A. Hampel, *Nucleic Acids Res.* **1994**, *22*, 1096–1100.
[8] (a) J. A. Grasby, K. Mersmann, M. Singh, M. J. Gait, *Biochemistry* **1995**, *34*, 4068–4074; (b) S. Schmidt, L. Beigelman, A. Karpeisky, N. Usman, U. S. Sørensen, M. J. Gait, *Nucleic Acids Res.* **1996**, *24*, 573–581.
[9] (a) P. A. Feldstein, G. Bruening, *Nucleic Acids Res.* **1993**, *21*, 1991–1998; (b) Y. Komatsu, M. Koizumi, H. Nakamura, E. Ohtsuka, *J. Am. Chem. Soc.* **1994**, *116*, 3692–3696; (c) Y. Komatsu, I. Kanzaki, M. Koizumi, E. Ohtsuka, *J. Mol. Biol.* **1995**, *252*, 296–304.
[10] S. E. Butcher, J. E. Heckman, J. M. Burke, *J. Biol. Chem.* **1995**, *270*, 29648–29652.
[11] (a) S. Schmidt, D. J. Earnshaw, S. Th. Sigurdsson, F. Eckstein, M. J. Gait, *Coll. Czech. Chem. Commun.* **1996**, *61*, 276–279; (b) D. J. Earnshaw, B. Masquida, S. Müller, S. Th. Sigurdsson, F. Eckstein, E. Westhof, M. J. Gait, *J. Mol. Biol.* **1997**, *274*, 197–212.
[12] J. F. Milligan, O. C. Uhlenbeck, *Meth. Enzymology* **1989**, *180*, 51–62.
[13] (a) S. L. Beaucage, R. P. Iyer, *Tetrahedron* **1992**, *48*, 2223–2311; (b) *ibid.*, **1993**, *49*, 6123–6194.
[14] (a) K. K. Ogilvie, E. A. Thompson, M. A. Quilliam, J. B. Westmore, *Tetrahedron Letts.*, **1974**, 2865–2868; (b) M. J. Damha, K. K. Ogilvie in 'Methods in Molecular Biology', (Ed. S. Agrawal), Humana Press, Totowa, NJ, **1993**, pp. 81–114.
[15] S. A. Scaringe, C. Francklyn, N. Usman, *Nucleic Acids Res.* **1990**, *18*, 5433–5441.

[16]  M. J. Gait, C. E. Pritchard, G. Slim, in *'Oligonucleotides and Analogues: A Practical Approach'*, (Ed. F. Eckstein), Oxford University Press, UK, **1991**, pp. 25–48.

[17]  F. Wincott, A. DiRenzo, C. Shaffer, S. Grimm, D. Tracz, C. Workman, D. Sweedler, C. Gonzalez, S. Scaringe, N. Usman, *Nucleic Acids Res.* **1995**, *23*, 2677–2684.

[18]  W. Gläsner, R. Merkl, S. Schmidt, D. Cech, H.-J. Fritz, *Biol. Chem. Hoppe-Seyler* **1992**, *373*, 1223–1225.

[19]  J. A. Esteban, A. R. Banerjee, J. M. Burke, *J. Biol. Chem.* **1997**, *272*, 13629– 13639.

[20]  (a) M. A. Rosen, D. J. Patel, *Biochemistry* **1993**, *32*, 6563–6581; (b) F. J. J. Overmars, J. A. Pikkemaat, H. H. Van den Elst Jacques, V. Boom, C. Altona, *J. Mol. Biol.* **1996**, *255*, 702–711.

# 6 Biosynthetic Pathways and Biochemistry

## 6.1 Neuropeptide Y: A molecule for hunger, stress and memory

*Annette G. Beck-Sickinger, Chiara Cabrele, and Richard Söll*

### 6.1.1 What is neuropeptide Y?

Neuropeptide Y (NPY) is a member of the pancreatic polypeptide hormone family, which also includes pancreatic polypeptide (PP), peptide YY (PYY) and anglerfish pancreatic peptide (PY). Features common to all members of this family are the chain length of 36 amino acids and the amidated carboxy-terminus. The polypeptides show a high homology in their primary structure, which is 69% between NPY and PYY and 50% between NPY and PP. With few exceptions, Pro is always present at positions 2, 5 and 8, Gly at position 9, Ala at position 12, Tyr at positions 20, 27 and 36, Thr at position 32 and Arg at positions 33 and 35 (Figure 6.1.1) [1]. NPY was first isolated from porcine brain in 1982 [2]. A three-dimensional conformation of NPY has been postulated by NMR, circular dichroism and molecular modeling based on the crystallographic data of the avian PP. These studies propose an N-terminal type II polyproline helix, involving the four proline residues at positions 2, 5, 8, and 13, and a C-terminal amphipathic $\alpha$-helix at positions 19–32. The two helices are joined by a $\beta$-turn of the residues 14–18. The last four C-terminal residues adopt a flexible loop structure.

NPY is widely distributed within the peripheral and central nervous systems and it is the most abundant neuropeptide in the brain [3]. NPY causes a long-lasting

```
NPY: YPSKPDNPGEDAPAEDMARYYSALRHYINLITRQRY
 PP: APLEPVYPGDNATPEQMAQYAADLRRYINMLTRPRY
```

**Figure 6.1.1.** Sequence of NPY and PP and model of a Y-receptor (G-protein-coupled receptor).

vasoconstriction in skeletal muscle, heart, kidney, and brain, whereas it has been shown to reduce local blood flow in a variety of vascular beds in different species. In vitro, NPY acts as a regulating factor, enhancing the action of other vasoconstrictor agents. Presynaptically, the hormone inhibits its own release as well as that of noradrenaline. Furthermore, NPY inhibits transmitter release from parasympathetic nerves in the heart, lower airways, and from airway and cardiac sensory nerves. NPY injections result in increasing memory performance in rats and in regulating the release of sexual hormones such as luteinizing hormone-releasing hormone. Finally, it should be emphasized that NPY plays a role in the pathophysiology of depression and anxiety, and is involved in controlling neural systems that stimulate eating behavior, that regulate levels of metabolic fuels, and that control energy expenditure (reviewed in [4]).

## 6.1.2    What are the molecular targets of neuropeptide Y?

The existence of specific NPY receptors could be demonstrated by replacement and competition tests using radioactive NPY analogs. All known NPY receptors belong to the large super-family of heptahelical G-protein-coupled receptors (Figure 6.1.1). They are designated by a capital Y in general and a lower case y for receptors without any known function. The different NPY receptor subtypes appear to use similar signal transduction pathways, although no distinct transduction pathway has been identified so far. In almost all cell types studied, NPY receptors act via pertussis toxin-sensitive G-proteins, *i.e.* members of the $G_i/G_o$ family: inhibition of adenylyl cyclase is found in almost every tissue and cell type, whereas inhibition of $Ca^{2+}$-channels, activation of $K^+$-channels and mobilization of $Ca^{2+}$ from intracellular stores is only found in some.

Five distinct NPY receptors have been cloned: $Y_1$, $Y_2$, $Y_4$, $Y_5$, and $y_6$. The sequences of $Y_1$-, $Y_4$- and $y_6$-receptors are more closely related to each other than to the $Y_2$- and $Y_5$-receptors. Furthermore $Y_2$- and $Y_5$-receptors are equally distantly

**Table 6.1.1.** Properties of the different NPY receptor subtypes.

| Receptor | $Y_1$ | $Y_2$ | $Y_4$ | $Y_5$ | $y_6$ |
|---|---|---|---|---|---|
| Ligands | NPY, [Pro$^{34}$] NPY | NPY, [Ahx$^{5-24}$] NPY | PP, [Pro$^{34}$] NPY | NPY, PP, [Pro$^{34}$] NPY | NPY |
| Amino acids | 384 | 381 | 375 | 455 | 371 |
| Signal transduction | cAMP inhibition $Ca^{2+}$ mobilisation | cAMP inhibition $Ca^{2+}$, $K^+$ channel | cAMP inhibition | cAMP inhibition $Ca^{2+}$ mobilisation | |
| Major occurence | periphery, hypothalamus | brain, hippocampus | intestine, colon | hypothalamus | not in human |
| Related action | vasoconstriction, anxiety, food intake? | memory, epilepsy, secretion | gastro-intestinal regulation | food intake? | |

related to one another as to the $Y_1/Y_4/y_6$-receptor group. A so-called $Y_3$-receptor has been described only pharmacologically [5].

## 6.1.3 What can we learn from structure–activity studies?

Neuropeptide Y binds to at least four different receptor subtypes. Each receptor mediates specific biological effects and is located in certain cell membranes. Ligands that bind selectively to certain receptors would be wonderful tools to localize each receptor by autoradiography with radioactive analogs, and to characterize the individual role of each receptor in vivo and under physiological conditions. Furthermore, selective ligands could be used as lead compounds for the development of drugs, that act specifically on one receptor subtype. How can we obtain subtype selective ligands? The different receptors usually recognize the ligand in a slightly different mode, for example either by binding the same ligand in a different conformation, different three-dimensional orientation, or by interacting with different side chains. In each case, it is important to understand the function of each part of the ligand, which best is performed by so-called structure–activity or structure–affinity studies: modified ligands are tested for their affinity and activity, and the biological information is correlated with the structural data. In order to understand the activity of bioactive peptides, segments (N- or C-terminal) are usually synthesized first and can lead to smaller, still biologically relevant peptides. In the case of NPY, for example the C-terminal dodecapeptide still maintains some activity at the $Y_2$-receptor [6]. The next steps frequently include so-called 'scans'. In the case of the 36-mer peptide NPY, 36 analogs have been required for the L-alanine scan with each position singly exchanged by L-alanine to obtain [Ala$^1$] NPY (Tyr at position 1 replaced by Ala), [Ala$^2$] NPY, etc. to [Ala$^{36}$] NPY [7]. In general, L-Ala scans lead to the identification of the important side chains of peptides, because Ala causes minimal influence on the conformation, but misses especially the properties of trifunctional amino acids. Further frequently performed scans of full-length peptides or of the biologically active peptide segment include D-amino acid scan (no change of charge, polarity, etc., but change of orientation of important side chains and of conformation), glycine scan (only spacer left), proline scan (looking for turns), and phenylalanine scan (role of hydrophobicity) [6, 8]. Furthermore, constrained analogs like cyclopeptides or peptides with conformationally restricting amino acids are required. Turn- or helix-inducing building blocks are also frequently used [9]. The structure of rigid, small peptides can be identified best with 2D-NMR techniques. In the case of NPY, the bioactive conformation of NPY at the $Y_2$-receptor could be identified with a C-terminal cyclic peptide, that showed full activity and of which the structure had been solved by 2D-NMR [10]. Conformational studies of larger and less constrained analogs can be obtained by circular dichroism spectroscopy. Although this technique does not provide the three-dimensional structure of the molecules, it is a rapid and convenient method which can be used to compare conformations of related peptides and help to distinguish between direct effects and indirect, conformationally induced ones [11].

### 6.1.4    Analogues of neuropeptide Y: Which amino acids are essential?

The contribution of each side chain of NPY to the binding was investigated by the systematic single exchange of each residue of NPY by L-Ala [7]. The four natural Ala residues at positions 12, 14, 18, and 23 were substituted by Gly. Furthermore, single amino acids were replaced by closely related ones. All peptides were tested for their affinity to the $Y_1$-, $Y_2$-, $Y_5$- and partly to the $Y_4$-receptor. Biological effects were investigated at the $Y_2$-receptor. It was striking, that the C-terminal pentapeptide was of major importance for the interaction with all receptors. Arg33 and Arg35 cannot be substituted, even not with Lys, without a complete loss of affinity at each receptor subtype. Because the sensitivity pattern of other positions was found to be completely different, we concluded that a different conformation of the peptide is required at each receptor subtype. At the $Y_1$-receptor, we find a very high sensitivity of Thr32 and Tyr36, whereas position 34 can be replaced by various amino acids, including Ala, Pro, and His. The exchange of Pro2 and Pro5 by Ala resulted in a 100-fold decrease of the $Y_1$-receptor affinity, probably because of the destabilization of the N-terminal polyproline helix. In the $\alpha$-helical segment, the most sensitive positions were Arg19, Tyr20, Leu24, Tyr27 and Ile31. Since these side chains are located at the hydrophobic face of the amphipathic $\alpha$-helix, they may also be involved in stabilizing the hairpin-like structure as well [6, 7].

The only amino acid of the C-terminal segment that could be replaced without a major loss of affinity to the $Y_5$-receptor is Thr32. In addition to [Ala$^{32}$] NPY, [D-Trp$^{32}$] NPY has been reported to maintain affinity [12]. At the $Y_2$-receptor, in addition to position 33 and 35, Gln34 is important and may not be replaced by Pro. This finding led to two NPY analogs, [Pro$^{34}$] NPY and [Leu$^{31}$, Pro$^{34}$] NPY, which bind to the $Y_1$-, $Y_4$-, and $Y_5$-receptors but not to the $Y_2$-receptor [13]. Exchange of single amino acids in the segment 1–18 caused only a minimal loss of activity at the $Y_2$-receptor, and suggests that this segment is not important for interaction. In the $\alpha$-helical region, the L-Ala scan analogs show only moderate effects, and this suggests that they play a role in conformational stabilization. Whereas [Pro$^{34}$] NPY is inactive at $Y_2$-receptors and as potent as NPY at $Y_1$- and $Y_5$-receptors, it showed enhanced affinity at $Y_4$-receptors. Arg33 is more affected by substitution than Arg35 or Tyr36 at the $Y_4$-receptor. For all members of the PP-family, however, the peptide acid is completely inactive.

### 6.1.5    Segments of the peptide – how do we get them active as well?

All N-terminal segments of NPY, such as NPY1–12 or NPY1–24, are completely inactive [14]. C-terminal segments maintain a receptor subtype-dependent affinity. N-terminally truncated segments, like NPY2–36 or NPY13–36, lose $Y_1$-receptor affinity, whereas they still bind to the $Y_5$- or $Y_2$-receptor, respectively.

The C-terminal decapeptide of NPY maintains some affinity at the $Y_2$-receptor. The exchange of Thr32 and Gln34 of NPY27–36 by Tyr and Leu, respectively, led to a 3700-fold enhancement [16]. Cyclization of a C-terminal fragment let to the smallest known full agonist at the human $Y_2$-receptor with nanomolar affinity:

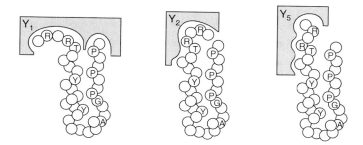

**Figure 6.1.2.** Hypothetical models of the interaction of NPY with the receptors $Y_1$, $Y_2$ and $Y_5$. The invariant positions in the NPY family are labeled.

cyclo[Lys$^{28}$-Glu$^{32}$]NPY Ac-25–36] [10]. With structure–affinity studies and the determination of the conformation of cyclic analogs of these molecules, the bioactive conformation of NPY at the $Y_2$-receptor could be described (Figure 6.1.2) [11].

Whereas C-terminal segments of NPY are inactive at the $Y_1$-, $Y_4$- and $Y_5$-receptors, modified analogs with high affinity for other receptor subtypes have also been obtained, like the very potent but rather unselective peptidic NPY antagonist, GW1229 (also known as 1229U91). This molecule was obtained by dimerization of the C-terminal nonapeptide of NPY with positions 30, 32, and 34 had been changed to Pro, Tyr, and Leu and positions 29 and 31 to Glu and diaminopropionic acid for dimerization [16].

## 6.1.6   Centrally truncated peptides: NPY analogs that miss the middle segment

Owing to the low affinity of N- and C-terminal segments, discontinuous analogs, which consist of a short N-terminal segment bound to a longer C-terminal segment via spacer, have been introduced with [Ahx$^{5-24}$] NPY [17]. This molecule contains 6-amino hexanoic acid to replace amino acids 5 to 24, thus connecting the N-terminal tetrapeptide and the C-terminal dodecapeptide (Figure 6.1.3). The molecules showed high affinity at the $Y_2$-receptor and full agonistic properties, but no activity at $Y_1$- and

**Figure 6.1.3.** Centrally truncated analogs of neuropeptide Y with high affinity to Y-receptors. 6-Amino-hexanoic acid (Ahx) has been used as a spacer to connect N- and C-terminal segments.

$Y_5$-receptors. It is much less 'sticky' than NPY itself, and accordingly is a very useful tool to study the $Y_2$-receptor-mediated activities, especially in vivo. Various scans, however showed, that no distinct amino acid of the N-terminal segment was required to obtain high biological activity; moreover, it serves to stabilize the bioactive conformation of the most important C-terminal pentapeptide [6].

By systematically increasing the length of the N- and the C-terminal segments, linear, discontinuous peptides were found that bind to the $Y_1$-receptor as well: $[Ahx^{8-20}]$ NPY and $[Ahx^{8-21}]$ NPY. It was then obvious that with a given N-terminal length, only one or two C-terminal peptides were optimal: $[Ahx^{7-22}]$ NPY, $[Ahx^{8-20}]$ NPY and $[Ahx^{9-18}]$ NPY [18]. Accordingly, zipper-like hydrophobic interaction between N- and C-terminus is more important for high $Y_1$-receptor activity than any specific amino acid either because N- and C-terminal segments (backbone groups more likely than side chains of the N-terminus) are important for high affinity, or because the hydrophobic interaction is required to stabilize the bioactive conformation of the C-terminal segment. By including Pro at position 34 to reduce the $Y_2$-receptor affinity, it was found that $[Ahx^{8-20}, Pro^{34}]$ NPY lacks $Y_2$-receptor affinity, but in addition to its high $Y_1$-receptor affinity, moderate $Y_5$-receptor affinity is obtained [19]. Accordingly, none of the so far NPY specific Y-receptor subtypes requires the loop part of NPY (amino acids 9 to 19) for direct receptor contact, and the role of this segment can only be speculated. Either it is required to form and stabilize the turn, that allows the N-terminal segment to interact with the C-terminus, or with its four acidic amino acids it is required to compensate the basic residues of the N- and C-terminal parts of NPY, that are required for receptor binding.

## 6.1.7    Why do we work with neuropeptide Y?

Neuropeptide Y is a most important molecule for all aspects of bioorganic research. It is the smallest 'protein' that is folded in solution, and accordingly is a great model to study folding by using synthetic analogs. Furthermore, the pancreatic polypeptide (PP) family is a unique ligand-receptor system with two ligands and five receptor subtypes to investigate selectivity, mode of binding, and activation of G-protein-coupled receptors. Last, but not least, NPY comprises important functions which make it a most interesting target in drug development, and which may help in the investigation of peptides in therapy.

## References

[1]  D. Larhammar, *Reg. Peptides* **1996**, *62*, 1–11.
[2]  K. Tatemoto, M. Carlquist, V. Mutt, *Nature* **1982**, *296*, 659–660.
[3]  W. F. Colmers, D. Bleakman, *Trends Neurosci.* **1994**, *17*, 373–379.
[4]  *Neuropeptide Y and Drug Development* (L. Grundemar and S. R. Bloom, eds.) **1997** Academic Press Limited, London UK.
[5]  M. C. Michel, A. G Beck-Sickinger., H. Cox, H. N. Doods, H. Herzog, D. Larhammer, R. Quirion, T. W. Schwartz, T. Westfall, *Pharmacol. Rev.* **1998**, 50, 143–150.

[6] A. G. Beck-Sickinger, G. Jung, *Biopolymers* **1995**, *37*, 123–142.

[7] A. G. Beck-Sickinger, H. A. Wieland, H. Wittneben, K. D. Willim, K. Rudolf, G. Jung, *Eur. J. Biochem.* **1994**, *225*, 947–958.

[8] B. Rist, M. Entzeroth, A. G. Beck-Sickinger, *J. Med. Chem.* **1998**, *41*, 117–123.

[9] A. G. Beck-Sickinger, in *Neuropeptide Protocols, Methods in Molecular Biology* (Irvine and C. H. Williams, eds.) **1996**, Humana Press Inc., Totowa NJ, *Vol. 73*, 61–73.

[10] B. Rist, O. Zerbe, N. Ingenhoven, L. Scapozza, C. Peers, P. F. T. Vaughan, R. L. McDonald, H. A. Wieland, A. G. Beck-Sickinger, *FEBS Lett.* **1996**, *394*, 169–173.

[11] B. Rist, N. Ingenhoven, L. Scapozza, G. Schnorrenberg, W. Gaida, H. A. Wieland, A. G. Beck-Sickinger, *Eur. J. Biochem.* **1997**, *247*, 1019–1028.

[12] A. Balasubramaniam, S. Sheriff, M. E. Johnson, M. Prabhakaran, Y. Huang, J. E. Fischer, W. T. Chance, *J. Med. Chem.* **1994**, *37*, 811–815.

[13] J. Fuhlendorff, U. Gether, L. Aakerlund, N. Langeland-Johansen, H. Thøgersen, S. G. Melberg, U. B. Olsen, O. Tharstrup, T. W. Schwartz, *Proc. Natl. Acad. Sci. USA* **1990**, *87*, 182–186.

[14] A. G. Beck-Sickinger, in *Neuropeptide Y and Drug Development* (L. Grundemar and S. R. Bloom, eds.) **1997**, Academic Press Limited, London UK, 107–126.

[16] A. J. Daniels, J. E. Matthews, R. J. Slepetis, M. Jansen, O. H. Viveros, A. Tadepalli, W. Harrington, D. Heyer, A. Landavazo, J. J. Leban, A. Spaltenstein, *Proc. Natl. Acad. Sci. USA* **1995**, *92*, 9067–9071.

[17] A. G. Beck, G. Jung, W. Gaida, H. Köppen, R. Lang, G. Schnorrenberg, *FEBS Lett.* **1989**, *244*, 119–122.

[18] B. Rist, H. A. Wieland, K. D. Willim, A. G. Beck-Sickinger, *J. Peptide Science* **1995**, *1*, 341–348.

[19] N. Ingenhoven, A. G. Beck-Sickinger, *J. Receptor & Signal Transd. Res.* **1997**, *17*, 407–418.

## 6.2    Biosynthetic studies on deoxysugars: Implications for enzyme-mediated synthesis of deoxyoligosaccharides

*Andreas Kirschning, Carsten Oelkers, Monika Ries, Andreas Schönberger, Sven-Eric Wohlert, and Jürgen Rohr*

### 6.2.1    Background

Deoxysugars are frequently found as components of oligosaccharides in glycoconjugates with antibacterial and antitumor activity [1]. They often play an important role in the mechanism of action, in particular for the DNA-interaction of anticancer drugs as well as in binding to RNA [2]. For instance, the cytostatic activity of the different landomycins [3], anti-tumor compounds of the angucycline group [4], directly depends on the length of the phenol-glycosidically linked deoxyoligosaccharide chains [5]. Thus, the member with the longest glycan moiety, landomycin A (**1**), exhibits the greatest anti-tumor activity. A unique feature of the hexasaccharide portion of **1** is its dimeric structure, each monomer consisting of a trisaccharide constructed of one 2,3,6-trideoxy hexose, named L-rhodinose (rho) (**6**), and two dideoxygenated D-olivose (oliv) units (**2**).

For the biosynthetic formation of deoxyhexoses it is generally accepted that deoxygenation occurs on nucleoside diphosphate intermediates [6], which are activated substrates for glycosyl transfer. The activation pathway always starts with a hexose-6-phosphate, which is transformed into the activated NDP-sugar in two enzymatic steps. This activation mechanism occurs prior to all deoxygenation steps, and therefore was assumed to exclude the activation of 6-deoxy-sugars. In our initial hypothesis, 2-deoxygenation takes place prior to 3-deoxygenation. Important intermediates, like activated 4-keto-2,6-dideoxy-D-glucose (**3**), L-aculose (**4**) and L-cinerulose (**5**) are depicted in Scheme 6.2.1.

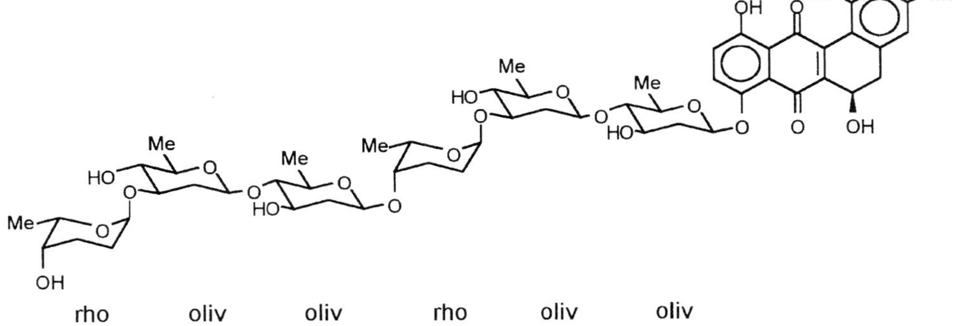

**Figure 6.2.1.** Structure of landomycin A (**1**).

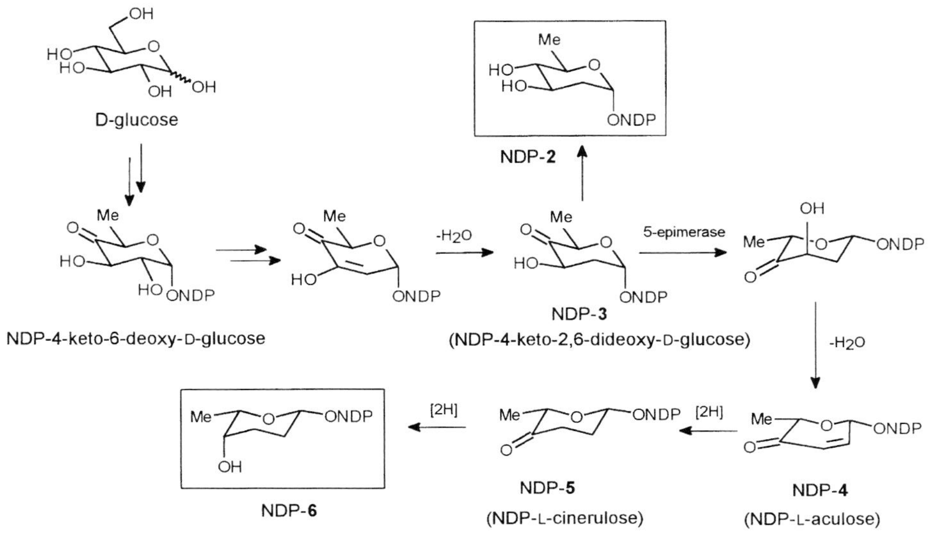

**Scheme 6.2.1.**  Proposed biosynthetic pathway for 2,6-di- and 2,3,6-trideoxyhexoses.

Apart from analyzing gene clusters which code for the enzymes of deoxysugar biosynthesis [7], the feeding of isotopically labeled hypothetical intermediates and NMR-analysis of the products obtained after fermentation is still an essential tool for unraveling a biosynthetic route. However, in the case of studying deoxysugar biosynthesis the availability of potential intermediates like the NDP-2-deoxysugar derivatives **NDP-3–NDP-5** is hampered by their extreme lability. In fact, currently no general chemical or enzymatic methods for the synthesis of 2-deoxygenated NDP-sugars are known [1a, 8]. Due to these properties, there is also a considerably high possibility that during the biosynthetic transformations of deoxysugars the acti-vated intermediates may be hydrolyzed to the parent hexoses. Thus, these sugars would be unable to participate in the final glycosyl transfer steps, if nature did not provide the possibility of a salvage pathway, best achieved by a direct activation. In fact, such a second activation pathway was first recognized for ʟ-fucose [9]. In order to prove that this activation pathway also operates in bacteria, we prepared iso-topically labeled non-activated deoxysugars in sufficient amounts, and these were employed in feeding experiments with *Streptomyces cyanogenus* (strain S-136, DSM 5087), the landomycin producer.

## 6.2.2   Synthesis of deuterated deoxysugars

In the first phase of this project, we devised efficient and flexible syntheses to various specifically labeled deoxysugars [10]. Dideuterated ʟ-rhodinose (2*S*, 3*R*)[2,3-$^2$H$_2$]-(**6**) and ʟ-cinerulose [2,3-$^2$H$_2$]-(**5**) were both prepared from di-*O*-acetyl-ʟ-rhamnal. (**7**) Ferrier rearrangement [11] and deacetylation afforded benzyl glycoside **8** which, after inversion of the configuration at C4, deuteration of the alkenic double bond and debenzylation of the anomeric center, afforded the target sugar [2,3-$^2$H$_2$]-**6** in 47% yield for six steps. Glycoside **8** also serves as starting material for ulose [2,3-$^2$H$_2$]-**5**. Dess-Martin periodinane-promoted oxidation [12] of the allylic hydroxy group, palladium-catalyzed deuteration in d$_1$-methanol and removal of the benzyl protection at C1 terminated the synthesis. The use of deuterated methanol in the reduction step is important for preventing solvent-promoted H/D exchange at C3. For the same reason, it was crucial to use DCl in D$_2$O for the deprotection of the anomeric center.

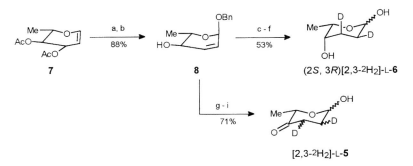

7                    8                    (2S, 3R)[2,3-$^2$H$_2$]-L-**6**

[2,3-$^2$H$_2$]-L-**5**

**Scheme 6.2.2.** Reagents and conditions: (a) BnOH, BF$_3$*OEt$_2$, CH$_2$Cl$_2$, -10°C, 30 min; (b) MeOH, NEt$_3$, H$_2$O (3:1:1), RT, 24 h; (c) PPH$_3$, DEAD, PhCO$_2$H, toluene, 0°C, RT, 30 min; (d) $^2$H$_2$, Pd/C, Et$_2$O, RT, 6 h; (e) MeOH, Na, RT, 3 days; (f) 0.5 M HCl, RT, 1 h; (g) Dess-Martin [12] periodinane oxidation; (h) $^2$H$_2$, Pd/C, MeO$^2$H, RT, 1.5 h; (i) 0.5N $^2$HCl, $^2$H$_2$O, RT, 1 h.

The key step for the preparation of [1-$^2$H]-D-olivose (**2**) involved deprotonation at C1 of 6-deoxy-D-glucal **9a** followed by deuteration of the intermediate carbanion. Acetoxyiodination using a diacetoxyiodine(I) anion [13] present in Et$_4$NI(OAc)$_2$ afforded **10** which was transformed into [1-$^2$H]-D-olivose (**2**) by radical removal of iodine at C-2 and removal of the silyl and acetyl protection under standard conditions.

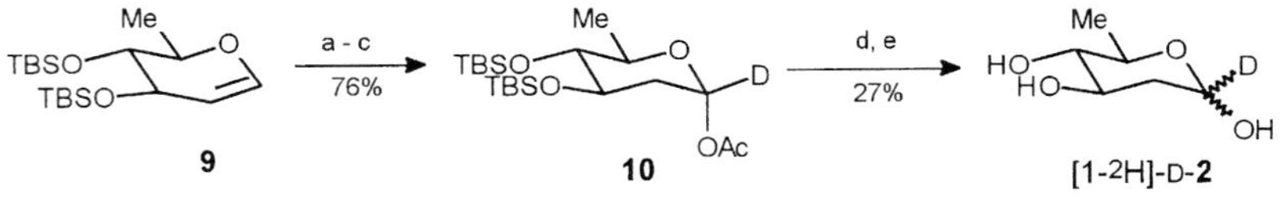

9                    10                    [1-$^2$H]-D-**2**

**Scheme 6.2.3.** Reagents and conditions: (a) $^t$BuLi (8 equiv), THF, -65 → -15°C, D$_2$O; (b) Et$_4$NI, PhI(OAc)$_2$, CH$_2$Cl$_2$, RT; (c) Bu$_3$SnH, AIBN, toluene, Δ; (d) TBAF, THF, RT; (e) Amberlyst A26 (H$^+$ form), MeOH.

As outlined in Scheme 6.2.4, the synthesis of [2,2,3,3-$^2$H$_4$]-D-rhodinose (**13**) relied on the preparation of key intermediate **12** which was prepared from optically pure *threo*-2,3-dihydroxybutanoic ester (**11**). Dibal-*H* promoted reduction to the corresponding aldehyde and Corey-Fuchs olefination set the stage for introducing the carbonyl functionality at C1 which was followed by exhaustive deuteration of the intermediate alkyne to give tetradeuterated ester **12**. Finally, the synthesis of $[2,2,3,3-^2H_4]$-D-rhodi-nose (**13**) was accomplished after acid-promoted cyclization and dibal-*H*-promoted reduction of the lactone carbonyl group.

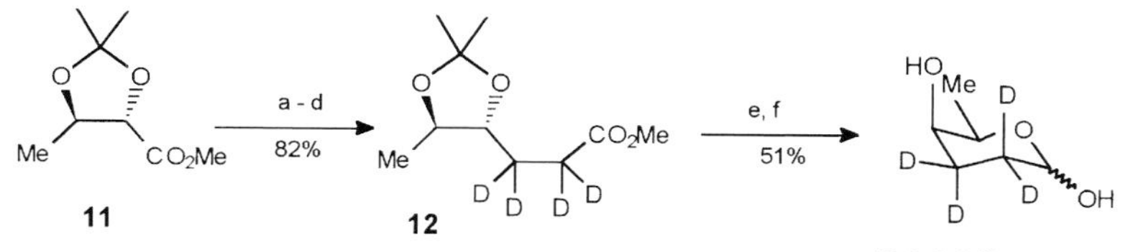

$[2,2,3,3-^2H_4]$-D-**13**

**Scheme 6.2.4.** Reagents and conditions: (a) dibal-*H*, $CH_2Cl_2$, $-78°C$, 1 h; (b) $PPh_3$, Zn, $CBr_4$, $CH_2Cl_2$, RT, 24 h; (c) BuLi (2.2 equiv), $-78°C \rightarrow RT$, 2.5 h, $ClCO_2CH_3$, $-78°C \rightarrow RT$; (d) $^2H_2$, Pd/C, MeOH, RT, 1.5 h; (e) 2N HCl, THF (1:2), RT, 12 h; (f) dibal-*H* (2.5 equiv), $CH_2Cl_2$, $-78°C$, 0.5 h.

## 6.2.3 Biosynthetic studies

The feeding of $[1-^2H]$-D-olivose (**2**) and $[2,3-^2H_2]$-L-cinerulose (**5**), respectively, to a growing culture of the landomycin A producer yielded landomycin A (**1**) in both cases which was analyzed by $^2H$-NMR. However, incorporation of **2** or **5** could not be detected. These results may have various implications. On one hand, biochemical activation of D-olivose (**2**), L-cinerulose (**5**) or a biological successor furnishing the NDP-derivatives does not happen. Alternatively, they were not taken up by the cell or were degraded prior to uptake. In fact, we found that cinerulose is a very labile hexose which is present as the corresponding acyclic aldehyde in a 50% portion [14].

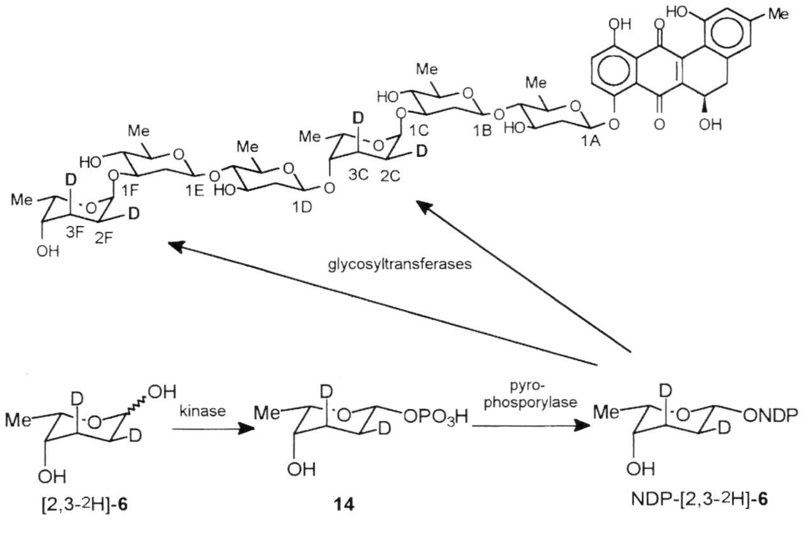

**Scheme 6.2.5.** Feeding experiment with $[2S,3R-^2H_2]$-L-rhodinose (**6**).

In the next step, feeding experiments in which synthetically prepared unlabeled L-rhodinose (**6**) was added to *S. cyanogenus* showed that **6** was consumed within 12 h and therefore seemed to be transported into the cells. The final proof of a successful incorporation of non-activated L-rhodinose into the L-rhodinose moeties of landomycin A **1** was obtained from feeding $(2S,3R)[2,3-^2H_2]$-L-(**6**) (Scheme 6.2.5). The $^2$H-NMR spectrum clearly showed $^2$H-enriched signals exclusively for 2C-H$_e$, 3C-H$_a$, 2F-H$_e$ and 3F-H$_a$ (Figure 6.2.2) [15]. Thus, the $^2$H-labeled L-rhodinose (**6**) must have been activated through an alternative pathway to the standard deoxygenation pathway. A control experiment, in which $[2,2,3,3-^2H_4]$-D-rhodinose (**13**) was fed, showed – as expected – no incorporation. We conclude that **6**, analogous to earlier studies on L-fucose [9] and 6-deoxy-D-glucose [16], is first converted into the corresponding L-rhodinose-1-phosphate **14** by a kinase and then transformed into NDP-**6** through a pyrophosporylase-mediated coupling reaction (Scheme 6.2.5). In fact, Bechthold and co-workers collected additional evidence that a short activation pathway is under operation, by analyzing the deoxysugar biosynthetic gene cluster of the landomycin producer [17].

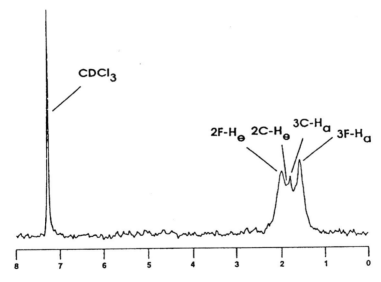

**Figure 6.2.2.** $^2$H-NMR spectrum of the isolated landomycin A (1) after feeding of $[2S,3R-^2H_2]$-L-rhodinose (**6**) to growing cultures of *Streptomyces cyanogenus* S-136. $^2$H signals were assigned based on $^1$H signals [2].

## 6.2.4   Conclusions

One can assume for organisms which produce deoxysugar-containing natural products that the short activation pathway is feasible for all deactivated intermediates of a certain biosynthetic deoxygenation cascade. Therefore, these results may have some fundamental impacts on further biosynthetic studies as well as for enzyme-mediated synthesis of deoxyoligosaccharides. Thus, the presence of a short activation pathway paves the way for more convenient biosynthetic investigations on deoxysugar biosynthesis cascades, since it should be possible to feed and incorporate isotope-labeled deoxysugars, i.e. non-activated analogs of the postulated intermediates of certain deoxygenation pathways. As another consequence of our findings, it will be advantageous to look for genes of such short activation pathways [6], since the corresponding enzymes, preferably the kinase and pyrophosporylase described in Scheme 6.2.5, should express a broad substrate flexibility and thus may be able to activate various deoxyhexoses, presumably at least all intermediates of a given deoxygenation cascade. Therefore, these enzymes may serve as powerful and new tools in the synthesis of deoxyoligosaccharides.

**Acknowledgements:** The authors thank the European Community (BIO4-CT96–0068), the Fonds der Chemischen Industrie, and particularly the Deutsche Forschungsgemeinschaft (grants Ro 676/5–1, 5–2, Ki 397/2–1 to 2–3 and SFB 416) for generous financial support. Dr. David O'Hagan and Dr. Ian McKeag, Department of Chemistry, University of Durham, UK, are gratefully acknowledged for excellent $^2$H-NMR spectra.

# References

[1] (a) A. Kirschning, A. Bechthold, J. Rohr, *Top. Curr. Chem.*, **1997**, *188*, 1–84; (b) S. J. Danishefsky, M. T. Bilodeau, *Angew. Chem.*, **1996**, *108*, 1483–1522, *Angew. Chem. Int. Ed. Engl.*, **1996**, *35*, 1380–1419; (c) F. Kennedy, C. A. White, *Bioactive Carbohydrates in Chemistry, Biochemistry, and Biology*, Ellis Horwood, Chichester, **1983**.

[2] (a) T. Hermann, E. Westhof, *Curr. Opin. Biotechnol.* **1998**, *9*, 66–73; (b) K. Michael, Y. Tor, *Chem. Eu. J.* **1998**, *4*, 2091–2098; (c) N. D. Pearson, C. D. Prescott, *Chem. Biol.* **1997**, *4*, 409–414.

[3] (a) T. Henkel, J. Rohr, J. Beale, L. Schwenen, *J. Antibiot.*, **1990**, *43*, 492–503; (b) S. Weber, C. Zolke, J. Rohr, J. Beale, *J. Org. Chem.*, **1994**, *59*, 4211–4214.

[4] J. Rohr, R. Thiericke, *Nat. Prod. Rep.*, **1992**, *9*, 103–137; (b) K. Krohn, J. Rohr, *Top. Curr. Chem.*, **1997**, *188*, 127–195.

[5] J. Rohr, A. Kirschning, S. Müller, E. Künzel, J. Frevert, unpublished results.

[6] (a) O. Gabriel, *Adv. Chem. Ser.*, **1973**, *117*, 387–410; (b) H.-w. Liu, J. S. Thorson, *Ann., Rev. Microbiol.*, **1994**, *48*, 223–256; (c) J. S. Thorson, H.-w. Liu, C. R. Hutchinson, *J. Am. Chem. Soc.*, **1993**, *115*, 6993–6994.

[7] L. Westrich, S. Domann, B. Faust, D. Bedford, D. Hopwood, A. Kirschning, A. Schönberger, U. Weißbach, J. Rohr, A. Bechthold, unpublished results.

[8] G. Srivastava, O. Hindsgaul, M. M. Palcic, *Carbohydr. Res.*, **1993**, *245*, 137–144.

[9] (a) J. G. Bekesi, R. J. Winzler, *J. Biol. Chem.*, **1967**, *242*, 3873–3879; (b) M. L. Reitman, I. S. Trowbridge, S. Kornfeld, *J. Biol. Chem.*, **1980**, *255*, 9900–9906.

[10] A. Kirschning, U. Hary, M. Ries, *Tetrahedron* **1995**, *51*, 2297–2304.

[11] R. J. Ferrier, D. M. Ciment, *J. Chem. Soc. (C)*, **1966**, 441–449; R. J. Ferrier, N. Prasad, *J. Chem. Soc (C)*, **1969**, 570–575.

[12] D. B. Dess, J. C. Martin, *J. Am. Chem. Soc.*, **1991**, *113*, 7277–7287.

[13]  (a) A. Kirschning, C. Plumeyer, L. Rose, *Chem. Commun.* **1998**, 33–34; (b) A. Kirschning, *Eu. J. Org. Chem.* **1998**, 2267–2274.

[14]  A. Schönberger, Diplomarbeit, Technische Universität Clausthal, **1997**.

[15]  J. Rohr, S.-E. Wohlert, C. Oelkers, A. Kirschning, M. Ries, *J. Chem. Soc., Chem. Commun.*, **1997**, 973–974.

[16]  S. K. Goda, M. Akhtar, *J. Antibiot.*, **1992**, *45*, 984–994.

[17]  The reader is kindly referred to Chapter 6.3 of this book.

# 6.3 Investigations on the biosynthesis of landomycin A

*Sven-Eric Wohlert, Andreas Bechthold, Claus Beninga, Thomas Henkel,*
*Meike Holzenkämpfer, Andreas Kirschning, Carsten Oelkers, Monika Ries,*
*Susanne Weber, Ulrike Weißbach, Lucia Westrich, and Jürgen Rohr*

## 6.3.1 Landomycins – an introduction

Over the past few years, the angucycline antibiotic landomycin A, produced by the actinomycete strain *Streptomyces cyanogenus*, has been intensively investigated. Landomycin A has a unique structure consisting of a decaketide aglycone and a hexadeoxysaccharide chain which is phenolglycosidically linked at C-8-O. Besides the major product landomycin A *S. cyanogenus* produces the related congeners landomycins B to D (Scheme 6.3.1). These differ from landomycin A mainly in the length of the sugar side chain, and in case of landomycin C also in the type of deoxysugars being used. The latter contains one D-amicetose unit, while the deoxysaccharide moieties of all other landomycins consist only of D-olivoses and of l-rhodinoses [1]. Later, a fifth member of this group of compounds, landomycin E (Scheme 6.3.1) was isolated from *Streptomyces globisporus*, and its structure elucidated. This natural product contains the same aglycone as all other landomycins, but only a trisaccharide chain which is exactly half the symmetrical hexadeoxysugar chain of landomycin A [2].

In various biological tests, landomycin A showed interesting anti-tumor activities, in particular against prostate cancer cell lines [3]. Its excellent anti-tumor activity was shown to result from interactions with the DNA, for which the long deoxysugar chain plays a major role [4]. The involvement of deoxysaccharide moieties in the mechanisms of action of natural anti-tumor drugs has been shown previously in several cases [5].

## 6.3.2 Biosynthetic studies on the aglycone moiety

All these findings justified an examination of the biosynthesis of landomycin A, with particular emphasis on its interesting hexadeoxysugar chain. Using mass spectrometry (FAB), various NMR experiments (proton, carbon, APT, H,H-COSY, C,H-COSY, COLOC, HMBC and INADEQUATE) and feeding experiments with $[1-^{13}C]$- and $[1,2-^{13}C_2]$ acetate it was possible to elucidate the structures of landomycins A to D [1b]. The results from the feeding experiments not only helped to assign all carbon atoms of the aglycone unit unambiguously, but also gave insight how the decaketide chain is incorporated into the landomycin aglycone (Scheme 6.3.2). Feeding of $[1-^{13}C, ^{18}O_2]$acetate, and fermentations under an $^{18}O_2$-enriched atmosphere as well as under reduced oxygen partial pressure, indicated the biogenetic origin of the different O-atoms of the aglycone moiety [1b, 6]. The experiments showed that the oxygens at C-7, 11 and 12 derive from atmospheric oxygen, and those at C-1and C-8 from acetate.

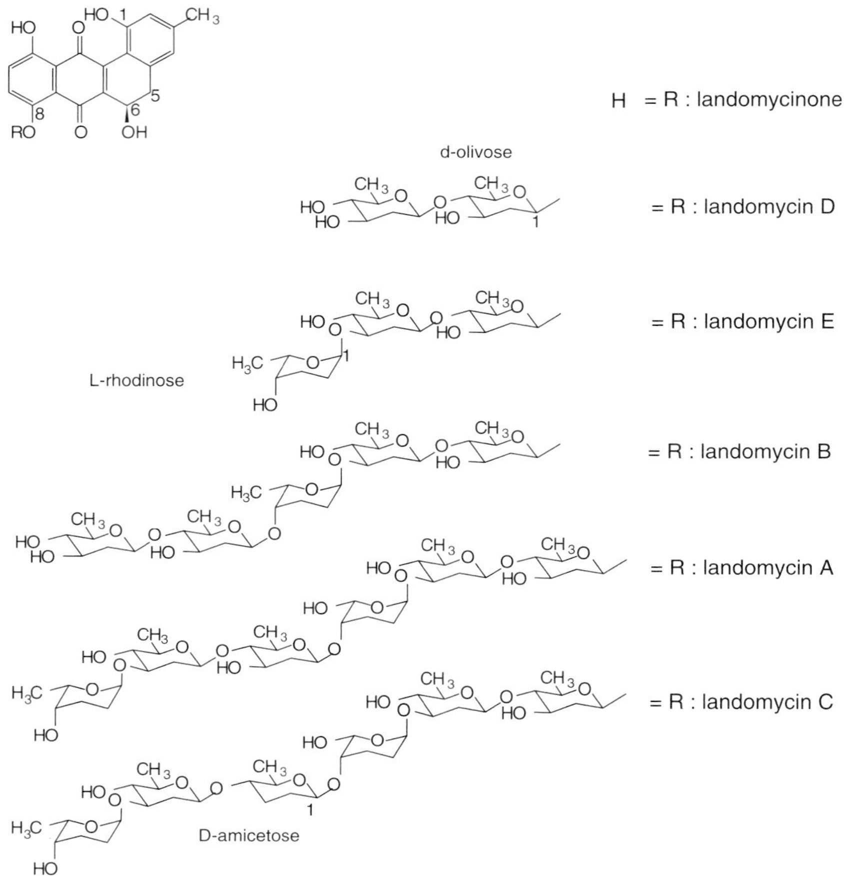

Scheme 6.3.1. Landomycins A to E.

Since the labeled acetate did not show any incorporation in the C-6 oxygen, and the experiment under $^{18}O_2$-enriched atmosphere yielded 5,6-anhydrolandomycin A instead of landomycin A, the biogenetic origin of C-6-O remained ambiguous. Either the fermentation under $^{18}O_2$-atmosphere, which was conducted in a closed system and may led to lack of oxygen, resulted in the production of the immediate precursor 5,6-anhydrolandomycin A, or the oxygen at 6-position derives from water. However, for the latter hypothesis no convincing chemical mechanism was available. Conducting a fermentation under willingly reduced oxygen supply conditions again resulted in the production of 5,6-anhydrolandomycin A instead of landomycin A. Thus, it was concluded that the C-6 oxygen is normally inserted at a late stage of the biosynthesis by an oxygenase which is not sufficiently supplied with oxygen under reduced oxygen partial pressure, and is therefore not able to finish the biosynthesis of landomycin A correctly.

### 6.3.3 Biosynthetic studies on the deoxyoligosaccharide chain

All following experiments were aimed at the biosynthesis of the deoxysugar chain. Initial feeding experiments with [1-$^{13}$C]-D-glucose confirmed that D-glucose was a precursor of all four D-olivose and of both L-rhodinose moieties in landomycin A (Scheme 6.3.2) [7].

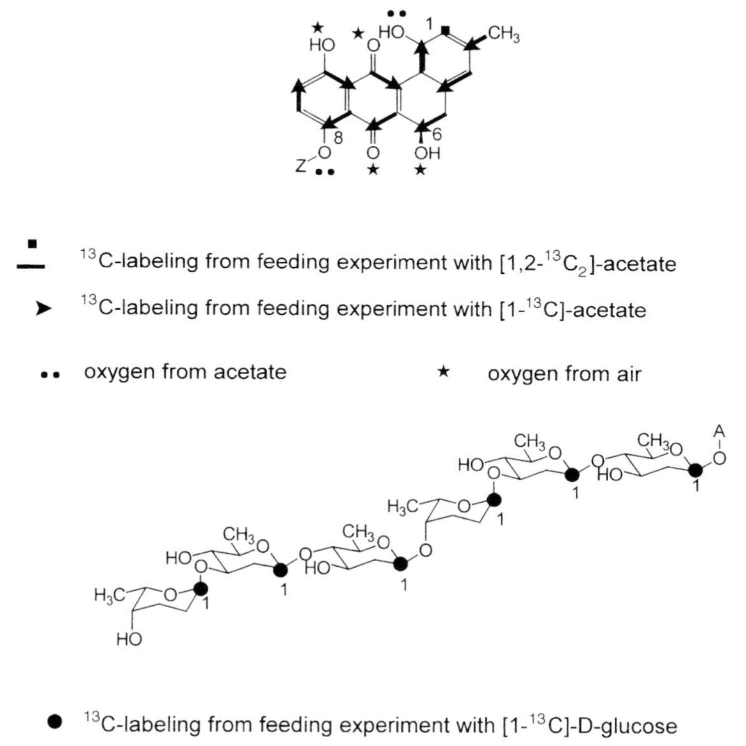

- $^{13}$C-labeling from feeding experiment with [1,2-$^{13}$C$_2$]-acetate

- $^{13}$C-labeling from feeding experiment with [1-$^{13}$C]-acetate

•• oxygen from acetate       ★   oxygen from air

●   $^{13}$C-labeling from feeding experiment with [1-$^{13}$C]-D-glucose

**Scheme 6.3.2.** Results of the $^{13}$C- and $^{18}$O-labeling experiments.

Since the structures of landomycins A, B, and D differ only in the length of their deoxysugar chains, a strong relation of their biosynthesis seemed to be obvious. Either landomycin A should be the precursor of landomycins B and D, the latter two resulting from the action of glycosylases, or, more likely, there is a linear construction of the landomycin A hexasaccharide chain. For this, landomycins D and B should be intermediates of the landomycin A biosynthesis. Landomycin C should also fit into such a biosynthetic scheme; however, being the only landomycin with an unusual D-amicetose moiety, it must branch off the main pathway. To differentiate between both above-mentioned alternatives, cross-feeding experiments using $^{14}$C-labeled landomycins were carried out [6, 8]. All $^{14}$C-labeled landomycins were obtained

through their biosyntheses by adding [1-[14]C]acetate to the growing culture of
*S. cyanogenus*. Each of the landomycins was isolated (the landomycinone through
hydrolysis from landomycin A) and then, in individual experiments, fed to growing
cultures of *S. cyanogenus*. Landomycins A, B, and D were again isolated from each
of these cultures, and their individual radioactivity was determined by scintillation
counting. The results showed that landomycinone was incorporated into all of the
landomycins, landomycin D into landomycin B and A, landomycin B only into lando-
mycin A, while landomycin A labeled none of the other landomycins. After the
discovery of landomycin E, this metabolite was produced in [14]C-labeled form and
fed to a growing culture of *S. cyanogenus*. It transpired [9] that landomycins B and
A, but not D, were labeled in this experiment. Thus, landomycin E fits well into the
biosynthetic scheme. The results of the cross-feeding experiments are outlined in
Scheme 6.3.3.

**Scheme 6.3.3.** Biosynthetic conversions of the landomycins.

Identification of the production curves of the different landomycins [8], the fact that
landomycin E cannot be detected in *S. cyanogenus* [9], and finding that only landomycins with two or four olivoses are produced, but none with three or just one olivose,
all led to the conclusion that the rhodinosyltransfer steps happen much more slowly in
the biosynthesis of the hexadeoxysugar chain than the olivosyltransfer steps. There is
also a possibility that the two olivoses are always transferred as a disaccharide unit.

## 6.3.4    Biosynthetic studies on the single deoxysugar building blocks

Following several feeding experiments with [1-[13]C]-D-glucose (the [13]C-labeled glucose
was fed in different growing stages), it was shown that D-olivose does not act as a
precursor of l-rhodinose, since the incorporation levels of the rhodinoses were in all
cases higher than those of the olivoses. This led to the conclusion that the pathways
of the two deoxysugars, although still strongly related, branched at one point [7]. It
is common knowledge that all deoxygenation steps in the biosynthesis of deoxysugars
are carried out on intermediates that are activated as nucleoside diphosphates. This
activation is needed for the glycosyl transfer step, but occurs at a very early biosynthetic stage, namely on glucose-1-phosphate which is converted into NDP
(=nucleosyldiphosphoryl)-D-glucose through the action of NDP glucose synthase,
also called NDP-glucose pyrophosphorylase. This activation includes a chemically
labile diphosphate bridge, and the loss of this activation would make the intermediate
useless for the organisms, at least for the immediate further use in the pathway (see

also Chapter 6.2). Thus, the investigation of deoxysugar biosynthetic pathways is extremely difficult, since proposed intermediates can only be fed already activated. However, the activation of deoxysugars as well as their pathway intermediates is still a very difficult issue, either by organic synthesis or by biological transformation. Nonetheless, there was some evidence in the literature for a 'short activation pathway' in some micro-organisms which enables them to reactivate intermediates that have lost their activation throughout the biosynthesis [10]. Such a 'short activation pathway' was shown to exist, for instance, in rats and mice as well as in one other *Streptomyces* species [11]. To prove that such a 'short activation pathway' also exists in *S. cyanogenus* feeding experiments with synthetic, non-activated, tritiated-([2$S$,3$R$-$^3$H$_2$]-L-rhodinose) and deuterium-labeled ([2$S$,3$R$-$^2$H$_2$]-L-rhodinose, [2,2,3,3-$^2$H$_4$]-D-rhodinose) deoxysugars were carried out [10, 12]. As expected in the case that a 'short activation pathway' exists, the correct isomer (in particular the [2$S$,3$R$-$^2$H$_2$]-L-rhodinose) was incorporated into landomycin A, while the wrong one (the labeled D-rhodinose) was not. This proved that *S. cyanogenus* indeed has a 'short activation pathway' for L-rhodinose, and this may be available for various other deoxysugars as well which could accidently arise from deactivation of intermediates of the deoxysugar biosynthetic pathway. There is some evidence from the analysis of the deoxysugar biosynthetic gene cluster of *Streptomyces cyanogenus*, in which a seemingly substrate-unspecific NDP-hexosepyrophosphorylase encoding gene was discovered (see also Chapter 6.4) [13]. Thus, the corresponding enzyme should also act on other deoxysugars. This is currently under investigation. One preliminary experiment, the feeding of [2,3-$^2$H$_2$]-L-cinerulose, failed to label the L-rhodinose moieties of landomycin A [13c]. The current interpretation of this result is that L-cinerulose is probably not a free intermediate of the deoxygenation cascade leading to L-rhodinose.

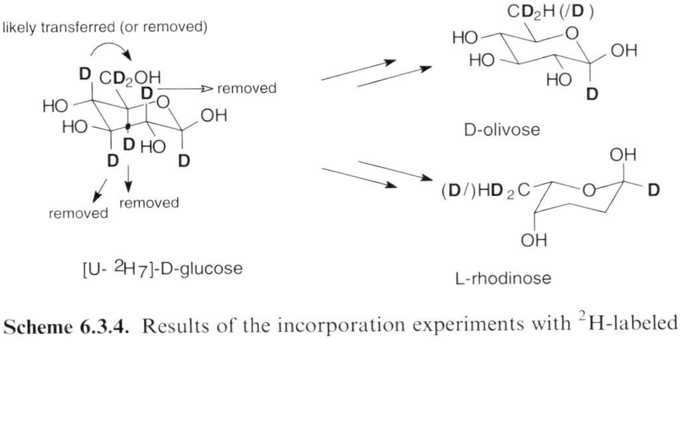

**Scheme 6.3.4.** Results of the incorporation experiments with $^2$H-labeled ethanol and D-glucose.

The biosynthesis of the deoxysugars from D-glucose requires several dehydration (deoxygenation), hydrogenation, and eventually dehydrogenation steps. While the removal of hydrogens requires either a dehydration step or a transfer of a hydrogen to $NAD(P)^+$, the insertion of a hydrogen is accordingly performed through the use of hydrogens coming from NAD(P)H [14]. To investigate which of the hydrogens of the precursor D-glucose is removed during the biosynthesis of the deoxysugar moieties of landomycin A, a feeding experiment using $[U-^2H_7]$-D-glucose was carried out. To 'see', where hydrogens are inserted, an experiment with $[1,1-^2H_2]$-ethanol was conducted [15], since these ethanol hydrogens are usually metabolized by ubiquitious

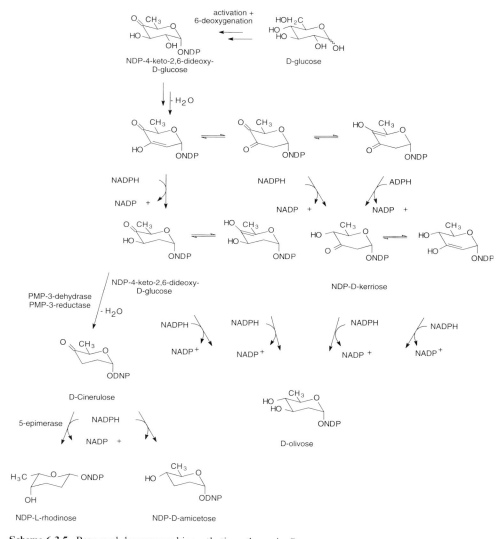

**Scheme 6.3.5.** Proposed deoxysugar biosynthetic pathway in *S. cyanogenus.*

alcohol dehydrogenases, and end up in NAD(P)H [14a,b]. The glucose-feeding experiment gave the expected relative high incorporation rate, and thus the modifications in both the D-olivoses and L-rhodinoses could be directly determined in landomycin A through $^2$H-NMR. As expected, the incorporation rate from the experiment with [1,1-$^2$H$_2$]-ethanol was much lower. Therefore, it was necessary to methanolyze the resulting landomycin A in order to aquire the $^2$H-NMR data on the methyl-D-olivoside. Unfortunately, it was not possible to analyze rhodinose this way, since both the L-rhodinose as well as methyl-L-rhodinoside are quite unstable. The result of the two feeding experiments are shown in Scheme 6.3.4.

For other micro-organisms it was shown for the 6-deoxygenation step which is catalyzed by an oxidoreductase, that 4-H is transferred to the 6-position, thus replacing the hydroxy group at the 6-position intramolecularly. Therefore, it is first transferred to NADP and then to the 6-position [5a, 16]. The feeding experiment with [1,1-$^2$H$_2$]-ethanol however showed that this is not (or only partially) the case in *S. cyanogenus*. The detected deuterium incorporation in 6-position (Scheme 6.3.4) of the D-olivose provided evidence that an exchange of the NAD(P)H formed during the oxidation of the 4-position and of other NAD(P)H must take place during the 6-deoxygenation step, *i.e.* the transfer from 4-H to the 6-position does not occur strictly intramolecularly, at least not in *S. cyanogenus*. Within the landomycin gene cluster several genes were found showing homologies to those already described [13, 17]. The described functions of those genes and results of biosynthetic studies for different natural products containing L-rhodinose as deoxysugar [17] support the hypothesis that there is a branched pathway leading to L-rhodinose and D-amicetose, respectively, possibly both from D-cinerulose. The first requires a 5-epimerization/reduction, the latter only a reduction. All the data so far aquired for the biosynthesis of the D-olivose and L-rhodinose moieties of landomycin A led to the hypothetical biosynthetic pathway shown in Scheme 6.3.5 [15].

**Acknowledgements:**   The authors would like to thank the Deutsche Forschungsgemeinschaft, the Fonds der Chemischen Industrie, the Deutsche Akademische Austauschdienst, the European Community and the Medical University of South Carolina for generous financial support. We also appreciate donations of chemicals from Hoechst AG (Frankfurt, Germany) and Merck (Rahway, NJ, USA).

# References

[1]  (a) T. Henkel, J. Rohr, J. M. Beale, L. Schwenen, *J. Antibiot.* **1990**, *43*, 492–503; (b) S. Weber, C. Zolke, J. Rohr, J. M. Beale, *J. Org. Chem.* **1994**, *59*, 4211–4214.

[2]  B. Matselyukh, I. Polishchuk, U. Weißbach, S.-E. Wohlert, J. Rohr, in preparation.

[3]  (a) National Cancer Institut, personal communications **1993–1998**; **(b)** M. R. Boyd, *Principles & Practices of Oncology* **1989**, *3*, 1.

[4]  (a) M. Zerlin, Hans-Knöll-Institut, Jena, personal communication, **1996**; (b) H. Decker, personal communication, **1996**; (c) H. Depenbrock, S. Bornschlegl, R. Peter, J. Rohr, P. Schmit, P. Schweighart, T. Block, J. Rastetter, A.-R. Hanauske, *Annals of Hematology* **1996**, *73*, A80.

[5]  (a) A. Kirschning, A. F.-W. Bechthold, J. Rohr, *Topics Curr. Chem.* **1997**, *188*, 1–83; (b) M. Hansen, L. H. Hurley, *J. Am. Chem Soc.* **1995**, *117*, 2421–2429; (c) D. Sun, M. Hansen, L. H. Hurley, *J. Am. Chem Soc.* **1995**, *117*, 2430–2440; (d) M. Hansen, S.-J. Lee, J. M. Cassady, L. H. Hurley, *J. Am. Chem. Soc.* **1996**, *118*, 5553–5561; (e) S. L. Walker, K. G. Valentine, D. Kahne, *J. Am. Chem. Soc.* **1990**, *112*, 6428–6429 ; (f) S. L. Walker, D. Kahne, *J. Am. Chem. Soc.* **1991**, *113*, 4716–4717 ; (g) S. L. Walker, J. Murnick, D. Kahne, *J. Am. Chem. Soc.* **1993**, *115*, 7954–7961; (h) K. C. Nicolaou, S. C. Tsay, T. Suzuki, G. F. Joyce, *J. Am. Chem. Soc.* **1992**, *114*, 7555–7557; (i) L. Gomez-Paloma, J. A. Smith, W. J. Chazin, K. C. Nicolaou, *J. Am. Chem. Soc.* **1994**, *116*, 3697–3708; (j) R. L. Halcomb, S. H. Boyer, S. J. Danishefsky, *Angew. Chem.* **1992**, *104*, 314–317; (k) J. Drak, N. Iwasawa, S. J. Danishefsky, D. Crothers, *Proc. Natl. Acad. Sci.* **1991**, *88*, 7464–7468.

[6]  U. Weißbach, *Diplomarbeit* **1996**, University of Göttingen.

[7]  S.-E. Wohlert, *Diplomarbeit* **1994**, University of Göttingen.

[8]  C. Beninga, *Diplomarbeit* **1994**, University of Göttingen.

[9]  M. Holzenkämpfer, *Diplomarbeit* **1998**, University of Göttingen.

[10]  J. Rohr, S.-E. Wohlert, C. Oelkers, A. Kirschning, M. Ries, *Chem. Commun.* **1997**, 973–974.

[11]  (a) J. G. Bekesi, R. J. Winzler, *J. Biol. Chem.* **1967**, *242*, 3873–3879; (b) M. L. Reitman, I. S. Townbridge, S. Kornfeld, *J. Biol. Chem.* **1980**, *255*, 9900–9906; (c) S.K. Goda, M. Akhtar, *J. Antibiot.* **1992**, *45*, 984–994.

[12]  (a) A. Kirschning, U. Hary, M. Ries, *Tetrahedron* **1995**, *51*, 2297–2304; (b) C. Oelkers, *Diplomarbeit* **1997**, University of Göttingen.

[13]  (a) A. Bechthold, personal communications **1997**; (b) L. Westrich, B. Faust, S. Domann, H. Decker, A. Bechthold, *Enzymology of Biosynthesis of Natural Products*, Poster No. 80, Berlin, Germany **1996**; (c) L. Westrich, S. Domann, B. Faust, D. Bedford, D.A. Hopwood, A. Bechthold, *FEMS Microbiology Letters*, **1999**, *170*, 381–387.

[14]  (a) H. G. Schlegel, *'Allgemeine Mikrobiologie'*, 7. Ed., Georg Thieme Verlag, Stuttgart, New York, **1992**; (b) L. Stryer, 'Biochemistry', 3rd Ed., W. H. Freeman & Co., New York, **1988**; (c) S. J. Gould, J. Guo, *J. Am. Chem. Soc.* **1992**, *114*, 10176–10181; (d) M. J. Schneider, F. S. Ungemach, H. P. Broquist, T. M. Harris, *J. Am. Chem. Soc.* **1982**, *104*, 6863–6864.

[15]  S.-E. Wohlert, *Dissertation* **1998**, University of Göttingen.

[16]  (a) O. Gabriel, *Carbohydr. Res.* **1968**, *6*, 111–117; (b) H. Zarkowsky, L. Glaser, *J. Biol. Chem.* **1992**, *244*, 4750–4756; (c) C. E. Snipes, G.-U. Brillinger, L. Sellers, L. Mascaro, H. G. Floss, *J. Biol Chem.* **1977**, *252*, 8113–8117.

[17]  (a) K. Ylihonko, J. Hakala, J. Niemi, J. Lundell, P. Mäntsälä, *Microbiology* **1994**, *140*, 1359–1365; (b) P. A. Pieper, Z. Guo, H.-W. Liu, *J. Am. Chem. Soc.* **1995**, *117*, 5158–5159; (c) C. R. Hutchinson, personal communication, **1997**.

# 6.4     Oligosacharide antibiotics: Perspective for combinatorial biosynthesis

*Andreas Bechthold, Jürgen Rohr*

Since many natural bioactive compounds are produced by actinomycetes, most of them by streptomyces species, scientists have focused their research on cloning biosynthetic genes of these organisms. Soon after the first biosynthetic genes had been isolated it became clear that in *Streptomyces* all biosynthetic genes necessary to synthesize a particular antibiotic are arranged together in one part (cluster) of the genome. Having compared sequences of genes of different clusters it became obvious that genes with similar function are highly conserved among *Streptomyces* and that biosynthetic genes could be used as hybridization probes to detect gene clusters in other strains [1]. More recently, researchers began to use biosynthetic genes to alter the structure of a compound or to combine genes from different organisms in oder to create new molecules. In particular Prof. Hopwood (Norwich, UK), Prof. Hutchinson (Madison, USA), Prof. Khosla (Stanford, USA), and all their co-workers succeeded in making large amounts of novel aromatic compounds after mixing polyketide biosynthetic genes. Thus a new technology named combinatorial biosynthesis was created [1–3]. Many antibiotics contain saccharides as important structural elements. Although in many cases the contribution of carbohydrate moieties on the mechanism of action of bioactive compounds has not been elucidated, sugars seem to be important recognition elements for the mechanism of action of the respective drug [4–5]. In 1994, we initiated an investigation on the genetic analysis of avilamycin, landomycin, and urdamycin biosynthesis, each of these antibiotics consisting of a polyketide moiety and a saccharide side chain. Our research was motivated by the belief that genes coding for proteins involved in the biosynthesis of these antibiotics can be used to create new saccharide antibiotics in the future.

## 6.4.1     Avilamycin A

Avilamycin A (**1**) (Scheme 6.4.1), produced by *Streptomyces viridochromogenes* Tü57 (*S. viridochromogenes* Tü57), is an oligosaccharide antibiotic and belongs to the orthosomycin group of antibiotics [6]. Avilamycin A, as well as other important members of the orthosomycins, contains a dichloroisoeverninic acid moiety and one or more orthoester linkages associated with carbohydrate residues [7]. The compound SCH27899 shows excellent activity against Gram-positive bacteria and is presently being tested for its possible use against human infectious diseases [8–9]. Avilamycin A inhibits the growth of Gram-positive bacteria, but the exact mode of action of the avilamycins is not known [10].

## 6.4.2     Landomycin and urdamycins

*Streptomyces cyanogenus* S136 is producing a complex of anti-cancer drugs (landomycins) comprising polyketide- and oligosaccharide moieties. The landomycins

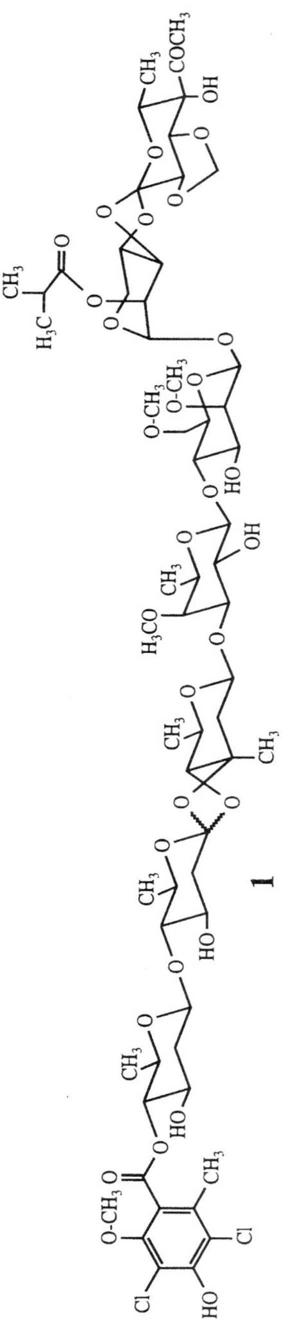

**1**

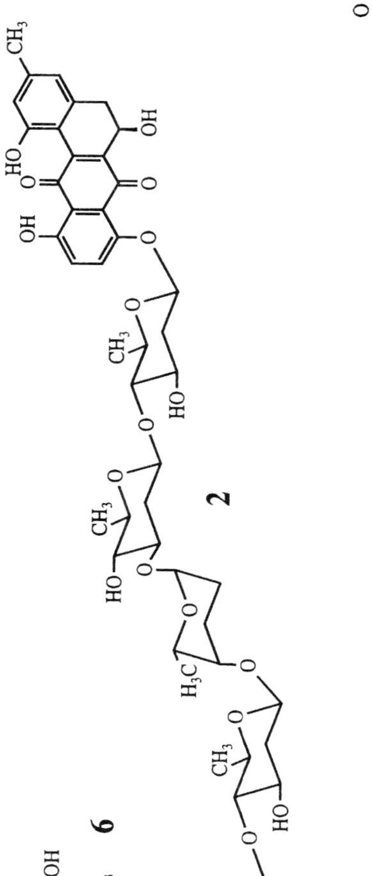

**2**

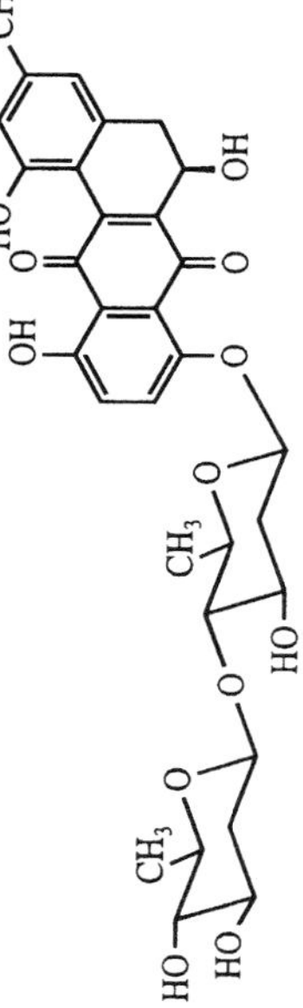

**3**

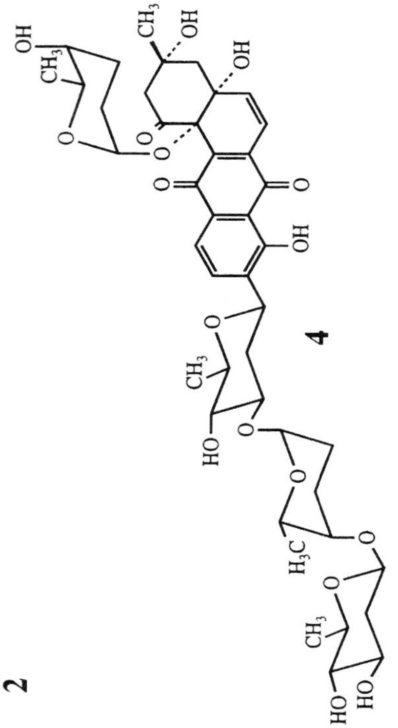

**4**

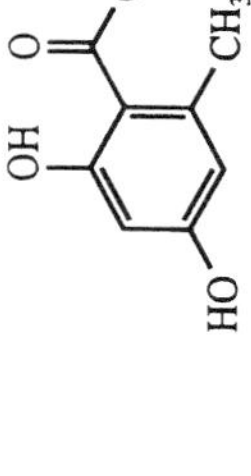

**6**

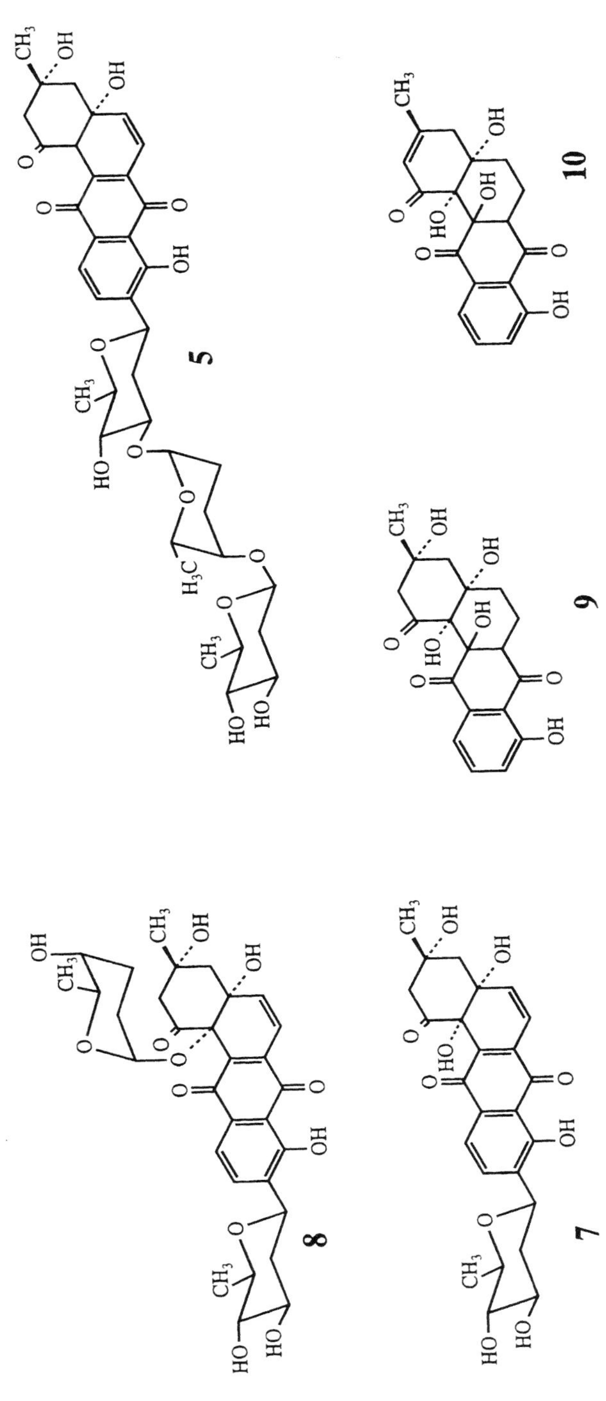

**Scheme 6.4.1.**  Chemical structures of compounds referred to in this paper.

belong to the angucycline class of antibiotics, the members differing in the length of a deoxysaccharide side chain which is phenol-glycosidically linked to a naphthazarine chromophore-containing aglycone. Landomycin A (**2**) (Scheme 6.4.1) is the largest member of this angucycline family and contains an unusual hexasaccharide chain consisting of the deoxysaccharides, D-olivose and L-rhodinose [11]. It has been shown that landomycin A inhibits the growth of tumor cells more strongly than does landomycin D (**3**), urdamycin A (**4**), urdamycin B (**5**), or several emycins (not glycosidated angucyclins) [12]. Urdamycins, produced by *S. fradiae* Tü2717, also belong to the angucyclin antibiotics and contain a saccharide side chain which is C–C-connected to the aglycone. In contrast to the landomycins, members of this group differ most in the structure of the polyketide moiety, but not in the length of the saccharide side chain [11, 13].

## 6.4.3    Results of investigations

### 6.4.3.1    Preparation and screening of cosmid libraries

The biosynthesis of deoxysugars of our antibiotics was thought to be made from a NDP-hexose, most likely from TDP-glucose. In most organisms that produce antibiotics containing 6-deoxyhexose moieties TDP-glucose is converted to TDP-4-keto-6-deoxyglucose by a TDP-glucose 4,6-dehydratase. Dehydratase genes are highly conserved in actinomycetes. We developed a PCR method which could be used for the amplification of DNA fragments from 4,6-dehydratase genes using genomic DNA of *S. fradiae* Tü2717, S. *cyanogenus* S136 and *S. viridochromogenes* Tü57 as template [14]. Gene cloning in cosmid vectors allows the isolation of DNA-fragments with a size of more than 40 kb in a single recombinant clone. Therefore, cosmid cloning is a very efficient method for the isolation of large biosynthetic gene clusters. Cosmid libraries from DNA of our *Streptomyces* strains were prepared. Colonies were probed by colony hybridization with the PCR fragments; in each case more than three colonies hybridizing to the probe were isolated [15–17].

### 6.4.3.2    Gene function

All three biosynthetic gene clusters mentioned above were characterized by sub-cloning, sequencing, expression, and gene inactivation experiments (Figure 6.4.1). More than 80 kb DNA have been sequenced. This chapter summarizes the most important insights into the possible function of genes found in the gene clusters of our antibiotics.

#### 6.4.3.2.1    Avilamycin

To test whether we had in fact cloned genes responsible for avilamycin biosynthesis, insertional inactivation experiments were carried out. For these experiments we

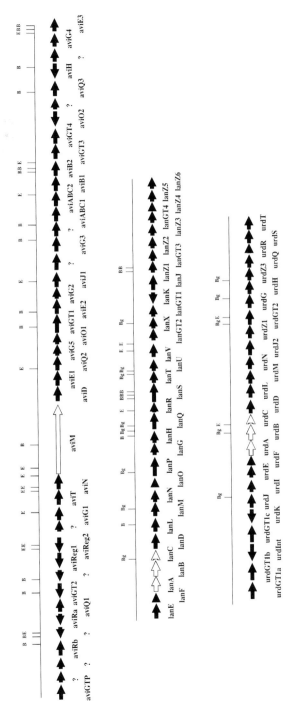

**Figure 6.4.1.** Organization of gene clusters for avilamycin A (I), landomycin A (II) and urdamycin A (III). Wedges, indicating open reading frames (genes) are orientated in the direction of gene transcription and are proportional to the size of the gene products. Genes coding for polyketid synthases are shown as white wedges. Restriction sites are indicated above the solid line (B, *Bam*HI; E, *Eco*RI). No *Bgl*II restriction sites are shown in the avilamycin cluster, no *Bam*HI sites in the urdamycin cluster.

inactivated the gene *aviE* (dehydratase gene) which we expected to be involved in early steps of the biosynthesis and whose inactivation would lead to a non-producing mutant. After transformation of the inactivation plasmid pDesery-, containing the erythromycin resistance gene, into *S. viridochromogenes* Tü57, erythromycin-resistant colonies were obtained. As shown by Southern hybridization, pDesery- had integrated into the chromosome at two different sites, both located around the dehydratase gene. None of these mutants was producing any avilamycin, indicating that we in fact had isolated the avilamycin biosynthetic gene cluster [17]. It was expected that the biosynthesis of the polyketid moiety of avilamycin is catalyzed by a set of enzymes, a ketosynthase (KS), an acyltransferase (AT), an acyl carrier protein (ACP), and a cyclase. Surprisingly, the deduced amino acid sequence of *aviM* showed homology to multifunctional proteins from fungi. Highest homology was found to a methylsalicylic acid synthase (MSAS) from *Penicillium patulum*. Most significantly, AviM contained three active sites (KS, AT, ACP) resembling the corresponding sites in MSAS. Expression of *aviM* in a heterologous *Streptomyces* strain (*S. lividans*) was leading to the accumulation of an aromatic compound which was identified as orsellinic acid (**6**). To our knowledge, AviM is the first multifunctional polyketide synthase isolated from bacteria which is able to produce an aromatic compound [17]. In many antibiotic biosynthetic gene clusters, resistance genes are linked to biosynthetic genes. DNA-fragments, randomly choosen, were expressed *in S. lividans*, which is an avilamycin-sensitive strain. DNA-fragments of resistant colonies were isolated. By this procedure, three different resistant genes were identified in the avilamycin biosynthetic gene cluster. The deduced amino acid sequence of *aviABC1* shows homology to ABC-transporters, which together with a transmembrane protein (AviABC2), participate in the export of drugs through ATP-dependent processes [18]. The function of the second gene (*aviRa*) is not known. A search of available protein databases with the deduced amino acid sequences failed to reveal significant similarities. AviRb is very similar to rRNA-methyl transferases, indicating that AviRb blocks avilamycin binding to rRNA by methylation. An additional transporter was detected in the cluster (*aviT*) by sequencing further DNA fragments. AviT shows homology to membrane proteins that mediate antibiotic efflux in a probably proton-dependent process [20]. For the biosynthesis of avilamycin, several enzymes involved in the formation of the sugar moieties, at least six glycosyltransferases, five methyltransferases and additional modifying enzymes are required. So far, eight deoxysugar biosynthetic genes (*aviQ1*, *aviQ2*, *aviQ3*, *aviD*, *aviE1*, *aviE2*, *aviE3*, and *aviJ*), four glycosyltransferase genes (*aviGT1*, *aviGT2*, *aviGT3*, and *aviGT2*), four putative sugar methyltransferase genes (*aviG1*, *aviG2*, *aviG3*, and *aviG5*) and two putative oxygenases genes (*aviO1* and *aviO2*) were detected. The function of AviQ was determined after expressing *aviQ1* in *E. coli*. The expressed enzyme showed UDP-glucose 4-epimerase activity, converting UDP-glucose to UDP-galactose. The function of further genes could be assigned just by comparing the deduced protein sequences with sequences of known proteins in databases. AviD and AviE1 are probably involved in the biosynthesis of the important key intermediate NDP-4-keto-6-deoxy-glucose, which can be converted to the 2,6-dideoxysugar components of avilamycin A, namely ring B (D-olivose), C (D-olivose), and D (D-evalose). AviQ2 and AviQ3 are similar to epimerases, AviE2 and AviE3 to NDP-glucose 4,6-dehydratases, and AviJ to oxidoreductases.

### 6.4.3.2.2 Landomycin and urdamycin

Formation of the polyketide-derived portion of both angucycline antibiotics seems to be governed by genes coding for iterative (type II) polyketide synthases (PKS). The *lanA*, *lanB*, and *lanC* genes are homologous to the *urdA*, *urdB*, and *urdC* genes and are the components of the PKS that create an uncyclized intermediate. This intermediate is reduced by LanD and UrdD. Two genes coding for cyclases (LanF, LanL and UrdF, UrdL) are found in both clusters. These proteins might catalyze the formation of cyclized intermediates which are modified by decarboxylation (LanP), oxygenation (LanE, LanM and UrdE, UrdM) and reduction (LanN, LanV, LanO, and LanZ4). Both polyketides undergo glycosylation with TDP-activated forms of D-olivose and L-rhodinose. These sugars are thought to be made from glucose-1-phosphate by several steps. After formation of TDP-glucose (LanG and UrdG), this activated sugar is converted to TDP-4-keto-6-deoxy-glucose by LanH and UrdH. Deoxygenation at position C-2 is probably catalyzed by LanS and UrdS respectively, leading to a hypothetical TDP-3,4-diketo-2,6-dideoxy-glucose intermediate which can be reduced at position 3 by LanT and UrdT. TDP-4-keto-2,6-dideoxy-glucose might be a central intermediate in the biosynthesis of both L-rhodinose and D-olivose. A reduction step at position C-4 is required to complete the biosynthesis of TDP-olivose. 3,5 epimerization (LanZ1, UrdZ1), dehydroxylation at position C-3 (LanQ, UrdQ), and reduction at position C-4 (lanZ3) would lead to the formation of TDP-rhodinose. Incorporation of labeled L-rhodinose into landomycin A has been described [21]. This experiment provides evidence for an alternative pathway leading to TDP-L-rhodinose (short activation pathway). In the landomycin biosynthetic gene cluster a second NDP-hexose synthetase was detected (lanZ2), which might be an important component of the second pathway to TDP-L-rhodinose. The design of the saccharide chains of landomycins and urdamycins is specified by glycosyltransferases. So far, four genes coding for glycosyltransferases have been found in both clusters (*lanGT1–4*, *urdGT1a-1c*, *urdGT2*). After expression of lanGT2 in *S. fradiae* Tü2717, accumulation of aquayamycin (**7**) and 100–2 (**8**) [11] was observed, both compounds were not detectable in the wild-type strain. It can be assumed that LanGT2 is binding either TDP-olivose or TDP-rhodinose, which is reducing the concentration of the activated sugar in the cell. The activity of glycosyltransferases of the urdamycin producer is affected, and this leads to the accumulation of intermediates of the biosynthesis. The functions of *urdGT1a*, *urdGT1c*, and *urdGT2* have been elucidated by gene inactivation experiments. *UrdGT2* mutants, containing an in-frame deletion, were accumulating urdamycin I (**9**) and urdamycin J (**10**), two non-glycosylated urdamycin derivatives. Therefore, *urdGT2* might code for a glycosyltransferase catalyzing the formation of a C–C bond. *UrdGT1a mutants* were producing aquayamycin, indicating that *urdGT1a* is coding for a 12β-rhodinosyltransferase. Mutants in which *urdGT1b*, *urdGT1c*, and urd*Int* had been deleted, were accumulating compound

100-2. UrdGT1b and UrdGT1c are probably involved in attaching D-olivose and L-rhodinose to compound 100/2.


### 6.4.3.3  Perspectives

More than 80 kb DNA have been cloned and sequenced in the biosynthetic gene clusters of the chosen oligosaccharide antibiotics. Although several aspects of the function of these genes await clarification, basic information is now available. It was not possible to produce new hybrid antibiotics just by expressing glycosyltransferase genes from the landomycin producer in *S. fradiae* Tü2717. Consequently, there is a need to focus on more complex approaches. By inactivating the glycosyltransferase genes of the urdamycin producer we created mutants which are able to synthesize activated deoxysugars but cannot assemble these sugars to form an oligosaccharide chain. These mutants will be used as recipients in which glycosyltransferase genes of *S. cyanogenus* S136 and *S. viridochromogenes* Tü57 can be expressed. This approach will be completed by expressing hybrid genes with altered substrate specificity. Combinatorial biosynthesis approaches may also be extended to deoxysugar biosynthetic genes. The assembling of mannose, evalose, or fucose molecules to the oligosaccharide chain of urdamycins is possible if *S. fradiae* Tü2717 is able to synthesize sufficient amounts of the desired sugar and to catalyze the connecting reaction. Co-expression of responsible biosynthetic genes and glycosyltransferase genes will provide support to reach this goal.


# References

[1]  D. A. Hopwood, *Chemical Reviews*, **1997**, *97*, 2465–2497.
[2]  C. R. Hutchinson, I. Fujii, *Annu. Rev. Microbiol.*, **1995**, *49*, 201–238.
[3]  C. Khosla, R. J. X. Zawada, *Trends Biotech*, **1996**, *14*, 335–342.
[4]  A. C. Weymouth-Wilson, *Nat. Prod. Rep.*, **1997**, *14*, 99–110.
[5]  A. Kirschning, A. Bechthold, J. Rohr, *Topics in Current Chemistry*, **1997**, *188*, 1–84.
[6]  F. Buzzetti, F. Eisenberg, H. N. Grant, W. Keller-Schierlein, W. Voser, H. Zähner, *Experientia*, **1968**, *24*, 320–323
[7]  D. E. Wright, *Tetrahedron*, **1979**, *35*, 1207–1237
[8]  M. G. Cormican, R.N. Jones, *Drugs*, **1996**, *51*, 6–12
[9]  C. Urban, N. Mariano, K. Mosinka-Snipas, C. Wadee, T. Chahrour, J. J. Rahal. J. *Antimicrob. Chemother.*, **1996**, *37*, 361–364.
[10]  H. Wolf, *FEBS Lett.*, **1973**, *36*, 181–186.
[11]  J. Rohr, R.Thiericke, *Nat. Prod. Rep.*, **1992**, *9*, 103–127.
[12]  H. Depenbrock, S. Bornschlegl, R. Peter, J. Rohr, P. Schmid, P. Schweighart, T. Block, J. Rastetter, A. R. Hanauske, *Ann. Hematol.*, **1996**, *73, Suppl. II*, A80/316.
[13]  H. Drautz, H. Zähner, J. Rohr, A. Zeek. *J. Antibiot.* **1986**, *39*, 1657–1669.
[14]  H. Decker, S. Gaisser, S. Pelzer, P. Schneider, L. Westrich, W. Wohlleben, A. Bechthold. *FEMS Microbiol. Lett.*, **1996**, *141*, 195–201.
[15]  L. Westrich, S. Gaisser, B. Reichenstein, A. Bechthold. *Biochemica*, **1997**, *101*, 30–32.
[16]  P. Schneider, Diploma thesis, Univerity of Tübingen, **1995**, 47–51.

[17] S. Gaisser, A. Trefzer, S. Stockert, A. Kirschning, A. Bechthold. *J. Bacteriol.*, **1997**, *179*, 6271–6278.

[18] C. Olano, A. M. Rodriguez, C. Mendez, J. Salas. *Mol. Microbiol.*, **1995**, *16*, 333–343.

[19] S. F. Haydock, J. A. Dowson, N. Dhillon, G. A. Roberts, J. Cortes, P. F. Leadlay. *Mol. Gen. Genet.*, **1991**, *230*, 120–128.

[20] D. A. Rouch, D. S. Cram, D. DiBernadino, T. G. Littlejohn, R. A. Skurray. *Mol. Microbiol.* **1990**, *4*, 2051–2062.

[21] J. Rohr, S. E. Wohlert, C. Oelkers, A. Kirschning, M. Ries. *Chem. Commun.*, **1997**, *10*, 973–974.

## 6.5    Biosynthesis of plant xanthones

*Ludger Beerhues, Wagner Barillas, Stefan Peters, and Werner Schmidt*

Plants produce an impressive diversity of natural products. The phenolic secondary metabolites include xanthones which are widely distributed in only two plant families, the Gentianaceae and the Hypericaceae [1]. Xanthones possess interesting pharmacological properties. For example, 2,3,4,5,6,7-hexahydroxyxanthone (Figure 6.5.1) exhibits strong anti-malarial activity by inhibiting heme polymerization in *Plasmodium falciparum* [2]. Psorospermin and gambogellic acid are cytotoxic and anti-tumor agents [3, 4], and subelliptenone F is an efficient inhibitor of topoisomerases I and II [5]. Rubraxanthone exhibits pronounced anti-bacterial activity against methicillin-resistant *Staphylococcus aureus* [6]. Bellidifolin is a selective and reversible inhibitor of monoamine oxidase A which is involved in neurotransmitter degradation [7]. Xanthones with anti-microbial properties have also been isolated from micro-organisms [8].

We study the biosynthesis of xanthones in plants, our investigations being performed with cell-suspension cultures of *Centaurium erythraea* (Gentianaceae) and *Hypericum androsaemum* (Hypericaceae).

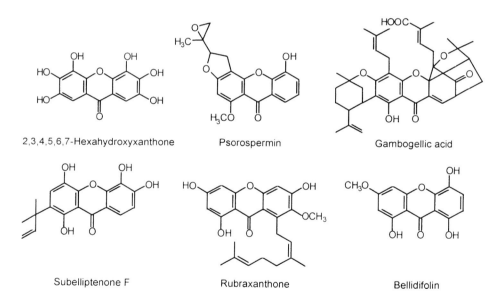

2,3,4,5,6,7-Hexahydroxyxanthone          Psorospermin          Gambogellic acid

Subelliptenone F          Rubraxanthone          Bellidifolin

**Figure 6.5.1.** Examples of pharmacologically active xanthones.

### 6.5.1    Xanthone biosynthesis in *Centaurium erythraea*

Cell cultures of *C. erythraea* contain 3,5,6,7,8-pentamethoxy-1-*O*-primeverosylxanthone (Figure 6.5.2) as the main constituent, and thus represent an attractive

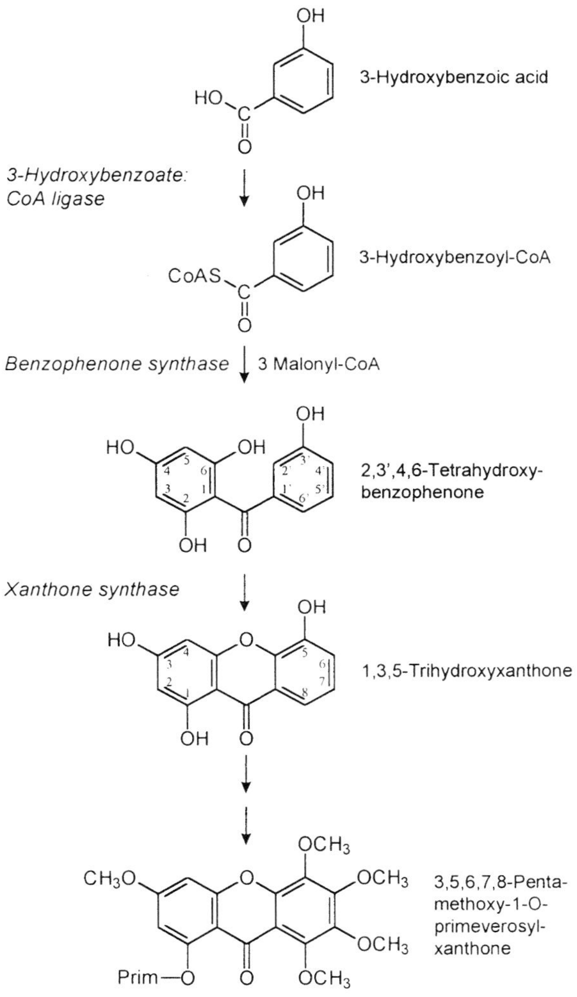

**Figure 6.5.2.** Proposed xanthone biosynthetic pathway in cell cultures of *Centaurium erythraea.*

model system to study xanthone biosynthesis [9]. Treatment of the cultured cells with various elicitors resulted in a differential accumulation of xanthones [10].

A central step in the xanthone biosynthetic pathway is the formation of the $C_{13}$ skeleton, *i.e.* an intermediate benzophenone (Figure 6.5.2). The involvement of this metabolite had previously been found in feeding experiments with intact plants [1]. The formation of the intermediate benzophenone is catalyzed by benzophenone synthase, which was detected for the first time in cultured cells of *C. erythraea* [11]. The enzyme condenses stepwise 3-hydroxybenzoyl-CoA with three molecules of malonyl-CoA to give 2,3',4,6-tetrahydroxybenzophenone. The underlying reaction mechanism is probably analogous to that of other plant polyketide synthases [12].

Besides 3-hydroxybenzoyl-CoA as the preferred substrate, benzophenone synthase converts benzoyl-CoA with half-maximal activity. Neither 2- nor 4-hydroxybenzoyl-CoA serve as substrates. At present, the benzophenone synthase cDNA is being isolated and heterologously expressed.

The preceding reaction in the biosynthetic route is the activation of 3-hydroxybenzoic acid, which is catalyzed by 3-hydroxybenzoate:CoA ligase (Figure 6.5.2) [13]. This enzyme was separated from 4-coumarate:CoA ligase by fractionated ammonium sulfate precipitation and hydrophobic interaction chromatography. 4-Coumarate:CoA ligase is the last enzyme of the general phenylpropanoid metabolism supplying substrates for lignin formation and other individual phenylpropanoid pathways [14]. The substrate specificities of the two CoA ligases are completely different [13]. 3-Hydroxybenzoate:CoA ligase activates only benzoic acids, with 3-hydroxybenzoic acid being the preferred substrate. It lacks affinity for cinnamic acids. These are activated by 4-coumarate:CoA ligase, with 4-coumaric acid being by far the most efficient substrate. Benzoic acids are not accepted by 4-coumarate:CoA ligase.

3-Hydroxybenzoate:CoA ligase has previously been detected in bacteria, but not yet been isolated and characterized [15]. Therefore, the enzyme from cultured *C. erythraea* cells was purified to apparent homogeneity, and its properties were studied [16]. The isolated enzyme protein is being internally digested and partially sequenced. In addition, we are studying the biosynthesis of 3-hydroxybenzoic acid. As yet, little is known about the formation of benzoic acids in plants.

The intermediate 2,3',4,6-tetrahydroxybenzophenone is subjected to regioselective cyclization (Figure 6.5.2) [17]. In *C. erythraea*, it is intramolecularly coupled to 1,3,5-trihydroxyxanthone, this ring closure being catalyzed by xanthone synthase which is, in contrast to benzophenone synthase and the CoA ligases, a cytochrome $P_{450}$ enzyme, probably a cytochrome $P_{450}$ oxidase. The activity of this membrane-bound enzyme is dependent on oxygen and NADPH and strongly reduced by established $P_{450}$ inhibitors. In addition, xanthone synthase is appreciably inhibited by carbon monoxide in the dark, and its activity is partly restored by illumination with white and blue light.

The reaction mechanism underlying the regioselective intramolecular cyclization of the benzophenone is likely to be an oxidative phenol coupling involving two one-electron oxidation steps (Figure 6.5.3) [17]. The first one-electron transfer and a deprotonation yield a phenoxy radical whose electrophilic attack at C-2' leads to the cyclization of the benzophenone. The intermediate resonance-stabilized hydroxy-cyclohexadienyl radical is converted by the loss of a further electron and proton to 1,3,5-trihydroxyxanthone. This mechanism is strongly favored by the *ortho-para*-directing 3-hydroxy group and supported by the substrate specificities of the preceding enzymes which convert most efficiently the 3-hydroxylated substrates.

## 6.5.2   Xanthone biosynthesis in *Hypericum androsaemum*

Cell cultures of *H. androsaemum* produce a broad spectrum of prenylated and/or glucosylated xanthones [18]. The accumulation of these compounds is strongly

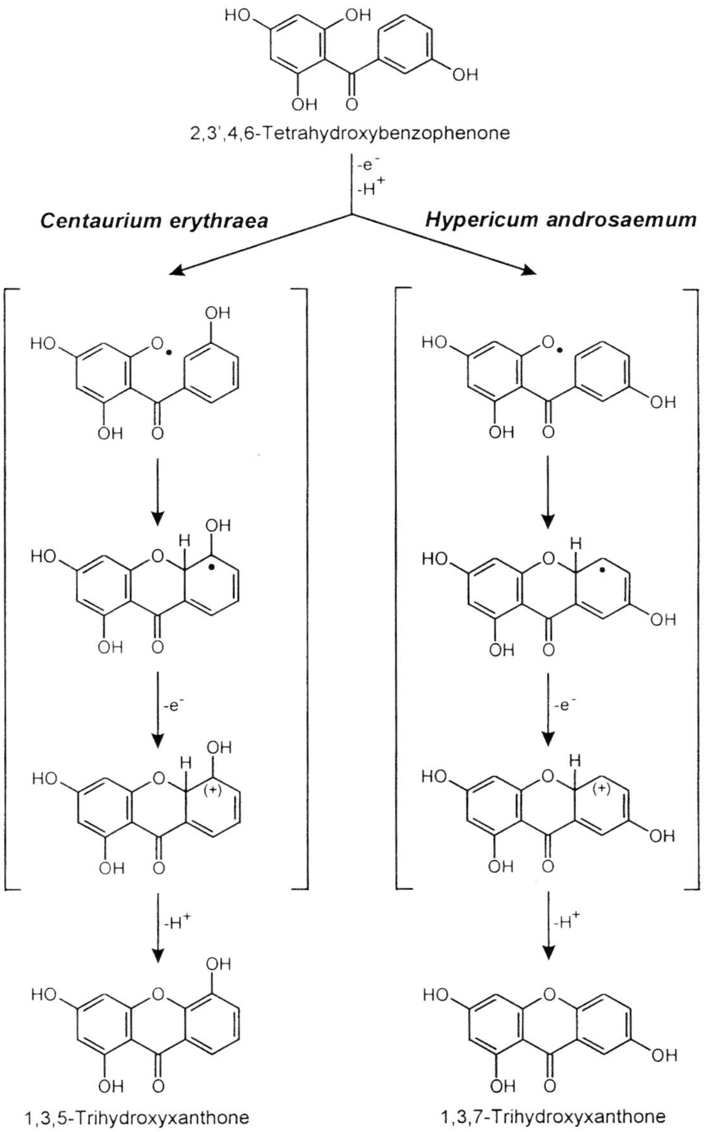

**Figure 6.5.3.** Postulated reaction mechanism underlying the regioselective oxidative phenol couplings of 2,3′,4,6-tetrahydroxybenzophenone in cultured cells of *Centaurium erythraea* and *Hypericum androsaemum.*

influenced by the composition of the culture medium and repressed by light. Our enzymologic studies were carried out with cells grown in a modified B5 medium in the dark.

The central enzyme reactions of xanthone biosynthesis were also detected in these cell cultures (Figure 6.5.4) [19]. Benzophenone synthase from *H. androsaemum* has,

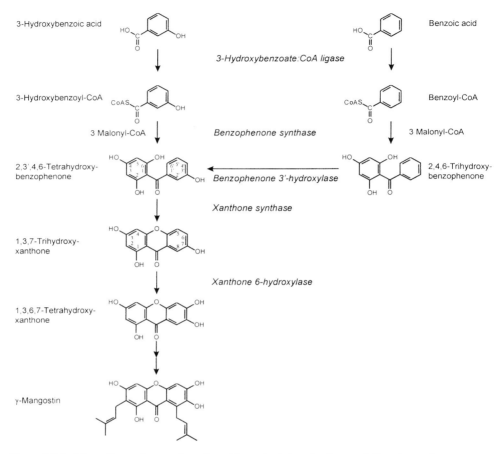

**Figure 6.5.4.** Alternative pathways of xanthone biosynthesis in cell cultures of *Hypericum adrosaemum.*

however, a different substrate specificity when compared to the enzyme from *C. erythraea*. While this is most active with 3-hydroxybenzoic acid, benzophenone synthase from *H. androsaemum* converts most efficiently benzoyl-CoA. 3-Hydroxy-benzoyl-CoA gives only half-maximal activity. Among the plant polyketide synthases, different substrate specificities have previously been observed with two stilbene synthases [20].

Both benzoyl-CoA and 3-hydroxybenzoyl-CoA are supplied by 3-hydroxyben-zoate:CoA ligase [19]. In *H. androsaemum,* this enzyme prefers 3-hydroxybenzoic acid and also acts relatively efficiently on benzoic acid. By comparison, in cultured *C. erythraea* cells, benzoic acid is only a poor substrate for 3-hydroxybenzoate:CoA ligase.

The sequential condensation of benzoyl-CoA with three molecules of malonyl-CoA yields 2,4,6-trihydroxybenzophenone (Figure 6.5.4). This intermediate is subsequently converted by benzophenone 3′-hydroxylase which was detected in the microsomal fraction [19]. The enzyme is a cytochrome $P_{450}$ mono-oxygenase, as shown by the studies described above for xanthone synthase. Benzophenone 3'-hydroxylase acts only on hydroxylated benzophenones of which 2,4,6-trihydroxybenzophenone is the most efficient substrate and is hydroxylated specifically in the 3'-position. The enzyme does not hydroxylate the 3-position of benzoic acid and cinnamic acid, and is thus not involved in the early steps of xanthone biosynthesis.

The substrate specificities of 3-hydroxybenzoate:CoA ligase and benzophenone synthase, as well as the occurrence of benzophenone 3'-hydroxylase, strongly suggest that alternative pathways lead to the formation of 2,3',4,6-tetrahydroxybenzophenone in cultured cells of *H. androsaemum* (Figure 6.5.4) [19]. The 3'-hydroxy group is introduced either at the benzophenone level or at an earlier, as yet unknown stage in the xanthone biosynthetic pathway.

In *C. erythraea*, 2,3',4,6-tetrahydroxybenzophenone is intramolecularly cyclized to 1,3,5-trihydroxyxanthone. By contrast, it is regioselectively coupled to the isomeric 1,3,7-trihydroxyxanthone in *H. androsaemum* (Figure 6.5.3) [17]. This oxidative phenol coupling occurs via the *ortho*-position to the 3'-hydroxy group in *C. erythraea*, however, via the *para*-position to the 3'-hydroxy group in *H. androsaemum*. This finding is in good agreement with the substitution patterns of the constituents accumulating in the two cell cultures. *C. erythraea* cells contain 5-oxygenated xanthones, whereas *H. androsaemum* cells accumulate mainly 7-oxygenated compounds [10, 18]. Both 1,3,5- and 1,3,7-trihydroxyxanthones are the precursors of the majority of plant xanthones [1]; thus, the regioselective cyclizations of 2,3',4,6-tetrahydroxybenzo-phenone represent an important branch point in plant xanthone biosynthesis. It will be interesting to clone and to analyze comparatively the xanthone synthases from the two cell cultures studied.

# References

[1]   G. J. Bennett, H. H. Lee, *Phytochemistry,* **1989**, *28*, 967–998.
[2]   M. V. Ignatushchenko, R. W. Winter, H. P. Bächinger, D. J. Hinrichs, M. K. Riscoe, *FEBS Lett.,* **1997**, *409*, 67–73.
[3]   J. M. Cassady, W. M. Baird, C. J. Chang, *J. Nat. Prod.,* **1990**, *53*, 23–41.
[4]   J. Asano, K. Chiba, M. Tada, T. Yoshii, *Phytochemistry,* **1996**, *41*, 815–820.
[5]   H. Tosa, M. Iinuma, T. Tanaka, H. Nozaki, S. Ikeda, K. Tsutsui, K. Tsutsui, M. Yamada, S. Fujimori, *Chem. Pharm. Bull.,* **1997**, *45*, 418–420.
[6]   M. Iinuma, H. Tosa, T. Tanaka, F. Asai, Y. Kobayashi, R. Shimano, K. Miyauchi, *J. Pharm. Pharmacol.,* **1996**, *48*, 861–865.
[7]   D. Schaufelberger, K. Hostettmann, *Planta medica,* **1988**, 219–221.
[8]   M. Chu, I. Truumees, R. Mierzwa, J. Terracciano, M. Patel, D. Loebenberg, J. J. Kaminski, P. Das, M. S. Puar, *J. Nat. Prod.,* **1997**, *60*, 525–528.
[9]   L. Beerhues, U. Berger, *Phytochemistry,* **1994**, *35*, 1227–1231.
[10]  L. Beerhues, U. Berger, *Planta,* **1995**, *197*, 608–612.
[11]  L. Beerhues, *FEBS Lett.,* **1996**, *383*, 264–266.
[12]  J. Schröder, *Trends Plant Sci.,* **1997**, *2*, 373–378.
[13]  W. Barillas, L. Beerhues, *Planta,* **1997**, *202*, 112–116.
[14]  K. Hahlbrock, D. Scheel, *Annu. Rev. Plant Physiol. Plant Mol. Biol.,* **1989**, *40*, 347–369.

[15]  U. Altenschmidt, B. Oswald, E. Steiner, H. Herrmann, G. Fuchs, *J. Bacteriol.,* **1993**, *175*, 4851–4858.

[16]  W. Barillas, L. Beerhues, unpublished results

[17]  S. Peters, W. Schmidt, L. Beerhues, *Planta,* **1998**, *204*, 64–69.

[18]  W. Schmidt, J. L. Wolfender, K. Hostettmann, L. Beerhues, unpublished results

[19]  W. Schmidt, L. Beerhues, *FEBS Lett.,* **1997**, *420*, 143–146.

[20]  R. Gehlert, A. Schöppner, H. Kindl, *Mol. Plant-Microbe Interact.,* **1990**, *3*, 444–449.

# 6.6 Screening of peptide libraries for the identification of mimotopes which crossreact with antibody epitopes

*Christoph Seidel, Michael Grol, Hans-Georg Batz,*
*Jens Schneider-Mergener, Ricardo Cortese, and Rob Meloen*

We have attempted to identify peptide sequences which mimic antibody epitope sequences by screening different peptide libraries. These sequences can be named mimotopes if the selected antibodies bind them with an affinity comparable to the affinity of the same antibodies to their natural epitopes. For the library screening, three well-defined monoclonal antibodies (mAbs) were selected, one against the protein troponin T (TnT) and two against creatine kinase MB (CK-MB). The selection contained one antibody against a linear epitope, one antibody against a conformational or assembled epitope, and one antibody against a mixed epitope form, which has both linear and conformational components. The following library types were used for screening: libraries of free soluble peptides, cellulose immobilized peptide libraries, and phage display libraries. We screened the libraries with different ELISA formats, BIAcore, and biopanning methods.

By screening the iterative/combinatorial, positional scanning, cellulose-bound and phage-displayed peptide libraries with three different antibodies, we could identify only the linear binding motifs of the epitopes, but no real mimotope. The shorter the linear part of the identified sequences, the lower was the affinity. The different libraries gave comparable results.

## 6.6.1 Peptide libraries

Since their first descriptions [1–3] peptide libraries got important in the field of therapeutic research. In the meantime, the use of library screening has become a potent tool for the identification of lead structures. Little attention has been directed to the identification of new specifiers for diagnostic applications by similar approaches. Specifiers are substances which react specifically and with high sensitivity with an analyte which is found in close relation with a defined clinical situation. In immunological diagnostic assays, antibodies play a key role because they may be either the specifiers or the analytes. They are involved in the diagnosis of infections as well as cancer, metabolic and other diseases. Our aim was to screen for linear structures which mimic conformational epitopes. Success may open the possibility of substituting a protein, or perhaps a specific part of a protein, by a low-molecular weight, well-defined linear peptide. The advantages of such peptide-based substances are the simple availability, the more precise means of characterization and, the avoidance of preparation from potentially infectious materials or other biomaterials which are not obtainable due to ethical reasons.

## 6.6.2    Materials and methods

### 6.6.2.1    Monoclonal antibody anti-Troponin T M7

The antibody was created against human heart muscle troponin T by classical hybidoma techniques (as all other antibodies described here). IgG was obtained by bioreactor fermentation, isolated by ammonium sulfate precipitation, and purified by an antibody specific protein A affinity chromatography. This antibody recognises the linear motif SLKDRIEKRRAE with an affinity of $1.8 \times 10^8$ mol/l determined by BIAcore (all following binding data were determined by BIAcore). The key sequence for binding is DRIEKR. The antigenic sequence was characterized by a PEPSCAN analysis [5]. A series of 12-mer peptides, each shifted in one-amino acid steps through the sequence of the human TnT protein, was synthesized. The reactivity of the overlapping peptides was tested in an ELISA format. The affinity constant Ka of the antibody binding to the whole protein is $1.0 \times 10^8$ mol/l.

### 6.6.2.2    Monoclonal antibody anti-CK-MB M-6.12.47

The antibody was created against human CK-MB. The IgG was obtained by bioreactor fermentation, isolated by ammonium sulfate precipitation, and purified by anion-exchange chromatography. The antigenic sequence was characterized by a PEPSCAN analysis [5] as above. The antibody recognizes the linear sequence motif KGKYY with low affinity. The affinity constant Ka of the antibody to the whole protein is $9.0 \times 10^8$ mol/l.

### 6.6.2.3    Monoclonal antibody anti-CK-MB M-7.4.5

The antibody was created against human CK-MB. It was obtained by bioreactor fermentation, isolated by ammonium sulfate precipitation and purified by anion-exchange chromatography. The PEPSCAN analysis and other methods (not published) showed reactivity only against the entire protein, but no reactivity against linear peptide parts. The affinity constant Ka of the antibody to the whole protein is $1.1 \times 10^8$ mol/l.

### 6.6.2.4    Free peptide model libraries for iterative and positional scanning processes [2, 4]

For the simulation of an iterative library screening, a model library consisting of free peptides was prepared. The six defined positions of the TnT epitope sequence DRIEKR were successively exchanged by a random (X) position (Table 6.6.1a). For the simulation of a positional scanning approach another synthetic library was prepared in which all positions but one were randomised (Table 6.6.1b). The peptide sublibrary mixtures were prepared by using an equimolar mixture of the protected natural amino acids on the X-positions in the course of classical solid phase peptide syntheses.

**Table 6.6.1.** Troponin T, M7 epitope (DRIEKR) model libraries for competitive, iterative (1a) and positional scanning (1b) library screening (ELISA with free peptides).

| Sublibrary 1a | Diversity | Sublibrary 1b | Diversity |
|---|---|---|---|
| DRIEKX | 18 | DXXXXX | $1.8 \times 10^6$ |
| DRIEXX | 324 | XRXXXX | $1.8 \times 10^6$ |
| DRIXXX | 5832 | XXIXXX | $1.8 \times 10^6$ |
| DRXXXX | 104976 | XXXEXX | $1.8 \times 10^6$ |
| EKXXXX | 104976 | XXXXKX | $1.8 \times 10^6$ |
| XXXXXX | $3.4 \times 10^7$ | XXXXXR | $1.8 \times 10^6$ |

X: Mixture of 18 natural amino acids except Tryptophan and Cystein; peptides are acetylated on the N-terminus and amidated on the C-terminus.

### 6.6.2.5 Cellulose membrane bound peptide libraries [3]

On 10 separate cellulose sheets, in 10 separate sections, peptide libraries of the common sequence $XXXXO_1B_1O_2B_2XXXX$ (X = amino acid mixture) were prepared by solid-phase-derived synthesis in the form of 36.100 single spots. Along the x co-ordinate (B2) and y co-ordinate (B1) of a single section of 361 ($19 \cdot 19$) single spots, 19 natural amino acids (except Cysteine) were permuted. On a section the amino acid position $O_2$ was defined, A or G for example or groups like the aromatic amino acids F,W,Y or the aliphatic amino acids I,V,L,M (see Table 6.6.2 and Figure 6.4.4). The permutation of position $O_1$ runs from sheet n to sheet $n + 1$ in the same manner as from section n to section $n + 1$. The spots were assayed with the monoclonals by ELISA-like techniques directly on the cellulose membrane. Visualization of the antibody-peptide binding was achieved with an anti-mouse $Fc\gamma$-alkaline phosphatase conjugate and a chromogenic substrate or, alternatively, with a horse-radish peroxidase conjugate with a chemoluminescent substrate. The 361 peptide mixtures are spotted in a quadratic $19 \cdot 19$ arrangement, the $10 \cdot 10$ sections are arranged in a $2 \cdot 5$ geometry

### 6.6.2.6 Phage display library

The phage display technique was first described by Smith et al. [1] and reviewed by R. Cortese et al. [6]. In brief, a randomized g-mer peptide was expressed as a fusion with the envelope protein pVIII of bacteriophage virion M13. The randomized fusion sequence in the framework of the outer surface protein sequence of the

**Table 6.6.2** Amino acids or amino acid groups of positions $O_1$ and $O_2$.

| Section/sheet no. | 1 | 2 | 3 | 4 | 5 | 6 | 7 | 8 | 9 | 10 |
|---|---|---|---|---|---|---|---|---|---|---|
| Amino acids | A | D, E | F, W, Y | G | H | I, V, L, M | K, R | N, Q | P | S, T |

phage virion is encoded by a randomized synthetic oligonucleotide sequence which is incorporated into the phage DNA. Theoretically, the library contains all 9-mer peptide sequences with a diversity of $5.0 \times 10^{11}$. The phage clones were tested with the antibodies by an in vitro selection process called biopanning. After affinity capturing, the phages were isolated, amplified, and submitted to further cycles of biopanning and amplification. The strong binders were characterized by DNA sequencing and ELISA testing.

### 6.6.2.7    Screening methods

The screening of the soluble peptide libraries was done in a competitive ELISA format. In the first step, a streptavidin-coated 96-well microtiter plate was incubated with the biotinylated human CK-MB or the synthetic TnT epitope. In a second step, the antibody was incubated together with the soluble sublibrary in increasing concentrations. Antibody (c = 1 nM) and the total number of all sublibrary molecules were in a molar ratio of $1:10^6$. The sublibraries were predissolved in 10% Dimethylsulfoxide to a concentration of 1.0 mg/ml and then diluted with Tris(hydroxymethyl)-aminomethane buffer, 40 mM, pH 7.5. The fraction of the antibody, which is not bound by the sublibrary is still bound to the surface of the microtiterplate. It was detected in the third step by incubating the wells with an anti-mouse Fc$\gamma$ horse radish peroxidase (HRPO) conjugate and a final chromogenic or chemoluminescence reaction [7].

### 6.6.2.8    Biosensor experiments [8]

The association constants Ka and reaction rates kon and koff were determined by the BIAcore system (BIAcore AB, Uppsala, Sweden). Running and dilution buffer was BIAcore ready-to-use system buffer. Biotinylated CK-MB, biotinylated synthetic TnT epitope and by the above experiments determined library peptides (in biotinylated form) were immobilized on a streptavidin-coated sensorchip. Antibody or its Fab fragment in a 125–1000 nM concentration range and then system buffer was injected for 3 min (flow rate 10 µl/min). The formation and dissociation of the solid phase bound antigen-antibody complexes were registered in real-time in the form of a sensorgram. The rate and dissociation constants were deduced with the BIAevaluation 2.1 software by non-linear curve adaptation. The comparison between IgG and Fab data allowed an estimation of avidity contributions.

## 6.6.3    Results

The concentrations above which the sublibraries 1a react distinctly in the competition with mAb TnT M7, are shown in Figure 6.6.1. Above about 0.1 µg/ml, a 50% signal reduction is reached by the target sequence DRIEKR. The mixture DRIXXX reaches

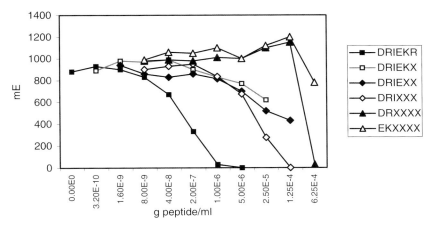

**Figure 6.6.1.** Competition of TnT MAK M7 and TnT-Bi reaction with model libraries of Table 6.6.1a in dependent on concentration (minus unspecific binding).

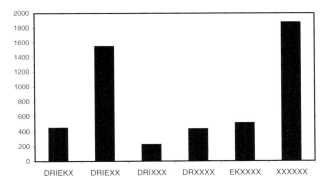

**Figure 6.6.2.** Competition of the TnT MAK M7 and TnT-Bi reaction with model libraries of Table 6.6.1a (Conc. 1.0 mg peptide mixture/ml).

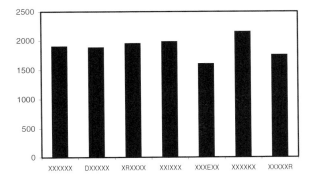

**Figure 6.6.3.** Competition of TnT MAK M7 and TnT-Bi reaction with model libraries of Table 6.6.1b (Conc. 1.0 mg peptide mixture/ml).

**Table 6.6.3.** Anti-creatine kinase-MB mAb-reactive sequences received by phage library screening.

| Antibody | Phageclone | Phage frame-**sequence**-phage frame* | Reactive sequence from PEPSCAN |
|---|---|---|---|
| MAK anti-CK-MB 7.4.5 | 2c | ...AEGEFC**STREIAWGA**CGDPAK... | no |
| MAK anti-CK-MB 6.12.47 | 2.1 | ....AEGEF**RGKFSLPIS**DPAK..... | LTGEFKGKYYPL |
|  | 2.4 | ....AEGEF**KGKFFLASQ**DPAK..... |  |
|  | 2.7 | ....AEGEF**QGQFYLQSW**DPAK..... |  |

*N-terminal phage frame sequence partially identical with creatine kinase K-chain primary structure (column 4).

this competition only above 20 μg/ml. The mixture DRIEXX, with one defined position more, reduces the signal by 50% only above 100 μg/ml. Because the solubility of the peptide mixtures is about 1 mg/ml, the experiments were done in the concentration range between 0.1 and 1.0 mg/ml.

The competing behavior of the mixtures at a concentration of 1 mg/ml is shown in Figure 6.6.2. The competition power of the mixtures declines in the following manner:

DRIXXX > DRXXXX = DRIEKX > EKXXXX > DRIEXX

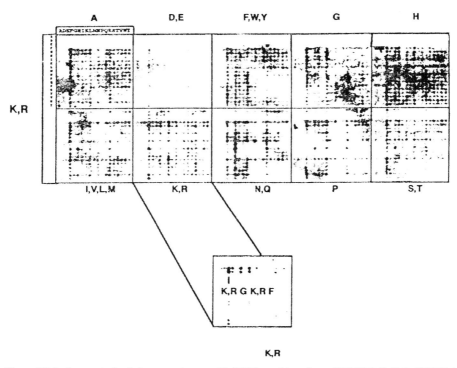

**Figure 6.6.4.** Segment of cellulose membrane with 36100 peptides of type $XXXXO_1B_1O_2B_2XXXX$; incubation with mAb anti-CK-MB 6.12.47; marker HRPO; chemoluminescence detection.

**Table 6.6.4.** BIAcore data of bivalent mAb 6.12.47 and monovalent Fab.

| Antibody form | Surface capture antigen | kon [l/mol × s] | koff [1/s] | Ka [mol/l] |
|---|---|---|---|---|
| IgG | Bi-CK-MB | $2.8 \times 10^5$ | $3.1 \times 10^{-4}$ | $9.0 \times 10^8$ |
| Fab | Bi-CK-MB | $3.1 \times 10^5$ | $6.4 \times 10^{-3}$ | $4.8 \times 10^7$ |
| IgG | Bi-Peptid-2.1 | $9.1 \times 10^3$ | $8.3 \times 10^{-4}$ | $1.1 \times 10^7$ |
| Fab | Bi-Peptid-2.1 | nd | nd | nd |
| IgG | Bi-Peptid-2.4 | $1.4 \times 10^4$ | $1.6 \times 10^{-3}$ | $8.3 \times 10^6$ |
| Fab | Bi-Peptid-2.4 | nd | nd | nd |
| IgG | Bi-Peptid-2.7 | $2.3 \times 10^4$ | $1.6 \times 10^{-3}$ | $1.5 \times 10^7$ |
| Fab | Bi-Peptid-2.7 | nd | nd | nd |

nd: not determinable, signal to low (<50 RU).

To simulate the iterative process, the model library 1b with only one defined position was tested in the same competition assay. Above a concentration of 1 mg/ml XXXEXX and XXXXXR (Figure 6.6.3) and above 0.33 mg/ml, DXXXXX showed a distinct competition.

To identify linear sequences which mimic conformational epitopes, several selection cycles were done with the phage display library and the CK-MB mAbs. The resulting sequences are shown in Table 6.6.3. In contrast to mAb 6.12.47, mAb 7.4.5 yielded neither with PEPSCAN, nor with the cellulose immobilized libraries, a specific binding sequence. The low-affinity phage sequence received with mAb 7.4.5 could not withstand control experiments. The cellulose membrane library delivered with mAb 6.12.47 the dominant consensus motif KGKF (Figure 6.6.4). Synthetic biotinylated control peptides of the clones 2.1, 2.4, and 2.7 showed with BIAcore and mAb 6.12.47 (IgG) comparable affinities to complete biotinylated CK-MB, but not with monomeric Fab fragments (Table 6.6.4).

## 6.6.4 Discussion

Houghten et al. [2] began the development of iterative library screening with libraries of the highest complexity and reduced the number of random positions by one in subsequent cycles. The percentage portion of the target sequence in the mixture increased correspondingly. Figure 6.6.2 shows however, that the competitive power of a mixture with four random positions is higher, than with only two. This results show, that the strategy of Houghten et al. does not always lead stringently to a target sequence.

A prerequisite for the success of a library screening is the concentration (Figure 6.6.1) of the soluble components, where the competition is most efficient and differentiation is most distinct. The concentration range is further limited by the solubility of the competitors. So, the mixture DRXXXX competes at a concentration of 0.5 mg/ml. The portion of target component DRIEKR in this mixture is only 5 ng/ml (calculated: 0.5 mg/ml devided by $18^4$). As seen in Figure 6.6.1, (■) the target sequence DRIEKR competes only above 100 ng/ml. Nevertheless, as competition takes place (∼) it becomes evident that not only the target sequence but the whole collective of all similar motifs compete together. Further, it is apparent, that the control sequence EKXXXX with the motif EK, which is shifted three positions to the N-terminus also competes. Individuals which contain parts of a target sequence in the 'wrong' position are also effective. This is another drawback of the iterative strategy.

The positional scanning method, which was also first published by Houghten et al. delineates in our example position 3 and 6. The example shows that this strategy was partially successful. Again, the success is dependent on the optimal concentration used in the test. Thus, position 1 could be delineated only at significant lower concentrations.

The search for sequences which mimic the conformational epitope of CK-MB mAb 7.4.5 with cellulose membrane and phage display libraries did not deliver any linear sequence that bind the mAb with sufficient affinity. Screening with the second mAb 6.12.47 led only to the short consensus motif K/RGKF, which is part of the primary component of the conformational epitope. Control syntheses showed that neighboring sequences did not contribute to the binding. The affinity data of the peptides in Table 6.6.4 appear only similar to the affinity of the natural protein. This apparent high affinity of the peptides is caused by the high density of the small peptides on the solid phase. This density allows a two-point binding with both Fab arms (avidity effect). This multivalent binding is not possible with a much larger protein such as CK-MB. If Fab fragments are used instead of whole IgG, the affinity drops so profoundly that it could not be determined by biosensor technology.

The experiments with the above-described peptide libraries showed that it was not possible to identify linear peptide mimotopes.

# References

[1]  S. F. Parmley, G. P. Smith, *Gene* **1988**, *37*, 305.
[2]  R. A. Houghten, C. Pinilla, S. E. Blondelle, J. R. Appel, C. T. Dooley und J. H. Cuervo, *Nature* **1991**, *354*, 84.
[3]  R. Frank, *Tetrahedron* **1992**, *48*, 9217.
[4]  C. Pinilla, J. R. Appel, P. Blanc, R. A. Houghten, *BioTechniques* **1992**, *13*, 901.
[5]  H. M. Geysen, R. H. Meloen, S. J. Barteling, *Proc. Natl. Acad. Sci. USA* **1984**, *81*, 3998.
[6]  R. Cortese, *Combinatorial libraries*, Walter De Gruyter, Berlin, New York **1996**.
[7]  2,2'-Azino-di-[3-ethylbenzthiazolinsulfonat(6)] (ABTS) from Roche Diagnostics GmbH.
[8]  F. Schindler, *BioTec Life Science* **1992**, *1*, 36.

# 6.7 Probes for DNA base flipping by DNA methyltransferases

*Birgit Holz and Elmar Weinhold*

## 6.7.1 DNA methyltransferases and the biological role of DNA methylation

In addition to the normal nucleobases adenine, thymine, guanine, and cytosine, the DNA of most organisms contains the methylated bases C5-methylcytosine (**1**) [1], N6-methyladenine (**2**) [2], or N4-methylcytosine (**3**) [3] (Figure 6.7.1).

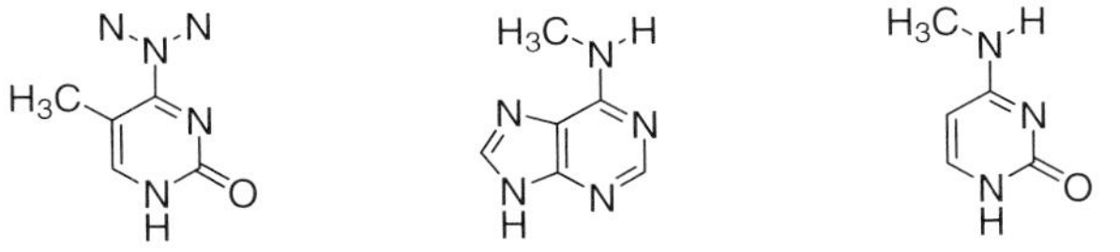

**Figure 6.7.1.** Methylated nucleobases found in the DNA of most organisms.

These methylated bases are formed by DNA methyltransferases (Mtases) which catalyze the transfer of the activated methyl group from the cofactor *S*-adenosyl-L-methionine (**4**, AdoMet) to the C5 carbon of cytosine, the N6 nitrogen of adenine, or the N4 nitrogen of cytosine within specific DNA sequences [4]. Most DNA Mtases recognize palindromic DNA sequences, which contain two symmetry-related target bases. After DNA replication, only the parental strand contains methylated bases (hemi-methylated DNA), and methylation of the daughter strand restores the fully methylated DNA (Scheme 6.7.1).

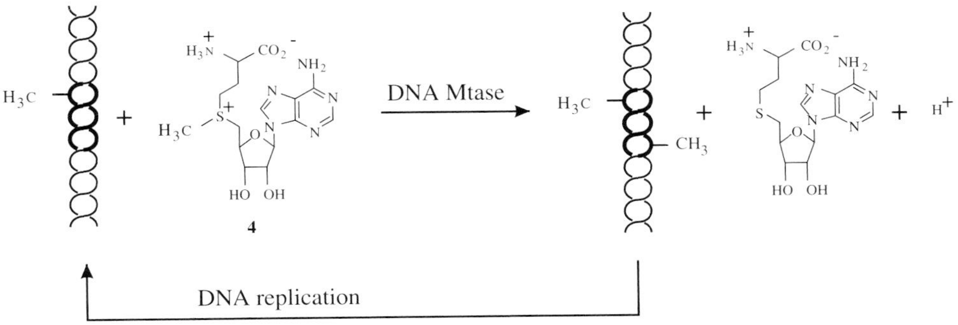

**Scheme 6.7.1.** Reaction catalyzed by DNA methyltransferase (Mtases).

Since DNA methylation is a postreplicative process and depends on the presence or regulation of DNA Mtases, a particular nucleotide sequence may exist in its fully methylated, its unmethylated, or transiently in its hemi-methylated form. Thus, DNA methylation can be regarded as an increase of the information content of DNA [5], which serves a wide variety of biological functions. In prokaryotes, DNA methylation is involved in protection of the host genome from endogenous restriction endonucleases, DNA mismatch repair after replication, regulation of gene expression, and DNA replication [6]. In eukaryotes, DNA methylation plays an important role in regulatory processes, such as regulation of gene expression, embryonic development, genomic imprinting, X-chromosome inactivation, and carcinogenesis [7].

## 6.7.2    DNA base flipping as observed in X-ray structures of DNA methyltransferases

For a long time it was unclear, how DNA Mtases reach their target bases, which are buried in the interior of the DNA helix. Only the three-dimensional structures of the C5-cytosine DNA Mtases M·*Hha*I [8] and M·*Hae*III [9] in complex with DNA showed that the target cytosines are completely rotated out of the DNA helix and placed in a cleft within each enzyme, where catalysis takes place. Thus, the Watson–Crick hydrogen bonds of the target cytosines to the partner guanines, and their stacking interactions with neighboring bases within the DNA helix are completely removed. The extrahelical conformation of the target cytosines is stabilized by numerous contacts to residues within the active sites. In addition, the unpaired guanines are hydrogen-bonded to protein residues. In the M·*Hae*III structure an additional rearrangement of the DNA is observed, in which the unpaired guanine forms a Watson–Crick-like hydrogen bond to the 3′-neighboring cytosine of the target base.

This novel mode of protein–DNA interaction was also suggested for the N6-adenine DNA Mtase M·*Taq*I and the N4-cytosine DNA Mtase M·*Pvu*II, based on their crystal structures in complex with the cofactor [10]. Both enzymes consist, like the C5-cytosine DNA Mtases, of two domains which form a positively charged cleft that is wide enough to accommodate double-stranded B-DNA. However, modeling of B-DNA into the structures showed that the distance between the target bases and the activated methyl group of the cofactor AdoMet is too large for a direct methyl group transfer. By rotating the target bases out of the DNA helix towards the cofactor, the distance between the methyl group acceptors and donor is significantly reduced, and hence a base flipping mechanism was proposed for M·*Taq*I and M·*Pvu*II.

## 6.7.3    2-Aminopurine in DNA as fluorescent probe for DNA base flipping

In general, X-ray crystallography of protein–DNA complexes provides the absolute proof for DNA base flipping. However, co-crystallization of proteins with their

DNA substrates is often cumbersome, and in some cases co-crystals might never be obtained. An alternative approach to detect base flipping makes use of the fluorescent adenine analog 2-aminopurine. The 2-aminopurine fluorescence is highly quenched in polynucleotides [11], mostly due to stacking interactions with neighboring bases, and is expected to increase dramatically if the 2-aminopurine base loses its stacking interactions with its neighboring bases due to base flipping. The incorporation of 2-aminopurine into oligodeoxynucleotides (ODNs) can be performed by solid-phase DNA synthesis using standard phosphoramidite chemistry [12].

### 6.7.3.1  Detection of DNA base flipping

In order to test whether duplex ODNs containing 2-aminopurine at the target position can be used as probes for DNA base flipping, we studied the C5-cytosine DNA Mtase M·*Hha*I as a paradigm for a base flipping enzyme [13]. M·*Hha*I recognizes the double-stranded DNA sequence 5′-GCGC-3′ and methylates the first cytosine (in bold). The 2-aminopurine fluorescence of the 37 base pair duplex ODN **5a** (Figure 6.7.2), in which the target cytosine is replaced by 2-aminopurine, is quenched about 100-fold. Upon titration of **5a** with M·*Hha*I the 2-aminopurine fluorescence intensity increases 54-fold. This result clearly demonstrates that duplex ODNs containing 2-aminopurine at the target site can serve as fluorescence probes for base flipping.

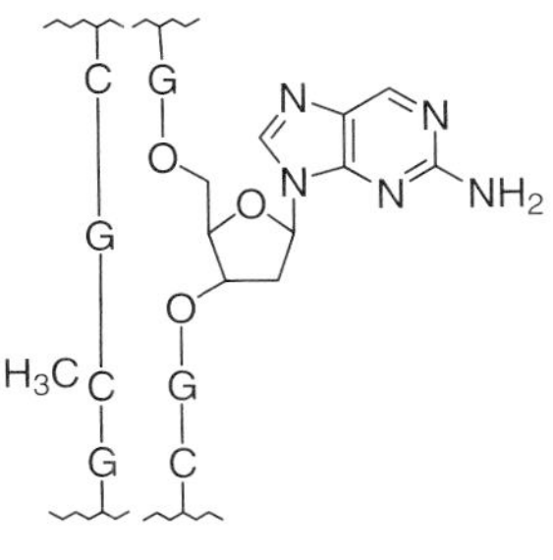

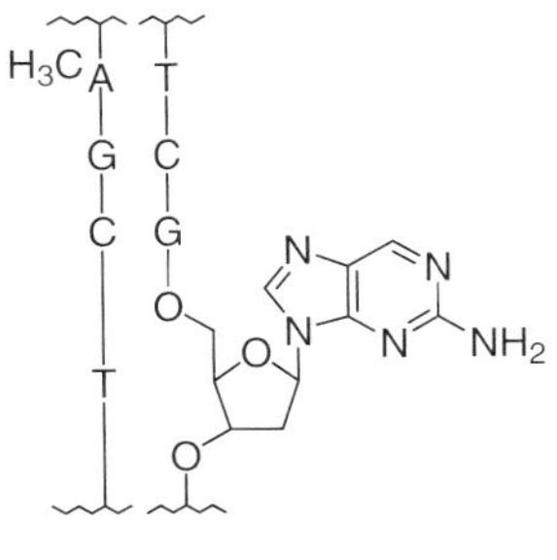

**Figure 6.7.2.** Duplex ODNs used for a fluorescence-based assay of DNA base flipping by the C5-cytosine DNA Mtase M·*Hha*I and the N6-adenine DNA Mtase M·*Taq*I. The duplex ODNs **5a** and **5b** contain 2-aminopurine at target positions within the recognition sequences of M·*Hha*I and M·*Taq*I, respectively.

We also tested the N6-adenine DNA Mtase M·*Taq*I to verify whether this enzyme also uses a base flipping mechanism [13]. M·*Taq*I catalyzes the methylation of adenine within the double-stranded DNA sequence 5′-TCGA-3′. Titration of the 36 base pair duplex ODN **5b** (Figure 6.7.2), in which the target adenine is replaced by 2-amino-purine, also showed a strongly enhanced (13-fold) 2-aminopurine fluorescence. Recently, fluorescence studies with duplex ODNs containing 2-aminopurine and the N6-adenine DNA Mtase M·*Eco*RI were reported [14]. Similar to our results with

M·*Taq*I, addition of stoichiometric amounts of M·*Eco*RI to a duplex ODN containing 2-aminopurine at the target position resulted in a 14-fold increase of the 2-amino-purine fluorescence. These findings with M·*Taq*I and M·*Eco*RI provide experimental evidence that base flipping similar to that described for the C5-cytosine DNA Mtases M·*Hha*I and M·*Hae*III also takes place in adenine-specific DNA Mtases. Since the active site residues of N6-adenine DNA Mtases and N4-cytosine DNA Mtases are highly conserved [15], a common base flipping mechanism for all DNA Mtases seems very likely.

### 6.7.3.2   Kinetics of DNA base flipping

In addition to the demonstration of base flipping by DNA Mtases, duplex ODNs containing 2-aminopurine at the target position can be used in stopped-flow experiments to monitor base flipping by DNA Mtases in real-time. These experiments provide kinetic and mechanistic information about the base flipping process mediated by DNA Mtases and complement structural information obtained by X-ray crystallography.

We performed stopped-flow measurements with the C5-cytosine DNA Mtase M·*Hha*I and the duplex ODN **5a** [16]. Rapid mixing of **5a** with an increasing excess of M·*Hha*I (pseudo first-order conditions) yielded observed rate constants $k_{obs}$ for the increase of the 2-aminopurine fluorescence, which were linearly dependent on the M·*Hha*I concentrations. From a plot of $k_{obs}$ against the enzyme concentration, a second-order rate constant of $3 \times 10^8 \, s^{-1}$ was calculated. Such a value is expected for a diffusion-controlled association between proteins and their substrates, and indicates that base flipping is very fast. However, it should be noted, that 2-amino-purine forms a mismatch with the partner guanine in the duplex ODN **5a** and thus the 2-aminopurine base should flip out of the DNA helix more easily, compared to the natural cytosine paired with guanine.

Since 2-aminopurine is an anolog of adenine, which can form Watson–Crick-like hydrogen bonds to thymine, this fluorescence probe should be more suitable for studies of the kinetics of base flipping by N6-adenine DNA Mtases. Recently, stopped-flow measurements with the N6-adenine DNA Mtase M·*Eco*RI and a duplex ODN containing 2-aminopurine at the target position were reported [17]. For the association process a biphasic 2-aminopurine fluorescence increase was observed. The fast phase showed an observed rate constant $k_{obs}$ of $21 \, s^{-1}$ and represented 75% of the total fluorescence change. Since $k_{obs}$ is more than 100-fold larger than the turnover number of the enzyme [18], base flipping is not rate-limiting for M·*Eco*RI. However, this value is close to the rate constant for methyl group transfer of $24 \, s^{-1}$, as determined in single-turnover experiments [19]. Thus, the rate-limiting step for M·*Eco*RI under steady-state kinetic conditions occurs after base flipping and methyl group transfer and is associated with the product release or a prior conformational change. The slower phase in the stopped-flow experiments yielded an observed rate constant $k_{obs}$ of $0.6 \, s^{-1}$ and represented 25% of the total fluorescence change. This second phase was attributed to a rearrangement of the extrahelical

2-aminopurine within the active site of the enzyme. Since this process is much slower than methyl group transfer, it probably does not lie on the reaction coordinate.

With the N6-adenine DNA Mtase M·*Taq*I and the duplex ODN **5b** we observed a monoexponential and enzyme concentration-independent 2-aminopurine fluorescence increase with an observed rate constant $k_{obs}$ of $20\,s^{-1}$ under pseudo first-order conditions [20]. The observed rate constant for base flipping is about 500-fold larger than the turnover number of M·*Taq*I, which again shows that base flipping is not rate-limiting. The concentration independence of $k_{obs}$ demonstrates that base flipping occurs after formation of a concentration-dependent initial collision complex. Thus, the enzyme does not recognize an extrahelical adenine within the recognition sequence. Rather, it initially binds to the recognition sequence and then promotes base flipping (Scheme 6.7.2). Recently, the equilibrium between the innerhelical conformation and extrahelical states of the target base was directly observed in a complex between M·*Hha*I and a duplex ODN containing 5-fluorocytosine at the target position by $^{19}$F-NMR spectroscopy [21]. However, DNA binding and base flipping may be accompanied by further structural changes of the DNA like rearrangement of base pairing, as seen in the C5-cytosine DNA Mtase M·*Hae*III [9], or DNA bending, as observed for the N6-adenine DNA Mtases M·*Eco*RI [22] and M·*Eco*RV [23].

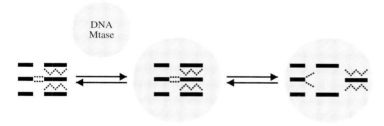

**Scheme 6.7.2.** Simple two-step binding mechanism for DNA Mtases in which base flipping occurs after initial collision complex formation. The solid lines represent the nucleobases within the DNA helix, and the dashed lines indicate interactions of the target base pair within the DNA and the DNA–Mtase complex.

The similarity between the observed rate constants for base flipping of M·*Eco*RI and M·*Taq*I is striking and suggests that the kinetics of base flipping mostly depend on the strength of the base pair that is opened, rather than on the DNA Mtases. The spontaneous opening of a base pair between 2-aminopurine and thymine in DNA was determined by iminoproton exchange measurements using NMR spectroscopy [24]. The reported rate constant of $700\,s^{-1}$ is about 35-fold larger than the base flipping rate constants observed with M·*Eco*RI and M·*Taq*I. Although the spontaneous base opening and base flipping by DNA Mtases could have different trajectories [17], the much faster spontaneous base pair opening due to thermal motions suggests that base flipping is not a DNA Mtase-catalyzed process. However, an important role of DNA Mtases is to stabilize the target base in an extrahelical conformation. They thereby alter the equilibrium between the inner- and the extrahelical target base, which normally lies far on the innerhelical side in DNA.

## 6.7.4    Tighter binding of modified DNA to DNA methyltransferases

### 6.7.4.1    Binding of duplex ODNs containing a mismatched target base

Recently, tighter binding of the C5-cytosine DNA Mtases M·*Hha*I [25, 26] and M·*Hpa*II [25] as well as the N6-adenine DNA Mtases M·*Eco*RV [27] and M·*Eco*RI [17] to duplex ODNs carrying mismatched bases or base analogs with reduced Watson–Crick hydrogen bonding potential at the target positions was observed. The reason for this tighter DNA binding was attributed to the thermodynamics of base flipping, which can be regarded as the sum of the energies needed to disrupt the base pair between the target base and the partner base and the energy gained by binding the extrahelical target base as well as the unpaired partner base by the DNA Mtases (Scheme 6.7.2). Thus, loosening the target base pair can shift the equilibrium between an inner- and extrahelical target base towards the extrahelical side. In a simple two-step binding mechanism, as shown in Scheme 6.7.2, the overall dissociation constant $K_D$ will be influenced by the equilibrium between the inner- and the extrahelical target base, and the more this equilibrium is shifted towards the extra-helical side in the DNA–Mtase complex, the higher will be the overall binding affinity.

### 6.7.4.2    Binding of duplex ODNs containing an abasic site

This mismatch base approach can be taken even further by deleting the target base completely. Duplex ODNs containing a deleted nucleotide or an abasic site at the target position were found to bind 45-fold, 10-fold, 7-fold and 4-fold tighter to M·*Hha*I [26], M·*Hpa*II [25], M·*Eco*RV [27] and M·*Eco*RI [17], respectively.

We have investigated binding of the N6-adenine DNA Mtase M·*Taq*I to the 36 base pair duplex ODNs **6a**, **6b** and **6c** containing 1,2-dideoxy-D-ribose (a stable abasic site) within or next to the 3′-TCGA-5′ recognition sequence of M·*Taq*I (Figure 6.7.3).

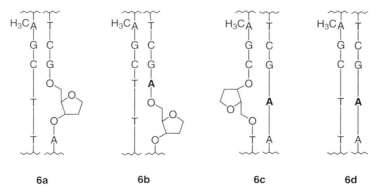

**6a**          **6b**          **6c**          **6d**

**Figure 6.7.3.** Duplex ODNs used in the gel mobility retardation assay with the N6-adenine DNA Mtase M·*Taq*I. The duplex ODNs **6a**, **6b** and **6c** contain 1,2-dideoxy-D-ribose (a stable abasic site) at the target position, 3′ to the target adenine and at the partner position within the recognition sequence of M·*Taq*I, respectively.

1,2-Dideoxy-D-ribose was incorporated into ODNs by solid-phase DNA synthesis using standard phosphoramidite chemistry [28], and binding to M·*Taq*I was analyzed after duplex formation in a gel mobility retardation assay [29]. The duplex ODN **6a** containing a stable abasic site at the target position binds to M·*Taq*I about 200-fold tighter than the regular duplex ODN **6d**. This result can be explained thermodynamically on the basis of a base flipping mechanism, as discussed above. Upon binding of **6a**, the enzyme does not need to pay the energetic cost to disrupt the target base pair and it cannot gain energy by binding an extrahelical target base, because this is absent in **6a**. However, it can bind the unpaired partner thymine directly, which should lead to an increased overall binding affinity. Similarly, a 60-fold tighter binding was observed for the duplex ODN **6b**, which contains the stable abasic site at the 3′-neighboring position of the target adenine. Here, it should be easier to flip the target adenine because the stacking interaction to its 3′-neighbor is missing. On the other hand, the extrahelical target adenine and the unpaired thymine could be bound as in the regular duplex ODN **6d**. This explains why a tighter overall binding results. With the duplex ODN **6c**, which contains the stable abasic site at the partner position, an almost unchanged overall binding affinity compared to **6d** was observed. In the duplex ODN **6c** it should be easier to flip the target adenine because the stabilizing Watson–Crick hydrogen bonds are missing. However, the enzyme can no longer bind the unpaired partner base because this is absent in **6c**. Thus, a higher, lower or even unchanged overall binding affinity is expected depending on the sum of the energies for these interactions.

## 6.7.5 Chemical detection of extrahelical bases in DNA–Mtase complexes

Very recently we described another method to detect extrahelical bases within DNA–Mtase complexes [30]. This method makes use of the observation that nucleobases within the DNA helix are less reactive towards chemical reagents than nucleobases in single-stranded DNA or free in solution. We found that the oxidation of thymine residues by potassium permanganate followed by piperidine-induced strand cleavage and separation of the cleaved strands by denaturing polyacrylamide gel electrophoresis provides a useful method to detect extrahelical thymine residues in DNA–Mtase complexes. In these experiments the target cytosine of the C5-cytosine DNA Mtase M·*Hha*I and the target adenine of the N6-adenine DNA Mtase M·*Taq*I were replaced by thymine residues. As discussed above, such a replacement of the natural target base by another base still leads to base flipping, as evidenced by the tighter binding of DNA Mtases to duplex ODNs containing a mismatched target base. Upon addition of M·*Hha*I and M·*Taq*I to these duplex ODNs, the reactivity of the targeted thymine residues towards potassium permanganate was clearly enhanced. This enhanced reactivity points directly to extrahelical thymine residues in the DNA–Mtase complexes. However, it should be noted that a hyper-reactivity of thymine residues in DNA–protein complexes does not *per se* prove base flipping, because other DNA-modifying enzymes that promote strand opening or melting of

DNA also lead to an enhanced reactivity of thymine residues towards potassium permanganate [31].

## 6.7.6    Other DNA-modifying enzymes using a DNA base flipping mechanism

In addition to DNA Mtases, the thymidine dimer excision repair enzyme endonuclease V [32], the uracil DNA glycosylase UDG [33], and the guanine-thymine/uracil mismatch specific DNA glycosylase MUG [34] were crystallized with DNA, and base or nucleotide flipping was observed in the crystal structures. While UDG and MUG flip their target nucleotides, endonuclease V was found to flip one of the partner adenines of the thymidine dimer. However, in all three cases nucleotide flipping allows the catalytic machinery to access the target site within the bulky DNA substrate. In addition, the crystal structures of several DNA-modifying enzymes in the absence of DNA were reported. The catalytic sites of O6-methylguanine DNA Mtase [35], $\beta$-glucosyl transferase [36], exonuclease III [37], photolyase [38], endonuclease III [39], DNA 3-methyladenine glycosylase II [40], and endonuclease HAP1 [41] were all found in the interior of the proteins. In each instance, nucleotide flipping appears to be necessary to bring the amino acid residues of the active site in close proximity to the target site. Thus, base or nucleotide flipping appears to be a widespread property of DNA-modifying enzymes and the probes developed for base flipping by DNA Mtases could also be useful for studying other DNA-modifying enzymes.

**Acknowledgements:** We are grateful to Roger Goody for his continuous support. This work was supported by a grant from the Deutsche Forschungsgemeinschaft and from the Volkswagen-Stiftung.

## References

[1]  R. D. Hotchkiss, *J. Biol. Chem.* **1948**, *168*, 315–332; G. R. Wyatt, *Biochem. J.* **1951**, *48*, 581–584.

[2]  D. B. Dunn, J. D. Smith, *Biochem. J.* **1958**, *68*, 627–636.

[3]  A. Janulaitis, S. Klimašauskas, M. Petrušyte, V. Butkus, *FEBS Lett.* **1983**, *161*, 131–134.

[4]  X. Cheng, *Annu. Rev. Biophys. Biomol. Struct.* **1995**, *24*, 293–318.

[5]  R. L. P. Adams, R. H. Burdon, *Molecular Biology of DNA Methylation*, Springer-Verlag, New York, **1985**.

[6]  J. Heitman, *Genetic Engineering* **1993**, *15*, 57–108; F. Barras, M. G. Marinus, *Trends Genet.* **1989**, *5*, 139–143.

[7]  J. P. Jost, H. P. Saluz, *DNA Methylation: Molecular Biology and Biological Significance*, Birkhäuser Verlag, Basel, **1993**.

[8]  S. Klimasauskas, S. Kumar, R. J. Roberts, X. Cheng, *Cell* **1994**, *76*, 357–369.

[9]  K. M. Reinisch, L. Chen, G. L. Verdine, W. N. Lipscomb, *Cell* **1995**, *82*, 143–153.

[10] J. Labahn, J. Granzin, G. Schluckebier, D. P. Robinson, W. E. Jack, I. Schildkraut, W. Saenger, *Proc. Natl. Acad. Sci. USA* **1994**, *91*, 10957–10961; W. Gong, M. O'Gara, R. M. Blumenthal, X. Cheng, *Nucleic Acids Res.* **1997**, *25*, 2702–2715.

[11] D. C. Ward, E. Reich, L. Stryer, *J. Biol. Chem.* **1969**, *244*, 1228–1237.

[12] L. W. McLaughlin, T. Leong, F. Benseler, N. Piel, *Nucleic Acids Res.* **1988**, *16*, 5631–5644; B. A. Connolly, *Methods Enzym.* **1992**, *211*, 36–53; S. Schmidt, D. Cech, *Nucleosides & Nucleotides* **1995**, *14*, 1445–1452; J. Fujimoto, Z. Nuesca, M. Mazurek, L. C. Sowers, *Nucleic Acids Res.* **1996**, *24*, 754–759.

[13] B. Holz, S. Klimasauskas, S. Serva, E. Weinhold, *Nucleic Acids Res.* **1998**, *26*, 1076–1083.

[14] B. W. Allan, N. O. Reich, *Biochemistry* **1996**, *35*, 14757–14762.

[15] T. Malone, R. M. Blumenthal, X. Cheng, *J. Mol. Biol.* **1995**, *253*, 618–632.

[16] S. Serva, S. Klimašauskas, E. Weinhold, *Biologija* **1997**, 9–12.

[17] B. W. Allan, J. M. Beechem, W. M. Lindstrom, N. O. Reich, *J. Biol. Chem.* **1998**, *273*, 2368–2373.

[18] N. O. Reich, N. Mashhoon, *Biochemistry* **1991**, *30*, 2933–2939.

[19] N. O. Reich, N. Mashhoon, *J. Biol. Chem.* **1993**, *268*, 9191–9193.

[20] B. Holz, H. Pues, J. Wölcke, E. Weinhold, *FASEB J.* **1997**, *11*, A1151.

[21] S. Klimašauskas, T. Szyperski, S. Serva, K. Wüthrich, *EMBO J.* **1998**, *17*, 317–324.

[22] S. Cal, B. A. Connolly, *J. Biol. Chem.* **1996**, *271*, 1008–1015.

[23] R. A. García, C. J. Bustamante, N. O. Reich, *Proc. Natl. Acad. Sci USA* **1996**, *93*, 7618–7622.

[24] P.-O. Lycksell, A. Gräslund, F. Claesens, L. W. McLaughlin, U. Larsson, R. Rigler, *Nucleic Acids Res.* **1987**, *15*, 9011–9025.

[25] A. S. Yang, J.-C. Shen, J.-M. Zingg, P. Mi, P. A. Jones, *Nucleic Acids Res.* **1995**, *23*, 1380–1387.

[26] S. Klimašauskas, R. J. Roberts, *Nucleic Acids Res.* **1995**, *23*, 1388–1395.

[27] S. Cal, B. A. Connolly, *J. Biol. Chem.* **1997**, *272*, 490–496.

[28] M. Takeshita, C.-N. Chang, F. Johnson, S. Will, A. P. Grollman, *J. Biol. Chem.* **1987**, *262*, 10171–10179.

[29] B. Holz, E. Weinhold, *Nucleosides & Nucleotides* **1999**, in press.

[30] S. Serva, E. Weinhold, R. J. Roberts, S. Klimašauskas, *Nucleic Acids Res.* **1998**, *26*, 3473–3479.

[31] W. Wang, M. Carey, J. D. Gralla, *Science* **1992**, *255*, 450–453; R. Visse, A. King, G. F. Moolenaar, N. Goosen, P. van de Putte, *Biochemistry* **1994**, *33*, 9881–9888; I. Konieczny, K. S. Doran, D. R. Helinski, A. Blasina, *J. Biol. Chem.* **1997**, *272*, 20173–20178.

[32] D. G. Vassylyev, T. Kashiwagi, Y. Mikami, M. Ariyoshi, S. Iwai, E. Ohtsuka, K. Morikawa, *Cell* **1995**, *83*, 773–782.

[33] R. Savva, K. McAuley-Hecht, T. Brown, L. Pearl, *Nature* **1995**, *373*, 487–493; G. Slupphaug, C. D. Mol, B. Kavli, A. S. Arvai, H. E. Krokan, J. A. Tainer, *Nature* **1996**, *384*, 87–92.

[34] T. E. Barrett, R. Savva, G. Panayotou, T. Barlow, T. Brown, J. Jiricny, L. H. Pearl, *Cell* **1998**, *92*, 117–129.

[35] M. H. Moore, J. M. Gulbis, E. J. Dodson, B. Demple, P. C. E. Moody, *EMBO J.* **1994**, *13*, 1495–1501.

[36] A. Vrielink, W. Rüger, H. P. C. Driessen, P. S. Freemont, *EMBO J.* **1994**, *13*, 3413–3422.

[37] C. D. Mol, C.-F. Kuo, M. M. Thayer, R. P. Cunningham, J. A. Tainer, *Nature* **1995**, *374*, 381–386.

[38] H.-W. Park, S.-T. Kim, A. Sancar, J. Deisenhofer, *Science* **1995**, *268*, 1866–1872.

[39] M. M. Thayer, H. Ahern, D. Xing, R. P. Cunningham, J. A. Tainer, *EMBO J.* **1995**, *14*, 4108–4120.

[40] Y. Yamagata, M. Kato, K. Odawara, Y. Tokuno, Y. Nakashima, N. Matsushima, K. Yasumura, K. Tomita, K. Ihara, Y. Fujii, Y. Nakabeppu, M. Sekiguchi, S. Fujii, *Cell* **1996**, *86*, 311–319; J. Labahn, O. D. Schärer, A. Long, K. Ezaz-Nikpay, G. L.Verdine, T. E. Ellenberger, *Cell* **1996**, *86*, 321–329.

[41] M. A. Gorman, S. Morera, D. G. Rothwell, E. de La Fortelle, C. D. Mol, J. A. Tainer, I. D. Hickson, P. S. Freemont, *EMBO J.* **1997**, *16*, 6548–6558.

# 6.8    Manipulating intracellular signal transduction

*Carsten Schultz, Andrew Schnaars, and Marco T. Rudolf*

Intracellular signal transduction is the field of life sciences (avoiding the more traditional and somewhat limiting assignment to biochemistry or cell biology) that deals with the transfer of information from the plasma membrane to the interior of the cell up to the point where the cell detectably responds to a given stimulus. The signaling cascades involve macromolecules, namely proteins and small molecules alike. Being chemists, it is not unexpected that our interest focuses on the small molecules. These signaling molecules are frequently called second messengers because they distribute information brought to the plasma membrane by hormones or neurotransmitters (first messengers) throughout the cytosol and/or the inner sheet of the plasma membrane. Surprisingly, the number of second messengers is rather limited. In fact, up to the early 1980s the whole second messenger concept, originally installed by Sutherland [1], based on the action of cyclic adenosine $3',5'$-monophosphate (cAMP) and – to a more disputed extent – cyclic guanosine $3',5'$-monophosphate (cGMP). The focus shifted to the inositol phosphates in 1983, when Streb and coworkers discovered the ability of *myo*-inositol 1,4,5-trisphosphate [$Ins(1,4,5)P_3$] to increase intracellular calcium levels by its release from internal stores [2]. More recently, inositol phosphate-containing phospholipids, the phosphoinositides, were suspected to mediate external signals, especially signals from growth factors [3]. There are others like diacylglycerol (DAG) and nitric oxide (NO), but apart from those, most second messengers share the feature of carrying negatively charged phosphates. Cell lysate analysis showed that in living cells a variety of around 20–30 inositol phosphates [4] and probably seven to eight different phosphoinositides were present [5]. Are they all intracellular messengers? If some of them are, what do they regulate? These are two of the questions we would like to answer in the future. Other aims involve the often-advertized hope that direct intracellular manipulation of second messenger levels, in by-passing the plasma membrane receptors, could serve as a therapeutic tool in the treatment of diseases.

To help answer these questions, chemists are able to provide tools for cell biology or biochemical experiments and eventually even therapeutics for medical treatment. As mentioned above, most intracellular messengers are phosphate esters. In the case of the inositol polyphosphates they have numerous negative charges, which prevent passive diffusion over the plasma membrane, thus allowing distinct signaling of single cells in large cell communities. If the goal is to manipulate the concentration of a single messenger by artificial delivery of the compound to the cytosol, these negative charges have to be masked by bioactivatable protecting groups.

## 6.8.1    Methodology

Bioactivatable masking groups have previously been used to convert carboxylic acids [6–8], phosphonates [9, 10], or phosphates [11] into uncharged, membrane-permeant derivatives. This methodology was frequently employed to create prodrugs with enhanced bioavailability [12]. Preferred masking groups were mostly acyloxymethyl esters. For the inositol polyphosphates, and in some cases also for cyclic nucleotides, we had to employ butyric esters to mask hydroxy groups. This appeared to be particularly important in cases where the hydroxy groups are located in a vicinal position to the masked phosphate triesters, because upon enzymatic hydrolysis the formation of unstable cyclic phosphate intermediates is likely to occur. This leads to the migration of the phosphate. Butyrylated inositol polyphosphate acetoxymethyl (AM) esters were lipophilic enough to be dissolved by toluene. On the other hand the compounds exhibited a moderate solubility in water. AM esters were shown to be sufficiently stable in extracellular medium with a half-life of several hours for most phosphate triesters [13]. However, endogenous intracellular enzymes rapidly hydrolyze the carboxylic ester, while the phosphate is instantly liberated by the loss of formaldehyde (Figure 6.8.1).

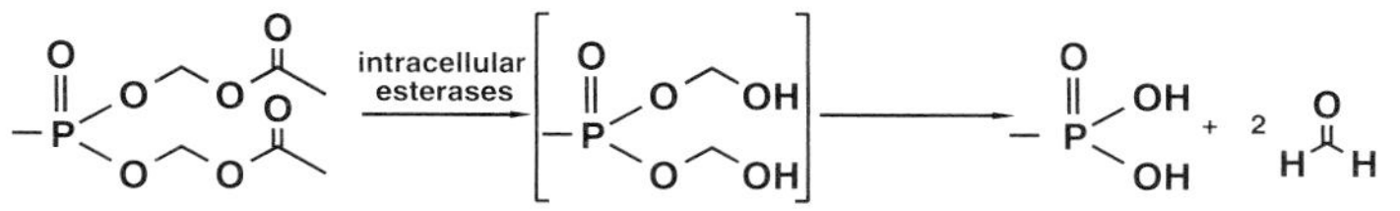

**Figure 6.8.1.** Acetoxymethyl esters are hydrolyzed under loss of acetic acid and formaldehyde.

## 6.8.2 Confirming known messenger functions

First attempts to apply the AM-ester methodology to second messengers concentrated on the cyclic nucleotides cAMP and cGMP, mainly because these compounds had only one negative charge on the phosphate and because a wide array of biological test sytems was available. The alkylation of $N^6,O^{2'}$-dibutyryl-cAMP ($Bt_2$cAMP) with acetoxymethyl bromide (AM-Br) in the presence of diisopropylethylamine gave $Bt_2$cAMP acetoxymethyl ester ($Bt_2$cAMP/AM; Figure 6.8.2) in good yields. $Bt_2$cAMP was chosen because of its known metabolic stability and because it exhibits a certain membrane-permeability itself. As expected, $Bt_2$cAMP/AM was several hundred-fold more potent than $Bt_2$cAMP in different applications such as the chloride secretion of epithelial cells or the aggregation of dye from fish melanophores [13].

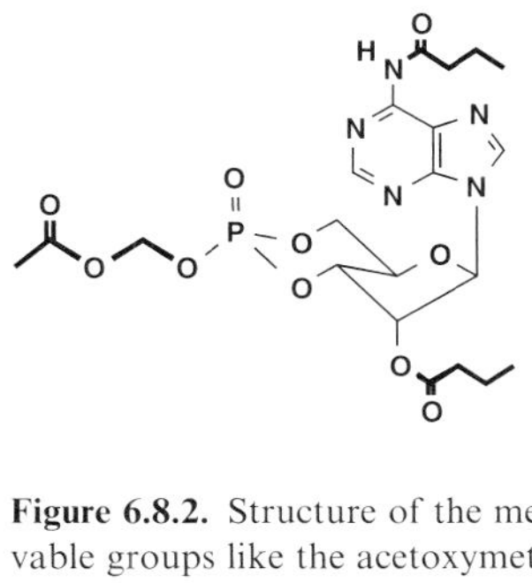

**Figure 6.8.2.** Structure of the membrane-permeant cAMP derivative $Bt_2$cAMP/AM. Enzymatically removable groups like the acetoxymethyl ester and the $O^{2'}$-butyrate are depicted in bold.

Threshhold concentrations for biological activity in the extracellular medium were as low as 10 nM [14]. Recent analysis of cell lysates from C6-glioma cells revealed that after incubation with 30 μM $Bt_2cAMP/AM$, the biologically active compound $N^6$-monobutyryl-cAMP accumulated inside the cell about 3.6-fold relative to the external concentration [15]. This shows that AM-esters and butyric esters are rapidly removed by endogenous esterases.

cAMP/AM, a derivative without butyrates was less potent but had the advantage of being able to generate cAMP itself, thus allowing a transient cAMP signal inside cells [16]. In the meantime, structurally modified cAMP derivatives have been synthesized and were successfully tested [17].

The synthesis of cGMP acetoxymethyl ester derivatives proved to be more difficult. The final product of cGMP/AM inhibited platelet aggregation with an $EC_{50}$ of 1 μM [18]. At this concentration the compound induced long-term-potentiation of rat hippocampal neurons [19].

The preparation of acetoxymethyl esters of the second messenger $Ins(1,4,5)P_3$ was more challenging. At that time $Ins(1,4,5)P_3$ (Figure 6.8.3a) represented the only inositol polyphosphate with an established biological function. First attempts required total synthesis from *myo*-inositol with regioselective introduction of the butyrates to cover three of the six hydroxy groups. The non-esterified OH-groups were phosphorylated and the phosphates were alkylated with AM-Br. The final product 2,3,6-tri-*O*-butyryl-*myo*-inositol 1,4,5-trisphosphate hexakis(acetoxymethyl) ester (Figure 6.8.3b) proved to be unreliable in its ability to elevate intracellular calcium concentrations [20]. This was probably due to the fact that the generation of intracellular $Ins(1,4,5)P_3$

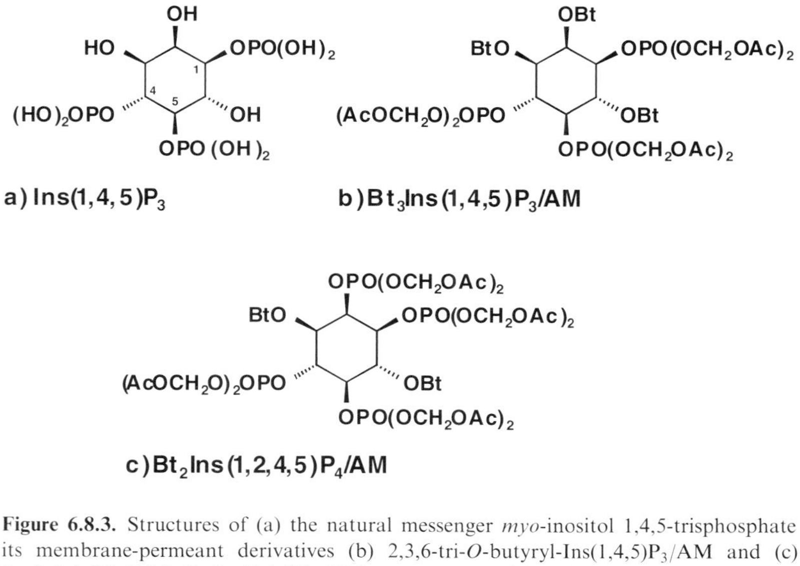

a) $Ins(1,4,5)P_3$          b) $Bt_3Ins(1,4,5)P_3/AM$

c) $Bt_2Ins(1,2,4,5)P_4/AM$

**Figure 6.8.3.** Structures of (a) the natural messenger *myo*-inositol 1,4,5-trisphosphate [$Ins(1,4,5)P_3$] and its membrane-permeant derivatives (b) 2,3,6-tri-*O*-butyryl-$Ins(1,4,5)P_3$/AM and (c) 3,6-di-*O*-butyryl-$Ins(1,2,4,5)P_4$/AM. $Bt_3Ins(1,4,5)P_3$/AM was unsuccessful probably due to fast intracellular metabolism (see text), but the additional 2-phosphate of $Bt_2Ins(1,2,4,5)P_4$/AM resulted in sufficient metabolic stability.

by intracellular esterases was slower than the metabolism of the generated Ins(1,4,5)P$_3$. Therefore, Wen-hong Lee from Roger Tsien's group prepared a derivative which refrained from the slowly hydrolyzing butyrates but carried propionoxymethyl (PM) groups instead of AM-groups to increase lipophilicity. This derivative was finally able to elevate calcium levels to a steady intracellular plateau [20]. Our own approach focused on the synthesis of a metabolically stable Ins(1,4,5)P$_3$ derivative. We therefore added another phosphate to the 2-OH-position of Ins(1,4,5)P$_3$. The resulting 3,6-di-*O*-butyryl-*myo*-inositol 1,2,4,5-tetrakisphosphate octakis(acetoxymethyl) ester (Figure 6.8.3c) was able to induce a long-lasting calcium plateau in PC12 cells [21]. Since the physiological signal is usually a calcium spike rather than a steadily elevated calcium level, Lee and co-workers introduced a photochemically removable nitrobenzyl group to the molecule. After incubation, the demasked Ins(1,4,5)P$_3$-derivative was shortly illuminated and by loss of the nitrobenzyl group the now biologically active compound gave a short and transient calcium spike. The introduction of several different time patterns of subsequent calcium spikes showed that a very particular activation of intracellular calcium levels was necessary to induce gene expression [22].

## 6.8.3 New messenger functions

When it became clear that the acyloxymethyl ester methodology would work for cyclic nucleotides and inositol phosphate alike, the door was opened for the intriguing possibility to study the yet unknown potential actions of the 20 or more inositol phosphates present in cells, in addition to Ins(1,4,5)P$_3$. Until recently, most of these were believed to be predominantly metabolites of the Ins(1,4,5)P$_3$ signaling pathway. In the past five years, experiments with radioactively labeled inositols showed that the higher phosphorylated inositol polyphosphates like some of the InsP$_4$s and all InsP$_5$s are formed too slowly from phosphatidylinositide breakdown and Ins(1,4,5)P$_3$ turnover (Figure 6.8.4), to be part of the same circuitry [23].

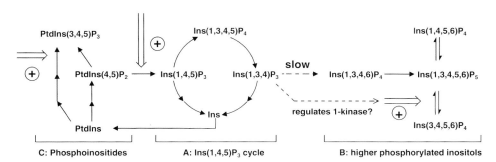

**Figure 6.8.4.** The intracellular orchestra of inositol phosphates: (A) The Ins(1,4,5)P$_3$ cycle known since the 1980s. The system of the higher phosphorylated inositol phosphates (B) is basically not metabolically connected to (A). (C) The phosphoinositides. PtdIns(3,4,5)P$_3$/AM accumulates after growth factor stimulation. Wide arrows show where extracellular agonists activate mass changes of the three distinct regulatory systems. For details on these signaling molecules, see text.

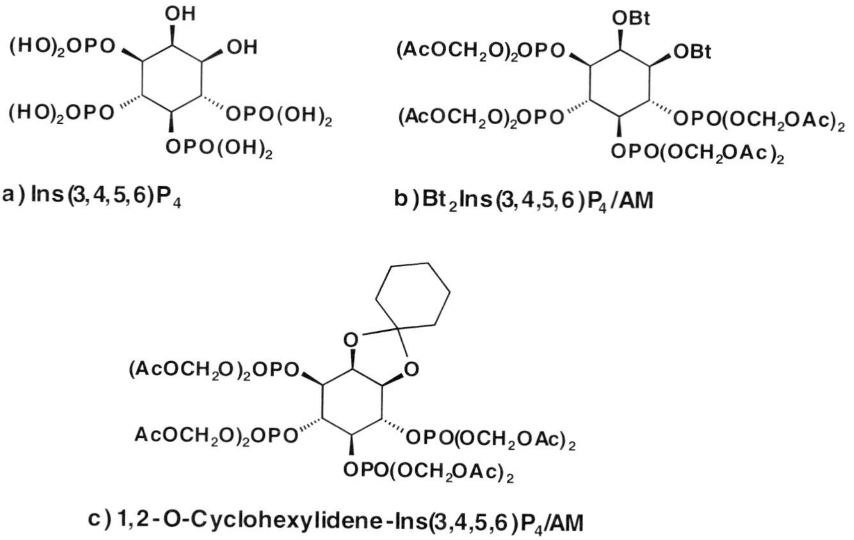

a) Ins(3,4,5,6)P$_4$

b) Bt$_2$Ins(3,4,5,6)P$_4$/AM

c) 1,2-O-Cyclohexylidene-Ins(3,4,5,6)P$_4$/AM

**Figure 6.8.5.** (a) The newly identified intracellular messenger *myo*-inositol 3,4,5,6-tetrakisphosphate [Ins(3,4,5,6)P$_4$]. Its membrane-permeant derivative 1,2-di-*O*-butyryl-Ins(3,4,5,6)P$_4$/AM (b) mimicked inhibitory effects of receptor activation for instance by carbachol on Cl$^-$ secretion of epithelial cells. (c) 1,2-*O*-Cyclohexylidene-Ins(3,4,5,6)P$_3$/AM represents an antagonist that blocks the inhibitory effect.

myo-*Inositol 3,4,5,6-tetrakisphosphate*. It has been known for quite some time that the stimulation of calcium-mediated chloride secretion (CaMCS) of epithelial cells by external agonists like carbachol or histamine could only be activated once in several hours. Because even artificial elevation of intracellular calcium by ionophores (thus circumventing receptor activation) was ineffective to activate CaMCS for a second time, it was proposed that: (i) a chemical messenger is preventing repeated CaMCS and (ii) this shut-off messenger is an inositol polyphosphate. Mass analysis from cell lysates revealed that one of them, *myo*-inositol 3,4,5,6-tetrakisphosphate [Ins(3,4,5,6)P$_4$; Figure 6.8.5a], had a sufficiently long life time to be a candidate [24].

We therefore prepared the membrane-permeant derivative Bt$_2$Ins(3,4,5,6)P$_4$/AM (Figure 6.8.5b) in a 10-step total synthesis. When cells of the human colon epithelial cell line T$_{84}$ were preincubated with Bt$_2$Ins(3,4,5,6)P$_4$/AM for 30 min the carbachol-induced Cl$^-$ secretion was greatly diminished (Figure 6.8.6). The enantiomeric derivative Bt$_2$Ins(1,4,5,6)P$_4$/AM was completely inactive. These results showed that Ins(3,4,5,6)P$_4$ acted as an intracellular shut-off messenger in epithelial cells [25]. Support was gained by the work form Debbie Nelson's group who performed whole-cell patch-clamp experiments and introduced Ins(3,4,5,6)P$_4$ to the cytosol *via* the patch pipette [26].

These results gave rise to a number of important questions. First, how is Ins(3,4,5,6)P$_4$ influencing Cl$^-$ secretion? Second, how is the formation of Ins(3,4,5,6)P$_4$ regulated? Clearly, Ins(3,4,5,6)P$_4$ was slowly building up after stimulation of the very same

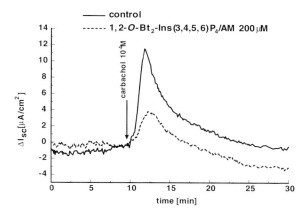

**Figure 6.8.6.** Preincubation with $Bt_2Ins(3,4,5,6)P_4$/AM for 30 min resulted in a greatly reduced secretion of $Cl^-$ when epithelial cells were stimulated with carbachol. Similar results helped to establish the shut-off messenger function of $Ins(3,4,5,6)P_4$. $I_{SC}$ = short circuit current which totally reflects the flux of $Cl^-$ from the basolateral to the apical side of the cells.

receptors which mediated the fast and transient rise in $Ins(1,4,5)P_3$ levels [25]. The pool from which $Ins(3,4,5,6)P_4$ is formed should be $Ins(1,3,4,5,6)P_5$, the most abundant inositol phosphate in most cells. How is the signal carried from the receptor to the phosphatase/kinase system that regulates $Ins(3,4,5,6)P_4$ levels? It appears now, that $Ins(1,3,4)P_3$ is regulating the formation of the shut-off messenger.

myo-*Inositol 1,3,4-trisphosphate [Ins(1,3,4)P_3].* $Ins(1,3,4)P_3$ is a major metabolite of $Ins(1,4,5)P_3$, formed after its phosphorylation to $Ins(1,3,4,5)P_4$ and subsequent dephosphorylation (Figure 6.8.4). *In-vitro* investigations of the two enzymes $Ins(1,3,4,5,6)P_5$ 1-phosphatase and $Ins(3,4,5,6)P_4$ 1-kinase by the group of Shears in Triangle Park showed that $Ins(1,3,4)P_3$ was a fairly good inhibitor of the 1-kinase. It was therefore proposed that elevated $Ins(1,3,4)P_3$ levels could regulate the $InsP_4$/$InsP_5$ equilibrium [27]. Our synthetic $Bt_3Ins(1,3,4)P_3$/AM was indeed able to reduce CaMCS of epithelial cells after stimulation with carbachol, just as we observed for the $Ins(3,4,5,6)P_4$-derivative [28]. Our hypothesis was further manifested by experiments with [³H]inositol-labeled AR4–2J pancreatoma cells performed by Shears' group: treatment with the membrane-permeant $Ins(1,3,4)P_3$-derivative increased levels of [³H]$Ins(3,4,5,6)P_4$ significantly [29]. It appears, that activation and shut-off signaling pathways are both mediated by $Ins(1,4,5)P_3$ in the first place. While the activation of for instance $Cl^-$ secretion depends on elevated calcium levels, the negative control involves the metabolite $Ins(1,3,4)P_3$, which regulates the formation of $Ins(3,4,5,6)P_4$. Although the $Ca^{2+}$/calmodulin-dependent protein kinase type II seems to be involved in $Ins(3,4,5,6)P_4$-signaling [26], its detailed way of function remains unclear at the moment. To us, this appears to be a major project to develop tools to illuminate this important facet in intracellular signaling.

myo-*Inositol 1,4,5,6-tetrakisphosphate.* As mentioned above, $Ins(1,4,5,6)P_4$, the enantiomer of $Ins(3,4,5,6)P_4$, is inactive in inhibiting $Cl^-$ secretion. However,

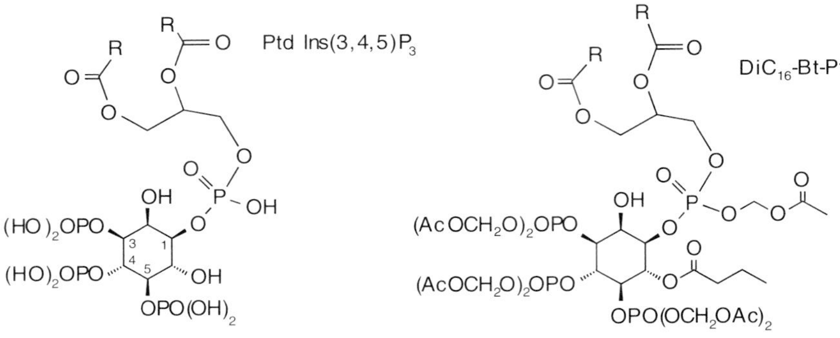

**Figure 6.8.7.** Structures of phosphatidylinositol 3,4,5-trisphosphate [PtdIns(3,4,5)P₃] and its membrane-permeant derivative 1′,2′-di-*O*-palmitoyl-6-*O*-butyryl-PtdIns(3,4,5)P₃/AM (R = C₁₅H₃₁). In the natural product, the fatty acids composition consists mostly of stearic and arachidonic acid (*sn2*-position).

when intestinal epithelial cells were infected with the enteric pathogen *Salmonella*, an invasive bacterium, Cl⁻ secretion is greatly increased causing severe diarrhea. At the same time, Ins(1,4,5,6)P₄ levels were found to be elevated manyfold [30]. It is currently unknown why this latter phenomenon occurs: is this induced by the bacteria? Does the infected cell try to defend itself? We provided the membrane-permeant derivative Bt₂Ins(1,4,5,6)P₄/AM to increase Ins(1,4,5,6)P₄ levels in non-infected epithelial cells, thus mimicking this particular effect induced by *Salmonella*. The surprising effect found was that the previously known inhibitory effect of growth factors like epidermal growth factor (EGF) on Cl⁻ secretion was partially blocked [30].

*Phosphatidylinositol 3,4,5-trisphosphate.* Many effects of growth factors are mediated via the activation of phosphatidyl inositol 3-kinase [5]. The enzyme phosphorylates phosphoinositides at the 3-position of the inositol moiety. The most prominent products to date are phosphatidylinositol 3,4,5-trisphosphate (PtdIns(3,4,5)P₃; Figure 6.8.7) and its dephosphorylated 3,4-bisphosphate [5, 31]. However, there was no definitive proof that these compounds function as intracellular messengers to mediate growth factor signaling, although many of the participating genes downstream have been identified and characterized [5]. Membrane-permeant derivatives should be able to help with this dilemma.

These compounds should be able to mimick growth factor activity, if for instance PtdIns(3,4,5)P₃ is an essential mediator in the signaling pathway. As described above, EGF inhibits Cl⁻ secretion of epithelial cells. We therefore synthesized 1′,2′-di-O-palmitoyl-6-O-butyryl-phosphatidyl-*myo*-inositol 3,4,5-trisphosphate heptakis-(acetoxymethyl) ester (DiC₁₆-Bt-PtdInsP₃/AM; Figure 6.8.7), a membrane-permeant derivative of PtdIns(3,4,5)P₃ [21]. The compound was inhibitory to epithelial Cl⁻ secretion similar to EGF [30]. Furthermore, this inhibitory effect was blocked by our membrane-permeant Ins(1,4,5,6)P₄ derivative (see above). Concerns that the observed effects were artifacts could be relieved by using similar membrane-permeant PtdIns(3,4,5)P₃ derivatives from friendly competitors, which were prepared by a totally different synthetic pathway [32]. In the meantime, Tsien's group showed that these compounds also mimic other known growth factor-mediated cellular responses such as glucose uptake of adipocytes (fat cells) [32]. Improved synthetic procedures will hopefully enable us and others to further manifest the function of PtdIns(3,4,5)P$_3$ and other phosphoinositides as intracellular messengers.

## 6.8.4 Drug development

Many publications describing the synthesis of inositol phosphates and their derivatives express the hope that the synthetic compounds are of use for medical treatment in the future [33]. Although this is certainly helpful to convince funding agencies, little work has been done to explore the possibilities of inositol phosphates as drugs. The lack of membrane-permeability was certainly one of the reasons since, after all, most inositol phosphates are acting inside cells. Our methodology to form membrane-permeant derivatives should help with this problem, although up to now the nature of the masking groups and in particular the formation of formaldehyde upon their hydrolysis is a caveat. Our interest concentrates on diseases involving epithelial cells, namely diarrhea and cystic fibrosis (CF). While in diarrhea Cl$^-$ secretion is greatly enhanced, cystic fibrosis patients suffer from chronically reduced Cl$^-$ secretion and subsequently a reduced water flux from the body to the lumen of the lungs and the gastrointestinal tract. This is due to a point mutation in the cAMP-regulated apical chloride channel [34]. Currently, there is no treatment for CF available which helps circumvent the molecular defect. Therefore, the average life-time of CF patients is now around 30 years (doubled in the last 20 years due to the efforts of physicians to relieve the symptoms), making CF the most abundant deadly inheritary disease in the western world. One of the ways discussed of treating CF patients is the activation of alternative chloride channels in the lung, in particular those activated by the calcium signaling pathway [34].

As we have seen above, this pathway can only be activated once in several hours due to the inhibitory effect of the intracellular messenger Ins(3,4,5,6)P$_4$. We therefore synthesized a moderately large number of Ins(3,4,5,6)P$_4$ derivatives and converted them to membrane-permeant forms [35]. One of the purposes was certainly the faint hope of finding an antagonist against Ins(3,4,5,6)P$_4$ which should enable us to block the inhibitory effect and allow repeated opening of calcium-activated chloride channels (Figure 6.8.8). Otherwise, systematic variations of the molecular structure should give information about the as yet unknown Ins(3,4,5,6)P$_4$ binding proteins. By modifying the two hydroxy groups of Ins(3,4,5,6)P$_4$ we were indeed fortunate to find two antagonists: 1,2-*O*-cyclohexylidene-Ins(3,4,5,6)P$_4$/AM (Figure 6.8.5) and 1-*O*-butyl-2-*O*-butyryl-Ins(3,4,5,6)P$_4$/AM. Both compounds did not inhibit chloride secretion as was shown for Bt$_2$Ins(3,4,5,6)P$_4$/AM. However, 25 min after a transient increase in chloride secretion induced by carbachol (which elevated natural Ins(3,4,5,6)P$_4$ levels) a second reponse to the calcium-mobilizing agent thapsigargin was fully restored compared to control experiments lacking the preincubation with the membrane-permeant antagonists (Figure 6.8.9) [36]. Since the epithelia of the

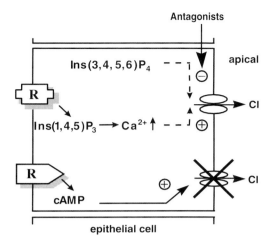

**Figure 6.8.8.** By blocking the inhibitory Ins(3,4,5,6)P₄ effect, antagonists should be able to allow repeated opening of calcium-activated chloride channels to compensate for the defective cAMP-regulated chloride channel, the cystic fibrosis transmembrane regulator (CFTR). Rs represent plasma membrane receptors. Their occupation leads to the formation of second messengers and subsequently to Cl⁻ secretion.

lung, where CF has its most dramatic effects, should be readily accessible to aerosols, future developments of antagonists with increased potency and penetration may become drug candidates for the treatment of CF. We are currently preparing small combinatorial libraries to find membrane-permeant antagonists with these improved properties.

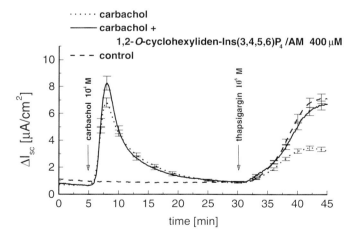

**Figure 6.8.9.** Membrane-permeant antagonists of Ins(3,4,5,6)P₄ allow a second activation of calcium-mediated chloride secretion after addition of carbachol (100 μM, first arrow). Thapsigargin, an ATPase-inhibitor that elevates intracellular calcium levels without triggering the receptor/Ins(1,4,5)P₃ pathway, was added where indicated by the second arrow (1 μM). $I_{SC}$ = short circuit current.

For the treatment of diarrhea, an inhibitory effect of membrane-permeant Ins(3,4,5,6)P$_4$ derivatives on chloride secretion is desirable. Derivatives like the acetoxymethyl ester of 2-deoxy-Ins(3,4,5,6)P$_4$ were shown to be partial agonists of Ins(3,4,5,6)P$_4$ [37]. These compounds could be a starting point for future developments in this direction. Our "*in vivo*" structure-activity results with the current set of derivatives show that the oxygen of the 1-hydroxy group of Ins(3,4,5,6)P$_4$ is essential for binding to the unknown binding proteins while the 2-hydroxy group offers more room for modifications. This information should help us to develop further tools to identify the binding proteins, for instance, affinity resins for chromatography or fluorescently labeled derivatives. Unfortunately, preliminary experiments with the latter were unsuccessful.

## 6.8.5   Conclusions, questions, and perspectives

We have shown that membrane-permeant derivatives of inositol phosphates and phosphoinositides are valuable tools to help identifying the previously unknown cellular functions of these compounds. Furthermore, treatment of cells with these derivatives allows a very specific manipulation of intracellular signaling molecules, thus disecting the strongly interacting signaling pathways. The question should be asked whether chemists will have to synthesize membrane-permeant derivatives of all natural inositol phosphates in the future, or whether one should concentrate only on currently fashionable isomers. From the standpoint of most chemists, the fifth or sixth total synthesis of just another structural isomer would be the time to change the project. Are the biological results important enough to justify only moderately interesting chemistry? Perhaps, the answer has to be found in the questions the scientist asked before starting the project. Some relief to this dilemma will arise from the synthesis of designed derivatives for drug development. We are currently synthesizing Ins(3,4,5,6)P$_4$ derivatives with added ring structures to the carbon backbone, which are expected to show improved properties to fulfil the criteria extracted from our structure-activity results.

An urgent need is the development of new masking groups to replace the acyloxymethyl esters. In particular, drug candidates should avoid the release of formaldehyde or other harmful by-products, but perhaps a finally successful derivative to treat CF will no longer have phosphates and therefore bioactivatable protecting groups will not be necessary. To be able to take more rational steps into this direction the Ins(3,4,5,6)P$_4$-binding proteins need to be identified, characterized, and purified. Most likely, the gene has to be cloned and the protein will need to be prepared by gene technology to produce sufficient amounts to perform *in-vitro* screening experiments. Moreover, our Ins(3,4,5,6)P$_4$ antagonists should be tested on tissue from CF patients.

Perhaps this project is an example to show that bioorganic chemistry is not limited to any particular discipline. Indeed, if performed with consequence, the chances are that one will need to work in aspects of chemistry, biochemistry, cell biology, molecular biology, and physiology – the life sciences.

# References

[1]   E. W. Sutherland, *Angew. Chem.* **1972**, *84*, 1117–1125; *Science* **1972**, *177*, 401–408.
[2]   H. Streb, R. F. Irvine, M. J. Berridge, I. Schulz, *Nature* **1983**, *306*, 67–69.
[3]   A. Ullrich, J. Schlessinger, *Cell* **1990**, *61*, 203–212.
[4]   G. W. Mayr, *Topics in Biochemistry*, Boehringer Mannheim **1988**.
[5]   A. Toker, L. C. Cantley, *Nature* **1997**, *387*, 673–676.
[6]   A. B. A. Jansen, T. J. Russell, *J. Chem. Soc.* **1965**, 2127–2132.
[7]   R. Y. Tsien, *Nature* **1981**, *290*, 527–528.
[8]   G. Grynkiewicz, M. Ponie, R. Y. Tsien, *J. Biol. Chem.* **1985**, *260*, 3440–3450.
[9]   R. P. Iyer, L. R. Phillips, J. A. Biddle, D. R. Thakker, W. Egan, S. Aoki, H. Mitsuga, *Tetrahedron Lett.* **1989**, *30*, 7141–7144.
[10]  D. Farquhar, S. Khan, M. C. Wilkerson, B. S. Anderson, *Tetrahedron Lett.* **1995**, *36*, 655–658.
[11]  J. K. Sastry, P. N. Nehete, S. Khan, B. J. Nowak, W. Plunkett, R. B. Arlinghaus, D. Farquhar, *Mol. Pharmacol.* **1992**, *41*, 441–445.
[12]  R. B. Silverman, *The organic chemistry of drug design and drug action, Chapter 8*, Academic Press, Inc., **1992**; *Medizinische Chemie, Chapter 8*, Verlag Chemie, Weinheim **1995**.
[13]  C. Schultz, M. Vajanaphanich, K. E. Barrett, P. J. Sammak, A. T. Harootunian, R. Y. Tsien, *J. Biol. Chem.* **1993**, *268*, 6316–6322.
[14]  M. Vajanaphanich, C. Schultz, A. E. Traynor-Kaplan, S. J. Pandol, R. Y. Tsien, K. E. Barrett, *J. Clin. Invest.* **1995**, *96*, 386–393.
[15]  M. Bartsch, *Diploma-Thesis*, University of Bremen **1997**.
[16]  C. Schultz, M. Vajanaphanich, H.-G. Genieser, B. Jastorff, K. E. Barrett, R. Y. Tsien, *Mol. Pharmacol.* **1994**, *46*, 702–708.
[17]  J. Kruppa, S. Keely, F. Schwede, C. Schultz, K. E. Barrett, B. Jastorff, *Bioorg. & Med. Chem. Lett.* **1997**, *7*, 945–948.
[18]  C. Schultz, L. Makings, R. Y. Tsien, unpublished results.
[19]  M. Zhuo, Y. Hu, C. Schultz, E. R. Kandel, R. D. Hawkins, *Nature* **1994**, *368*, 635–639.
[20]  W-h. Li, C. Schultz, J. Llopis, R. Y. Tsien, *Tetrahedron* **1997**, *53*, 12017–12040.
[21]  C. Schultz, M. T. Rudolf, H. Gillandt, A. E. Traynor-Kaplan. In: *Phosphoinositides* (ed. K. S. Bruzik), *Am. Chem. Soc., Symp. Series* **1999**, *718*, 232–243.
[22]  W-h. Li, J. Llopis, M. Whitney, G. Zlokarnik, R. Y. Tsien, *Nature* **1998**, *392*, 936–941.
[23]  F. S. Menniti, K. G. Oliver, J. W. Putney, Jr., S. B. Shears, *Trends Biochem. Sci.* **1993**, *18*, 53–56.
[24]  U. Kachintorn, M. Vajanaphanich, K. E. Barrett, A. E. Traynor-Kaplan, *Am. J. Physiol.* **1993**, *264*, C671–C676.
[25]  M. Vajanaphanich, C. Schultz, M. T. Rudolf, M. Wasserman, P. Enyedi, A. Craxton, S. B. Shears, R. Y. Tsien, K. E. Barrett, A. E. Traynor-Kaplan, *Nature* **1994**, *371*, 711–714.
[26]  W. Xie, M. A. Kaetzel, K. S. Bruzik, J. R. Dedman, S. B. Shears, D. J. Nelson, *J. Biol. Chem.* **1996**, *271*, 14092–14097.
[27]  Z. Tan, K. S. Bruzik, S. B. Shears, *J. Biol. Chem.* **1997**, *272*, 2285–2290.
[28]  M. T. Rudolf, A. E. Traynor-Kaplan, C. Schultz, *Bioorg. & Med. Chem. Lett.* **1998**, *8*, 1857–1860.
[29]  X. Yang, M. T. Rudolf, M. A. Carew, M. Yoshida, V. Nevreter, A. M. Riley, S.-K. Chung, K. S. Bruzik, B. V. L. Potter, C. Schultz, S. B. Shears, *J. Biol. Chem.* **1999**, *274*, 18973–18980.
[30]  L. Eckmann, M. T. Rudolf, A. Ptasznik, C. Schultz, T. Jiang, N. Wolfson, R. Y. Tsien, J. Fierer, S. B. Shears, M. F. Kagnoff, A. E. Traynor-Kaplan, *Proc. Natl. Acad. Sci. USA* **1997**, *94*, 14456–14460.
[31]  R. Kapeller, L. C. Cantley, *BioEssays*, **1994**, *16*, 565–576.
[32]  T. Jiang, G. Sweeney, M. T. Rudolf, A. Klip, A. E. Traynor-Kaplan, R. Y. Tsien, *J. Biol. Chem.* **1998**, *273*, 11017–11024.
[33]  B. V. L. Potter, D. Lampe, *Angew. Chem.* **1995**, *107*, 2085–2125; *Angew. Chem., Int. Ed. Engl.* **1995**, *34*, 1933–1972.

[34] M. J. Welsh, A. E. Smith, *Spektr. d. Wiss.* **1996**, 32–39.

[35] S. Roemer, C. Stadler, M. T. Rudolf, B. Jastorff, C. Schultz, *J. Chem. Soc., Perkin Trans. I* **1996**, 1683–1694.

[36] M. T. Rudolf, *PhD Thesis,* University of Bremen, Wissenschaft & Technik Verlag, Berlin **1998**.

[37] M. T. Rudolf, W-h. Li, N. Wolfson, A. E. Traynor-Kaplan, C. Schultz, *J. Med. Chem.* **1998**, *41*, 3635–3644.

# 7 Physical and Analytical Methods

## 7.1    Adventures with atomic force microscopy

*Chris Abell, Rachel McKendry, and Trevor Rayment*

Much of organic chemistry has been focused on the synthesis of novel compounds, understanding the mechanisms of reactions, and structure determination (generally of monomeric molecules). At the same time, the subject has been rooted in experimental protocols which have involved the preparation of at least milligram amounts of material, generally in a glass round-bottomed flask. It is clear that this regime is now undergoing a revolution.

This revolution is being driven from many directions, not least by the radical concepts underpinning the whole area of combinatorial chemistry, and by a merging of interests with biochemists on macromolecular structure. Another contributing factor is the freedom of scale given by the higher resolution and greater sensitivity of conventional analytical techniques such as NMR spectroscopy and mass spectrometry. However, detection of *atto*moles by mass spectrometry seems rather insensitive compared to the opportunities presented by techniques that enable us to look at and manipulate single molecules.

A key development in this respect was the invention of the scanning tunnelling microscope (STM) by Binnig and Rohrer in 1982, in which individual surface atoms were imaged with unprecedented resolution, in real space [1]. For this discovery, Binnig and Rohrer were awarded with the Nobel Prize for Physics in 1986. In the same year this group also described the atomic force microscope [2]. This instrument has proved to have many advantages for imaging all kinds of samples. However, its potential extends far beyond this role, most notably to the study of intermolecular interactions. This potential led us to set up an interdisciplinary collaboration to find new applications for atomic force microscopy (AFM).

Our thesis was that by combining our skills in organic synthesis, bioorganic chemistry, and scanning probe methodology we would be able to ask new questions. Undertaking scientific investigations in this way is exciting because the collaborators have different value systems and recognize different goals as being important. What actually emerges may be something that neither party can fully assess, largely because it is in new and unknown territory. However, if the participating scientists find it exciting, it is likely that the broader scientific community will also. In this article we briefly describe the ideas behind AFM and highlight the achievements of other groups who have pioneered the technique. We then describe our own first steps in what is proving to be a very rewarding scientific adventure.

## 7.1.1   Atomic force microscopy (AFM)

AFM is one of a family of techniques based on mechanically scanning a probe across a sample surface to investigate properties of the surface. Unlike STM, it does not rely on electron tunneling to probe the properties of a conductive sample, but instead measures the forces between the tip and sample, using the equivalent of a sense of touch. Images are obtained by scanning a very sharp silicon nitride tip across a surface. The deflection of a cantilever is monitored by focusing a laser beam off the free end of the cantilever onto a multiple segment photodiode known as a position-sensitive detector. The sample is scanned point-by-point, line-by-line until a two-dimensional map, or image is created (Scheme 7.1.1).

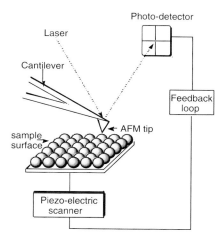

**Scheme 7.1.1.** The basic components of the atomic force microscope.

The AFM has a theoretical force sensitivity of $10^{-14}$ N, a lateral resolution of 0.1 nm, and the capability of obtaining atomic resolution of metals and even organic compounds. Samples can be imaged in a vacuum, air or in liquid [3]. The sample is mounted on a piezoelectric scanner which can translate the sample in the x, y, and z direction. In contact mode, the tip actually touches the sample surface, and the sample is rastered beneath the tip with a constant force. This probes the repulsive force which arises from overlapping electron orbitals between the tip and sample. Because these forces are short range and strongly distance-dependent, contact AFM achieves high spatial resolution. However, large pressures are generated at the surface which can damage or deform the sample. This is a particularly important concern when imaging biological specimens. In tapping mode, the tip is brought into contact with the sample, immediately retracted and then moved laterally to a different position where a new approach is made. In this way lateral forces and 'dragging' are minimized [4].

In addition to imaging samples, AFM can also provide information about the forces which exist between the tip and sample. It is possible measure the force felt by the cantilever as the probe tip is brought close to and even indented into the sample surface, and then pulled away. This technique, termed force spectroscopy, can be used to measure the attractive or repulsive forces between the probe tip and the sample surface, providing information about the local chemical and mechanical properties like adhesion and elasticity.

AFM can also be used to study the frictional forces between the tip and a surface at an atomic scale (nanotribology). Indeed, AFM has unique advantages for the study of friction as it simultaneously measures topography, and normal and lateral forces, the three most important quantities in any tribological process. As the tip is rastered across sample surface in contact mode, surface features result in changes in the signal detected by the upper and lower parts of a quadrant detector. At the same time, lateral torsion due to friction results in different signals to the left and right side of the detector. A two-dimensional friction force image can be generated, where high friction corresponds to lighter areas and low friction to darker regions on the image.

## 7.1.2  Biological imaging

AFM has become a very powerful technique for imaging biological samples. One reason for its success is the ability to image samples of native specimens in physiological buffers, under ambient conditions. This is in contrast to samples imaged by scanning and transmission electron microscopy, which generally require extensive preparation, such as staining or metal coating. However, in order to withstand the very large tip forces, biological specimens must be rigidly fixed to a suitable, flat substrate. This has been achieved by simply air-drying macromolecules to substrates such as mica [5] or gold [6]. In general it is better to image samples in tapping mode in liquids [7]. This greatly reduces the lateral forces exerted by the tip on the sample and so improves image resolution.

AFM can be used to obtain images very quickly, in the order of seconds, and so many biological processes can be observed in real-time, e.g. the polymerization of fibrin, the basic component of blood clots [8].

The ability to generate images at nanometer scale resolution of nucleic acids has broad and exciting implications, way beyond the studies that have so far been reported. The first highly reproducible images of DNA were obtained in 1991 [9]. Since then, AFM has become an important technique for providing structural information about DNA and DNA–protein complexes. For example, AFM has been used to resolve the pitch of DNA [10], and to show *Eco*RI endonuclease bound to its nucleotide recognition sequence on individual plasmids [11]. The presence of kinks in free circular DNA was demonstrated by the direct visualization *in situ* [12]. It was also shown that small DNA circles change from smoothly bent to abruptly kinked shapes on the addition of $Zn^{2+}$. The bending of DNA fragments upon binding to Cro repressor protein has also been observed [13]. Fluid tapping mode AFM was used to observe *Escherichia coli* RNA polymerase transcribing two different linear-stranded DNA templates [14], the transcription process being observed by monitoring the translocation of the DNA template by RNA polymerase on addition of ribonucleoside 5-triphosphates (NTPs) in a series of real-time sequential images.

There have been numerous studies where proteins have been imaged. The best of these studies reveal details of the quaternary structures [15], although in many others the protein is rather poorly defined. It has been proposed that it may be better to image biological samples at cryogenic temperatures [16]. Under these conditions hydrogen bonding should be strengthened, giving a more stable structure, better able to withstand probing by the AFM tip. Cryogenic imaging of two types of immunoglobulins, IgG and IgM immobilized on a mica substrate at 90 K have been reported [16]. We have recently used AFM to determine the structure of novel macromolecular assemblies where proteins are organized at defined positions along a DNA scaffold [17].

## 7.1.3 Chemical force microscopy (CFM)

The ability to do synthetic chemistry opens up a whole new universe of applications for AFM, and makes the instrument an exciting tool for the organic (as well as the physical) chemist. When a freshly prepared gold surface is exposed to a solution of alkyl thiols, the thiols spontaneously chemisorb into the gold to form a self-assembled monolayer (SAM) [18]. This presents a uniform surface made up of the methyl groups at the other end of the thiols. By using thiols which terminate in other functional groups, *e.g.* carboxyl, amino etc., it is possible to generate correspondingly functionalized surfaces. In the same way, the AFM tip can be pre-coated with gold onto which a SAM can be formed. These coatings can be characterized by simple contact angle measurements, X-ray photoelectron spectroscopy, and vibrational spectroscopy. Once the surface and the tip have been modified, then the AFM can be used to measure the interaction between the coating on the tip and the coating on the surface – a technique referred to as chemical force microscopy. This has been applied to hydrophobic and hydrophilic interactions [19], the binding between biotin and streptavidin [20], and the binding between DNA nucleotide bases [21].

We have shown that the quality of the force measurements obtained in CFM experiments depend in part on the quality of the underlying gold surface. In some early CFM studies, the distribution of the data was rather large. However, this was for samples where the gold surface was simply prepared by thermal evaporation of gold at room temperature. We found that this forms surfaces which were essentially poly-crystalline in nature. We prepared gold films on freshly cleaved mica at a pressure of less then $1 \times 10^{-8}$ mbar by thermal evaporation in a Balzer's UHV chamber. The chamber and samples were prebaked at 160°C and 340°C respectively overnight. After deposition, the samples were annealed for at least 5 h, a procedure which gave smoother surfaces that were predominantly Au (111). In CFM experiments, such surfaces give a narrower distribution of adhesion forces [22].

In order to make comparative measurements between a given tip and two or more surfaces it is preferable to have the surfaces as part of the same sample. This can be used to provide an internal standard, to overcome problems with variability between samples, or simply to speed up data collection. To achieve this we use the microcontact printing methodology developed by Kumar and Whitesides [23]. In this way we can routinely generate a patterned surface comprising alternating SAMs from two different thiols which run in parallel lines 2 μm in width across the surface.

### 7.1.3.1  Chiral discrimination by chemical force microscopy

There have been questions raised about what is actually detected in a CFM experiment [24]. In particular, could different interactions measured between the tip and SAMs of two different compounds simply be due to either differential packing or differential solvation of the two SAMs? In considering the implications of this problem, we decided to see if CFM could discriminate between two SAMs that differed only in the chirality of the group presented at the end of thiol. The SAM made using an *R*-thiol should pack identically to that made with the *S*-thiol. Similarly, there should be no difference is solvation of the two surfaces using an achiral solvent.

Mandelic acid was chosen as the chiral end group for the thiols forming the SAM (Scheme 7.1.2) because it has a single chiral center with four distinct groups attached (Ph, H, OH, CO₂H). In order to see any discrimination between two chiral surfaces, it

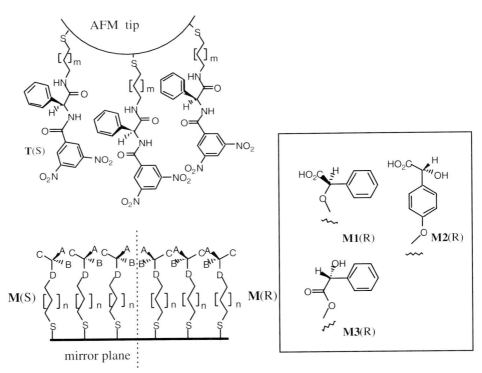

**Scheme 7.1.2.**  The experimental set up to demonstrate chiral discrimination by CFM. The AFM tip was coated with a self-assembled monolayer terminating in a phenylglycine amide.The surface was coated with alternating monolayers terminating in the different enantiomers of mandelic acid, attached through either the alcohol, aromatic ring, or acid (inset).

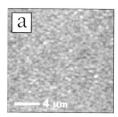 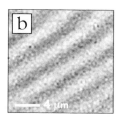

**Figure 7.1.1** The topographic image (a) and friction map (b) obtained in the chiral discrimination experiment. The tracks of alternating chiral surfaces are at 45 degrees. In (b) the lighter areas correspond to regions of higher friction.

was also necessary to chirally functionalize the tip of the AFM. This was done using a phenylglycine derivative, so that we could compare the results obtained by CFM with discrimination seen when analogous compounds were put down a Pirkle (phenylglycine) HPLC column.

The topographic map of the surface (Figure 7.1.1a) obtained when the chiral tip was dragged across a surface of alternating tracks of R- and S-mandelic acid-terminated thiol showed no distinction between the tracks. This was to be expected as the tracks correspond to monolayers that differ only in their chirality. In contrast, the friction map (Figure 7.1.1b) showed an alternation in the frictional force. The differences in frictional force are assumed to correspond to the differences in the strengths of the transient diastereomeric complexes formed between the chiral molecule on the tip (say the R-configuration) and either the R- or S-thiol on the surface. This conclusion was supported by force distance measurements which showed quantitative discrimination between the two enantiomeric surfaces [25].

The question of substrate orientation was also addressed by attaching the mandelic acid to the surface with a thiol linker through the OH (M1), the phenol ring (M2), and the acid (M3). For surfaces made of either M1 and M2 the R-surface showed stronger adhesion forces to the S-tip, than the R-tip. The racemic tip gave an intermediate result. When the thiol was attached through the acid (M3), the adhesion forces were much weaker, and showed a reversal of discrimination. The adhesion forces correlated with the retention times of closely analogous compounds on a Pirkle column. The S-enantiomers corresponding to M1 and M2 eluted more slowly than the R-enantiomers on an R-column, whereas the R-enantiomer corresponding to M3 eluted more slowly, albeit the retention time for M3 was much less than for M1 and M2.

These were fundamentally important experiments because they established that CFM could distinguish between subtly different surfaces where packing and solvation must be identical. The results open up ways to further our understanding of the relationship between adhesion, friction, and wear in nanotribology.

## 7.1.4   Future directions

Our interest in using AFM is expanding rapidly. On a fundamental level we want to have a better picture of what happens when the AFM tip addresses a surface, and to understand the relationship between adhesion and friction. On an applied level, our ambitions are rather recklessly unbounded. New projects are directed towards such diverse goals as using AFM to screen combinatorial libraries of small molecule inhibitors, using the AFM tip as a molecular pencil for writing on surfaces, and developing a whole new synthetic chemistry on surfaces. This is science which excites us!

**Acknowledgements:** We gratefully acknowledge support from the EPSRC and Zeneca.

# References

[1]   G. Binnig, H. Rohrer, C. Gerber, *Phys. Rev. Lett.* **1982**, *49*, 57–61.

[2]   G. Binnig, C. F. Quate, C. Gerber, *Phys. Rev. Lett.* **1986**, *56*, 930–933.

[3]   D. L. Worcester, R. G. Miller, P. J. Bryant, *J. Microsc.* **1988**, *152*, 817–821.

[4]   M. H. Jericho, B. L. Blackford, D. C. Dahn, *J. Appl. Phys* **1989**, *65*, 5237–5239.

[5]   E. A. G. Chernoff, D. A. Chernoff, *J. Vac. Sci. Tech. A*, **1992**, *10*, 596–599; Y. L. Lyubchenko, P. I. Oden, D. Lampner, S. M. Lindsay, K. A. Dunker, *Nucleic Acids Res.* **1993**, *21*, 1117–1123.

[6]   S. M. Lindsay, T. Thundat, L. Nagahara, U. Knipping, R. L. Rill, *Science* **1989**, *244*, 1063–1064.

[7]   P. K. Hansma, J. P. Cleveland, M. Radamacher, D. A. Walters, P. E. Hillner, M. Bezanilla, M. Fritz, D. Vie, H. G. Hansma, C. B. Prater, J. Massie, L. Fukunaga, J. Gurley, V. Elings, *Appl. Phys. Lett.* **1994**, *64*, 1738–1740.

[8]   B. Drake, C. B. Prater, A. L. Weisenhorn, S. A. C. Gould, T. A. Albretch, C. F. Quate, D. S. Cannell, H. G. Hansma, P. K. Hansma, *Science* **1989**, *243*, 1586–1589.

[9]   H. G. Hansma, J. Vesenka, C. Siegerist, G. Kelderman, H. Morrett, R. L. Sinsheimer, V. Elings, C. Bustamante, P. K. Hansma, *Science* **1992**, *256*, 1180–1184.

[10]  H. G. Hansma, D. E. Laney, M. Bezanilla, R. L. Sinsheimer, P. K. Hansma, *Biophys. J.* **1995**, *68*, 1672–1677; J. X. Mou, D. M. Czajkowsky, Y. Y. Zhang, Z. F. Shao, *F. E. B. S. Lett* **1995**, *371*, 279–282.

[11]  D. P. Allison, P. S. Kerper, M. J. Doktycz, T. Thundat, P. Modrich, F. W. Larimer, D. K. Johnson, P. R. Hoyt, M. L. Mucenski, R. J. Warmack, *Genomics* **1997**, *41*, 379–389.

[12]  W. H. Han, S. M. Lindsay, M. Dlakic, R. E. Harington, *Nature* **1997**, *386*, 563.

[13]  D. A. Erie, G. L. Yang, H. C. Schultz, C. Bustamante, *Science* **1994**, *266*, 1562–1566.

[14]  S. Kasas, N. H. Thompson, B. L. Smith, H. G. Hansma, X. S. Zhu, M. Guthold, C. Bustamante, E. T. Kool, M. Kashlev, P. K. Hansma, *Biochem*istry **1997**, *36*, 461–468.

[15]  F. A. Schabert, C. Henn and A. Engel, *Science* **1995**, *268*, 92–94; D. J. Muller, C. A. Schoenenberger, F. Schabert , A. Engel, *J. Struct. Biol.* **1997**, *119*, 149–157.

[16]  Z. Shao, J. Mou, D. M. Czajkowsky, J. Yang, J. Yuan, *Adv. Phys.* **1996**, *45*, 1–86.

[17]  K. Barnes, J. Tomkins, R. McKendry, J. Blacker, C. Abell, unpublished results

[18]  R. G. Nuzzo, D. L. Allara, *J. Am. Chem. Soc.* **1983**, *105*, 4481–4483; M. D. Porter, T. B. Bright, D. L. Allara , C. E. D. Chidsey, *J. Am. Chem. Soc.* **1987**, *109*, 3559–3568.

[19]  C. D. Frisbie, L. F. Rozsnyai, A. Noy, M. S. Wrighton, C. M. Lieber, *Science* **1994**, *265*, 2071–2074.

[20]  G. U. Lee , D. A. Kidwell, R. Colton, *Langmuir* **1994**, *10*, 354–357; M. Ludwig, W. Dettmann, H. E Gaub, *Biophys. J.* **1997**, *72*, 445–448.

[21]  T. Boland, B. D. Ratner, *Proc. Natl. Acad. Sci. USA* **1995**, *92*, 5297–5301.

[22]  R. McKendry, M.-E. Theoclitou, C. Abell, T. Rayment, *Langmuir* **1998**, *14*, 2846–2849.

[23]  A. Kumar, G. M. Whitesides, *Appl. Phys. Lett.* **1993**, *63*, 2002–2004.

[24]  S. K. Sinniah, A. B. Steel, C. J. Miller, R. E. Reuttrobey, *J. Am. Chem. Soc.* **1996**, *118*, 8925–8931.

[25]  R. McKendry, M.-E. Theoclitou, T. Rayment, C. Abell, *Nature* **1998**, *391*, 556–557.

# 7.2    Marine natural products: New ways in the constitutional assignment

*Matthias Köck and Jochen Junker*

Marine natural products chemistry has attracted much interest during the past decade in the search for new lead compounds for pharmaceutical applications. In contrast to terrestrial natural products chemistry, marine natural products chemistry is a relatively new area of research. It was initiated by the work of Bergmann in 1950 [1], and the first comprehensive review article was published in 1973 [2]. The marine environment has great potential as a source for finding new natural products, as: (i) 71% of the earth's surface is covered by oceans; (ii) the biodiversity of the marine environment is larger than its terrestrial counterpart, which is of fundamental interest for the diversity of the secondary metabolites (almost all animal phyla are represented in the oceans [3]); and (iii) only 8% of all known natural products are from marine sources [4]. However, as yet no pharmacon from marine sources is in clinical use. At the moment, the most promising compounds for pharmaceutical application are the anti-tumor agents bryostatin 1 (**1**), isolated from the bryozoan *Bugula neritina* [5], and depsipeptide aplidine (**2**, dehydrodidemnin B) [6], isolated from the ascidian *Aplidium albicans*. Both compounds are in clinical phase II. Aplidine has replaced didemnin B (isolated from the ascidian *Trididemnum solidum* [7]) in clinical phase II trials. Aplidine is in comparison to didemnin B more active and has no cardiotoxicity [8]. Two other very promising anti-tumor agents are in clinical phase I: dolastatin 10 (**3**) isolated from the sea hare *Dolabella auricularia* [9], and ecteinascidin 743 (**4**) isolated from the ascidian *Ecteinascidia turbinata* [10]. It can, however, take a long time from the first discovery of an active compound to its use as a drug. For example, the extract of *Taxus brevifolia* was shown in 1962 to be cytotoxic against KB cells, but the active constituent taxol, was first isolated only in 1971 [11] and finally introduced as a pharmacon in 1993.

## 7.2.1    Constitutional assignment by NMR spectroscopy

In order to develop the potential of marine natural products, a reliable structure elucidation of the isolated compounds is essential. For non-crystalline compounds, the elucidation of their structure relies on NMR spectroscopy, which is the only method by which information with atomic level resolution in solution can be obtained. Usually, the NMR investigations depend on the knowledge of the molecular formula which is derived from high-resolution mass spectrometry. Other methods such as IR and UV spectroscopy can only distinguish among structural alternatives.

The general approach for the constitutional assignment of natural products using 2D NMR spectroscopy is as follows: (1) the COSY experiment [12] is used for the proton assignment, while in case of overlapping proton signals, a TOCSY experiment [13] can be of help; (2) the carbons adjacent to the protons are assigned by a HMQC [14] or a HSQC [15] experiment; while in case of overlapping signals, these experiments can be combined with a COSY or a TOCSY step resulting in HMQC/HSQC-COSY [16] or HMQC/HSQC-TOCSY [17]; and (3) the HMBC experiment [18] is used for the assignment of quaternary carbons and furthermore to connect spin systems through quaternary carbons via heteronuclear long-range correlations ($^2J_{CH}$ and $^3J_{CH}$). The procedure described uses only proton-based experiments, which is not sufficient for all kinds of compounds. Polycyclic aromatic compounds, for example, have a relatively small number of protons and therefore the information obtained from the HMBC experiment may not be sufficient for an unambiguous structure elucidation. Another general disadvantage is that the HMBC experiment cannot distinguish between $^2J_{CH}$ and $^3J_{CH}$ correlations.

The impact of the information contained in different NMR experiments is illustrated in Figure 7.2.1. These three natural products document which kind of

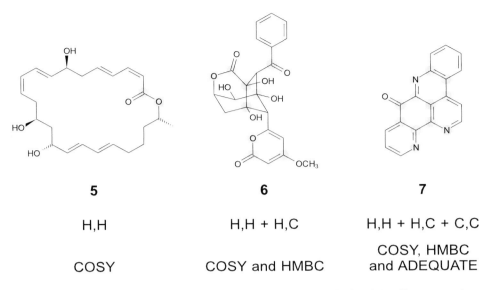

**5**

H,H

COSY

**6**

H,H + H,C

COSY and HMBC

**7**

H,H + H,C + C,C

COSY, HMBC
and ADEQUATE

**Figure 7.2.1.** Structures of macrolactin A (**5**), enterocin (**6**) and ascididemin (**7**). In all structures the non-protonated atoms are indicated by gray circles. Below the structures the correlations, and respectively the NMR experiments, are given which are necessary for an unambiguous assignment of the constitution.

NMR experiments are necessary to assign the constitution of different classes of compounds. Macrolactin A (**5**) [19] is an example of a compound with a small number of non-protonated atoms (indicated by gray circles). The constitution of such compounds can be solved unambiguously using H,H correlations (COSY, Relayed COSY [20], TOCSY) if the proton signals are not severely overlapped and if the compound appears only as one conformer in solution. For the second example enterocin (**6**) [21], the number of non-protonated atoms is substantially larger and therefore H,C besides H,H correlations are necessary to assign the constitution unambiguously. The third example is ascididemin (**7**) [22], which has a large number of adjacent quaternary centers. Thus, H,H and H,C correlations are not sufficient for an unambiguous assignment of the constitution. In this case C,C correlations are required.

The structure elucidation of proton-poor marine natural products, as shown before, represents a challenge to NMR spectroscopy. This is documented for two additional examples (pyridoacridines) from the literature. Bromoleptoclinidinone (**8**, for R = Br) was first isolated by Schmitz et al. in 1987 [23]. One year later, Kobayashi et al. published the structure of ascididemin (**8**, for R = H) [22]. The two structures differed by the position of one nitrogen, as highlighted in Figure 7.2.2. Schmitz et al. revised the structure of bromoleptoclinidinone according to ascididemin in 1989 [24]. The first 2D NMR studies of neocalliactine acetate (**9**) [25] by Cimino et al. in 1987 proposed four possible structures [26], which were reduced to two by Schmitz et al. in 1991 [27]. The final structural proof was achieved by the total synthesis of neocalliactine acetate in 1992 [28].

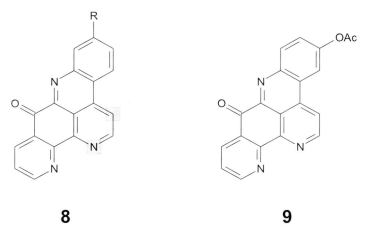

**8**                                    **9**

**Figure 7.2.2.** Structural formulae of 2-bromoleptoclinidinon (**8**, R = Br), ascididemin (**8**, R = H), and neocalliactine acetate (**9**). The gray squares in the left formula indicate the misassigned centers of 2-bromo-leptoclinidinon in the originally published structure (exchange of C and N).

### 7.2.1.1 New NMR experiments for the constitutional assignment: ADEQUATE

The INADEQUATE experiment [29], which yields C,C correlations, is usually too insensitive to be of practical use for natural product samples since only milligram quantities are available. Therefore, proton-detected INADEQUATE type experiments with increased sensitivity are required for the structure elucidation of proton-poor compounds. The experiments must be carried out with proton excitation and detection in order to gain the maximum sensitivity as $\gamma_H = 4\gamma_C$. Thus, the ADEQUATE (Adequate Sensitivity Double-Quantum Spectroscopy) experiments were introduced as a sensitive method to observe C,C correlations [30]. They represent a combination of a HSQC or a HMBC step followed by the evolution of $^1J_{CC}$ or $^nJ_{CC}$ couplings. NMR experiments using C,C correlations with proton excitation and detection had already been published in 1993 [31] and 1995 [32] for $^{13}$C-enriched compounds as well as a 2 M sample of sucrose. These kind of C,C correlation experiments applied to compounds with natural abundance of $^{13}$C were only made possible with the introduction of pulsed field gradients [33] to suppress the undesired magnetization and select only $^{13}$C,$^{13}$C pairs that contain the information of interest [34].

The pulse sequence of the HSQC-type ADEQUATE experiment is shown in Figure 7.2.3a. The desired correlation of the second transfer ($^1J_{CC}$ or $^nJ_{CC}$) is dependent on the length of the delay $\Delta_2$. The pulse sequence starts with an INEPT transfer ($\Delta_1 = 1/4J_{CH}$) for the magnetization transfer from protons to carbons (part A in Figure 7.2.3a). In the delay $\Delta_2$ $(1/4J_{CC})^nJ_{CC}$ couplings evolve which results in a term $4H_zC_{1x}C_{2z}$ (part B in Figure 7.2.3a). The $90°$ pulse converts the carbon anti-phase magnetization in a double-quantum (DQ) term which evolves during $t_1$ (part C in Figure 7.2.3a). Further terms, especially for molecules with only one $^{13}$C atom, as $2H_zC_{1x}$, are suppressed by the first gradient (G1) in the evolution time of the double-quantum coherence which selects $2H_zC_1^+C_2^+$ (part D in Figure 7.2.3a). After the evolution time $t_1$, the $^{13}C,^{13}C$ double-quantum coherence is converted into $^{13}C$ single-quantum (SQ) coherence (part E in Figure 7.2.3a). The maximum transfer from DQ ($2H_zC_1^+C_2^+$) to SQ ($4H_zC_1^+C_{2z}$, $4H_zC_1^-C_{2z}$) coherence is not achieved with a 90° pulse, but with a 60° or a 120° pulse [35]. The gradient G2 selects the $^{13}C$-SQ coherence. The other pulses on $^1H$ and $^{13}C$ achieve a polarization transfer from $4H_zC_1^+C_{2z}$ ($^{13}C$-SQ) to H$^-$ ($^1H$-SQ) using COS-HSQC [36] which includes the reverse

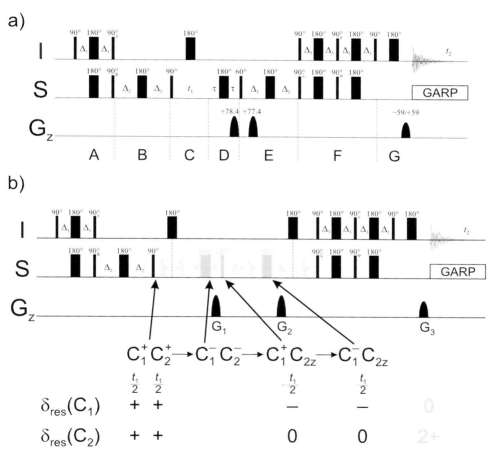

**Figure 7.2.3.** (a) Pulse sequence of the HSQC-type ADEQUATE experiment in the non-refocused version. Depending on the length of the $\Delta_2$ delay, the experiment is a 1,1- (5 ms) or a 1,$n$-ADEQUATE (30 ms). A detailed description of the pulse sequence is given in the text (divided in the parts A to G). The phases for the carbon pulses are: $\phi = $ x, $-$x; $\zeta = $ y, $-$y, x, $-$x; $\psi = -$y, y, $-$x, x. The gradient strength is given in the pulse sequence above the pulsed field gradients. (b) Pulse sequence of the HSQC-type ADEQUATE experiment in the $\omega_1$-refocused version. To refocus the chemical shift of C-1, the $t_1$ time is implemented in the $\Delta_2$ delay. The explicit terms after the generation of the double-quantum term as well as the resulting chemical shift evolution of C-1 and C-2 are given.

INEPT transfer and sensitivity enhancement [37] (part F in Figure 7.2.3a). The last gradient (G3) directly before the acquisition time selects proton single-quantum magnetization (part G in Figure 7.2.3a). The gradients must fulfil the condition $(2G1 + G2)/G3 = \gamma_H/\gamma_C = 4$ to achieve an echo or anti-echo.

The resulting spectrum obtained with the pulse sequence shown in Figure 7.2.3a has double-quantum frequencies in $F_1$ as it is also known from the INADEQUATE experiment. Thus, the experiment cannot be analyzed in the same way as an HMBC spectrum. It would be desirable to obtain a spectrum with single-quantum frequencies of $^{13}C$ in $F_1$. Therefore, the chemical shift of C-1 in the DQ-term (C-1 + C-2) has to be refocused during the evolution time $t_1$. In order to achieve this the $t_1$ time is also implemented in the C,C refocusing delay ($\Delta_2$, see Figure 7.2.3b, gray part of the pulse sequence). The chemical shift evolution of C-1 and C-2 is given for all $t_1$ periods in the pulse sequence (see Figure 7.2.3b). In this constant-time evolution, the homonuclear $^{13}C,^{13}C$ coupling is refocused, the heteronuclear coupling is decoupled and the chemical shift evolution of C-1 during the evolution of the double-quantum term is refocused. The resulting chemical shift for C-1 is 0 as required, whereas the chemical shift of C-2 evolved.

The ADEQUATE experiments can be run with an HSQC ($^1J_{CH}$) or HMBC ($^{2/3}J_{CH}$) for the transfer from protons to carbons and a $^1J_{CC}$ or $^nJ_{CC}$ transfer. Altogether, four ADEQUATE experiments are possible which allow the observation of correlations over two, four or six bonds, starting from a proton. The resulting correlations are shown schematically in Figure 7.2.4. The first experiment is the 1,1-ADEQUATE. The first 1 stands for the $^1J_{CH}$ transfer, which is achieved by an HSQC step; the second 1 stands for a $^1J_{CC}$ transfer. The resulting correlation is equivalent to a $^2J_{CH}$ coupling and could be used for the distinction of $^2J_{CH}$ and $^3J_{CH}$ correlations in the HMBC experiment. Furthermore, the observation of two-bond H,C correlations in olefinic or aromatic systems are possible which are known to be of low intensity in the HMBC experiment ($^2J_{CH}$ couplings). The second experiment is the 1,n-ADEQUATE, which combines a HSQC step with a $^nJ_{CC}$ correlation. For $n = 3$ the resulting correlation is a pseudo $^4J_{CH}$ correlation and therefore an additional tool for assigning the constitution. The n,1-ADEQUATE which combines a HMBC step ($^nJ_{CH}$) and a $^1J_{CC}$ transfer, the resulting correlation is also a pseudo $^4J_{CH}$ correlation. If both $^3J$ transfers are combined in an m,n-ADEQUATE, correlations over six bonds can be observed.

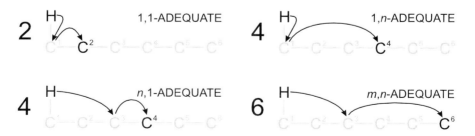

**Figure 7.2.4.** Schematic representation of the correlations (for n, m = 3) obtained with the four ADEQUATE experiments. The bold number indicates the number of bonds covered by the experiment.

**7.2.1.2 Cocon – Translation of NMR correlation data into molecular constitutions**

The general strategy for the constitutional assignment of natural products using NMR spectroscopy is to assign the signals of the NMR active nuclei (usually $^1$H, $^{13}$C) and if the resulting structure is in accordance with the correlation data, the structure is supposed to be solved. Using this procedure the question remains over whether the structure is really unambiguously solved. In general, the answer to this question must be no. One should not be satisfied to find one solution for a given data set; more constitutional possibilities must be considered. That this is a necessary approach was already pointed out by John Faulkner [38] in 1992: "Rather than defining a structure that can be shown to fit the spectral data, it is best to examine many possible structures (we would say all) and treat each proposed structure as a hypothesis that cannot be proved but can only be disproved by incompatible data." Therefore, high-quality NMR data and a computer-based structure generator are necessary in order to fulfil the required approach. The three examples in Figure 7.2.1 highlight that the NMR measurements must be adapted to the complexity of the problem.

We have introduced the computer program Cocon (**Con**stitutions from **Con**nectivities) [39] which allows chemists to discuss comprehensively the entirety of constitutions which are in accordance with the NMR correlation data and the molecular formula [40]. The program is able to use all different types of NMR correlation data and it optionally obeys rules on $^{13}$C NMR chemical shifts [41] which, intentionally, are defined very coarsely. There are two Cocon analyses possible: the first needs the hybridization state (atom type) of every atom, in the second the definition of the atom types is not necessary because the possible hybridization states are permutated [42]. Cocon has two general applications: (i) analysis of structural proposals which have already been made (discussion of alternatives); and (ii) de novo structure elucidation. A $^{13}$C NMR chemical shift calculation (SpecEdit) [43] can be applied after a Cocon calculation in order to get a ranking or a selection of the constitutions [44].

## 7.2.2   Constitutional assignment of marine natural products

In the following, the combination of different NMR experiments and Cocon for the constitutional assignment of compounds from marine organisms is exemplified for three cases. The first example is cadinadiene (**10**), a sesquiterpene from an unidentified octocoral from Western Australia which serves here as a model for a proton-rich compound [45, 46]. Cadinadiene is contained in this organism as 0.2% of dry weight. For the structural elucidation, COSY, HSQC and HMBC experiment were carried out. Already from the 1D spectra of **10**, the following structural elements can be concluded: (i) the carbon chemical shifts indicated that there is no heteroatom included in this compound and therefore it could be identified as a sesquiterpene just by counting the number of carbon resonances; (ii) from the 1D proton and carbon spectrum, as well as DEPT [47], it could be concluded that **10** contains three methyl groups which are adjacent to a non-olefinic methin group; (iii) the peak intensities of the 1D $^{13}$C spectrum indicated that two of the four olefinic carbons are quaternary;

(iv) one of the protonated carbon signals is shifted to high field (below 110 ppm), indicating an olefinic methylene group, which leads to an exocyclic double bond. Altogether there are three $-CH_3$, four $>CH_2$, four $>CH-$, one $=CH_2$, one $=CH-$ and two $=C<$ groups, resulting in a molecular formula of $C_{15}H_{24}$ (Table 7.2.1). Two of the four degrees of unsaturation are the double bonds, and because no hetero-atoms are included, the remaining two degrees come from to ring closures. One of the double bonds must be located at the ring junction, because otherwise two protons or a methyl group bound to a double bond would be expected, which was not observed.

**Table 7.2.1.** Chemical shifts of cadinadiene (**10**).

| No. | $\delta(^1H)$ [ppm] | $\delta(^{13}C)$ [ppm] | HMBC | Possible atom types[*] |
|---|---|---|---|---|
| 1 | 0.87 | 14.9 | 7, 9, 10 | $-CH_3$ |
| 2 | 0.93 | 21.3 | 3, 6, 11 | $-CH_3$ |
| 3 | 0.76 | 21.8 | 2, 6, 11 | $-CH_3$ |
| 4 | 1.60/1.25 | 22.8 | 11, 15 | $>CH_2$ |
| 5 | 1.65/1.50 | 27.2 | 8, 10, 14 | $>CH_2$ |
| 6 | 1.75 | 27.3 | 4, 11 | $>CH-$ |
| 7 | 1.72/1.27 | 29.3 | 1, 4, 9, 11 | $>CH_2$ |
| 8 | 2.25/2.10 | 29.6 | 5, 10, 12, 14 | $>CH_2$ |
| 9 | 1.88 | 34.9 | 7, 10 | $>CH-$ |
| 10 | 2.28 | 37.0 | 5, 9, 13, 15 | $>CH-$ |
| 11 | 1.60 | 51.2 | 4, 13, 15 | $>CH-$, $\equiv CH$ |
| 12 | 4.54 | 107.9 | 8, 13, 14 | $>CH_2$, $=CH_2$ |
| 13 | 5.90 | 126.5 | 8, 10, 11, 12, 14 | $=CH-$ |
| 14 | – | 144.2 | – | $=C<$ |
| 15 | – | 144.5 | – | $=C<$ |

[*] proposed by COCON.

By using the COSY correlations (20 were obtained experimentally), COCON generated three possible constitutions for **10** when bonds between atoms not showing a COSY correlation are forbidden (see Figure 7.2.5). If this restriction was not used, 12

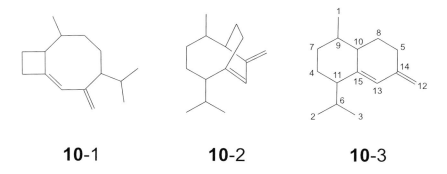

**10**-1                **10**-2                **10**-3

**Figure 7.2.5.** The three constitutional proposals of cadinadiene (**10**) generated by COCON using only COSY data. The correct structure is **10**-3.

constitutions were generated. For further analysis, we used $^{13}$C NMR chemical shift prediction and energy arguments. With the help of the computer program SpecEdit [43], all $^{13}$C-NMR chemical shifts for every constitution generated by COCON were calculated. The averaged $^{13}$C chemical shift deviation over all carbon atoms ($\Delta[\delta(^{13}C)]$) is 3.40 ppm for **10**-1, 3.07 ppm for **10**-2, and 2.47 ppm for **10**-3. Proposal **10**-2 violates Bredt's rule which is not automatically investigated by COCON [48]. The total energies were calculated with the program HyperChem (**10**-1, 66 kcal/mol; **10**-2, 148 kcal/mol; **10**-3, 26 kcal/mol). In both cases **10**-3, which is the correct constitution, is preferred. Thirty-seven HMBC correlations were obtained, which is 57% of the correlations expected theoretically for **10**-3 [49]. Using the HMBC data, only constitution **10**-3 remains. In principle, the HMBC experiment is already oversized for this problem. If the hybridization states are not defined prior to the COCON calculation, the same number of possible constitutions is generated [50].

The second example is oroidin (**11**), a marine alkaloid which was first isolated in 1971 from the sponge *Agelas oroides* [50]. The sample investigated here was isolated from *Agelas clathrodes* in which it is contained as 0.4% of dry weight [42]. The NMR experiments COSY, HSQC ($^{13}$C and $^{15}$N), HMBC ($^{13}$C and $^{15}$N) and 1,1-ADEQUATE were necessary for an unambiguous assignment of the constitution. Four data sets which differ by the number of HMBC correlations were obtained for **11** (see Table 7.2.2). To judge the effect of the 1,1-ADEQUATE data, columns 1 and 2 in Table 7.2.2 must be compared. These numbers indicate that the 1,1-ADEQUATE correlations have a great impact on the structure elucidation. An even more dramatic effect is obtained with the $^{1}$H,$^{15}$N-HMBC correlations for **11** (compare columns 1 and 3 in Table 7.2.2) [52].

**11**

**Table 7.2.2.** COCON results for oroidin (**11**).

| Data set | No. of HMBC correlations | col 1 1100 | col 2 1110 | col 3 1101 | col 4 1111 |
|---|---|---|---|---|---|
| | | *NMR correlations*[*] | | | |
| A | 9 (29%) | 234 336 | 10 597 | — | — |
| B | 18 (58%) | 27 142 | 1 655 | 10 | 3 |
| C | 23 (74%) | 690 | 33 | 10 | 3 |
| D | 26 (84%) | 60 | 10 | 6 | 2 |

[*] 1100 stands for COSY and HMBC correlations, 1110 for COSY, HMBC and 1,1-ADEQUATE, 1101 for COSY, HMBC ($^{13}$C) and HMBC ($^{15}$N).

Here, we want to focus on the importance of the quality of the HMBC data set. If the data set has only a relatively small number of HMBC correlations (data set A) [53], the 1,1-ADEQUATE correlations do not help very much because there are still more than 10 000 possible constitutions. In contrast, with the well-defined data set D only two solutions are obtained. The two final possible constitutions differ only in the substitution pattern of the pyrrole part and can easily be distinguished by a $^{13}$C chemical shift calculation [54]. The results for oroidin (**11**) without predefined hybridization states are given in reference [42].

The final example is 5,6-dihydro lamellarin H (**12**-1), a polycyclic aromatic compound isolated from an ascidian of the genus *Didemnum*. It is contained as 0.2% of dry weight. The lamellarins were first described by Faulkner et al. in 1985 [55]. The complexity of this constitutional problem is shown by the number of non-protonated atoms, indicated by the gray circles in **12**-1 (see Figure 7.2.6). The following NMR experiments were acquired: COSY, HSQC, HMBC, 1,1-ADEQUATE and 1,*n*-/*n*,1-ADEQUATE. Thirty-five HMBC cross-peaks were observed for **12**-1. The theoretical maximum for the assigned constitution of **12**-1 is 41. The six missing cross-peaks in the HMBC spectrum are all $^{2}J_{CH}$ correlations which were observed in the 1,1-ADEQUATE. Thus, the experimental data set for **12**-1 equals the theoretical one in respect to the data used by Cocon (COSY, HMBC, 1,1-ADEQUATE). With these data, Cocon generated 1448 structural proposals, which indicates that more NMR data is required. By using only six long-range ADEQUATE correlations (pseudo $^{4}J_{CH}$, from 1,*n* and *n*,1), the number of constitutions is reduced to 60. The subsequent calculation of the carbon chemical shifts (SpecEdit) of the Cocon structures proposed two constitutions with chemical shift deviation $\Delta[\delta(^{13}C)]$ of 4.50 ppm for **12**-1 and $\Delta[\delta(^{13}C)]$ of 5.80 ppm for **12**-2 (see Figure 7.2.6). The total energies are 48 kcal/mol for **12**-1 and 122 kcal/mol for **12**-2 which violates Bredt's rule. In both cases **12**-1 is preferred, which is the correct constitution.

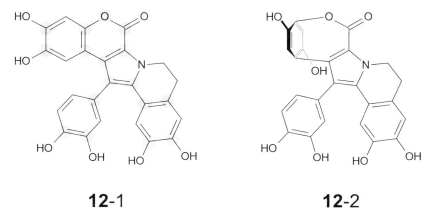

**12**-1                            **12**-2

**Figure 7.2.6.** The two structures of lamellarin (**12**) with the lowest chemical shift deviation calculated by SpecEdit. Using all experimental correlation data, Cocon generated 60 possible structures. The subsequent calculation of the carbon chemical shifts was used to rank the structural proposals.

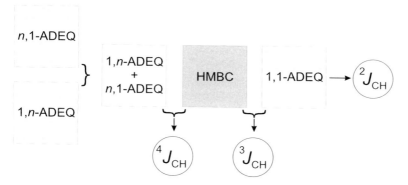

**Figure 7.2.7.** General approach for the constitutional assignment of proton-poor compounds. The refocused versions of the ADEQUATE experiments allow an analysis as known from the HMBC. Therefore, the assignment of the cross-peaks to the corresponding $J$ coupling can be carried out by overlaying the spectra.

## 7.2.3  Summary and outlook

The presented structure elucidations prove that experiments such as $^1H,^{15}N$-HMBC, and 1,1-ADEQUATE are very important for an unambiguous assignment of the constitution. Furthermore, a NMR-based structure generator like Cocon is necessary to discuss alternative structural proposals. The subsequent calculation of the carbon chemical shifts is very useful for the selection among the generated structures.

A general strategy for the constitutional assignment of proton-poor compounds is presented in Figure 7.2.7. The HMBC is the central element. The 1,1-ADEQUATE yields $^2J_{CH}$ correlations, by comparison of the HMBC and the 1,1-ADEQUATE spectrum the $^3J_{CH}$ correlations can be obtained. The comparison of the HMBC with the sum of the 1,$n$- and $n$,1-ADEQUATE spectra yields $^4J_{CH}$ correlations. These correlation data serve as input for the NMR-based structure generator Cocon. If the HMBC spectrum contains $^4J_{CH}$ correlations, there is a possibility of distinguishing between $^3J_{CH}$ and $^4J_{CH}$ using a refocused 1,1-ADEQUATE and a non-refocused $n$,1-ADEQUATE [56].

**Acknowledgements:** This research was financially supported by the Fonds der Chemischen Industrie (Li 149/4 and Sachkostenzuschuß) and the Deutsche Forschungsgemeinschaft (Ko 1314/3–1 and 3–2). We acknowledge the support of Prof. Dr. Christian Griesinger. We further thank Prof. Dr. William Fenical (San Diego) for the collaboration with cadinadiene and lamellarin, Dr. Heonjoong Kang (San Diego) for the collaboration with lamellarin, Dr. Bernd Reif, Prof. Dr. Christian Griesinger and Dr. Rainer Kerssebaum (Bruker, Karlsruhe) for the collaboration with the ADEQUATE experiment, and Dr. Thomas Lindel (Heidelberg) for the collaboration with oroidin and Cocon.

# References

[1]  W. Bergmann, R. J. Feeney, *J. Am. Chem. Soc.* **1950**, *72*, 6463–6465 and *J. Org. Chem.* **1951**, *16*, 981–987.

[2]  P. Scheuer, *Chemistry of Marine Natural Products*, Academic Press, New York, **1973**.

[3]  J. L. Gould, W. T. Keeton, Biological Science, 6th Edition, W. W. Norton & Company, New York, **1996**.

[4]  MarinLit – A database of the literature on marine natural products, J. W. Blunt, M. H. G. Munro, University of Canterbury, Christchurch, New Zealand.

[5]  G. R. Pettit, C. L. Herald, D. L. Doubek, D. L. Herald, *J. Am. Chem. Soc.* **1982**, *104*, 6846–6848.

[6]  K. L. Rinehart, A. M. Lithgow-Bertelloni, PCT Int. Pat. Appl., WO 91.04985, Apr. 18, 1991; GB Appl. 89/22,026, Sept.29, 1989; *Chem. Abstr.* **1991**, *115*, 248086q.

[7]  (a) K. L. Rinehart, Jr., J. B. Gloer, J. C. Cook, Jr., S. A. Mizsak, T. A. Scahill, *J. Am. Chem Soc.* **1981**, *103*, 1857–1859; (b) K. L. Rinehart, Jr., J. B. Gloer, R. G. Hughes, Jr., H. E. Renis, F. A. McGovern, E. B. Swynenberg, D. A. Stringfellow, S. L. Kuentzel, L. H. Li, *Science* **1981**, *212*, 933–935.

[8]  R. Sakai, K. L. Rinehart, V. Kishore, B. Kundu, G. Faircloth, J. B. Gloer, J. R. Carney, M. Namikoshi, F. Sun, R. G. Hughes, Jr., D. G. Gràvalos, T. G. de Quesada, G. R. Wilson, R. M. Heid, *J. Med. Chem.* **1996**, *39*, 2819–2834.

[9]  G. R. Pettit, Y. Kamano, C. L. Herald, A. A. Tuinman, F. E. Boettner, H. Kizu, J. M. Schmidt, L. Baczynskyj, K. B. Tomer, R. J. Bontems, *J. Am. Chem. Soc.* **1987**, *109*, 6883–6885.

[10]  K. L. Rinehart, Jr., T. G. Holt, N. L. Fregeau, J. G. Stroh, P. A. Keifer, F. Sun, L. H. Li, D. G. Martin, *J. Org. Chem.* **1990**, *55*, 4512–4515.

[11]  M. C. Wani, H. L. Taylor, M. E. Wall, P. Coggon, A. T. McPhail, *J. Am. Chem. Soc.* **1971**, *93*, 2325–2327.

[12]  W. P. Aue, E. Bartholdi, R. R. Ernst, *J. Chem. Phys.* **1976**, *64*, 2229–2246.

[13]  (a) L. Braunschweiler, R. R. Ernst, *J. Magn. Reson.* **1983**, *53*, 521–528; (b) A. Bax, R. A. Byrd, A. Aszalos, *J. Am. Chem. Soc.* **1984**, *106*, 7632–7633; (c) A. Bax, D. G. Davis, *J. Magn. Reson.* **1985**, *65*, 355–360.

[14]  (a) L. Müller, *J. Am. Chem. Soc.* **1979**, *101*, 4481–4484; (b) M. R. Bendall, D. T. Pegg, D. M. Doddrell, *J. Magn. Reson.* **1983**, *52*, 81–117; (c) A. Bax, R. H. Griffey, B. L. Hawkins, *J. Am. Chem. Soc.* **1983**, *105*, 7188–7190 and *J. Magn. Reson.* **1983**, *55*, 301–315.

[15]  G. Bodenhausen, D. J. Ruben, *Chem. Phys. Lett.* **1980**, *69*, 185–189.

[16]  J. D. Forman-Kay, A. M. Gronenborn, L. E. Kay, P. T. Wingfield, G. M. Clore, *Biochemistry* **1990**, *29*, 1566–1572.

[17]  L. Lerner, A. Bax, *J. Magn. Reson.* **1986**, *69*, 375–380.

[18]  A. Bax, M. F. Summers, *J. Am. Chem. Soc.* **1986**, *108*, 2093–2094.

[19]  K. Gustafson, M. Roman, W. Fenical, *J. Am. Chem. Soc.* **1989**, *111*, 7519–7524.

[20]  G. Eich, G. Bodenhausen, R. R. Ernst, *J. Am. Chem. Soc.* **1982**, *104*, 3731–3732.

[21]  (a) N. Miyairi, H.-I. Sakai, T. Konomi, H. Imanaka, *J. Antibiotics* **1976**, *29*, 227–235; (b) Y. Tokuma, N. Miyairi, Y. Morimoto, *J. Antibiotics* **1976**, *29*, 1114–1116.

[22]  J. Kobayashi, J. Cheng, H. Nakamura, Y. Ohizumi, Y. Hirata, T. Sasaki, T. Ohta, S. Nozoe, *Tetrahedron Lett.* **1988**, *29*, 1177–1180.

[23]  S. J. Bloor, F. J. Schmitz, *J. Am. Chem. Soc.* **1987**, *109*, 6134–6136.

[24]  F. S. de Guzman, F. J. Schmitz, *Tetrahedron Lett.* **1989**, *30*, 1069–1070.

[25]  First isolation: E. Lederer, C. Teissier, C. Huttrer, *Bull. Soc. Chim. Fr.* **1940**, *7*, 608–615.

[26]  G. Cimino, A. Crispino, S. de Rosa, S. de Stefano, M. Gavagnin, G. Sodano, *Tetrahedron* **1987**, *43*, 4023–4030.

[27]  F. J. Schmitz, F. S. DeGuzman, M. B. Hossain, D. van der Helm, *J. Org. Chem.* **1991**, *56*, 804–808.

[28]  F. Bracher, *Liebigs Ann. Chem.* **1992**, 1205–1207.

[29] A. Bax, R. Freeman, T. A. Frankiel, *J. Am. Chem. Soc.* **1981**, *103*, 2102–2104.

[30] (a) B. Reif, M. Köck, R. Kerssebaum, H. Kang, W. Fenical, C. Griesinger, *J. Magn. Reson.* **1996**, *A118*, 282–285; (b) M. Köck, B. Reif, W. Fenical, C. Griesinger, *Tetrahedron Lett.* **1996**, *37*, 363–366.

[31] (a) Y. Q. Gosser, K. P. Howard, J. H. Prestegard, *J. Magn. Reson.* **1993**, *B101*, 126–133; (b) J. Chung, J. R. Tolman, K. P. Howard, J. H. Prestegard, *J. Magn. Reson.* **1993**, *B102*, 137–147; (c) T. K. Pratum, B. S. Moore, *J. Magn. Reson.* **1993**, *B102*, 91–97.

[32] J. Weigelt, G. Otting, *J. Magn. Reson.* **1995**, *A113*, 128–130.

[33] R. E. Hurd, B. K. John, *J. Magn. Reson.* **1991**, *91*, 648–653.

[34] Therefore, these kinds of experiments are very insensitive in comparison to HSQC. When using a micro probe, the minimum amount of material is about 15 mg for a compound with a molecular weight of 500. A micro probe is a 2.5 mm probe (diameter) which has a sensitivity improvement of a factor of 1.7 in comparison to the 5 mm probe with a coil length of 16 mm. A cryogenic probe offers a sensitivity enhancement of a factor of 4.

[35] T. H. Mareci, R. Freeman, *J. Magn. Reson.* **1982**, *48*, 158–163.

[36] (a) J. Schleucher, M. Sattler, C. Griesinger, *Angew. Chem.* **1993**, *105*, 1518–1521; *Angew. Chem. Int. Ed. Engl.* **1993**, *32*, 1489–1491; (b) M. Sattler, M. Schwendinger, J. Schleucher, C. Griesinger, *J. Biomol. NMR* **1995**, *5*, 11–22.

[37] A. G. Palmer III, J. Cavanagh, P. E. Wright, M. Rance, *J. Magn. Reson.* **1991**, *93*, 151–170.

[38] D. J. Faulkner, in *Marine Biotechnology, Vol. 1*; D. H. Attaway, O. R. Zaborsky, Eds.; Plenum Press: New York **1993**, 459–474.

[39] T. Lindel, J. Junker, M. Köck, *J. Mol. Mod.* **1997**, *3*, 364–368.

[40] The high calculation speed of the COCON program is based mainly on the integrated evaluation of HMBC information simultaneously with the generation of constitutions. Another important fact is that constitutionally equivalent isomers are detected at the earliest possible stage in the program.

[41] E. Pretsch, T. Clerc, J. Seibl, W. Simon, *Tables of Spectral Data for Structure Determination of Organic Compounds*, 2nd ed.; Springer-Verlag, New York, **1989**.

[42] T. Lindel, J. Junker, M. Köck, *Eur. J. Org. Chem.* **1999**, 573–577.

[43] W. Maier in *Computer-Enhanced Analytical Spectroscopy, Volume 4* (Ed.: C. L. Wilkins), Plenum Press, New York, London, **1993**, 37–55.

[44] M. Köck, J. Junker, W. Maier, M. Will, T. Lindel, *Eur. J. Org. Chem.* **1999**, 579–586.

[45] (a) N. H. Andersen, D. D. Syrdal, B. M. Lawrence, S. J. Terhune, J. W. Hogg, *Phytochemistry* **1973**, *12*, 827–833; (b) S. J. Terhune, J. W. Hogg, B. M. Lawrence, *Phytochemistry* **1974**, *13*, 1183–1185; (c) M. Köck, K. M. Jenkins, W. Fenical, unpublished results.

[46] The methanol/dichloromethane extract of the coral was solvent-partitioned with trimethylpentane, ethyl acetate, water, and *n*-butanol. The trimethylpentane-soluble part was separated by silica flash column chromatography with a trimethylpentane/ethyl acetate gradient, resulting in 13 fractions. Fraction 1 (100% trimethylpentane) yielded the fraction with cadinadiene (**10**). The stereochemistry of **10** was not assigned because the constitution was already known. The correct name of **10** is cadina-4(14),5-diene, other names are known in the literature for the different stereoisomers. The numbering of the structure used here is according to the decreasing $^{13}$C chemical shifts.

[47] (a) D. M. Doddrell, D. T. Pegg, M. R. Bendall, *J. Magn. Reson.* **1982**, *48*, 323–327; (b) D. T. Pegg, M. R. Bendall, *J. Magn. Reson.* **1983**, *55*, 114–127.

[48] A method to introduce Bredt's rule into a structure generator was published by J.-M. Nuzillard, *J. Chem. Inf. Comp. Sci.* **1994**, *34*, 723–724.

[49] Nine further HMBC correlations with low intensity were not used because the occurence of $^4J$ correlations cannot be excluded.

[50] (a) S. Forenza, L. Minale, R. Riccio, E. Fattorusso, *J. Chem. Soc., Chem. Commun.* **1971**, 1129–1130; (b) E. E. Garcia, L. E. Benjamin, R. I. Fryer, *J. Chem. Soc., Chem. Commun.* **1973**, 78–79; (c) R. P. Walker, D. J. Faulkner, D. van Engen, J. Clardy, *J. Am. Chem. Soc.* **1981**, *103*, 6772–6773.

[51]  Two possible atom types are generated for C-11 and C-12, while all other atoms had only one possible atom type. Only one out of four atom type combinations yielded solutions because of an odd number of sp or $sp^2$ atoms.

[52]  M. Köck, J. Junker, T. Lindel, *Org. Lett.* submitted.

[53]  Four different data sets for oroidin (**11**) were used for the COCON calculations. Data set A was measured at 500 MHz in DMSO-$d_6$, data set B (600 MHz, DMSO-$d_6$), data set C (600 MHz, chloroform-$d_1$/DMSO-$d_6$ 3:2), and D (600 MHz, chloroform-$d_1$/DMSO-$d_6$ 3:2 and the correlations obtained for 4,5-dibromopyrrole-2-carboxylic acid).

[54]  H.-O. Kalinowski, S. Berger, S. Braun, $^{13}$C-NMR-Spektroskopie, Thieme Verlag, Stuttgart, **1984**.

[55]  (a) R. J. Anderson, D. J. Faulkner, H. Cun-heng, G. D. Van Duyne, J. Clardy, *J. Am. Chem. Soc.* **1985**, *107*, 5492–5495; (b) N. Lindquist, W. Fenical, G. D. Van Duyne, J. Clardy, *J. Org. Chem.* **1988**, *53*, 4570–4574.

[56]  M. Köck, B. Reif, C. Griesinger, to be published.

# 7.3 Structural studies of intermolecular interactions by NMR spectroscopy

*Gerd Gemmecker*

The use of multidimensional NMR spectroscopy for the structure determination of small organic molecules is well established in the scientific community. In the last decade, the range of such studies has been extended stepwise to larger biomacro-molecules, mainly proteins and nucleic acids. Structure determinations of proteins up to ca. 20–30 kDa are now possible in a more or less routine manner, thanks to recent progress in protein expression, isotopic labeling and NMR methodology. Meanwhile, it is also accepted that the resolution of NMR protein structures is generally comparable to that of X-ray structures, and in recent years NMR studies have contributed every fourth structure newly deposited in the Brookhaven Protein Data Bank [1].

However, NMR structure determination is not just a mere competition or, better, an alternative to X-ray diffraction in the mentioned size range. It also offers a wealth of possibilities to study intermolecular interactions, reactions, and unstable intermediates under conditions that are not suited for crystallization, but closely resemble the physiological environment. In addition, NMR spectroscopy also allows us to gain information about the dynamic behavior of the systems under study, thus leading us from the often merely static definition of 'structure' towards a much more realistic view of biologically active molecules and their interactions.

## 7.3.1 Binding site mapping

Protein complexes with high-affinity ligands, *e.g.*, enzyme inhibitors, generally behave like single molecules in NMR and X-ray studies and thus pose no special problems, except for their larger size and possible changes in solubility. Under normal protein NMR conditions (*i.e.*, sample concentrations in the order of ca. 1 mM), binding constants in the nanomolar range lead to negligible concentrations of free protein and ligand under stoichiometric conditions.

However, chemical and biological processes often also include weaker inter-molecular interactions leading to a dynamic equilibrium between free and bound states of the components involved. For a dissociation constant in the micro- or milli-molar range and a stoichiometric ratio of protein and ligand (ca. 1 mM each), the system will exist in an equilibrium of the protein-ligand complex and significant amounts of the free molecules, in slow or fast exchange with the complex. It is obvious that under such conditions the NMR spectrum of the system will be far more complex and difficult to analyze than for the isolated complex alone. This problem can be resolved by differential isotopic labeling of the components, *i.e.*, only one of the components is enriched with a stable, NMR active, but naturally rare isotope. For biomacromolecules the isotopes $^2$H, $^{13}$C, and $^{15}$N are usually employed for this purpose, by means of appropriately isotope-labeled culture media in the bacterial overexpression of the protein. Isotope-filters in the NMR experiments then allow us to selectively observe either the labeled or the unlabeled component, with the selected component practically quantitatively transferred into the complexed state by adding an excess of the non-observed component (which is suppressed anyway in the NMR measurement).

## 7.3.2    The bacterial phosphotransferase system

In the following, some applications of the aforementioned techniques pertaining to the phosphotransferase system (PTS) will be shown. The bacterial phosphoenol pyruvate-dependent phosphotransferase system combines the cytosolic uptake of $C_6$ sugars and sugar alcohols with the phosphorylation of its substrates and regulatory functions [2, 3]. While the transporter proper is formed by the transmembrane IIC subunit, the other sugar-specific domains IIA and IIB, as well as the sugar-unspecific proteins Enzyme I and HPr, act as a phosphorylation cascade (Figure 7.3.1).

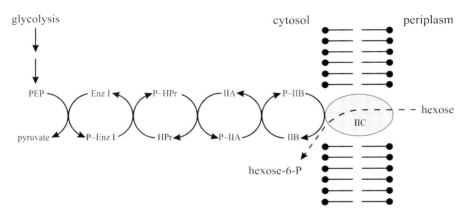

**Figure 7.3.1.**  Schematic organization of the bacterial phosphotransferase system (PTS).

The two hydrophilic subunits IIA and IIB are, in nature, realized either as separate proteins or as domains in various combinations, also together with the transmembrane subunit IIC containing the sugar binding site. During the phosphate transfer from phosphoenol pyruvate (PEP), all PTS proteins (except for IIC) are transiently phosphorylated at a histidine or cysteine residue. Each bacterium contains several different sugar-specific parts which also differ between organisms and have been classified into various families, according to sequence homology [4].

In addition to their importance for the bacterial metabolism, the highly specific protein–protein interactions involved in the phosphate transfer between the PTS proteins make them ideal objects for NMR studies of intermolecular interactions.

For the glucose transporter of *E. coli*, several NMR and X-ray studies of the soluble IIA$^{\text{Glc}}$ protein have already been conducted in the unphosphorylated and also in the

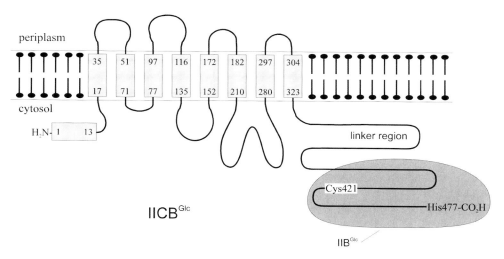

**Figure 7.3.2.** Topology of the IICB$^{Glc}$ protein of *E. coli*. The C-terminal IIB$^{Glc}$ subdomain (shaded area) can be overexpressed separately from the membrane-spanning IIC domain and linker region.

phosphorylated state [5, 6]. The IIB$^{Glc}$ domain of the *E. coli* glucose transporter, however, is covalently linked to the C-terminus of the transmembrane domain (IICB$^{Glc}$), and thus is not accessible to high-resolution NMR (Figure 7.3.2). However, it could be shown that the IIB$^{Glc}$ domain can be overexpressed separately, folding into a stable, soluble and fully functional protein of ca. 11 kDa molecular weight [7].

A high-resolution 3D structure of IIB$^{Glc}$ could be determined by multidimensional NMR spectroscopy, yielding a new fold consisting of a four-stranded antiparallel $\beta$-sheet and three $\alpha$-helices located on one side of the $\beta$-sheet (Figure 7.3.3) [8]. Our next interest focused on the structural effects of IIB$^{Glc}$ phosphorylation. From mutation studies it had been postulated that IIB$^{Glc}$ is transiently phosphorylated at the sulfhydryl group of Cys35 (= Cys421 in the original IICB$^{Glc}$ protein) – in contrast to all other known PTS proteins, which are *N*-phosphorylated at a histidine side chain. In order to bring IIB$^{Glc}$ into the phosphorylated state and keep it phosphorylated for an extended time period, the unlabeled PTS phosphorylation cascade (Enzyme I, HPr and IIA$^{Glc}$) was added in catalytic amounts to a solution of uniformly $^{15}$N labeled IIB$^{Glc}$. An excess of PEP allowed us to keep IIB$^{Glc}$ quantitatively phosphorylated for a four-day period, sufficient to run 2D- and 3D-NMR spectra [9].

The Cys-phosphorylation was confirmed by the $^{31}$P chemical shift of 13 ppm, characteristic for a phosphate thioester. While IIB$^{Glc}$ phosphorylation was complete within 15 min, hydrolysis occured with a half-life time of ca. 4 h in aqueous solution at pH 6, after depletion of the PEP reservoir. A comparison of the fingerprint $^{1}$H,$^{15}$N-HSQC amide spectra and the NOE patterns of phosphorylated and unphosphorylated IIB$^{Glc}$ gave strong evidence that the global fold of the protein was not changed by phosphorylation. However, while the $\beta$-turn region containing Cys35 showed signs of slow conformational flexibility in the unphosphorylated state, it

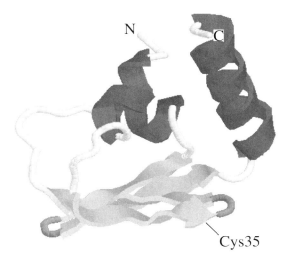

**Figure 7.3.3.** Cartoon representation of the 3D structure of IIB$^{Glc}$, with the active site residue Cys35 located in position i of a $\beta$-turn.

seemed to become quite rigid after phosphorylation. We assume that this rigidity arises from salt bridges between the phosphate group and the side chains of Arg38 and Arg40, both of which are highly conserved among the IIB domains of the glucose PTS family (Figure 7.3.4).

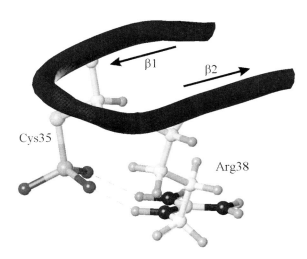

**Figure 7.3.4.** Model of the IIB$^{Glc}$ phosphorylation site. In the phosphorylated state, the phosphate group bound to Cys35 could form a salt bridge with the Arg38 guanidinium group, thus bringing the formerly flexible $\beta$-turn between strands 1 and 2 into a rigid conformation.

## 7.3.3 Mapping protein–protein interactions by NMR

The IIB$^{Glc}$ domain is involved in two intermolecular interactions: the phosphate transfer from IIA$^{Glc}$ to IIB$^{Glc}$, and the phosphorylation of the sugar in the ternary complex with the membrane-bound IIC$^{Glc}$ domain. While IIC$^{Glc}$ cannot be brought into solution without denaturation, both the IIA and IIB domains are soluble hydrophilic proteins accessible to high-resolution NMR studies. However, two problems exist that make a direct study of the 3D structure of the IIA–IIB complex impossible. The first is the sheer size of the complex (ca. 30 kDa) and the complexity of the spectra from 269 amino acids; however, with recent isotopic labeling methods (including fractionated deuteration) and NMR techniques, such a task should still be within the range of NMR structure determination [10].

The second problem arises from the (bio)chemical properties of the two proteins involved. In a sample with typical NMR concentrations (ca. 0.1–1 mM) of the two proteins, in the proper stoichiometric ratio 1:1, the IIA–IIB complex forms only partially, in slow equilibrium with an almost equal amount of the free proteins IIA$^{Glc}$ and IIB$^{Glc}$. The NMR spectra show separate overlapping signal sets for each of the three species, rendering the spectra far too complex for interpretation.

Generally, such an equilibrium of free and complexed species can be observed for micro- to millimolar $K_d$ values. In order to obtain almost quantitative complex formation for typical NMR concentrations, the binding constant of the proteins would have to be in the nanomolar range. However, one of the two components can often be forced into a completely (*i.e.*, >95%) complexed state by adding an excess of the other complex partner. For example, a two-fold excess of IIA$^{Glc}$ (ca. 0.5 mM concentration) was enough to bring the uncomplexed fraction of IIB$^{Glc}$ below the detection limit (corresponding to a $K_d$ of ca. 30 μM), and vice versa.

Of course, the signals of the component in excess (in a mixture of free and complexed state) have to be suppressed in the NMR spectra. This can be readily achieved by differential isotopic labeling of the two components, in several different ways. Thus, the component to be suppressed can be deuterated (*i.e.*, depleted of the natural NMR active isotope $^1$H), or the component to be observed can be labeled with an NMR active isotope of low natural abundance ($^{15}$N or $^{13}$C), or both approaches could be combined.

In our case we chose to uniformly label only one component with $^{15}$N, which allowed us to observe it selectively in $^{15}$N-filtered spectra. A comparison of the simple 2D $^1$H,$^{15}$N-correlation spectra of each component already indicates the regions of the protein involved in complex formation, which display significant signal shifts (Figure 7.3.5).

However, the majority of the signals of both proteins do not shift upon complex formation, indicating that their global fold is preserved (this can be further confirmed by comparing other structure-sensitive parameters, *e.g.*, NOE patterns). In such a case, the interaction sites of the components in the complex can be found by simply mapping the observed signal shifts onto the known 3D structures of the (uncomplexed) proteins.

For both IIA$^{Glc}$ and IIB$^{Glc}$ one obtains binding sites consisting of a single surface patch centered around the two phosphorylation sites, His90 of IIA$^{Glc}$ and Cys35 of IIB$^{Glc}$ (Figures 7.3.6 and 7.3.7).

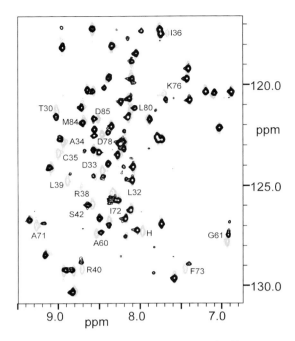

**Figure 7.3.5.** Comparison of a section of the $^1$H,$^{15}$N-correlation spectra of IIB$^{Glc}$ in the free state (light gray) and in the complex with unlabeled IIA$^{Glc}$ (black). The sequential assignment of free IIB$^{Glc}$ is given for all peaks that show a significant shift upon complexation.

The two interaction sites display a complementary structure: while His90 of IIA$^{Glc}$ is located in a hydrophobic depression of the protein surface, Cys35 of IIB$^{Glc}$ is lying on a hydrophobic protrusion. Both contact sites contain mainly non-polar residues, so that hydrophobic interactions will play an important part

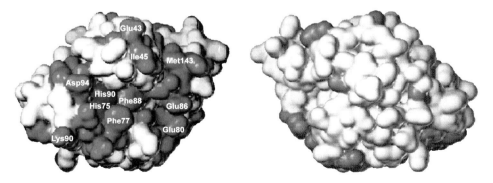

**Figure 7.3.6.** Contact site of IIA$^{Glc}$ for the complexation with IIB$^{Glc}$ (left). Residues with significantly shifted amide signals are shown in dark; the protein moiety opposite the phosphorylation site His90 (right) is not affected. (Reproduced with permission from [9]. Copyright 1997 American Chemical Society)

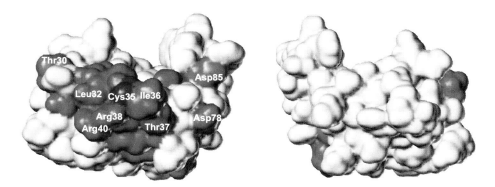

**Figure 7.3.7.** Contact site of IIB$^{Glc}$ for the complexation with IIA$^{Glc}$. (Reproduced with permission from [9]. Copyright 1997 American Chemical Society) Again, the residues with significantly shifted amide signals (shown in dark) are centered around the phosphorylation site Cys35 (left), while the opposite part of the protein is not affected (right).

in the IIA$^{Glc}$–IIB$^{Glc}$ interaction. However, apart from the phosphorylation sites themselves, some other polar amino acids are also located in the binding surfaces, *e.g.*, Lys69, Glu86, and Ser141 of IIA$^{Glc}$, and Gln70 of IIB$^{Glc}$. All these are highly conserved among the glucose family PTS proteins, and it must be assumed that they are involved in polar interactions essential for the specificity of the inter-molecular interaction.

## 7.3.4 SAR-by-NMR

In the example of the IIA$^{Glc}$–IIB$^{Glc}$ complex, the exchange between the free and complexed state was slow on the NMR time scale ($k_{off} <$ ca.10 Hz), resulting in separate signal sets for both states and the need to add an excess of the unlabeled compound to suppress the signals of free labeled protein.

Especially in the case of low binding constants, the exchange between free and complexed state is usually much faster, so that only *one* averaged signal set will be observed for the labeled compound, even if only partially complexed. Thus, the effects of ligand binding to a protein (*i.e.*, amide signal shifts in the binding site) can already be detected when only a small percentage of the protein exists in the complexed state. Under these conditions, even binding constants in the 10–100 millimolar range are sufficient to cause observable protein signal shifts, even for ligand concentrations as low as ca. 10 mM.

This possibility to detect relatively weak binding interactions and, at the same time, also identify the binding site, has prompted Fesik and co-workers to develop a technique named SAR-by-NMR (SAR = structure–activity relationships) [11, 12]. By acquiring $^1$H,$^{15}$N correlation spectra of a $^{15}$N-labeled protein in the presence of a mixture of ca. 10 small organic compounds and searching for signals shifted relative

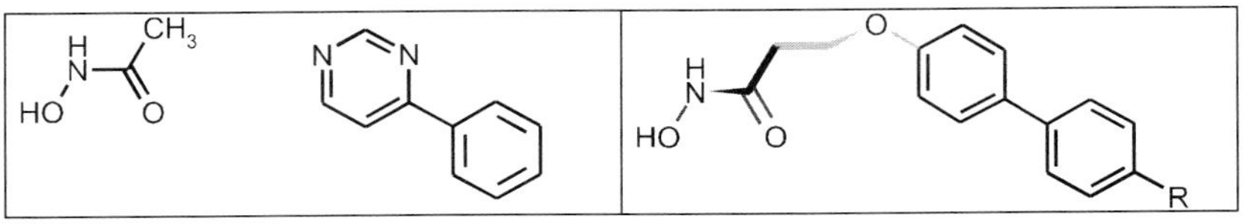

**Figure 7.3.8.** Example for SAR-by-NMR. From a library with small organic compounds, acetohydroxamic acid and 4-phenylpyrimidine were found to weakly bind to stromelysin ($K_d$ 17 mM and 20 µM, respectively) at adjacent sites (left). Introduction of a linker (shown in grey) and systematic modifications resulted in the development within a few months of a highly potent inhibitor ($IC_{50} = 25$ nM) of stromelysin and gelatinase A (right) [14].

to the free protein, they are able to screen up to 1000 compounds per day for millimolar binding affinities.

In addition, the information about the individual ligands' binding sites at the protein can be used to find ligands interacting with slightly different, but adjacent interaction sites. These weakly binding ligands are then connected to a single molecule with an appropriate linker, yielding new high-affinity ligands that can be useful leads for drug development (Figure 7.3.8). The practicability of this approach – combining screening methods with structure-based rational drug design – has already been demonstrated in several examples [13, 14].

**Acknowledgments:** The results presented here could not have been obtained without the efforts of Matthias Eberstadt and Simona Golic Grdadolnik, in co-operation with Prof. Bernhard Erni and Andreas Buhr (University of Berne) and Prof. Kessler (TU München). Financial support from the DFG, the SFB 369, the Fonds der Chemischen Industrie and the Dr.-Ing. Leonhard-Lorenz-Stiftung (München) is gratefully acknowledged.

# References

[1]  W. A. Hendrickson, K. Wüthrich, *Macromolecular Structures*, Current Biology, London, **1995**.

[2]  P. W. Postma, J. W. Lengeler, G. R. Jacobson, *Microbiol. Rev.* **1993**, *57*, 543–594.

[3]  M. H. Saier Jr., S. Chauvaux, J. Deutscher, J. Reizer, J.-J. Ye, *Trends Biochem. Sci.* **1995**, *20*, 267–271.

[4]  B. Erni, *Int. Rev. Cytol.* **1992**, *137A*, 127–148.

[5]  J. G. Pelton, D. A. Torchia, N. D. Meadow, S. Roseman, *Biochemistry* **1992**, *31*, 5215–5224.

[6]  J. G. Pelton, D. A. Torchia, N. D. Meadow, S. Roseman, *Protein Sci.* **1993**, *2*, 543–558.

[7]  A. Buhr, K. Flükiger, B. Erni, *J. Biol. Chem.* **1994**, 269, 23437–23443.

[8]  M. Eberstadt, S. Golic Grdadolnik, G. Gemmecker, H. Kessler, A. Buhr, B. Erni, *Biochemistry* **1996**, *35* (35), 11286–11292.

[9]  G. Gemmecker, M. Eberstadt, S. Golic Grdadolnik, H. Kessler, A. Buhr, B. Erni, *Biochemistry* **1997**, *36* (24), 7408–7417.

[10]  M. Sattler, S. W. Fesik, *Structure* **1996**, *4* (11), 1245–1249.

[11]  S. B. Shuker, P. J. Hajduk, R. P. Meadows, S. W. Fesik, *Science* **1996**, *274*, 1531–1534.

[12]  H. Kessler, *Angew. Chemie Int. Ed. Engl.* **1997**, *36* (8), 829–831.
[13]  P. J. Hajduk, J. Dinges, G. F. Miknis, M. Merlock, T. Middleton, D. J. Kempf, D. A. Egan, K. A. Walter, T. S. Robins, S. B. Shuker, T. F. Holzman, S. W. Fesik, *J. Med. Chem.* **1997**, *40*, 3144.
[14]  P. J. Hajduk, R. P. Meadows, S. W. Fesik, *Science* **1997**, *278*, 497–499.

## 7.4     Tailor-made experimental building blocks for NMR studies of bioorganic compounds

*Steffen J. Glaser*

The characterization of the configuration and conformation of molecules is an important aspect of chemisty in general, and of bioorganic chemistry in particular, where detailed structural data form the basis for an understanding of the biological function of a molecule. Nuclear magnetic resonance (NMR) provides powerful tools to determine structural as well as dynamical parameters in solution [1, 2]. However, the study of biological macromolecules, such as proteins, DNA, and RNA poses a number of problems. The NMR spectra of these molecules contain hundreds of resonances that are relatively broad and lead to severe overlap. In addition, the molecules are often only available in limited quantities or are stable in solution only in relatively low concentrations by usual NMR standards. Hence, the most important difficulties are the limited resolution and sensitivity of NMR spectra. Progress in technological, experimental, and synthetic techniques have considerably improved the situation during the past years. These developments include hardware improvements, such as stronger magnets and more sensitive NMR probes, new approaches for the uniform or specific incorporation of $^{13}$C, $^{15}$N, or $^{2}$H labels in the molecules of interest using chemical or biological synthesis, and the design of new multi-dimensional NMR experiments. At first sight, the number of published NMR experiments that are of potential use in bioorganic chemistry is overwhelming. However, a closer inspection reveals that all these experiments are combinations of a relatively small number of experimental building blocks. One of the goals of our research is the development of new basic building blocks with tailor-made properties.

### 7.4.1     Computer-aided design of NMR experiments

Whereas the most simple building blocks consist of a single radiofrequency pulse element, such as a 90° or 180° pulse, there are also a number of important experimental building blocks that consist of hundreds or even thousands of well-defined radiofrequency pulses. In high-resolution NMR, these so-called multiple-pulse sequences are frequently used for the transfer of magnetization between coupled spins or for the elimination of couplings [1]. With the help of simulation programs it is possible to predict with high accuracy the result of any given NMR pulse sequence applied to a spin system that is characteristic for the class of investigated compounds, such as the sugar moieties in DNA and RNA or the individual amino acid moieties in peptides and proteins. Furthermore, multiple-pulse sequences with desired properties can be optimized from scratch. This is possible if the extent to which these properties are met by a given sequence is translated into a figure of merit that reflects the fitness of the sequence [3, 4]. Hence, the computer-aided design of multiple-pulse sequences is a powerful tool for the development of novel experimental building blocks with superior characteristics.

With the help of our simulation program SIMONE [3] we developed a number of new experimental building blocks that are tailor-made for the NMR spectroscopy of biological macromolecules in solution. These building blocks include multiple-pulse sequences for the efficient correlation of resonances in two-dimensional NMR spectra to help in the resonance assignment of isotopically labeled [4–10] and unlabeled compounds [3, 4, 11, 12], experiments for the determination of homonuclear and heteronuclear coupling constants in peptides, proteins and RNA [4, 13–18], robust pulse sequences for the measurement of inter-nuclear distances [19, 20], and band-selective heteronuclear decoupling sequences [21]. The goal of this contribution is to illustrate the underlying design principles of some of the new building blocks and to point out practical examples. In addition, a new theoretical tool is discussed that makes it possible for the first time to decide whether a given experimental building block is optimal with respect to sensitivity or if there is still room for improvement. The emphasis will be on basic principles in order to make it possible to judge if these or related tailor-made solutions are applicable to the specific compounds of the reader. For experimental and theoretical details, the reader is referred to the original literature.

## 7.4.2 Tailor-made experiments for the correlation of resonances

With increasing magnetic fields the absolute spectral width (measured in units of Hertz; Hz) also increases. In principle, the amplitude of the radiofrequency pulses should also be increased in proportion in order to cover the same relative spectral width (measured in units of parts per million; ppm). However, the amplitude of the radiofrequency field is not only restricted by hardware limitations of available spectrometers but also by restrictions due to sample heating by the applied radiofrequency power. Hence, one goal for the development of new multiple-pulse sequences is to increase the covered bandwidth relative to the applied radiofrequency amplitude [3–5]. However, in contrast to the development of such broadband sequences it can also be of advantage to create band-selective experimental building blocks. This is exemplified for the case of homonuclear $J$-correlation spectra of peptides and proteins [12]. With the exception of proline (no $H_\alpha$ proton) and glycine (two $H_\alpha$ protons), every amino acid residue is expected to create a single correlation peak between the amide proton $H^N$ and $H_\alpha$ due to the vicinal $^3J(H^N,H_\alpha)$ coupling. The corresponding set of cross-peaks provides a fingerprint that is characteristic for each peptide or protein. Fingerprint signals can be acquired using COSY (*c*orrelated *s*pectroscop*y*) [22] or TOCSY (*to*tal *c*orrelation *s*pectroscop*y*) [4, 23–27] experiments. However, it is often not possible to detect the complete set of $H^N$–$H_\alpha$ fingerprint signals because of the limited sensitivity of these experiments. Experiments with improved sensitivity and the potential of spectral editing to resolve overlap in the fingerprint region are highly desirable. Whereas for two coupled spins the TOCSY experiment is inherently more sensitive than the COSY experiment [28], in larger spin systems the TOCSY experiment transfers magnetization between all spins that are part of the J-coupling network (see Figure 7.4.1a). The resulting dilution of magnetization in the spin system results in a reduced intensity of the cross-peaks of interest between $H^N$ and $H_\alpha$. These signals can be enhanced if the transfer of magnetization within the

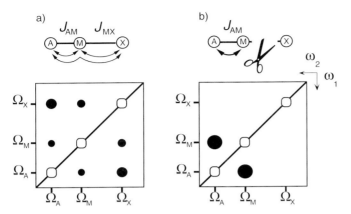

**Figure 7.4.1.** Magnetization transfer in a system consisting of three coupled spins (A, M and X) during the mixing period of TOCSY (a) and TACSY (b) experiments (top) and a schematic representation of the corresponding two-dimensional spectra (bottom). TOCSY spectra contain cross-peaks (solid circles) between all resonances on the diagonal whereas TACSY spectra yield selected cross-peaks with increased intensity.

spin system is controlled in the mixing period of the experiment. This can be achieved if some spins are excluded from the transfer process by effectively reducing the size of a given spin system or a class of spin systems (see Figure 7.4.1b). This class of experiments is called *tailored correlation spectroscopy* (TACSY) [29] because, in contrast to TOCSY experiments, only selected resonances are correlated by cross-peaks in the corresponding two-dimensional spectra (Figure 7.4.1b). This can be achieved if two spins A and M (*e.g.* $H^N$ and $H_\alpha$) can be manipulated in a different way than a third spin X (*e.g.* $H_\beta$), which is possible for several amino acid types (Ala, Arg, Glu, Gln, Ile, Lys, Met, and Val) for which the typical chemical shifts of the $H_\beta$ resonances are well separated from the $H^N$ and $H_\alpha$ resonances. The CABBY-1 sequence [12] was developed specifically for this application. This not only provides increased intensities of the $H^N$–$H_\alpha$ fingerprint signals of these amino acid residues compared to conventional experiments but also makes it possible to edit the fingerprint region based on amino acid type and on the size of the $^3J(H^N,H_\alpha)$ coupling constants. The TACSY principle has also been successfully applied to the transfer of magnetization between $H'_3$, $H'_4$, $H'_5$, and $H''_5$ protons in DNA where tailor-made multiple-pulse sequences make it possible to avoid a loss of magnetization to the $H'_2$, $H''_2$ and $H'_1$ protons [30]. In addition, tailor-made multiple-pulse sequences were developed for the correlation of $C_\alpha$ and $C'$ and for the correlation of $C_\beta$ and aromatic carbon resonances in $^{13}$C-labeled proteins [6].

## 7.4.3   Tailor-made experiments for the determination of coupling constants

TOCSY and TACSY building blocks not only provide efficient means to establish correlations between coupled resonances in order to help in the assignment of complex

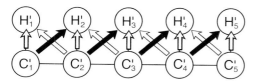

**Figure 7.4.2.** Schematic representation of the $^1$H and $^{13}$C spins of the ribose moiety of RNA. The directed TOCSY building block makes it possible to create selectively the 'forward directed' $C'_i$-$H'_{i+1}$ correlations (black arrows) whereas the $C'_i$-$H'_i$ and $C'_i$-$H'_{i-1}$ correlations (white arrows) are suppressed.

spectra, but also provide tools for the quantitative determination of $J$ coupling constants which yield valuable structural information. The E.TACSY (*exclusive tailored correlation spectroscopy*) experiment was used to determine $^3J(C', H_\beta)$ coupling constants in $^{13}$C-labeled proteins [13, 14]. It is based on tailor-made multiple-pulse sequences that not only restrict magnetization transfer to a defined subset of the spin system (*e.g.* $C_\alpha$ and $C_\beta$), but also leave the spin state of another set of spins untouched (*e.g.* $C'$). A novel selection principle was used for the determination of $^3J(H'_1, H'_2)$, $^3J(H'_2, H'_3)$, $^3J(H'_3, H'_4)$ and $^3J(H'_4, H'_5)$ coupling constants in $^{13}$C-labeled RNA. The most important problem for the NMR spectroscopy of RNA is the severe overlap of resonances. Even heteronuclear two- or three-dimensional experiments often do not provide the required resolution. In particular, conventional experiments cannot discriminate between different directions of coherence transfer (see white and black arrows in Figure 7.4.2). The computer-aided development of new experimental building blocks for the *directed* transfer of magnetization (see black arrows in Fig. 7.4.2) avoids undesired transfers and leads to drastically simplified spectra in which all redundant cross-peaks are eliminated. This technique made it possible to measure almost all proton-proton coupling constants of the ribose rings in a 19-mer RNA hairpin [15–17]. This directed TOCSY experiment makes use of previously usually disregarded coherences that are always created as a by-product during the transfer of magnetization in TOCSY or TACSY experiments [4].

These additional coherences (zero-quantum coherences) have also been put to good use in a new experimental approach for the quantitative determination of $^3J(H^N, H_\alpha)$ coupling constants in unlabeled peptides and proteins [18]. If for a given coupling constant (*e.g.* $^3J(H^N, H_\alpha)$) both in-phase and antiphase multiplets are available, it is possible to extract the corresponding coupling constant, even if it is smaller than the linewidth [31]. The SIAM experiment [18] allows the *s*imultaneous acquisition of *i*n-phase and *a*ntiphase *m*ultiplets with markedly improved sensitivity and resolution compared to conventional approaches. The experiment is based on the fact that zero-quantum coherence between two spins (*e.g.* $H^N$ and $H_\alpha$) can be turned into double-quantum coherence if a selective 180° pulse can be applied to one of the two spins. This is always possible if the two spins have well-separated chemical shift ranges, such as the $H^N$ and $H_\alpha$ resonances. The created double-quantum coherence can easily be separated from the transfered magnetization by using standard phase-cycling procedures and it is possible to extract from a single experimental data set two different spectra:

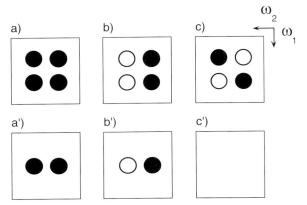

**Figure 7.4.3.** Schematic representation of characteristic cross-peak multiplet components in 2D-NMR spectra. (a) in-phase splitting in $\omega_1$ and $\omega_2$ (*e.g.* TOCSY and ZQF-SIAM); (b) in-phase splitting in $\omega_1$ and antiphase splitting in $\omega_2$ (DQF-SIAM); (c) antiphase splitting in $\omega_1$ and $\omega_2$ (*e.g.* DQF-COSY). Positive signals are white, negative signals are black. (a′), (b′) and (c′) show the corresponding multiplets if $\omega_1$ decoupling is used. In the case of (a′) and (b′), this leads to a signal enhancement and to an improved resolution in the $\omega_1$ dimension, whereas in (c′) the signals cancel.

1.  A zero-quantum filtered (ZQF) SIAM spectrum with in-phase splittings of the multiplets in the $\omega_2$ dimension (see Figure 7.4.3a).
2.  A double-quantum filtered (DQF) SIAM spectrum with antiphase splittings of the multiplets in the $\omega_2$ dimension (see Figure 7.4.3b).

As both spectra have in-phase splittings in $\omega_1$, it is possible to further increase the resolution and the sensitivity by $\omega_1$ decoupling [32] (c.f. Fig. 3a′ and 3b′). This is not possible in a conventional COSY [22] or DQF-COSY [33] experiment with antiphase splittings both in $\omega_1$ and $\omega_2$ (Figure 7.4.3c) where $\omega_1$ decoupling would erase the entire cross-peak (Figure 7.4.3c′). In Figure 7.4.4, a small section (containing

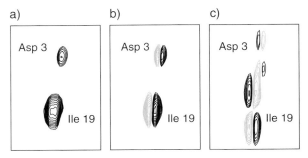

**Figure 7.4.4.** Section of the fingerprint region of a ZQF-SIAM with $\omega_1$ decoupling (a), a DQF-SIAM with $\omega_1$ decoupling (b) and a conventional DQF-COSY spectrum without $\omega_1$ decoupling (c) of the protein BPTI at 600 MHz [18]. Positive and negative signals are shown using black and gray contour lines, respectively.

cross-peaks of Asp3 and Ile19) of the fingerprint region of the protein BPTI is shown. This demonstrates the superior sensitivity and resolution of the DQF-SIAM spectrum (Figure 7.4.4b) compared to a conventional DQF-COSY spectrum (Figure 7.4.4c) which was used previously to extract antiphase multiplets. As pointed out above, the same experimental data-set provides (without any additional measurement time) a ZQF-SIAM spectrum (Figure 7.4.4a) from which the complementary in-phase multiplets can be extracted that are necessary to quantitatively determine the desired $^3J(H^N, H_\alpha)$ coupling constants. The SIAM method is not restricted to the determination of $^3J(H^N, H_\alpha)$ couplings or to applications to peptides or proteins. A necessary condition is that the two mutually coupled spins of interest (here $H^N$ and $H_\alpha$) have different chemical shifts. Furthermore, it must be ensured that in large spin systems the pure in-phase and antiphase multiplets are not contaminated by other signals. A sufficient (but not necessary) condition for this is that the detected spin (here $H^N$) has no further coupling partners. Hence, the SIAM method complements the powerful E.COSY technique [34] for which a third common coupling partner is mandatory.

## 7.4.4  A new criterion for optimal sensitivity

In addition to optimal resolution, one of the most important design criteria for applications in bioorganic chemistry is the maximum possible sensitivity of an experiment. Until recently, there was no general method for the prediction of the theoretical optimum for a given experiment, even if relaxation can be neglected. The development of a gradient-based approach [35] to determine the theoretical upper bound for the transfer efficiency of any experimental building block makes it possible to predict whether a given experimental implementation of a NMR experiment achieves the theoretical limit, or if it can be improved by a new experimental building block. It could be shown that the efficiency of some commonly used experimental building blocks can be improved by up to 40% [35]. This motivated the computer-aided development of new pulse sequence elements that actually reach the theoretical optimum [36]. In summary, the combination of new theoretical insights with the computer-aided design of NMR experiments makes it possible to create tailor-made experimental building blocks with superior characteristics that enhance the power of high resolution NMR to solve structural and dynamical problems in the field of bioorganic chemistry.

**Acknowledgments:** I would like to thank the current and former members of my group for their dedication, enthusiasm and contributions: Gerd Hauser, Frank Kramer, Burkhard Luy, Raimund Marx, Thomas Prasch, Jens Quant, Timo Reiss, Oliver Schedletzky, and Peter Schmidt. I am also grateful to Teresa Carlomagno, John Marino, Niels C. Nielsen, Michael Sattler, Harald Schwalbe, Jürgen Schleucher, Thomas Schulte-Herbrüggen, Malte Sieveking, Ole W. Sørensen, and in particular to Christian Griesinger for stimulating discussions and collaborations. I acknowledge support by the DFG (Gl 203/1 and Gl 203/2) and the Fonds der

Chemischen Industrie. NMR Experiments were performed at the Large Scale Facility for Biomolecular NMR (ERB CT 950034).

# References

[1]  R. R. Ernst, G. Bodenhausen, A. Wokaun, *Principles of Nuclear Magnetic Resonance in One and Two Dimensions*, Clarendon Press, Oxford **1987**.

[2]  W. R. Croasmun, R. M. K. Carlson, Eds., *Two-Dimensional NMR Spectroscopy*, VCH Publishers, Inc., New York **1994**.

[3]  S. J. Glaser, G. P. Drobny in *Advances in Magnetic Resonance, Vol. 14* (Ed.: W. S. Warren), Academic Press, San Diego, **1990**, pp. 35–58.

[4]  S. J. Glaser, J. J. Quant in *Advances in Magnetic and Optical Resonance, Vol. 19* (Ed.: W. S. Warren), Academic Press, San Diego, **1996**, pp. 59–252.

[5]  M. G. Schwendinger, J. Quant, S. J. Glaser, C. Griesinger, *J. Magn. Reson. B* **1994**, *111*, 115–120.

[6]  T. Carlomagno, M. Maurer, M. Sattler, M. G. Schwendinger, S. J. Glaser, C. Griesinger, *J. Biomol. NMR* **1996**, *8*, 161–170.

[7]  T. Carlomagno, B. Luy, S. J. Glaser, *J. Magn. Reson.* **1997**, *126*, 110–119.

[8]  J. Schleucher, M. Schwendinger, M. Sattler, P. Schmidt, O. Schedletzky, S. J. Glaser, O. W. Sørensen, C. Griesinger, *J. Biomol. NMR* **1994**, *4*, 301–306.

[9]  M. Sattler, P. Schmidt, J. Schleucher, O. Schedletzky, S. J. Glaser, C. Griesinger, *J. Magn. Reson. B* **1995**, *108*, 235–242.

[10] M. Sattler, J. Schleucher, O. Schedletzky, S. J. Glaser, C. Griesinger, N. C. Nielsen, O. W. Sørensen, *J. Magn. Reson. A* **1996**, *119*, 171–179.

[11] U. Kerssebaum, R. Markert, J. Quant, W. Bermel, S. J. Glaser and C. Griesinger, *J. Magn. Reson.* **1992**, *99*, 184–191.

[12] J. Quant, T. Prasch, S. Ihringer, S. J. Glaser, *J. Magn. Reson. B* **1995**, *106*, 116–121.

[13] P. Schmidt, H. Schwalbe, S. J. Glaser, C. Griesinger, *J. Magn. Reson. B* **1993**, *101*, 328–332.

[14] D. Abramovich, S. Vega, J. Quant, S. J. Glaser, *J. Magn. Reson. A* **1995**, *115*, 222–229.

[15] H. Schwalbe, J. P. Marino, S. J. Glaser, C. Griesinger, *J. Am. Chem. Soc.* **1995**, *117*, 7251–7252.

[16] S. J. Glaser, H. Schwalbe, J. P. Marino, C. Griesinger, *J. Magn. Reson. B* **1996**, *112*, 160–180.

[17] J. P. Marino, H. Schwalbe, S. J. Glaser, C. Griesinger, *J. Am. Chem. Soc.* **1996**, *118*, 4388–4395.

[18] T. Prasch, P. Gröschke, S. J. Glaser, *Angew. Chem. Int. Ed.* **1998**, *37*, 802–806.

[19] J. Schleucher, J. Quant, S. J. Glaser, C. Griesinger, *J. Magn. Reson. A.* **1995**, *112*, 144–151.

[20] J. Schleucher, J. Quant, S. J. Glaser, C. Griesinger in *Encyclopedia of Nuclear Magnetic Resonance* (Ed.: D. M. Grant, R. K. Harris), John Wiley, **1996**, pp. 4789–4804.

[21] U. Eggenberger, P. Schmidt, M. Sattler, S. J. Glaser, C. Griesinger, *J. Magn. Reson.* **1992**, *100*, 604–610.

[22] W. P. Aue, E. Bartholdi, R. R. Ernst, *J. Chem. Phys.* **1976**, *64*, 2229–2246.

[23] L. Braunschweiler, R. R. Ernst, *J. Magn. Reson.* **1983**, *53*, 521–528.

[24] A. Bax, D. G. Davis, *J. Magn. Reson.* **1985**, *65*, 355–360.

[25] J. Listerud, S. J. Glaser, G. P. Drobny, *Mol. Phys.* **1993** *78*, 629–658.

[26] O. Schedletzky, S. J. Glaser, *J. Magn. Reson. A* **1996**, *123*, 174–180.

[27] O. Schedletzky, B. Luy, S. J. Glaser, *J. Magn. Reson.* **1998**, *130*, 27–32.

[28] J. Briand, R. R. Ernst, *J. Magn. Reson. A* **1993**, *104*, 54–62.

[29] S. J. Glaser, *J. Magn. Reson. A* **1993**, *103*, 283–301.

[30] T. Prasch, *Entwicklung neuer NMR-spektroskopischer Methoden und deren Anwendungen auf Biomakromoleküle*, Thesis, J. W. Goethe-Universität, Frankfurt **1998**.

[31] J. Titman, J. Keeler, *J. Magn. Reson.* **1990**, *89*, 640–646.

[32] R. Brüschweiler, C. Griesinger, O. W. Sørensen, R. R. Ernst, *J. Magn. Reson.* **1988**, *78*, 178–185.

[33] D. Neuhaus, G. Wagner, M. Vasak, J. H. R. Kägi, K. Wüthrich, *Eur. J. Biochem.* **1985**, *151*, 257–273.

[34] C. Griesinger, O. W. Sørensen, R. R. Ernst, *J. Am. Chem. Soc.* **1985**, *107*, 6394–6396.

[35] S. J. Glaser, T. Schulte-Herbrüggen, M. Sieveking, O. Schedletzky, N. C. Nielsen, O. W. Sørensen, C. Griesinger, *Science* **1998**, *280*, 421–424.

[36] T. Untidt, T. Schulte-Herbrüggen, B. Luy, S. J. Glaser, C. Griesinger, O. W. Sørensen, N. C. Nielsen, *Mol. Phys.* **1998**, *95*, 787–796.

## 7.5    NMR Techniques for the investigation of carbohydrate-protein interactions

*Thomas Weimar*

### 7.5.1    Carbohydrate–protein complexes

The steadily increasing number of oligosaccharide structures found in biological systems illustrates the importance of these compounds and their conjugates in nature. Besides their role as energy storage molecules or structural materials, oligosaccharides are widely involved in biological recognition phenomena such as cell targeting and differentiation, tumor invasion and metastasis, immune response, or bacterial and viral adhesion to host cells [1]. Despite the knowledge of the importance of oligosaccharides in these biological events, there is still a lack of understanding how proteins recognize carbohydrates. This is mainly due to the fact that carbohydrate-protein complexes display dissociation constants ($K_D$) most often in a range between $10^{-2}$ to $10^{-5}$ M. These weak interactions yield a low occupancy of the ligand in the binding site which complicates crystallographic investigations of carbohydrate–protein complexes.

NMR spectroscopy has long been the method of choise to study carbohydrate structure and conformation in solution [2], and is, as will be described in this chapter, also particularly useful in deriving the protein bound conformation of carbohydrate ligands. With experiments performed in the liquid state, NMR can closely reproduce the natural environment of carbohydrate–protein systems. Therefore, NMR spectroscopy can help to close the gap between crystallographic data of carbohydrate-binding proteins on one hand and thermodynamic and kinetic investigations of carbohydrate–protein complexes performed with microcalorimetry or surface plasmon resonance on the other.

### 7.5.2    Carbohydrate NMR

NMR spectroscopy of carbohydrates can be grouped in two categories; structure determination, and conformational analysis. The structure of an oligosaccharide is defined by its sequence (and branching) which relates best to the definition of the primary structure of a protein, whereas the relative orientation of the monosaccharide units define the conformation of a carbohydrate. Larger oligosaccharide structures are still most often isolated from biological sources, making an isotope enrichment usually impossible. Therefore, the structure determination of oligosaccharides by NMR uses $^1$H and $^{13}$C chemical shift and correlation informations of the carbohydrate resonances together with qualitative distance information from NOE (*N*uclear *O*verhauser *E*nhancement) experiments [3, 4].

When performing a conformational analysis of oligosaccharides, the majority of the experimental data stem from nuclear Overhauser experiments either in the laboratory (NOE) [4] or rotating frame [4] (*R*otating *F*rame *O*verhauser *E*nhancement; ROE). From these experiments interproton distances are derived which are used to determine models of the three-dimensional structure of the molecules. In principle, NOE experiments are easy to perform (in particular, the 2D-NOESY experiment is as simple to set up as a COSY experiment), but are subject to a variety of potential pitfalls which can lead to artifacts in the acquired spectra. The most important prerequisite for high-quality NOE spectra of small- to medium-sized molecules is the stability of the spectrometer and the temperature in the probe [4]. In the past years, the manufacturers have addressed these machine-dependent aspects successfully by improving the temperature control of the probe, the phase and amplitude stability of the pulse-generating components, and the sensitivity of the probe and the receiver. Also, the introduction of actively shielded pulsed field gradients into the spectrometers offers many new ways to improve the quality of the spectra by selecting coherence transfer pathways or suppressing artifacts [5]. When working with a modern spectrometer [6] and a carefully prepared sample, it no longer seems to make much difference whether to run two-dimensional NOESY experiments, or one-dimensional variants [7] with shaped pulse excitation/inversion such as 1D-NOESY [8], 1D-transient [4, 9] and double pulsed field gradient spin-echo NOE experiments [10]. Nowadays, with all these experiments the detection and quantification of NOEs <1% is possible.

Although the quality of the spectrometers has improved, the conformational analysis of oligosaccharides is still not a routine task. First, due to their intrinsic structure, oligosaccharides show only very few interglycosidic NOE effects which define the conformation(s) of a specific glycosidic linkage. Second, at proton resonance frequencies of 500 or 600 MHz the correlation time dependence of the NOE places oligosaccharides often in a region of the function (see Figure 7.5.1) where the effects are very weak or disappear completely. ROE experiments do not have this limitation but, in addition to the distance dependence, ROE data depend also on the offset to the spin locking field [4]. The appearance of TOCSY peaks in ROE experiments can also complicate the analysis of the spectra [4]. Third, glycosidic linkages have a rather flexible nature, leading to an ensemble of rapidly interconverting oligosaccharide conformations. Compared to this interconversion, NMR is a slow technique, producing time-averaged observables like coupling constants or NOE effects. Consequently, a conformational analysis of these molecules has to take different conformations into account. This is either done by Boltzmann-averaging of conformations from potential energy calculations [11] or Molecular-dynamics [12] or Monte-Carlo simulations [13].

## 7.5.3  NMR experiments of carbohydrate–protein complexes – Transferred NOEs

The above-mentioned high dissociation constants of carbohydrate–protein complexes most often result in rapid exchange between the free state and the protein-bound state of

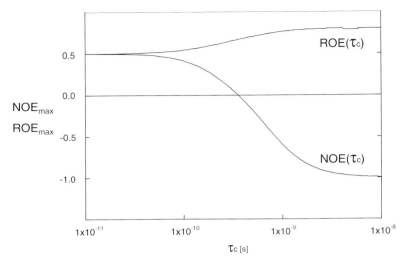

**Figure 7.5.1.** Plots of the maximum homonuclear NOE and ROE enhancements for a two-spin system at a magnetic field strength of 500 MHz as a function of the correlation time ($\tau_c$). Small molecules tumble rapidly in solution and display short correlation times, whereas proteins tumble slow and display long correlation times. The crossover point of the NOE function depends also on the magnetic field strength ($\omega_0$). The NOE enhancement becomes zero at $\omega_0\tau_c = (5/4)^{1/2} \sim 1.12$.

an oligosaccharide ligand. This has influence on NMR observables such as coupling constants, chemical shifts or ROE effects since, in addition to the conformational average of the free oligosaccharide, the bound conformation of the carbohydrate also partly affects the measured parameters. Unfortunately, the contribution of the bound conformation to such an averaged observable is usually small so that no information about the bound state of the oligosaccharide can be extracted from these parameters. The only NMR observable which benefits from this fast exchange is the NOE effect. Whenever the dissociation rate constant ($k_d$ or $k_{off}$) of a complex is fast compared to the $T_1$-relaxation time scale of NMR spectroscopy, the observed NOE effects in such a mixture are the average between the NOEs of the free and the bound ligand. Due to the pronounced dependence upon correlation time and hence the size of a molecule (see Figure 7.5.1), the NOE effects of free di- to hexasaccharides have little or no influence on the observed averaged NOEs. The bound carbohydrate on the other side has a correlation time which is defined by the size of the protein leading to strong (and negative) NOE effects for the complexed form of the ligand. Therefore, the observed, averaged (free and bound) NOE effects in such an experiment have a negative sign [14] and represent the protein-bound conformation of the ligand.

In this way, the ligand resonances are labeled with the correlation time of the protein-bound state and transfer the information of this conformation into the free state. This exchange phenomenon was first described by Balaram et al. and consequently named transferred NOE [15] (Figure 7.5.2). As summarized recently [16], the nature of the interactions between carbohydrates and proteins make these

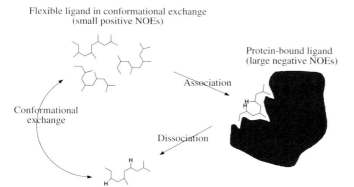

Flexible ligand in conformational exchange
(small positive NOEs)

Protein-bound ligand
(large negative NOEs)

Association

Conformational
exchange

Dissociation

**Figure 7.5.2.** Schematic representation of the trNOE concept. The dissociation rate constant of the complex has to be fast compared to the $T_1$-relaxation time scale of NMR spectroscopy to observe trNOEs. In that case, the close distance between the two marked protons in the protein-bound form of the ligand will give an averaged negative NOE enhancement.

systems an excellent target for trNOE experiments. In the following, two examples of NMR investigations of carbohydrate–protein complexes based on transferred NOE experiments will be given.

## 7.5.4   Case 1: Using transferred NOE experiments to solve the bound conformation of a *Streptococcus* Group A trisaccharide-antigen [17]

After an immunologically significant infection with *Streptococcus pyogenes* (Group A), a small but significant portion of patients develop *Streptococcus*-related rheumatic diseases such as rheumatic fever, heart-valvular disease, glomerulonephritis, and other rheumatic disorders [18]. These patients show an autoimmune (antibody) cross-reactivity between the surface antigens of the bacterium and antigens of cardiac and muscle tissue. Such a cross-reactivity with heart tissue is known for antibodies against the *Streptococcus* Group A cell-wall polysaccharide [19]. This cell-wall polysaccharide consists of a poly-$\alpha$-L-rhamnopyranosyl backbone composed of alternating $(1 \rightarrow 2)$ and $(1 \rightarrow 3)$ linkages of which $\beta$-$N$-acetyl-D-glucosamine residues are attached at the 3-position of every second rhamnose residue (Figure 7.5.3). The minimum unit of this polysaccharide which shows good inhibition of the anti-*Streptococcus* mouse monoclonal antibody, Strep 9, is the synthetic trisaccharide propyl 3-$O$-(2-aceta-mido-2-deoxy-$\beta$-D-glucopyranosyl)-2-$O$-($\alpha$-L-rhamnopyranosyl)-$\alpha$-L-rhamnopyrano-side (**1**) (Figure 7.5.3). The dissociation constant of the complex of this trisaccharide and the monoclonal antibody Strep 9 is in the order of $10^{-4}$ M. Based on the assumption that the association rate constant is diffusion controlled or slower, this complex must have a dissociation rate constant which permits a trNOE investigation of the antibody-bound conformation of the trisaccharide antigen.

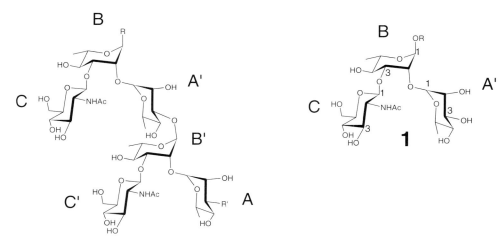

**Figure 7.5.3.** Representation of the *Streptococcus* Group A cell-wall polysaccharide (left) and trisaccharide **1** (right) used for the trNOE experiments. The numbering of the monosaccharide units and some protons are given. The anomeric center of **1** is protected with a propyl group.

### 7.5.4.1 T$_{1\rho}$-filtered trNOE experiments

In the sample of the antibody Strep 9 and ligand **1**, two sorts of linewidths appear. The NMR resonances of the ligand in fast exchange with the protein are narrow, whereas the protein resonances are very broad (slower correlation time; Figure 7.5.4). Although, trNOE experiments are typically performed with a 10- to 20-fold excess of ligand, the majority of NMR resonances in such a sample stem from the protein in the mixture. The trisaccharide/antibody sample used here contained 6 mg antibody (80 nmol binding sites) and 1.2 mg of the antigen (ligand : antibody ratio, 14 : 1) in ~500 µl phosphate-buffered D$_2$O solution. A very convenient tool to suppress the protein background signals in such a sample is a T$_{1\rho}$ filter [20], which can be built into a standard NOESY pulse-sequence as well as into a normal 1D experiment as a short spin lock pulse after the first 90° pulse. This T$_{1\rho}$ filter takes advantage of the fast T$_{1\rho}$ relaxation of the protein resonances and relaxes these signals within some milliseconds very effectively prior to acquisition (Figure 7.5.4). An alternative for a T$_{1\rho}$ filter is a T$_2$ filter [21], but such a T$_2$ filter has the disadvantage that couplings which evolve during the delays might disturb the spectrum.

### 7.5.4.2 Spin diffusion in trNOE spectra

Figure 7.5.5A shows a part of a 2D trNOE spectrum of the antibody/trisaccharide sample at the position of the anomeric protons. One recognizes the large amount of cross-peaks, although many of these are very small. Without going into detail it is

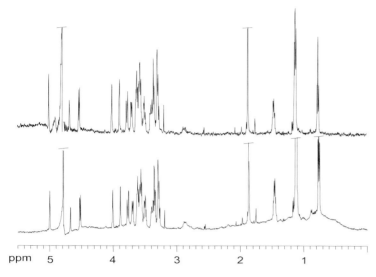

**Figure 7.5.4.** Effect of a $T_{1\rho}$ filter in one-dimensional experiments. Lower spectrum: Four experiments have been added with the standard pulse-acquire sequence. Upper spectrum: Two experiments have been added with a $T_{1\rho}$ filter sequence (10 ms spin lock after the 90° pulse). This short spin lock is sufficient to relax all protein signals in this experiment.

easy to understand that this large number of cross-peaks for a bound trisaccharide has to be questioned. Due to the slow reorientation of these molecules, a potential source of artifacts in (tr)NOE experiments is spin diffusion [22, 23], which can produce cross-peaks between distant protons through a secondary (and unwanted) magnetization transfer [4]. In NOE-type experiments these spin diffusion cross-peaks have the same sign as regular trNOEs, but cannot be considered in a structure determination since they lead to an underestimation of distances. The usual way to minimize the effect of spin diffusion is to measure NOE effects at short mixing times [4, 22]; however, since trNOEs are often rather weak, mixing times between 30 and 100 ms are often not applicable. Longer mixing times on the other hand increase the risk of spin diffusion in the spectra.

It has been shown by different groups [17, 21, 22, 24, 25] that spin diffusion in trNOE experiments can be addressed with ROESY [4] experiments. The cross-peaks in these NOE experiments in the rotating frame have different behavior as NOE effects in that every magnetization transfer gives rise to a change in sign of the observed effects. Cross-peaks arising exclusively from spin diffusion have, therefore, the same sign as the diagonal signals, making them easy to differentiate from first-order (tr)ROE effects, which would have the opposite sign [4]. A lack of cross-peaks in the (tr)ROE spectra is then due either to a lack of magnetization transfer or to a cancellation of (tr)ROE- and spin diffusion effects [24]. Figure 7.5.5B demonstrates the use of ROE experiments to uncover spin diffusion artifacts in the trNOE spectra of the antibody/trisaccharide sample. Here, all small effects are not observed in the

A

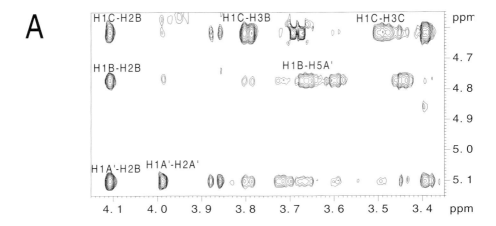

B

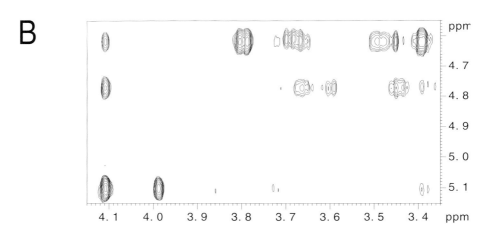

**Figure 7.5.5.** Comparison of a trNOE spectrum (A) with a trROE spectrum (B) of the trisaccharide/ antibody complex at the position of the anomeric protons. All effects which have disappeared in the trROE spectrum arise through spin diffusion.

ROE experiments and consequently they have to be omitted in the determination of the bound conformation of the ligand.

The integration of the remaining trNOE effects for antibody-bound **1** finally yielded interglycosidic distances (H1A'-H2B (2.30 Å), H1B-H5A' (2.51 Å), H1C-H3B (2.35 Å), and H1C-H2B (3.30 Å)) which were used in simulated annealing calculations to define the bound conformation of the ligand. During these calculations it turned out that, in addition to the NOE-derived restraints, repulsive restraints defining non-observable trNOEs between the proton pairs H1A'-H2C and H1-H6$_{proS(R)}$C had also to be included [17].

# Global minimum energy conformation

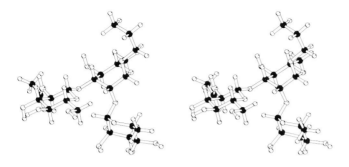

# Antibody bound conformation

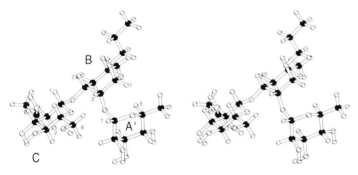

**Figure 7.5.6.** Stereo plots of the global minimum energy conformation and the antibody-bound conformation of trisaccharide **1**, respectively. The numbering of the protons is indicated in the latter plot. Plots have been produced with Molscript [35].

   The conformation of the ligand which, ultimately, satisfied all experimental observations differs from the global minimum energy conformation by the rotation of the $\psi$ angle at the $\alpha$–$(1 \rightarrow 2)$ linkage from a +*gauche* into a −*gauche* orientation (compare Figure 7.5.6). This finding has implications for the interactions of the bacterial cell-wall polysaccharide in complex with the antibody. If the global minimum energy conformations of the free trisaccharide and other oligomer portions of the polysaccharide [26] are also representative for the conformation of the whole cell-wall polysaccharide, complexation by the antibody induces a conformational change of the polysaccharide, at least at the $\alpha$–$(1 \rightarrow 2)$ linkage. However, if the complexed form of the trisaccharide resembles the conformation of the polysaccharide, no conformational change during binding of the polysaccharide is necessary.

### 7.5.5    Case 2: An investigation of the complex of glucoamylase with a maltose heteroanalog inhibitor [27]

Analogs of carbohydrates which are carbohydrase inhibitors offer a means of studying events in the binding- and catalytic mechanism of the enzymes. An important industrial enzyme in this respect is glucoamylase (1,4-α-D-glucan glucohydrolase; EC 3.2.1.3) from *Aspergillus niger*. This enzyme is widely used to degrade starch to glucose syrup. Since heteroanalogs of maltose in which the ring oxygen and/or the interglycosidic oxygen atoms are replaced by sulfur or nitrogen [27, 28] have inhibition constants ($K_i$) for glucoamylase in the milli- to micromolar range [27, 28], trNOE experiments can be used to derive the bound conformation of the inhibitors. The model for the bound conformation can then be docked into the catalytic site of the X-ray structure of the enzyme [29] to study the complex and the carbohydrate–protein interactions therein.

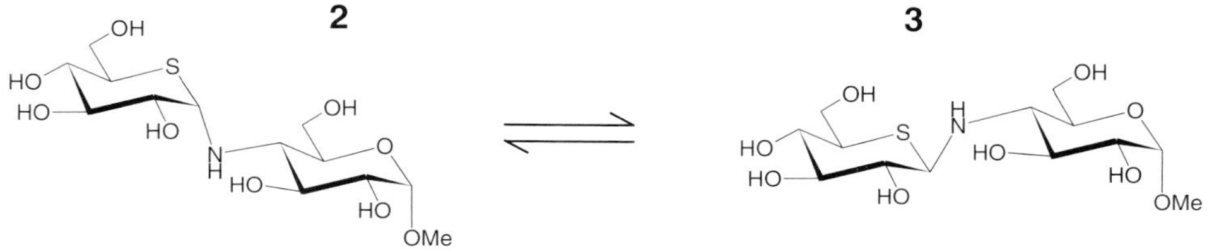

**Figure 7.5.7.** Structural formula of 5'-thio-4-*N*-α-maltoside (**2**) and 5'-thio-4-*N*-α-cellobioside (**3**).

The heteroanalog methyl 5'-thio-4-*N*-α-maltoside (**2**) (Figure 7.5.7) is a fairly potent inhibitor of glucoamylase [27], The disadvantage of this compound being that it anomerizes to give a mixture with the corresponding cellobioside analog 5'-thio-4-*N*-α-cellobisoside (**3**) [27] (Figure 7.5.7). Due to the slow anomerization process (within days) the NMR spectra show two different compounds. For this reason, the first step in the analysis of the experimental data of the complex is to decide if the enzyme differentiates between the two heteoanalogs and, if so, which compound it binds. The question if a mixture contains a bioactive component can be answered quickly and elegantly by acquiring and inspecting a NOE spectrum of the sample in the presence of the protein (glucoamylase). A bioactive compound should display trNOEs (negative cross-peaks), whereas small non-binding compounds show positive cross-peaks. Figure 7.5.8 displays positive and negative levels of a portion of such a NOE spectrum of a sample of the heteroanalogs **2** and **3** and glucoamylase. It is obvious that only some negative trNOE peaks are observable. A more detailed analysis of the cross-peaks reveals that **2** is the only bioactive compound in this NMR sample [27]. In the meantime, this bioaffinity-NMR concept which is based on the trNOE methodology has been further developed and used successfully to screen mixtures for biological activity [30]. With three-dimensional experiments based on this concept it was also shown that the identification and the structural analysis of a bioactive component in a complex mixture is possible [31].

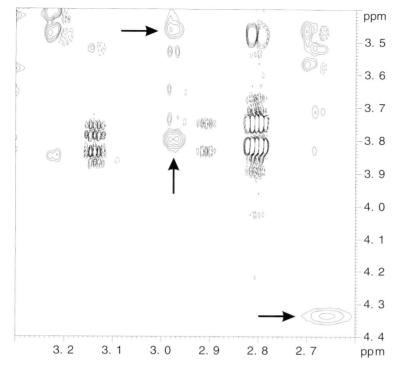

**Figure 7.5.8.** (Tr)NOE spectrum of the mixture of the heteroanalogs **2** and **3** and glucoamylase. Positive (interrupted lines) and negative levels (solid lines) are shown. The trNOE effects are marked with arrows and belong to **2**. All other cross-peaks are either positive NOEs or zero-quantum artifacts (COSY-like peaks) which stem from compound **3**.

Having defined **2** as the bioactive compound, the question of the conformation in the enzyme-bound state can be followed further. It is well known for maltose [32] as well as for maltose analogs [33] that these molecules mainly populate two conformational families at the $\alpha-(1 \rightarrow 4)$ glycosidic linkage. These conformational families (see Figure 7.5.9) interconvert through a rotation around the $\psi$ angle from $\sim -25°$ to $\sim 180°$ with the majority of conformations in the region of the global minimum energy conformation ($\psi \sim -25°$). Both conformational families show specific NOE effects across the glycosidic linkage (Figures 7.5.9 and 7.5.10) which can be used to qualitatively define the conformational region of the enzyme-bound inhibitor **2**. Comparing slices of the NOE experiment of free **2** with slices from the trNOE spectrum of the bound compound at the position of the proton H1′, the trNOE spectrum does not show the effects H1′-H3 and H1′-H5, which suggests that the inhibitor is bound in a conformation from the global minimum energy family [27] (Figures 7.5.9 and 7.5.10). A more detailed analysis of trNOE effects of **2** and a second maltose heteroanalog derived the bound conformation of both inhibitors which were then docked into the high-resolution X-ray structure [29] of the enzyme. The obtained

Global Minimum                                    Local Minimum

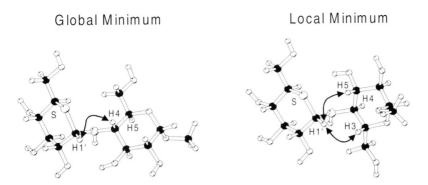

**Figure 7.5.9.** Ball and stick models (Molscript) [35] of the global and the local minimum energy conformations of **2**. The expected interglycosidic NOEs in each conformation are indicated.

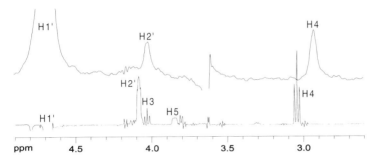

**Figure 7.5.10.** Comparison of cross-sections of trNOE and NOE spectra of the bound (upper spectrum) and free (lower spectrum) form of inhibitor **2**. The spectrum of the free inhibitor shows the interglycosidic NOE effects H1'-H3, H1'-H4 and H1'-H5. In the trNOE spectrum the effects H1'-H3 and H1'-H5 are missing (compare Figure 7.5.9). Shift differences between the resonances in the two spectra are a result of the different experimental temperatures.

models of the enzyme inhibitor complexes could then very successfully be used to probe carbohydrate–protein interactions in these systems [34].

### 7.5.6  Conclusions

For protein ligand complexes in fast exchange on the NMR $T_1$-relaxation time-scale, trNOE-based experiments offer a potent means of determining the conformation of the bound ligands. The method does not depend on the size of the protein, and thus very large proteins may be used in these experiments. Especially when the structure of the protein is known, the protein ligand complex can be modeled with the knowledge of the bound conformation of the carbohydrate ligand after which the models can be used for further analysis of protein ligand interactions.

The correlation time dependence of trNOEs also makes these experiments a valuable tool when complex mixtures, for example from combinatorial synthesis or biological isolates, must be screened for biological activity. In one single experiment, such a mixture can be examined for its protein-binding affinities.

**Acknowledgements:** This work was supported by Grants from the Deutsche Forschungsgemeinschaft and Fonds der Chemischen Industrie. I am indebted to Dr. B. Svensson (Carlsberg Laboratory, Copenhagen, Denmark) and Prof. B. M. Pinto (Simon Fraser University, Burnaby, B.C., Canada) for providing the material for the investigations described here and to Prof. Th. Peters (Medical University of Lübeck, Lübeck, Germany) for allowing me access to the computer and NMR facilities in Lübeck.

# References

[1]  (a) A. Varki, *Glycobiology* **1993**, *3*, 97–130; (b) R. A. Dwek, *Science* **1995**, *269*, 1234–1235.

[2]  (a) C. A. Bush, *Curr. Opin. Struct. Biol.* **1992**, *2*, 655–660; (b) R. J. Woods, *ibid.* **1995**, *5*, 591–598.

[3]  J. Dabrowski in *Two-Dimensional NMR Spectroscopy. Applications for Chemists and Biochemists* (Eds.: W. R. Croasmun, R. M. K. Carlson), VCH Publishers, New York, **1994**.

[4]  D. Neuhaus and M. Williamson, *The Nuclear Overhauser Effect in Structural and Conformational Analysis*, VCH Publishers, New York, **1989**.

[5]  (a) J. Keeler, R. T. Clowes, A. L. Davis, E. D. Laue, *Methods Enzymol.* **1994**, *239*, 145–207; (b) L. E. Kay, *Curr. Opin. Struct. Biol.* **1995**, *5*, 674–681.

[6]  Coming from the world of Bruker spectrometers, I define a modern spectrometer as a machine of the D-series or the equivalent from other manufacturers.

[7]  (a) H. Kessler, S. Mronga, G. Gemmeker, *Magn. Reson. Chem.* **1992**, *29*, 527–557; (b) K. Scott, J. Keeler, Q. N. Van, A. J. Shaka, *J. Magn. Reson.* **1997**, *125*, 302–324.

[8]  (a) H. Kessler, H. Oschkinat, C. Griesinger, *J. Magn. Reson.* **1986**, *70*, 106–133; (b) L. Emsley, G. Bodenhausen, *ibid.* **1989**, *82*, 211–221.

[9]  (a) T. Weimar, B. Meyer, T. Peters, *J. Biomol. NMR* **1993**, *3*, 399–414; (b) T. Peters, T. Weimar, *ibid.* **1994**, *4*, 97–116.

[10]  K. Scott, J. Stonehouse, T. L. Hwang, J. Keeler, A. J. Shaka, *J. Am. Chem. Soc.* **1995**, *117*, 4199–4200.

[11]  (a) K. Bock, H. Lönn, T. Peters, *Carbohydr. Res.* **1990**, *198*, 375–380; (b) C. A. Bush and P. Cagas in *Advances in Biophsical Chemisty*, Vol. 2 (Ed.: C. A. Bush), JAI Press, Stamford, CT, **1991**.

[12]  (a) R. J. Woods in *Reviews in Computational Chemistry*, Vol. 9 (Eds. K. B. Lipkowitz, D. B. Boyd), VCH Publishers, New York, **1996**. (b) K. H. Ott, B. Meyer, *Carbohydr. Res.* **1996**, *281*, 11–34.

[13]  (a) T. Peters, B. Meyer, R. Stuike-Prill, R. Somorjai, J.-R. Brisson, *Carbohydr. Res.* **1993**, *238*, 49–73; (b) T. Weimar, T. Peters, S. Pérez, A. Imberty, *Theochem. (J. Molec. Struc.)* **1997**, *395/396*, 297–311.

[14]  Historically, the definition of the sign of NOE effects stems from one-dimensional steady-state experiments of small molecules. In modern transient NOE experiments, like two-dimensional NOESY, the sign is defined relative to the diagonal signals in the 2D spectrum with signals having the same sign as the diagonal being negative, and signals having the opposite site as the diagonal being positive.

[15]  (a) P. Balaram, A. A. Bothner-By, J. Dadok, *J. Am. Chem. Soc.* **1972**, *94*, 4015–4017; (b) P. Balaram, A. A. Bothner-By, E. Breslow, *ibid.*, 4017–4018. (c) J. P. Albrand, B. Birdsall, J. Feeney, G. C. K. Roberts, A. S. V. Burgen, *Int. J. Biolog. Macromol.* **1979**, *1*, 37–41.

[16] (a) T. Peters, B. M. Pinto, *Curr. Opin. Struct. Biol.* **1996**, *6*, 710–720; (b) A. Poveda, J. L. Asensio, J. F. Espinosa, M. Martin-Pastor, J. F. Cañada, J. Jiménez-Barbero, *J. Mol. Graphics Mod.* **1997**, *15*, 9–17.

[17] T. Weimar, S. J. Harris, J. B. Pitner, K. Bock, B. M. Pinto, *Biochemistry* **1995**, *34*, 13672–13681.

[18] (a) A. L. Bisno in Nonsuppurative Poststreptococcal Sequelae: Rheumatic Fever and Glomerulonephritis. *In Principles and Practice of Infectious Diseases*, 2nd Ed. (Eds.: G. L. Mandell, R. G. Douglas, J. E. Bennett), Wiley, New York, **1985**; (b) J. Rotta in *Towards Better Carbohydrate Vaccines* (Eds.: R. Bell, G. Torrigiani), John Wiley & Sons, New York, **1987**.

[19] (a) I. J. Goldstein, B. Halpern, L. Roberts, *Nature* **1967**, *213*, 44–47; (b) I. J. Goldstein, P. Rebeyrotte, J. Parlebas, B. Halpern, *Nature* **1968**, *219*, 86–868.

[20] T. Scherf, J. Anglister, *Biophys. J.* **1993**, *64*, 754–61.

[21] S. R. Arepalli, C. P. J. Glaudemans, G. D. Jr. Daves, P. Kovác, A. Bax, *J. Magn. Reson. B* **1995**, *106*, 195–198.

[22] L. Y. Lian, I. L. Barsukov, M. J. Sutcliffe, K. H. Sze, G. C. K. Roberts, *Methods Enzymol.* **1994**, *239*, 657–700.

[23] (a) A. P. Campbell, B. D. Sykes, *J. Magn. Reson.* **1991**, 93, 77–92; (b) F. Ni, H. A. Scheraga, *Acc. Chem. Res.* **1994**, *27*, 257–264.

[24] M. E. Perlman, D. G. Davis, G. W. Koszalka, J. V. Tuttle, R. E. London, *Biochemistry* **1994**, *33*, 7547–7559.

[25] J. L. Asensio, J. F. Cañada, J. Jiménez-Barbero, *Eur. J. Biochem* **1995**, *233*, 618–630.

[26] (a) U. C. Kreis, V. Varma, B. M. Pinto, *Int. J. Biol. Macromol.* **1995**, *17*, 117–130; (b) R. Struike-Prill, B. M. Pinto, *Carbohydr. Res.* **1995**, *279*, 59–73.

[27] J. S. Andrews, T. Weimar, T. P. Frandsen, B. Svensson, B. M. Pinto, *J. Am. Chem. Soc.* **1995**, *117*, 10799–10804.

[28] S. Mehta, J. S. Andrews, B. D. Johnston, B. Svensson, B. M. Pinto, *J. Am. Chem. Soc.* **1995**, *117*, 9783–9790.

[29] A. E. Aleshin, B. B. Stoffer, L. M. Firsov, B. Svensson, R. B. Honzatko, *Biochemistry* **1996**, 35, 8319–8328.

[30] (a) B. Meyer, T. Weimar, T. Peters, *Eur. J. Biochem.* **1997**, *246*, 705–709; (b) D. Henrichsen, B. Ernst, J. L. Magnani, W.-T. Wang, B. Meyer, T. Peters, *Angew. Chem. Int. Ed.*, **1999**, 38, 98–102.

[31] L. Herfurth, T. Weimar, B. Meyer, T. Peters, presented at the Franco-German NMR Conference II, Obernai, France, **1998**; L. Herfurth, T. Weimar, B. Meyer, T. Peters, manuscript in preparation.

[32] A. S. Sashkov, G. M. Lipkind, and N. K. Kochetkov, *Carbohydr. Res.* **1986**, *147*, 175–182.

[33] (a) K. Bock, J. O. Duus, and S. Refn, *Carbohydr. Res.* **1994**, *253*, 51–67; (b) T. Weimar, U. C. Kreis, J. S. Andrews, B. M. Pinto, *Carbohydr. Res.*, **1999**, *315*, 222–233.

[34] T. Weimar, B. Stoffer, B. Svensson, B. M. Pinto, presented at the XIX[th]. International Carbohydrate Symposium, San Diego, CA, USA, **1998**; T. Weimar, B. Stoffer, B. Svensson, B. M. Pinto, manuscript in preparation.

[35] P. J. Kraulis, *J. Appl. Crystallogr.* **1991**, *24*, 946–950.

# 7.6     Access to structural diversity *via* chemical screening

*Susanne Grabley and Ralf Thiericke*

In order to take advantage of modern drug discovery strategies like the so-called high-throughput screening (HTS), access to high sample numbers covering a broad range of low-molecular mass diversity is essential. Today, compound collections from both, combinatorial chemistry [1], and exploiting structural diversity derived from natural sources contribute to improved lead discovery. In particular, low-molecular mass natural products from bacteria, fungi, plants, and invertebrates either from terrestrial or marine environments bear unique structural diversity [2].

In comparison to target-directed screening attempts, chemical screening strategies applied to concentrates from natural sources (*e.g.* micro-organisms and plants) were found to be efficient supplemental and alternative methods, especially with the aim to discover new secondary metabolites [3]. Chemical screening using thin-layer chromatography (TLC) and various staining reagents offers the opportunity to visualize a nearly complete picture of a produced secondary metabolite pattern (metabolic finger-print) [4]. The chromatographic behaviour of metabolites on TLC-plates, as well as their chemical reactivity towards staining reagents under defined reaction conditions being the selection criteria resulted in a number of new secondary metabolites displaying a wide range of structurally different compounds out of nearly every class known so far [3]. In a following step, the purified natural products are passed through different biological assay systems for lead discovery purposes. The availability of both, sufficient amounts of pure metabolites, and a broad range of biological tests are the critical points for success. As predicted, a high percentage of the secondary metabolites coming out of chemical screening progams showed biological effects and in defined cases gave reason for further, more detailed biological studies towards new lead structures.

This article provides information about the different chemical screening approaches and focuses on both, the metabolic diversity accessible, and methodical aspects.

## 7.6.1     The chemical screening method

Chemical screening is based on the analysis of the chemical reactivity of secondary metabolites towards defined staining reagents using TLC [5]. In order to apply chemical screening in a reproducible way, standardized procedures have to be used. Concentrated extracts from natural sources are applied on TLC-plates and chromatographed with different solvent systems (separation step). After drying the chemical reactivity of the metabolites is analyzed by making use of defined chemical reagents sprayed directly onto the TLC plates (detection step). Selection criteria are both, the chromatographic parameters, and the colorization behaviour after spraying and heating, using defined and reproducible conditions.

As an example, the chemical screening method is described for the screening of secondary metabolites from microbial sources. Using standardized conditions and culture media, the strains are cultivated in shake flasks. Mycelium and culture broth are separated, defined extracts (50-fold concentrated) are prepared from both the culture filtrate (adsorption chromatography on Amberlite XAD-16; elution with methanol/water) and the mycelium (extraction with organic solvents like acetone). In a routine procedure, the obtained concentrates are analyzed by TLC using silica gel plates with different solvent systems (*e.g.* CHCl$_3$/MeOH, 9:1, and 1-butanol/ acetic acid/water, 4:1:5, upper phase), visual detection under the support of UV-extinction (254 and 366 nm) and colorization reactions obtained by staining with different reagents (*e.g.* anisaldehyde-, orcinol-, blue tetrazolium and Ehrlich's reagent) [27]. This procedure mainly focuses on the chemical behavior and reactivity of the

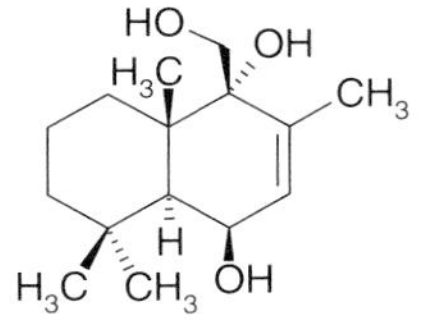

6-epi-Albrassitriol (**1**)

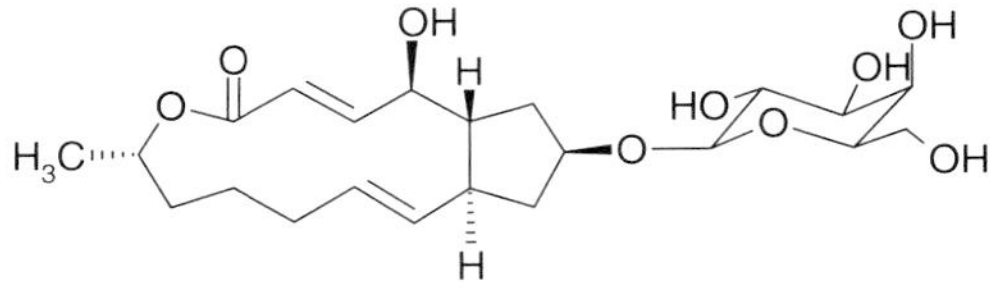

7-O-β-D-Galactosyl-brefeldin A (**2**)

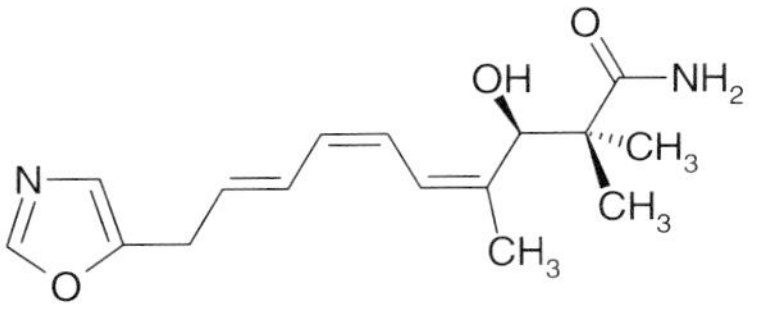

Inthomycin A (**3**)

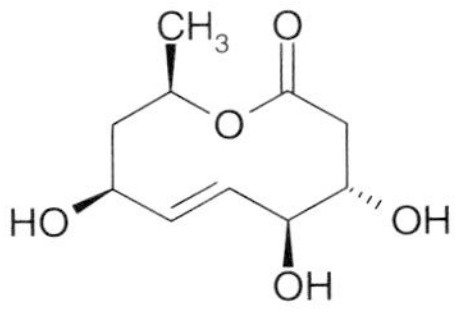

Decarestrictine D (**4**)

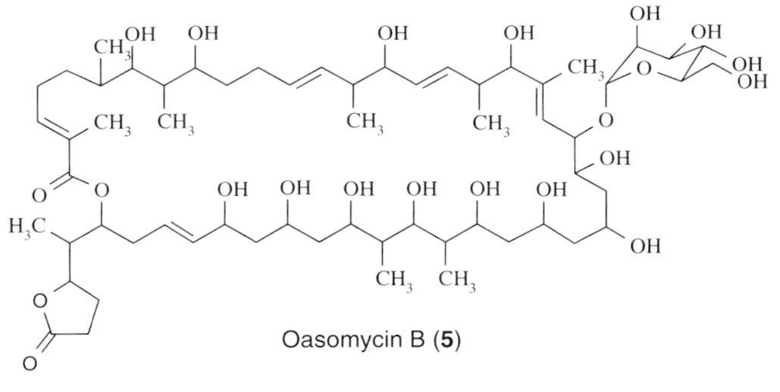

Oasomycin B (**5**)

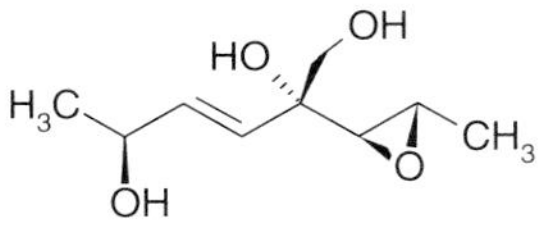

Aspinonene (**6**)

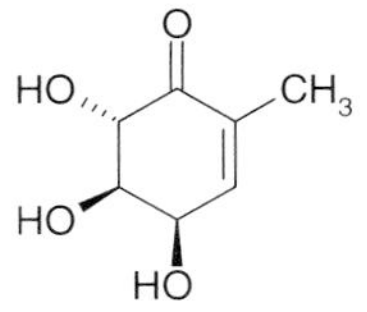

Gabosine A (**7**)

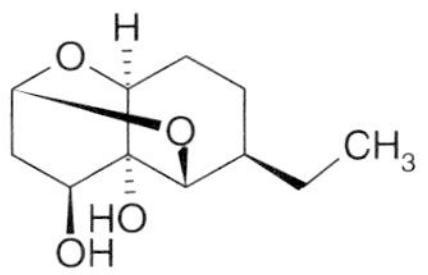

Agistatine B (**8**)

**Scheme 7.6.1.** Selected secondary metabolites detected by chemical screening approaches.

natural products and offers the opportunity to visualize a nearly complete picture of a microbial secondary metabolite pattern produced by each strain (metabolic fingerprint). On the basis of reference substances, spots on the chromatograms can be classified to be a constituent of the nutrient broth, a frequently formed and thus widely distributed metabolite, or a strain-specific compound which merits further attention. It is possible to adopt chemical screening to various organisms such as microalgae, myxobacteria, coryneforme and nocardiforme bacteria, or other so-called neglected genera of actinomycetes and plants.

## 7.6.2    New secondary metabolites from chemical screening

Based on the detailed analysis of the secondary metabolite pattern produced by each organism, a number of new secondary metabolites have been discovered. The structural diversity found is tremendous. In Scheme 7.6.1 some examples of new secondary metabolites from microbial sources are depicted. Pointing to the molecular masses as a distinguishing criterion, the majority of new secondary metabolites described from chemical screening exhibit masses lower than 500 Da, but in a number of cases metabolites with molecular masses up to 1200 Da were found. A broader overview of the structural diversity is given in reference [3].

According to the classification system of secondary metabolites by Berdy et al. [6] which represents compounds from classical biological screening, chemical screening has been proven to detect metabolites of all structural classes described. While biological screening on the basis of more or less anti-infective screening models obviously results in compounds of the carbohydrate-, macrocycle-, quinoid- and amino acid classes, compounds from chemical screening are predominantly *N/S*- and *O*-heterocycles, alicycles, aromatic, and aliphatic compounds. However, depending on the selection criteria used in chemical screening approaches, this result can be shifted towards other distributions (Scheme 7.6.2).

Based on our acquired expertise in qualifying the spots obtained via the TLC method, about 50% of all metabolites selected, isolated, and structurally classified appeared to be new [4]. However, these findings correlate strongly to the quality of the isolation protocols and the microbiologically based selection criteria leading to the strains, as well as to the routine cultivation conditions and ingredients of the culture media.

## 7.6.3    Hydrophilic-chemical screening

Several successful attempts for a further development of the basic chemical screening from the methodical point of view have been achieved. A broad variety of different TLC plates, staining reagents and solvent systems were tested and applied in routine screenings. It has been found that the screening procedure can be adopted to various objectives. As an example, the chemical screening was optimized for a screening towards highly hydrophilic metabolites (HPTLC-silica gel plates; chloroform/methanol/2-propanol/ammonia (25%)/water 25:30:30:0.4:25 vol%; staining reagents

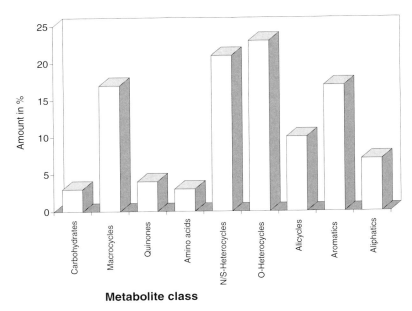

**Scheme 7.6.2.** Distribution of metabolites discovered from chemical screening according to the structural classification by *Berdy*. [6] (Data from a BMBF-supported joint-project of the Hoechst AG and A. Zeeck [3].)

anisaldehyde/$H_2SO_4$ and naphthoresorcin/$H_2SO_4$). In a first attempt with neglected genera of actinomycetes and different streptomycetes, this extension of the routine chemical screening resulted predominantly in the discovery of oligosaccharides, with about 70% referring to new compounds [7]. The hydrophilic-chemical screening completes the chemical screening in the hydrophilic region of secondary metabolites.

## 7.6.4 Biomolecular-chemical screening

In addition, chemical screening has been further developed in the direction of a biomolecular-chemical screening which allows the advantages of TLC and colorization by the use of staining reagents to be combined with binding studies of a secondary metabolite to biomacromolecules like DNA. On the basis of this method binding of secondary metabolites to DNA can be tested directly on TLC ($RP_{18}$-silica gel plates) with the advantage of a subsequent application of staining reagents. Results of the TLC-based biomolecular-chemical screening method have been verified independently by DNA melting curves and studies with reflectometric interference spectroscopy (RIFS; in cooperation with Prof. Dr. G. Gauglitz, University of Tübingen) [8]. Pure compounds can easily be examined by one-dimensional TLC, in which DNA is applied together with the test compound at the starting spot. Analysis is performed by differences in $R_f$-values in comparison to the reference. DNA-binding properties for a number of natural products have been discovered with this method [9].

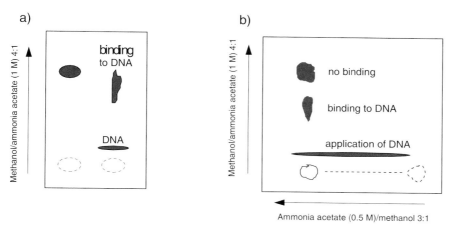

**Scheme 7.6.3.** Biomolecular-chemical screening [9, 10]:
(a)  1D TLC for the analysis of pure compounds. Chromatographic behavior without DNA (left), and with DNA (right).
(b)  2D TLC for the analysis of crude extracts from natural sources.

For screening of crude extracts, two-dimensional TLC is necessary. In the first dimension the metabolites are separated, while in the second dimension binding studies are performed in the same way as described for pure compounds (Scheme 7.6.3). Several new secondary metabolites (e.g. (2*R*,4*R*)-4-hydroxy-2-(1,3-pentadienyl)-piperidine (**9**) and 3,4-seco-4,23-hydroxyolean-12-ene-22-on-3-carboxylic acid (**10**)) have been discovered in the first screening attempts [9, 1].

## 7.6.5    Classification of microbial strain collections

As a consequence of the detailed analysis of the results obtained from the chemical screening of microbial extracts, it seems reasonable to comment on the secondary

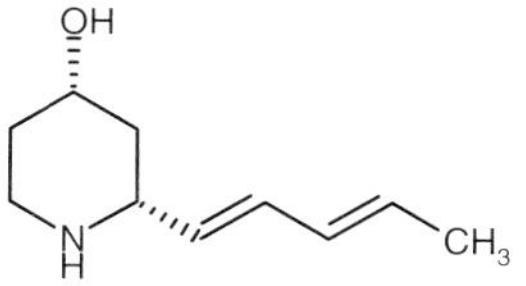

(2*R*,4*R*)-4-Hydroxy-2-(1,3-pentadienyl)-
piperidine (**9**)

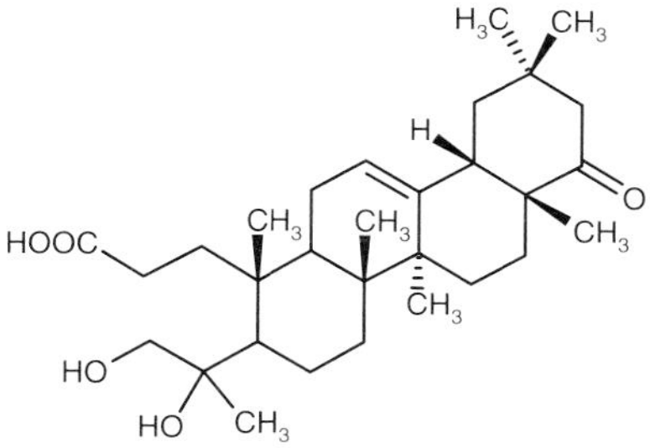

3,4-Seco-4,23-hydroxyolean-12-ene-
-22-on-3-carboxylic acid (**10**)

**Scheme 7.6.4.** Examples of new secondary metabolites from a biomolecular-chemical screening.

metabolism of the analyzed strains. With regard to the metabolic fingerprints visualized after fermentation on different media using various harvesting times, the TLC-based method can also be used advantageously for both the detection of so-called 'talented' strains, and for profiling microbial strain collections, especially as a first and fundamental step of efficiently applied biological high-throughput assays. For example, based on metabolic fingerprints, microbial isolates from the genus streptomyces can be classified in [4]:

(i)  **non-producing organisms**, which gave no indication of the formation of secondary metabolites up to a detection limit of 0.1 to 1 mg/l culture broth (though non-producing organisms can produce biologically active secondary metabolites in small amounts);

(ii)  **organisms of narrow productivity** which produce one or two secondary metabolites as main products with a restricted dependence on the alteration of the culture conditions; and

(iii) **talented organisms** which produce a large variety of structurally diverse secondary metabolites, usually in a stronger dependence to the applied cultivation conditions.

According to this terminology, from our experience nearly one third of the screened strains belong to the category of non-producing organisms, while about 1% could be highlighted as talented strains.

An integration of the analytical potency of the chemical screening method allows both qualifying microbial strain collections with the possibility to determine non-producing organisms, and a valuation of hits from primary high-throughput screenings with crude extracts from natural sources. One advantage is that the rapid recognition and dereplication possibilities of TLC analysis, coupled with carefully examined staining reagents, could add a form of chemical typification to the information from biological secondary screenings. Thus, a hitlist for further, more detailed studies can be created efficiently. It is expected that the critical period, from the primary screening to first promising lead compounds, can be shortened.

## 7.6.6  Biological activities

In contrast to any biological screening, the chemical screening approaches a priori possess no or only little (*e.g.* the biomolecular-chemical screening) correlation to a defined biological activity. However, the strategy of chemical screening is based on the hypothesis that all new secondary metabolites could bear interesting biological activities for various future applications. Striking examples are the decanolides of the decarestrictine family in which decarestrictine D (**4**) [11] was shown to be an efficient inhibitor of de novo cholesterol biosynthesis in vitro and in vivo, [12] or the obscurolides (*e.g.* obscurolide A$_2$ (**11**)), being a new class of phosphodiesterase inhibitors [13–15]. The oxazole-trienes of the inthomycin type (*e.g.* inthomycin A = phthoxazolin A (**3**)) bear significant inhibitory effects on cellulose biosynthesis and may be suitable as herbicides [16, 17]. Elloramycin A (**16**) exhibits significant antibacterial activity [18]. Cytotoxic effects against murine L-1210 leukemia cells in both,

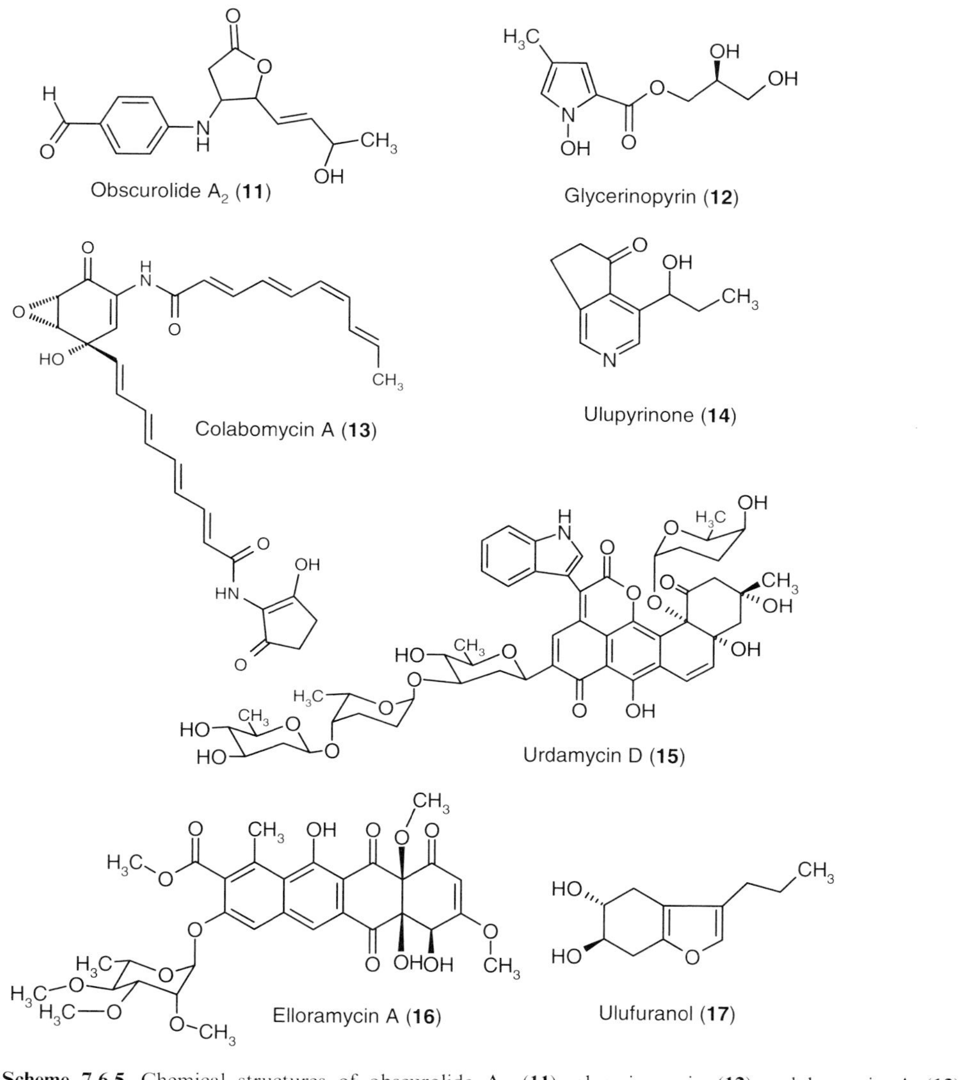

**Scheme 7.6.5.** Chemical structures of obscurolide A$_2$ (**11**), glycerinopyrin (**12**), colabomycin A (**13**), ulupyrinone (**14**), urdamycin D (**15**), elloramycin A (**16**), and ulufuranol (**17**).

proliferation, and stem cell assays were observed for colabomycin A (**13**) [19] and urdamycin A (**15**) [20]. The 34-membered marginolactone amycin B bears antibacterial and antifungal activity [21], while the 42-membered marginolactone oasomycin B (**5**) appears to be a weak inhibitor of de novo cholesterol biosynthesis, as observed in a HEP-G2 cell-based assay [22].

In contrast, on rare occasions no significant biological activity has been reported for new secondary metabolites from chemical screening. It is obvious that these metabolites are predominantly low-molecular mass compounds smaller than 250 Da, *e.g.* glycerinopyrin (**12**) [23], the carba-sugars of the gabosine family like gabosine A, (**7**) [5], the pyranacetals of the agistatine group like agistatine B (**8**) [24], the heteroaromates ulupyrinone (**14**), and ulufuranol (**17**) [25], or the multifunctional aspinonene (**6**). [26] Due to these unique structures, it may be expected that success in finding biological properties might correlate to a broader biological testing program.

## 7.6.7  Discussion

The potential of the chemical screening approaches lies in the possibility of obtaining access to outstanding structural resources from nature. This allows us to build up a collection of both pure new and already known natural products which, advantageously, can be integrated into target-oriented biological screening programs. It must be stressed that chemical screening programs end up with pure compounds. As traditional natural product screening is done by testing crude extracts, followed by the crucial work of back-tracking the active compounds from the 'hit'-extracts, much experience is required to exclude both false-positive results and doublettes. Considerable efforts are required to obtain sufficient quantities of raw material for the purpose of reproduction, isolation, structure elucidation and subsequent verification of the biological activity of a metabolite. The complete process was proved to be highly time- and capacity-consuming; thus it seems preferable to perform target-oriented screenings with pure natural products rather than with crude extracts.

Compound collections of natural origin with substantial structural diversity contribute to improved lead discovery and efficiently support synthetic libraries (*e.g.* from classical or combinatorial synthesis). It should be highlighted that the chemical screening approaches resulted in a number of new secondary metabolites in good yields which, advantageously, can be used as unique templates for multiple parallel synthesis either on solid- or in liquid phase. Thus, it is possible to strengthen the generation of structural diversity *via* 'combinatorial' synthesis, and additionally combine chemical screening with high-throughput screening for more efficient drug discovery strategies.

**Acknowledgements:** The authors would like to thank all their colleagues and co-workers who are engaged in chemical screening approaches. We are indepted to Prof. Dr. H. Zähner and Prof. Dr. A. Zeeck for all the inspiration and discussions over the years.

## References

[1]  S. R. Wilson, A. W. Czarnik (Eds.) *Combinatorial Chemistry. Synthesis and Application.* John Wiley & Sons, New York, **1997**.

[2]  S. Grabley, R. Thiericke (Eds.), *Drug Discovery from Nature*, Springer Verlag, Heidelberg, **1999**.

[3]  S. Grabley, R. Thiericke, A. Zeeck in *Drug Discovery from Nature*, pp. 124–148, Springer Verlag, Heidelberg, **1999**.

[4]  K. Burkhardt, H.-P. Fiedler, S. Grabley, R. Thiericke, A. Zeeck, *J. Antibiot.* **1996**, *49*, 432.

[5]  S. Bach, S. Breiding-Mack, S. Grabley, P. Hammann, K. Hütter, R. Thiericke, H. Uhr, J. Wink, A. Zeeck, *Liebigs Ann. Chem.* **1993**, 241.

[6]  J. E. Berdy, A. Aszalos, M. Bostian, K. L. McNitt, *Handbook of Antibiotic Compounds*, CRC-Press, Boca Raton, FL, **1980–1987**, Vol. I–XIV.

[7]  M. Hajek, 'Eine neue Screening-Strategie für hydrophile Naturstoffe sowie der Einsatz des Transkriptionsassays Myc-Max in der automatisierten Wirkstoffsuche', PhD-Thesis, University of Jena (Germany), **1997**.

[8]  J. Piechler, A. Brecht, G. Gauglitz, M. Zerlin, C. Maul, R. Thiericke, S. Grabley, *Anal. Biochem.* **1997**, *249*, 94.

[9]  C. Maul, 'Biomolekular-Chemisches Screening: Etablierung einer neuartigen Strategie für die Wirkstoffsuche sowie Isolierung und Strukturaufklärung neuer Naturstoffe', PhD-Thesis, University of Jena (Germany), **1997**.

[10]  A. Maier, C. Maul, M. Zerlin, S. Grabley, R. Thiericke, *J. Antibiot.* submitted.

[11]  A. Göhrt, A. Zeeck, K. Hütter, R. Kirsch, H. Kluge, R. Thiericke, *J. Antibiot.* **1992**, *45*, 66.

[12]  S. Grabley, E. Granzer, K. Hütter, D. Ludwig, M. Mayer, R. Thiericke, G. Till, J. Wink, S. Philipps, A. Zeeck, *J. Antibiot.* **1992**, *45*, 56.

[13]  H. Hoff, H. Drautz, H.-P. Fiedler, H. Zähner, J. E. Schultz, W. Keller-Schierlein, S. Philipps, M. Ritzau, A. Zeeck, *J. Antibiot.* **1992**, *45*, 1096.

[14]  M. Ritzau, S. Philipps, A. Zeeck, H. Hoff, H. Zähner, *J. Antibiot.* **1993**, *46*, 1625.

[15]  M. Zerlin, R. Thiericke, *Nat. Prod. Lett.* **1996**, *8*, 163.

[16]  T. Henkel, A. Zeeck, *Liebigs Ann. Chem.* **1991**, 367.

[17]  Y. Tanaka, I. Kanaya, Y. Takahashi, M. Shinose, H. Tanaka, S. Omura, *J. Antibiot.* **1993**, *46*, 1208.

[18]  H. Drautz, P. Reuschenbach, H. Zähner, J. Rohr, A. Zeeck, *J. Antibiot.* **1985**, *38*, 1292.

[19]  R. Grote, A. Zeeck, H. Drautz, H. Zähner, *J. Antibiot.* **1988**, *41*, 1178.

[20]  J. Rohr, R. Thiericke, *Nat. Prod. Rep.* **1992**, *9*, 103.

[21]  S. Grabley, P. Hammann, W. Raether, J. Wink, *J. Antibiot.* **1990**, *43*, 639.

[22]  S. Grabley, G. Kretzschmar, M. Mayer, S. Philipps, R. Thiericke, J. Wink, A. Zeeck, *Liebigs Ann. Chem.* **1993**, 573.

[23]  M. Schönewolf, S. Grabley, K. Hütter, R. Machinek, J. Wink, A. Zeeck, J. Rohr, *Liebigs Ann. Chem.* **1991**, 77.

[24]  A. Göhrt, S. Grabley, R. Thiericke, A. Zeeck, *Liebigs Ann. Chem.* **1996**, 627.

[25]  P. Henne, S. Grabley, R. Thiericke, A. Zeeck, *Liebigs Ann./Recueil* **1997**, 937.

[26]  J. Fuchser, S. Grabley, M. Noltemeyer, S. Philipps, R. Thiericke, A. Zeeck, *Liebigs Ann. Chem.* **1994**, 831.

[27]  P. Henne, R. Thiericke, S. Grabley, K. Hütter, J. Wink, E. Jurkiewicz, A. Zeeck, *Liebigs Ann. Chem.* **1993**, 565.

## 7.7    Carbohydrate – protein interaction studies by various NMR methods and computational calculations

*Hans-Christian Siebert*

When a protein such as a lectin recognizes the distinct sugar residues of an oligosaccharide chain, knowledge about the primary structure of the ligand alone does not suffice for detailed understanding of these interaction processes. After the elaboration of analytical methods to reliably unravel the sequence, the next challenge is posed by the necessity to describe the conformation and structural flexibility of oligosaccharides in order to understand their ligand properties on a submolecular level. In the case where the structure of two linked hexopyranose rings is carefully inspected, this problem turns out to be less complicated than initially thought. The main source of spatial alterations for saccharides resides in the bonds of the glycosidic linkage due to the rather rigid ring structure with their pronounced energetic preference for the chair conformation. As shown in Figure 7.7.1, both dihedral torsion angles $\Phi$ and $\Psi$ of these two exocyclic bonds can be independently varied, thereby altering the relative positions of the two rings.

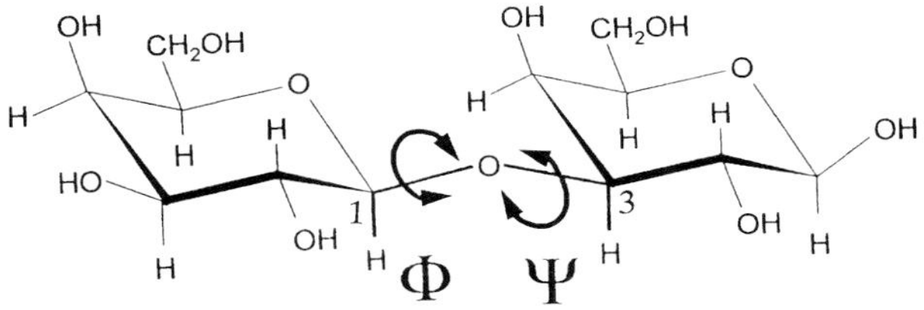

**Figure 7.7.1.** Representation of the disaccharide Gal$\beta$1–3Gal with the glycosidic angles $\Phi/\Psi$.

The questions immediately arise: which of the angles positions will be energetically preferred for individual oligosaccharides; and to what extent conformational flexibility will exist and whether binding of the ligand to a receptor may distort predominant angle positions. A solution of these problems, whose acquisition will have far-reaching consequences for drug design, exists in a combined application of computer modeling, X-ray crystallography, and NMR spectroscopy [1, 2]. All three methods have their limitations but together they work in a complementary manner and are therefore an excellent modus operandi to study the structural basis of the sugar code, *i.e.* carbohydrate–protein interactions on an atomic scale.

### 7.7.1    Unbound oligosaccharides

The two torsion angles $\Phi$ and $\Psi$ define the conformational space for any disaccharide (Figure 7.7.1). Drawing the analogy to topographical maps, the area of a $\Psi, \Psi$-representation receives its third dimension not by altitude, but by the energy content

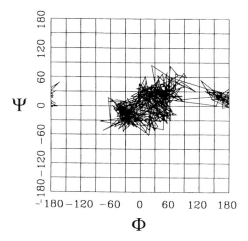

**Figure 7.7.2.** MD simulation of the conformation of Galβ1–2Gal'β1-R at the start coordinates of Ψ, Ψ (180°/0°) using the Consistence Valence Force Field (CVFF) with the explicit inclusion of solvent molecules at 300 K and a simulation time of 1000 ps [6].

of each conformation, defined by a certain Ψ, Ψ-combination. The total potential energy for any Ψ, Ψ-set can be calculated, if the individual energetic contributions are adequately taken into consideration.

Molecular dynamics calculations can be compared to a series of snapshots of a molecule after distinct periods of time. The selection of this interval is determined by the velocity of vibrations. To follow closely stepwise alterations of any aspect of conformation, the snapshot should be recorded before, *e.g.* one C–H stretch vibration is completed and the next vibration is already under way. Consequently, periods of 1 fs $(10^{-15}$ s) are chosen to be able to establish trajectories of Ψ, Ψ-alterations. They are sufficient to account for the behavior of the molecule during the period of individual stretch vibrations which are measured to last approximately 10 fs. Examples of trajectories for the two torsion angles of a disaccharide are shown in Figure 7.7.2 for the disaccharide Galβ1–2Gal', simulated explicitly in water molecules.

The two local energy-minimum conformations in the central low-energy valley are mainly populated. Remarkably, this calculation includes the presence of solvent molecules, which generally damp the frequency of oscillations and transitions [1, 3]. In view of biological systems with water, their consideration is essential, although the number of equations of motions automatically increases. If the size of the calculations exceeds a reasonable computational time to complete the run, the effect of solvent molecules is approximated by choosing a certain value for the dielectric constant $\varepsilon$. Although this procedure certainly is a simplification of the studied complex system, these approximations are generally approved of, until the problem of the enormous size of calculation steps to be carried out in a few hours is solved. Since this parameter change also is of relevance for molecular mechanics calculations, its impact warrants attention in this field. Molecular mechanics calculations and molecular dynamics simulations at various dielectric constants (between 1 and 80) are also important because often one does not actually know its value in or in the near of the binding pocket. This value can differ significantly from that of water ($\varepsilon_{water} = 80$) [4].

Evidently, oligosaccharides cannot superficially be referred to as rigid molecules. For this reason, the concept of the possibility for an ensemble of energetically favorable conformers as ligands is increasingly appreciated [2, 3, 5]. This knowledge has far-reaching consequences for the understanding of protein–carbohydrate interaction. Detailed calculations for any sugar compound are necessary, because a low-energy conformer can in principle be the object of selection by a receptor. Lectins, which are carbohydrate-binding proteins without immunologic origin and enzymatic activity and which play important roles in various regulation processes, possess in many cases rather shallow binding pockets. Therefore, in some cases parts of an oligosaccharide chain are still flexible, even if it is bound to a lectin [6].

In the free state, the populations are often equally distributed at least between two conformational states. Especially, when low-energy conformations are positioned in different sections of a rather extended low-energy-level valley, frequent transitions between two positions separated only by a minor barrier can occur, as was also recently shown for disaccharide ligands of galactoside-binding lectins [6, 7]. This situation calls the question to be answered about the spatial parameters of a flexible ligand after complex formation with a receptor. Before this issue will be addressed, it should not be overlooked that our reasoning on flexibility up to this point exclusively rests on computer-assisted calculations. Their results need to be ascertained by experimental input.

An indication for the inherent flexibility is given by the often-noted difficulty to crystallize complex carbohydrates. Even if successful, crystallography cannot reflect the dynamic behavior of molecules in solution. Therefore, spectroscopic techniques which afford the capacity to monitor molecules in solution are essential. In NMR spectroscopy, through-space dipolar interactions between protons acting as a transmitter–receiver pair are a valuable source of information, in case their spatial vicinity does not exceed a distance of 3.5 Å for a finite time period when oligosaccharides are measured [1, 8, 9]. These contacts can be intra- and inter-residual. In order to collect information about $\Psi, \Psi$-values, only inter-residual contacts are useful, which qualify to draw conclusions on the $\Psi, \Psi$-features, as indicated in the right part of Figure 7.7.3.

For a completely rigid molecule, the measurement of these nuclear Overhauser effects (NOE) allows us unequivocally to translate the signal intensities into internuclear distances, employing the spectroscopic signal as a molecular ruler. For a rigid molecule, no further complications affect the interpretation. However, the inherent flexibility with rapid fluctuations hampers such straightforward derivation of conclusions in most cases. The representation of the NMR-derived information as a single (virtual) conformation can be misleading. It is essential to bear in mind that individual conformers will contribute to the overall signal, because the comparatively long time frame of the measurement cannot distinguish between the frequently and rapidly changing positions. The lack of ability to carry out snapshots in the pico-second-range with the NMR-spectrometer causes inevitable time- and ensemble-averaging over all contributing conformers [1, 8, 10, 11]. Since the number

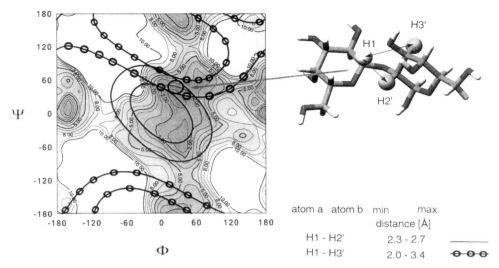

Ψ

Φ

| atom a atom b | min | max |
| --- | --- | --- |
| | distance [Å] | |
| H1 - H2' | 2.3 - 2.7 | ———— |
| H1 - H3' | 2.0 - 3.4 | o-o-o |

**Figure 7.7.3.** Superposition of the distance map of Gal$\beta$1–2Gal'$\beta$1-R in complex with the galactoside-specific mistletoe lectin *Viscum album* agglutinin (VAA) and the RAMM-derived energy profile of the same disaccharide in the free state at $\varepsilon = 4$ (calculation of energy levels from 0 to 41.86 kJ, i.e. 0–10 kcal/mol). The two measured interresidual NOE contacts are: (o-o-o): Gal H1–Gal H3': 2.3 Å (lower limit), 2.7 Å (upper limit, measured NOE distance); (———): Gal H1–Gal H2': 2.0 Å, 3.4 Å. The area of overlap in the Ψ, Ψ-diagram defines the set of angle combinations in which both constraints are satisfied in one conformation [6].

inter-residual contacts in oligosaccharides is generally small, the availability of only one or few distance constraint(s) and the problem of averaging preclude an immediate depiction of an actual conformation. However, at least the problem of averaging is manageable.

To generate a NOE, two protons come into a dipolar contact over a fixed distance or over two or more individual distances which are averaged during the measurement. Taking flexibility into account, the probably averaged signal can originate from, *e.g.* two distances. This assumption is the basis of the distance-mapping approach [1, 3, 5]. The actual presence of an inter-residual dipolar contact is attributed to two distances, which yield the measurable distance upon averaging. The lower limit for the distance of 2.0 Å is given by the van der Waals radii. A lower limit of, *e.g.* 2.3 Å results from other intramolecular steric hindrances. An upper limit will not exceed 3.5 Å in case of oligosaccharides, because the strength of the through-space magnetization transfer fades away with an $r^{-6}$-dependence on the distance between sender and receiver protons. Upper and lower limits encircle a section of the Φ and Ψ conformational-space by contour lines. This set of limited areas defines an ensemble of Ψ, Ψ-combinations which are consistent with the distance constraint(s) derived from the NMR experiment. The principal idea is drawn in the left parts of Figure 7.7.3. Weighting of this total population of Ψ, Ψ-combinations can be performed by combining the illustrations of the mapping approach with the results of molecular mechanics calculations. If a second distance constraint is available, the size of the population with compatible $\Psi$, $\Psi$-combinations in the area of overlap is drastically reduced, as is clearly apparent in the left parts of Figure 7.7.3. Again, accounting of energy levels in this marked area is conducive to infer conformational parameters of a disaccharide. Thus, NMR-derived distance constraints are helpful to deduce whether conformational energy valleys can be populated in solution. If they do not fit into the area delimited by a set of contour lines in the distance-mapping approach, the probability for the occurrence of the corresponding $\Psi$, $\Psi$-combinations is negligible. These constraints can also be assessed experimentally for ligands associating with a receptor and rapidly exchanging between solution and the binding site. The monitoring of transferred nuclear Overhauser effects (trNOE) in receptor/ligand-mixtures at subsaturational receptor concentrations provides a means to characterize spatial features of the bound ligand [12].

## 7.7.2    Ligand–lectin complexes

Without knowledge of the binding site architecture, the discussed structural informations are incomplete. A complete elucidation of the ligand's conformation in the binding pocket by NMR spectroscopy would require the collection of the entire set of NOE-data from the complexed molecules in technically demanding two- or more-dimensional NMR-techniques [11, 13]. The linebroadening of signals which must be ascribed to individual protons and which increases for bigger molecules limits the scope of this sophisticated method to rather small proteins (40 kDa). Under these circumstances it is not surprising that the smallest known lectins, *i.e.* hevein and pseudo-hevein (two isolectins from tree rubber tree) with its 43 amino acids, have so far been the prime target for this approach [14–17]. The aromatic part of a 2D-TOCSY-spectrum of pseudo-hevein is shown in Figure 7.7.4 (unpublished result).

Although astounding progress is currently being made in this area, and isotope-enriched receptors and ligands can simplify the task, the combined application of X-ray crystallography, special NMR-methods and molecular modeling techniques is still sure to be a major source of topological information. One such special NMR-method which especially addresses the involvement of tyrosine-, tryptophan-, and histidine-residues in saccharide binding is the CIDNP (Chemically Induced Dynamic Nuclear Polarization) technique.

## 7.7.3    CIDNP experiments

The CIDNP-method can be used in many aspects when studying biomolecules. CIDNP was previously used to study the non-specific and specific interaction between lac-repressor headpiece and DNA [18, 19]. (For reviews, see [20, 21].) The method can also be used for the analysis of denatured states of lysozyme [22] and intact glycoproteins in comparison with the deglycosylated [23] and desialylated [24, 25] forms in solution. CIDNP experiments have also successfully been used to explore

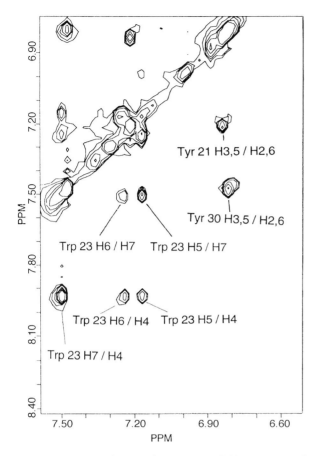

**Figure 7.7.4.** Aromatic part of a 600 MHz TOCSY spectrum of pseudo-hevein.

the function of the aromatic amino acids tyrosine, tryptophan, and histidine at a lectin surface in respect to carbohydrate binding. [15, 26, 27]

Many plant and animal lectins harbor these amino acid residues as a part of their binding pocket architecture [28–30]. The presence of aromatic amino acids in the binding site of various lectins raises the question why such hydrophophic amino acids are important for the binding of the hydrophilic carbohydrates. The CIDNP-technique addresses this question and has to be considered as a powerful tool in glycosciences since the CIDNP-reactive amino acids often mediate hydrophobic interactions between the unpolar parts of a saccharide and its receptor. In this respect, CIDNP is a sensitive spectroscopic method which focuses on distinct structural aspects of oligosaccharide–protein interactions, thereby supplementing the topological data obtained by other techniques.

Briefly, the CIDNP technique can be described as follows: A flavin dye is added to a protein solution. During one part of the CIDNP-experiment the sample is irradiated

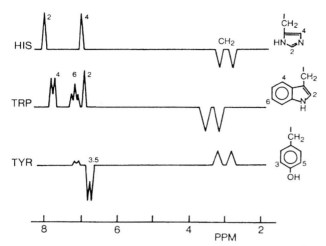

**Figure 7.7.5.** Schematic representation of the laser light-induced $^1$H photo CIDNP effects observed for the amino acids histidine, tryptophan, and tyrosine.

by an argon-ion laser. The photo-excited dye reacts reversibly with surface-exposed amino acid side chains of histidine, tyrosine, and tryptophan, thereby generating protein–dye radical pairs. Nuclear spin polarization is obtained from back-reactions of the radical pairs. By alternately recording light and dark free induction decays, and subtracting the resulting spectra, the CIDNP difference spectrum is created, containing only lines of polarized residues. The tyrosine CIDNP signal appears as an emission line in negative direction; CIDNP signals of histidine and tryptophan residues occure as absorption lines in positive direction (Figure 7.7.5).

The tyrosine CIDNP effect corresponds to the spin-density distribution of the intermediate phenoxy radical with strong emissssion at the 3, 5 protons and weak positive enhancement at the 2, 6 protons. The CIDNP spectra of tryptophan correspond with an intermediate radical with positive spin density at the 2, 3, 4, and 6 positions of the indole ring and very small spin density at the 5 and 7 positions, leading to absorption spectra (positive direction). The CIDNP difference spectrum of histidine at neutral pH is quite strong with the $\beta$-CH$_2$ group in emission and the 2 and 4-protons showing enhanced absorption (Figure 7.7.5). In case a CIDNP-reactive amino acid residue constitutes the binding site architecture of a lectin, the corresponding CIDNP-singal will be partly or entirely be suppressed by the addition of a specific ligand.

The role of Tyr-, His-, and Trp-residues can be studied exemplarily in the small lectins hevein, pseudo-hevein and the B-domain of wheat germ agglutinin (WGA) which consists of 43 amino acid residues [15–17]. All hevein-like lectins have a binding specificity for *N*-acetylglucosamine oligomers. The CIDNP-signals of the corresponding amino acids in the binding pocket show a significantly weakened intensity after complexation with a ligand, as paradigmatically shown for hevein and pseudo-hevein (Figure 7.7.6 a–d). The involvement of Trp-NH-protons in (GlcNAc)$_n$-binding can be studied directly by comparison of an 1D-proton-NMR spectrum of native hevein with one of a hevein–ligand complex (Figure 7.7.7a and b).

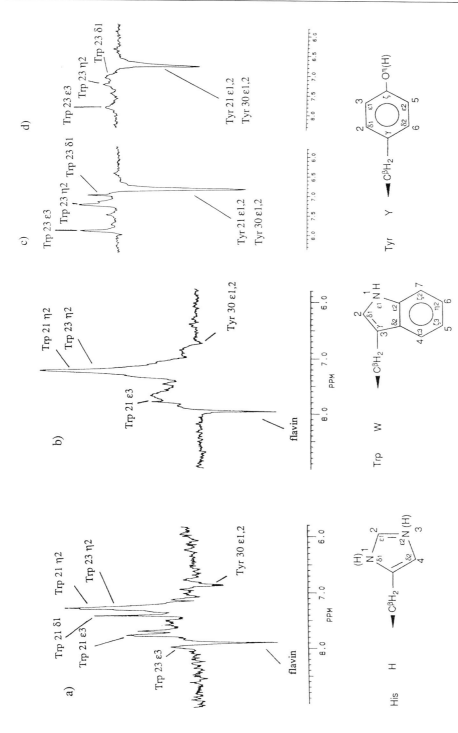

**Figure 7.7.6.** Laser photo CIDNP difference spectra of: (a) hevein: (b) a hevein-*N,N*,-diacetylchitobiose complex: (c) pseudo-hevein: (d) a pseudo-hevein-*N,N*,-diacetylchitobiose complex.

a)

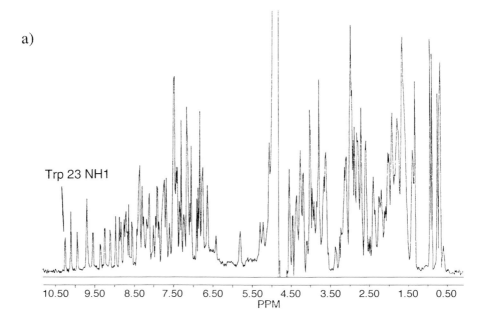

Trp 23 NH1

b)

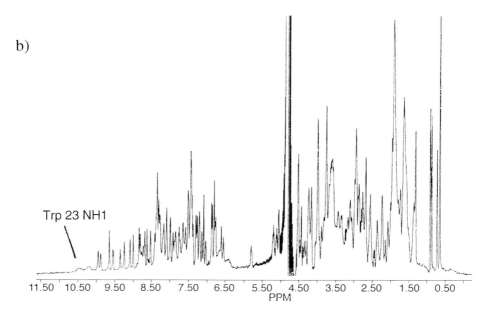

Trp 23 NH1

**Figure 7.7.7.** (a) One-dimensional spectrum of hevein in $H_2O$, (b) One-dimensional spectrum of a hevein-$N,N$,-diacetylchitobiosecomplex in $H_2O$.

The lectin *Urtica dioica* agglutinin (UDA) from the stinging nettle and WGA also have an affinity to *N*-acetylglucosamine oligomers. UDA consists of two, and WGA of four, hevein-like domains. Beside its affinity to *N*-acetylglucosamine oligomers, WGA also binds sialic acid-containing oligosaccharides. In all mentioned examples, the corresponding signals of Tyr-, His-, and Trp-residues of the binding pocket are affected by the addition of *N*-acetylglucosamine oligomers. Different structural aspects of these interactions can also be studied by two-dimensional NMR-methods. [11, 31, 32] Interestingly, complexation of UDA with a small amount of (GlcNAc)$_3$ leads to an altered CIDNP spectrum in which a new small signal has occurred. This finding argues in favor for a small conformational change in the binding pocket during ligand binding [15]. Such conformational changes caused by ligand binding or site-directed mutagenesis can also be elegantly detected by the CIDNP-method [26, 27].

In order to obtain further information about the architecture of the hevein-domains, the CIDNP-data were correlated with X-ray crystallographic coordinates from WGA [33] and NMR-obtained conformations from hevein [14–17]. Comparison of CIDNP-derived surface accessibilities with computational calculated values on the basis of refined X-ray or NMR structures could be used for structural refinement. The accessibilities of the tyrosine, tryptophan, and histidine residues of the studied proteins are as follows (with a dot density of 1 and a sphere radius of 1.5 Å): Buried tyrosines have a surface accessibility below 30 Å$^2$, partly buried tyrosine residues have accessibilities from 30 to 80 Å$^2$, while a value above 80 Å$^2$ is calculated for the tyrosines which are considered to be the surface-exposed and causing intense CIDNP signals. Corresponding values are estimated for tryptophan and histidine residues [15]. As can be seen from these data [15] and Figure 7.7.6 (a–d), Tyr 30 retains its partial surface accessibility in pseudo-hevein, leading only to a minor contribution to the huge Tyr-signal in the CIDNP-difference spectrum. The major contribution to this signals stems from Tyr23, which has a surface accessibility comparable with that of Trp23 in hevein. CIDNP- and modeling-data indicate that the architecture of the binding pockets of hevein, pseudo-hevein and the B-domain of WGA are similar [15]. A model of a hevein–(GlcNAc)$_2$ complex and a stereoscopic representation of pseudo-hevein are shown in Figure 7.7.8 a and b.

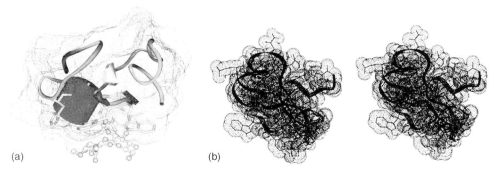

(a)                                                    (b)

**Figure 7.7.8.** (a) Model of a hevein-*N*,*N*′-diacetylchitobiose complex. (b) Model of pseudo-hevein (in stereo) – the binding pocket is located in the lower left part of the figure.

## 7.7.4    Conclusion

The sugar code is well established and the emphasis of research has shifted from detection to delineation of structural and functional details of lectin–carbohydrate interactions and their elicited signaling processes [2, 3]. Such investigations have uncovered a veritable degree of conformational flexibility of ligands and are currently unraveling intrinsic details about their molecular get-together with a lectin. Combinations of NMR spectroscopy and X-ray crystallography with sophisticated computer calculations encompassing molecular mechanics, molecular dynamics, homology modeling and docking algorithms are pertinent to comprehend how the sugar language is decoded. Moving forward from this firm foundation, it is reasonable to expect proper guidance from this panel of methods on the way to an optimized rational marker or drug design with clinical implications. Improvements in NMR-spectroscopic techniques [34] and in the analysis of complex molecular dynamics simulations [35] are underlining these assumptions.

## References

[1]  H.-C. Siebert, C.-W. von der Lieth, M. Gilleron, G. Reuter, J. Wittmann, J. F. G. Vliegenthart, H.-J. Gabius, Carbohydrate–protein interaction, in *Glycosciences: Status and Perspectives* (H.-J. Gabius and S. Gabius, eds.), Chapman & Hall, London–Weinheim, **1997**, pp. 291–310.

[2]  H. J. Gabius, H.-J., *Pharm. Res.* **1998**, *15*, 24–31.

[3]  C.-W. von der Lieth, T. Kozár, W. E. Hull, *J. Mol. Struct.* **1997**, *395/396*, 225–244.

[4]  S. C. Harvey, *Proteins* **1989**, *5*, 78–92.

[5]  H.-C. Siebert, G. Reuter, R. Schauer, C.-W. von der Lieth, J. Dabrowski, *Biochemistry*, **1992**, *31*, 6962–6971.

[6]  M. Gilleron, H.-C. Siebert, H. Kaltner, C.-W. von der Lieth, T. Kozár, K.M. Halkes, E. Y. Korchagina, N. V. Bovin, H.-J. Gabius, J. F. G. Vliegenthart, *Eur. J. Biochem.* **1998**, *252*, 416–427.

[7]  H.-C. Siebert, M. Gilleron, H. Kaltner, C.-W. von der Lieth, T. Kozár, N. V. Bovin, E. Y. Korchagina, J. F. G. Vliegenthart, H.-J. Gabius, *Biochem. Biophys. Res. Commun.* **1996** *219*, 205–212.

[8]  J. P. Carver, *Pure & Appl. Chem.* **1993**, *65*, 763–770.

[9]  S. W. Homans, Three-dimensional structures of oligosaccharides explored by NMR and computer calculation, in *Glycoproteins* (J. Montreuil, J. F. G. Vliegenthart and H. Schachter, eds.), Elsevier, Amsterdam, **1995**, pp. 67–86.

[10] O. Jardetzky, *Biochim. Biophys. Acta* **1980**, *621:* 227–232.

[11] H.-C. Siebert, R. Kaptein, J. F. G. Vliegenthart, Study of oligosaccharide-lectin interaction by various nuclear magnetic resonance (NMR) techniques and computational methods, in *Lectins and Glycobiology* (H.-J. Gabius, S. Gabius, eds.) Springer Verlag, Heidelberg, New York, **1993**, pp. 105–116.

[12] F. Ni, *Progr. NMR Spectr.* 1994, *26*, 517–606.

[13] W. E. Hull, Experimental aspects of two-dimensional NMR, in *Two-dimensional NMR-spectroscopy: applications for chemists and biochemists* (W. R. Croasmun and R. M. K. Carlson, eds.), VCH Publishers Inc., New York, **1994**, pp. 67–456.

[14] N. H. Andersen, B. Cao, A. Rodriguez-Romero, B. Arreguin, *Biochemistry* **1993**, *32*, 1407–1422.

[15] H.-C. Siebert, C.-W. von der Lieth, R. Kaptein, J. J. Beintema, K. Dijkstra, N. van Nuland, U. M. S. Soedjanaatmadja, A. Rice, J. F. G. Vliegenthart, C. S. Wright, H.-J. Gabius, *Proteins* **1997**, *28*, 268–284.

[16] J. L. Asensio, F. J. Cañada, M. Bruix, A. Rodríguez-Romero, J. Jiménez-Barbero, *Eur. J. Biochem.* **1995**, *230*, 621–633.

[17] J. L. Asensio, F. J. Cañada, M. Bruix, C. Gonzáles, N. Khiar, A. Rodríguez-Romero, J. Jiménez-Barbero, *Glycobiology* **1998**, *8*, 569–577.

[18] F. Buck, H. Rüterjans, R. Kaptein, K. Beyreuter, *Proc. Natl. Acad. Sci. USA* **1980** 77, 5145–5148.

[19] S. Stob, R. M. Scheek, R. Boelens, R. Kaptein *FEBS Lett.* **1988**, *239*, 99–104.

[20] R. Kaptein, Photo CIDNP studies of proteins, in *Biological Magnetic Resonance 4* (L. J. Berliner, ed.), Plenum Press, New York, **1982**, pp. 145–191.

[21] P. J. Hore, R. W. Broadhurst, *Progr. NMR Spectroscopy* **1993**, *25*, 345–402.

[22] R. W. Broadhurst, C. M. Dobson, P. J. Hore, S. E. Radford, M. L. Rees, *Biochemistry* **1991**, *30*, 405–412.

[23] K. Hård, J. P. Kamerling, J. F. G. Vliegenthart, *Cabohydr. Res.* **1992**, *236*, 315–320.

[24] H.-C. Siebert, S. André, G. Reuter, H.-J. Gabius, *FEBS Lett.* **1995**, *371*, 13–16.

[25] H.-C. Siebert, S. André, G. Reuter, R. Kaptein, J. F. G. Vliegenthart, H.-J. Gabius *Glycoconjugate J.* **1997**, *14*, 945–949.

[26] H.-C. Siebert, E. Tajkhorshid, C. W. von der Lieth, R. G. Kleineidam, S. Kruse, R. Schauer, R. Kaptein, J. F. G. Vliegenthart, H.-J.Gabius *J. Mol. Model.* **1996**, *2*, 446–455.

[27] H.-C. Siebert, R. Adar, R. Arango, M. Burchert, H. Kaltner, G. Kayser, E. Tajkhorshid, C.-W. von der Lieth, R. Kaptein, N. Sharon, J. F. G. Vliegenthart, H.-J. Gabius, *Eur. J. Biochem.* **1997**, *249*, 27–38.

[28] N. Sharon, *Trends Biochem. Sci.* **1993**, *18*, 221–226.

[29] J. M. Rini, *Annu. Rev. Biophys. Biomol. Struct.* **1995**, *24*, 551–577.

[30] H.-J. Gabius, *Eur. J. Biochem.* **1997**, *243*, 543–576.

[31] K. Hom, M. Gochin, W. J. Peumans, N. Shine, *FEBS Lett.* **1995**, *361*, 157–161.

[32] H.-C. Siebert, R. Kaptein, J. J. Beintema, U. M. S. Soedjanaatmadja, C. S. Wright, A. Rice, R. G. Kleineidam, S. Kruse, R. Schauer, P. J. W. Pouwels, J. P. Kamerling, H.-J. Gabius, J. F. G. Vliegenthart, *Glycoconjugate J.* **1997**, *14*, 531–534.

[33] C. S. Wright, *J. Mol. Biol.* **1989**, *209*, 475–487.

[34] K. Wüthrich, *Nature Struct. Biol.* **1998**, *July*, 492–495.

[35] H.-C. Siebert, C.-W. von der Lieth, X. Dong, G. Reuter, R. Schauer, H.-J. Gabius, J. F. G. Vliegenthart, *Glycobiology* **1996**, *6*, 561–572.

# Subject Index